Lecture Notes in Computer Scie

T0238098

Commenced Publication in 1973
Founding and Former Series Editors:
Gerhard Goos, Juris Hartmanis, and Jan van Leeuwen

Klaus Jansen   Sanjeev Khanna
José D.P. Rolim   Dana Ron (Eds.)

Approximation, Randomization, and Combinatorial Optimization

Algorithms and Techniques

7th International Workshop on Approximation Algorithms
for Combinatorial Optimization Problems, APPROX 2004
and 8th International Workshop on Randomization
and Computation, RANDOM 2004
Cambridge, MA, USA, August 22-24, 2004
Proceedings

Volume Editors

Klaus Jansen
Christian-Albrechts-University of Kiel
Institute of Computer Science and Applied Mathematics
Olshausenstr. 40, 24098 Kiel, Germany
E-mail: kj@informatik.uni-kiel.de

Sanjeev Khanna
University of Pennsylvania, Dept. of Computer and Information Science
3330 Walnut Street, Philadelphia, PA 19104, USA
E-mail: sanjeev@cis.upenn.edu

José D.P. Rolim
University of Geneva, Computer Science Department
24 rue Général Dufour, 1211 Geneva 4, Switzerland
E-mail: jose.rolim@cui.unige.ch

Dana Ron
Harvard University, Radcliffe Institute for Advanced Study
10 Garden St, Cambridge, MA 02138, USA
E-mail: danar@eecs.harvard.edu

Library of Congress Control Number: 2004110448

CR Subject Classification (1998): F.2, G.2, G.1

ISSN 0302-9743
ISBN 3-540-22894-2 Springer Berlin Heidelberg New York

Springer is a part of Springer Science+Business Media

springeronline.com

© Springer-Verlag Berlin Heidelberg 2004
Printed in Germany

Typesetting: Camera-ready by author, data conversion by Olgun Computergrafik
Printed on acid-free paper   SPIN: 11311911   06/3142   5 4 3 2 1 0

Foreword

This volume contains the papers presented at the *7th International Workshop on Approximation Algorithms for Combinatorial Optimization Problems* (APPROX 2004) and the *8th International Workshop on Randomization and Computation* (RANDOM 2004), which took place concurrently at Harvard University, Cambridge, on August 22–24, 2004. APPROX focuses on algorithmic and complexity issues surrounding the development of efficient approximate solutions to computationally hard problems, and this year's workshop was the seventh in the series after Aalborg (1998), Berkeley (1999), Saarbrücken (2000), Berkeley (2001), Rome (2002), and Princeton (2003). RANDOM is concerned with applications of randomness to computational and combinatorial problems, and this year's workshop was the eighth in the series following Bologna (1997), Barcelona (1998), Berkeley (1999), Geneva (2000), Berkeley (2001), Harvard (2002), and Princeton (2003).

Topics of interest for APPROX and RANDOM are: design and analysis of approximation algorithms, inapproximability results, approximation classes, on-line problems, small space and data streaming algorithms, sub-linear time algorithms, embeddings and metric space methods in approximation, math progamming in approximation algorithms, coloring and partitioning, cuts and connectivity, geometric problems, network design and routing, packing and covering, scheduling, game theory, design and analysis of randomized algorithms, randomized complexity theory, pseudorandomness and derandomization, random combinatorial structures, random walks/Markov chains, expander graphs and randomness extractors, probabilistic proof systems, random projections and embeddings, error-correcting codes, average-case analysis, property testing, computational learning theory, and other applications of approximation and randomness.

The volume contains 19+18 contributed papers, selected by the two program committees from 54+33 submissions received in response to the call for papers.

We would like to thank all of the authors who submitted papers, the members of the program committees

APPROX 2004

Chandra Chekuri, Bell Laboratories
Lisa Fleischer, Carnegie Mellon U. and IBM T.J. Watson
Sudipto Guha, U. of Pennsylvania
Sanjeev Khanna, U. of Pennsylvania (Chair)
Rajmohan Rajaraman, Northeastern U.
Tim Roughgarden, UC Berkeley
Baruch Schieber, IBM T.J. Watson
Martin Skutella, Max Planck Institute
Dan Spielman, MIT
Luca Trevisan, UC Berkeley
Mihalis Yannakakis, Columbia U.
Neal Young, UC Riverside

RANDOM 2004

Noga Alon, Tel Aviv U.
Amos Beimel, Ben Gurion U.
Peter Bro Miltersen, U. of Aarhus
Funda Ergun, Case Western Reserve U.
Uri Feige, Weizmann Institute
Leslie Ann Goldberg, U. of Warwick
Russell Impagliazzo, UC San Diego
Adam Kalai, Toyota Technological Institute
Satish Rao, UC Berkeley
Dana Ron, Tel Aviv U. and Radcliffe Institute, Harvard (Chair)
Rocco Servedio, Columbia U.
Neal Young, UC Riverside

and the external subreferees Udo Adamy, Dorit Aharonov, Ernst Althaus, Javed Aslam, Nikhil Bansal, Therese Beidl, Petra Berenbrink, Joan Boyar, Gruia Calinescu, Kamalika Chaudhuri, Bernard Chazelle, Jiangzhuo Chen, Marek Chrobak, Colin Cooper, Erik Demaine, Nikhil Devanur, Thomas Erlebach, Martin Farach-Colton, Sándor Fekete, Tom Friedetzky, Rajiv Gandhi, Mordecai Golin, Oded Goldreich, Alex Hall, Refael Hassin, Stefan Hougardy, Piotr Indyk, Wojtek Jawor, Mark Jerrum, Lujun Jia, Valentine Kabanets, Marek Karpinski, Tali Kaufman, Julia Kempe, Claire Kenyon, Tracy Kimbrel, Robert Kleinberg, Adam Klivans, Jochen Könemann, Stavros Kolliopoulos, Madhular Korupolu, Robert Krauthgamer, Kofi Laing, Matt Levine, Moshe Lewenstein, Guolong Lin, Yi-Kai Liu, Stefano Lonardi, Marco Lübbecke, Mohammad Mahdian, Russ Martin, Daniele Micciancio, Mike Molloy, Cris Moore, Elchanan Mossel, Ian Munro, Assaf Naor, Moni Naor, Kirk Pruhs, Harald Räcke, Balaji Raghavachari, Vijaya Ramachandran, Dana Randall, April Rasala, Mauricio Resende, Gaby Scalosub, Guido Schäfer, Amir Shpilka, Dana Shapira, Riccardo Silvestri, Dan Spielman, Nicolas Stier, Leen Stougie, Martin Strauss, Maxim Sviridenko, Hisao Tamaki, Prasad Tetali, Santosh Vempala, Eric Vigoda, Anil Vullikanti, Dan Wang, Enav Weinreb, Gerhard Woeginger, Nick Wormald, Alex Zelikovsky and Yan Zhang.

We gratefully acknowledge support from the Radcliffe Institute for Advanced Study and the Computer Science Department of Harvard University, Cambridge, MA, the Institute of Computer Science of the Christian-Albrechts-Universität zu Kiel and the Department of Computer Science of the University of Geneva. We also thank Ute Iaquinto and Parvaneh Karimi Massouleh for their help.

August 2004   Sanjeev Khanna and Dana Ron, Program Chairs
                    Klaus Jansen and José D.P. Rolim, Workshop Chairs

Table of Contents

Contributed Talks of APPROX

Contributed Talks of RANDOM

Designing Networks with Existing Traffic to Support Fast Restoration

Mansoor Alicherry[1], Randeep Bhatia[1], and Yung-Chun (Justin) Wan[2]

[1] Bell Labs, Lucent Technologies, Murray Hill, NJ 07974
{mansoor,randeep}@research.bell-labs.com
[2] Department of Computer Science, University of Maryland, College Park, MD 20742
ycwan@cs.umd.edu

Abstract. We study a problem motivated by a scheme for supporting fast restoration in MPLS and optical networks. In this local restoration scheme detour paths are set-up a priori and network resources are pre-reserved exclusively for carrying rerouted traffic under network failures. (i.e. they do not carry any traffic under normal working conditions). The detours are such that failed links can be bypassed locally from the first node that is upstream from the failures. This local bypass activation from the first detection point for failures along with the dedication of network resources for handling failures permits very fast recovery times, a critical requirement for these networks. By allowing sharing of the dedicated resources among different detours the local restoration scheme results in efficient utilization of the pre-reserved network capacity.

In this paper we are interested in the problem of dedicating the least amount of the currently available network capacity for protection, while guaranteeing fast restoration to the existing traffic along with any traffic that may be admitted in the future. We show that the problem is NP-hard, and give a 2-approximation algorithm for the problem. We also show that the integrality gap of a natural relaxation of our problem is $\Omega(n)$, thus establishing that any LP-based approach using this relaxation cannot yield a better approximation algorithm for our problem.

1 Introduction

Dynamic provisioning of bandwidth guaranteed paths with fast restoration capability is an important network service feature for the emerging Multi-Protocol Label Switched (MPLS) networks [7] and optical mesh networks [18]. The fast restoration capabilities are required in order to provide the needed reliability for services such as packetized voice, critical VPN traffic, etc. Traditionally ring based SONET [11] networks have offered 50ms restoration to bandwidth guaranteed services, using pre-reserved spare protection capacity and pre-planned protection paths. Pre-planning protection in rings has been especially attractive, because of the availability of exactly one backup path between any two nodes, leading to very simple and fast automatic protection switching mechanisms. However in ring based SONET networks these advantages come at the cost of reserving at least half the total capacity for protection.

A local restoration scheme [12], [16], [20] is proposed to provide fast restoration in mesh based MPLS and optical networks. In this scheme which is also referred to as

K. Jansen et al. (Eds.): APPROX and RANDOM 2004, LNCS 3122, pp. 1–12, 2004.

link restoration the traffic on each link e of the network is protected by a detour path that does not include link e. Upon failure of any link e, any traffic on e is switched to its detour path. Thus, link restoration provides a local mechanism to route around a failure. In this restoration scheme the restoration capacity of the pre-setup detours is not used under normal no-failure conditions (except possibly by low priority preemptible traffic). Local restoration when used in conjunction with advanced reservation of the restoration capacities and pre-setup detours results in low restoration latency. Pre-provisioned link restoration also results in operational simplicity since the detours have to be only provisioned once for a given network topology and since the online connection routing can now be done oblivious to the reliability requirements, using only the resources that are not reserved for restoration.

An important consideration for any fast restoration scheme is to minimize the network resources dedicated for restoration and hence to maximize the proportion of network resources available for carrying traffic under normal working conditions. In general, link restoration provides guaranteed protection against only single link failures hence the reserved restoration capacity may be shared among the different pre-setup detours (since at most one detour may carry restored traffic at any given time). Thus pre-provisioned link restoration scheme offers the promise of fast restoration recovery for just a small fraction of the total capacity reserved for restoration, due to the high degree of restoration capacity sharing that is possible in mesh networks.

In many situations one would like to support fast restoration on existing networks without disturbing the existing traffic, meaning that the restoration scheme can only use up to the current available network capacity (link capacity minus existing traffic capacity) for protecting the existing traffic and any new traffic. Note that existing traffic makes the problem harder. A simpler polynomial-time 2-approximation algorithm in the absence of existing traffic was presented in [1] (see Related Work).

In this paper we are interested in the optimization problem of dedicating the least amount of the currently available network capacity for protection, while guaranteeing fast restoration to the existing traffic along with any traffic that may be admitted in the future, for the pre-provisioned link restoration scheme. Specifically we are interested in partitioning the available link capacities into working and protection, such that the latter is dedicated for restoration and the former is available to carry any current or new traffic, with the objective of guaranteeing link restoration for minimal total protection capacity. Note that in a network with a static topology this problem may need be solved only once, since the solution remains feasible even as the admitted traffic pattern changes. However, the solution may not stay optimal over time as the admitted traffic pattern changes, and may be recomputed occasionally to ensure efficient utilization of network resources. Also changes in network topology (which are common but not frequent) may require recomputing the solution, since the old solution may not even guarantee link restoration.

1.1 Problem Definition

Given a undirected network $G = (V, E)$, with link capacities u_e and existing traffic W_e on link $e \in E$ the problem is to partition the capacities on link e into a working capacity w_e and a protection capacity p_e (s.t. $w_e + p_e = u_e$) such that

- The total protection capacity $\sum_{e \in E} p_e$ is minimized.
- For every link $e = (u, v)$, in the network $G - e$ obtained by removing link e from G, there exists a path P_e (detour for link e) between nodes u and v, such that every link e' on P_e satisfies $p_{e'} \geq w_e$.
- The working capacity w_e is at least the amount of the existing traffic W_e.
- In case such a partition is not feasible on G, output an empty solution.

In other words, on link e, p_e capacity is reserved for carrying restored traffic during failures and at most w_e ($w_e \geq W_e$) traffic is carried during normal working conditions. Thus on the failure of link e, at most w_e traffic on it is rerouted over the pre-setup detour path P_e using only the reserved capacities on the links on P_e.

Note that given the p_e and w_e values for all links e in any feasible solution, the detour paths P_e can be easily computed. Hence we do not include the computation of the detour paths P_e in the statement of the problem.

1.2 Our Contribution

We show that given an instance of the problem, it can be determined in polynomial time, using a fast and efficient algorithm, if the problem has a feasible solution. However, we show that computing an optimal solution for feasible instances is NP-hard. Moreover, we present a simple and efficient algorithm that computes a solution to the given feasible instance in which the total protection capacity reserved is guaranteed to be within two times the protection capacity reserved in any optimal solution. We also show that the integrality gap of a natural relaxation of our problem is $\Omega(n)$, thus establishing that any LP-based approach using this relaxation cannot yield a better approximation algorithm for our problem.

2 Related Work

The main approaches for supporting pre-provisioned link restoration scheme in mesh networks are based on identifying ring structures. Once the set of rings are identified then pre-planned restoration schemes as in SONET [11] are employed. In some of these approaches the network is designed in term of rings [17] or by partially using rings [9]. Thus, these schemes are only applicable to constrained topologies. In some other of these approaches each link is covered by a cycle leading to a cycle cover for the network [9]. Each of these cycles is then provisioned with enough protection capacity to cover the links that belong to it. On the failure of the link the working traffic is rerouted over the protection capacities in the surviving links of the covering cycle. There are two drawbacks of these approaches: first the amount of pre-provisioned protection capacity can be significant and second it is hard to find the smallest cycle cover of a given network [19]. An improvement to these schemes is those based on the notion of p-cycle [10]. Here the main idea is that a cycle can be used to protect not just the links on the cycle but also the chords (spokes) of the cycle, thus showing that far fewer cycles (than in a cycle cover) may be sufficient for providing full protection. An algorithm to minimize the total spare capacity, based on solving an integer program over all possible cycles is given in [10]. To the best of our knowledge no fast approximation algorithms for this problem are known. An alternative to cycle covers, intended to overcome the difficulty of finding good covers, is to cover every link in a network with exactly two cycles [8]. A set of cycles that meets this requirement is called a double cycle cover [13]. For planar graphs, double cycle covers can be found in polynomial-time. For non-planar graphs, it is conjectured that double cycle covers exist, and they are typically found quickly in practice. However, even for double cycle cover based protection schemes, the required pre-provisioned protection capacity can be significant.

Non-ring based approaches to link restoration on mesh networks include generalized loop-back [14], [15], where the main idea is to select a digraph, called the primary, such that the conjugate digraph, called the secondary, can be used to carry the switched traffic for any link failure in the primary. Chekuri *et al.* [5] consider the problem of adding protection capacity to the links of a given network (primary) carrying working traffic, at minimum cost, so that the resulting network is capable of supporting link protection for a given set of links, where the protection is provided to the working traffic on the primary network. In their model no limit is imposed on the total capacities of the links, and they provide a 4-approximation algorithm when all links in the original primary network have uniform bandwidth (carrying the same amount of working traffic) and they provide a 10.87-approximation algorithm for the general case. In addition they also provide a $O(\log n)$-approximation algorithm for the problem of jointly designing the primary and protected networks, given a demand matrix for the working traffic.

Our previous work [1] considers the problem under the assumption that there is no existing working traffic on any link ($W_e = 0$). We also allow the rerouted traffic to be split on at most two detours in the event of a failure. We show that the optimization problem is NP-hard and provide a 2-approximation algorithm. We also give a lower bound for the problem when there is no restriction on splitting of rerouted traffic.

All the schemes mentioned earlier assume that protection is provided for a single link failure. Choi *et al.* [6] present a heuristic for protecting against two link failures, based on link restoration. The problem of survivable network design has also been extensively studied [4, 3]. Most of the work here has focused on obtaining strong relaxations to be used in cutting plane methods.

3   Algorithm

Recall that u_e is the total capacity of link e, and W_e is the amount of existing traffic in link e. We define $u_e - W_e$ to be the *maximum protection capacity available* on the link e.

Our algorithm (Algorithm A) creates a solution by first computing a maximum spanning tree T_A based on the maximum protection capacity available on the links. It sets the protection capacity of any link e on the maximum spanning tree T_A to the maximum protection capacity available on link e. For any non-tree (T_A) link e (called *cross link*), it initially sets its working capacity equal to the minimum protection capacity assigned to the links on the unique path in T_A between the endpoints of link e. It then selects a few of the cross links and *boosts up* their protection capacities. The main idea is to protect the cross links by using only links in the tree T_A. Also for each tree link e its detour consists of all but one tree link, and the boosting up of some cross links is used to ensure that the cross link on this detour has enough protection to protect e.

The detailed algorithm is given in Algorithm A. We denote the working and protection capacity assigned by Algorithm A on link e by w_e^A and p_e^A respectively ($w_e^A + p_e^A = u_e$). Conceptually the algorithm (as well as the analysis) consists of two phases – finding basic protection capacity in all links, and then boosting the protection capacity for some cross links. These two steps are combined into one in the algorithm, by setting $w_{e_i}^A$ of cross link e_i to $\max(\min(p, u_{e_i} - w), W_{e_i})$ in line 18. p is the minimum protection capacity of the unique path in the tree, which can protect this cross link. w is the maximum working capacity of a link in the unique path, which is still unprotected. So we need to assign at least w protection, but cannot have a working capacity of more than p in this cross link. Also, since we have to support all the existing traffic, $w_{e_i}^A$ has to be at least W_{e_i}.

Algorithm 1 *Algorithm A*

Let $\{e_1, e_2, \ldots, e_m\}$ be the links sorted in decreasing order of maximum available protection capacities ($u_e - W_e$).
$T_A = \phi$
for $i = 1, \ldots, m$ {
    if ($T_A \cup \{e_i\}$ does not form a cycle) {
        $T_A = T_A \cup \{e_i\}$
        if ($W_{e_i} > u_{e_i}/2$)
            No solution exists.
        $w_{e_i}^A = W_{e_i}$
        $p_{e_i}^A = u_{e_i} \quad w_{e_i}^A$
        Mark e_i as unprotected
    } else {
        Let $P = \{e_{j_1}, e_{j_2}, \ldots, e_{j_k}\}$ be the unique path in T_A connecting the endpoints of e_i.
        Let M be the links in P which are marked as unprotected.
        $w = \max_{e \in M} w_e^A \quad$ ($w = 0$ if $M = \emptyset$)
        $p = \min_{e \in P} p_e^A$
        if ($p < W_{e_i}$)
            No solution exists.
        $w_{e_i}^A = \max(\min(p, u_{e_i} - w), W_{e_i})$
        $p_{e_i}^A = u_{e_i} - w_{e_i}^A$
        if ($w > p_{e_i}^A$)
            No solution exists.
        Unmark edges in M.
    }
}
if any link e_i is marked
    No solution exists.

3.1 Correctness

Lemma 1. *The solution (if any) returned by Algorithm A is a feasible solution.*

Proof. It is easy to see that the amount of working traffic of every link (w_e^A) in any solution returned by Algorithm A, is always at least the amount of existing traffic (W_e).

Now we show that for every link $e = (u, v)$, there is a backup path P_e in $G - e$ between nodes u and v, such that every link e' on P_e satisfies $p^A_{e'} \geq w^A_e$.

Case 1, for all $e \in T_A$: Note that $G-e$ is connected, because otherwise link e would stay marked at the end of the algorithm in which case no solution is returned by Algorithm A. Thus, there is at least one cross link $e_c = (u_c, v_c)$, which, together with the unique path in T_A between nodes u_c and v_c (excluding link e), forms a backup path for e. Without loss of generality, let e_c be the first cross link considered by the algorithm, such that adding it to T_A results in a cycle C containing link e. Consider the path $P_e = C \setminus \{e\}$. The link e_c has been assigned the least protection by Algorithm A among the links on P_e. Note that link e is marked (hence in M) at the time when e_c is considered by the algorithm. Thus when link e_c is considered by the algorithm we must have $p^A_{e_c} \geq w \geq W_e = w^A_e$ implying that P_e is a valid backup (detour) for link e.

Case 2, for all $e \notin T_A$: The backup path P_e for link e is the unique path in T_A connecting e, which always exists. The links on this path have enough protection because the algorithm sets the working traffic w^A_e of link e to at most $\min_{e' \in P_e} p^A_{e'}$.

Lemma 2. *If a feasible solution exists, the algorithm will return a solution.*

Proof. The algorithm will not return a solution in 4 cases.

Case 1, there exists a link $e \in T_A$, $W_e > u_e/2$: If a tree link e has $W_e > u_e/2$, then no solution exists. This is because, the maximum protection capacity available on e ($u_e - W_e$) is strictly less than $u_e/2$. If there was a solution, then there must exist a path P_e in $G - e$ between the end points of e all whose links e' have maximum protection capacity available $u_{e'} - W_{e'} > u_e/2$. But then T_A is not a maximum spanning tree based on the maximum protection capacity available on the links, a contradiction.

Case 2, there exists a link $e_c \notin T_A$, $p < W_e$: The proof for this case uses arguments similar to case 1.

Case 3, there exists a link $e_c \notin T_A$, $w > p^A_{e_c}$: Note that in this case $M \neq \emptyset$. Let $e_c = (u_c, v_c)$ and let P_{e_c} be the unique path in T_A connecting nodes u_c and v_c. Let $e \in M$ be a marked link on the path P_{e_c} with $w = W_e = w^A_e$, at the time when e_c is considered by the algorithm. Let P_e be a feasible detour for link e. Thus the maximum protection capacity available on all links e' on P_e is $u_{e'} - W_{e'} \geq w$. Also $P_e \cup \{e\}$ forms a cycle in G, thus at least one link on P_e is a cross link for T_A, that forms a cycle containing link e when added to T_A. This link must have been considered before link e_c since it has strictly more maximum protection capacity available on it. Hence link e must already be marked before link e_c is considered, a contradiction.

Case 4, there is an unmarked link $e = (u, v)$ at the end: In this case in $G - e$ nodes u and v are not connected. Hence, no detour is possible for link e.

From the above two lemmas, we have the following theorem.

Theorem 1. *Algorithm A is correct.*

Remark: For each link e in T_A, we can lower its protection capacity until lowering it further would require decreasing the working capacity of some other links, or would make the working capacity of link e so large that there is no feasible detour for it. This trimming may reduce the total protection capacity however, it has no implication on our analysis of the worst-case approximation ratio for the algorithm.

4   Analysis

In this section we show that A is a 2-approximation algorithm. First we define some notations. Let p_e^{OPT} be the protection capacity of link e in an optimal solution OPT, and w_e^{OPT} be the working capacity of link e in OPT, thus $w_e^{OPT} = u_e - p_e^{OPT}$. To compare the solution of Algorithm A with OPT, we construct a maximum spanning tree T_{OPT} based on the protection capacities p_e^{OPT}.

We partition the links in the network into 4 sets: $E_{AO} = \{e | e \in T_A, e \in T_{OPT}\}$, $E_{\bar{A}\bar{O}} = \{e | e \notin T_A, e \notin T_{OPT}\}$, $E_{A\bar{O}} = \{e | e \in T_A, e \notin T_{OPT}\}$, and $E_{\bar{A}O} = \{e | e \notin T_A, e \in T_{OPT}\}$. We define p_E^{OPT} to be $\sum_{e \in E} p_e^{OPT}$ (sum of protection capacities in OPT over all links in E), p_E^A to be $\sum_{e \in E} p_e^A$ (sum of protection capacities in our solution over all links in E), and u_E to be $\sum_{e \in E} u_e$. For a link $e \notin T_A$, we use B_e^A to represent the unique path in T_A connecting the endpoints of e. Similarly for $e \notin T_{OPT}$, we use B_e^{OPT} to represent the unique path in T_{OPT} connecting the endpoints of e. Since both T_A and T_{OPT} are spanning trees, both trees have the same number of edges. It is easy to see that $|E_{A\bar{O}}| = |E_{\bar{A}O}|$.

For a cross link e (not in T_A), we assume that the Algorithm A assigns a basic "Level-1" protection capacity of $p_e^{L1} = \max(0, u_e - \min_{e' \in B_e^A} p_{e'}^A)$ and a working capacity of $w_e^{L1} = \min(u_e, \min_{e' \in B_e^A} p_{e'}^A)$. The cross links e with $p_e^{L1} = p_e^A$ are *non-boosted* cross links. Note that the protection capacity of the remaining cross links e are *boosted* from p_e^{L1} to $W_{e'}$ by Algorithm A, where e' is the tree link with the largest existing traffic ($W_{o'}$), among all tree links protected by e (which are in M when link e is considered by the algorithm). Also note that there are at most $n - 2$ such *boosted* cross links, each protecting a different link in T_A. It is easy to see that $p_E^A \leq \sum_{e \in T_A} p_e^A + \sum_{e \notin T_A} p_e^{L1} + \sum_{e \in T_A}' W_e$.

The total protection capacity used by Algorithm A is thus at most $p_{E_{AO}}^A + p_{E_{A\bar{O}}}^A + p_{E_{\bar{A}O}}^{L1} + p_{E_{\bar{A}\bar{O}}}^{L1} + W_{E_{AO}} + W_{E_{A\bar{O}}}$.

Lemma 3. *For all links $e_{A\bar{O}} \in E_{A\bar{O}}$, $w_{e_{A\bar{O}}}^{OPT} \leq p_{e'}^{OPT}$, for any $e' \in B_{e_{A\bar{O}}}^{OPT}$.*

Proof. For all links $e_{A\bar{O}} \in E_{A\bar{O}}$, (note that $e_{A\bar{O}}$ is a cross link for T_{OPT})

$$w_{e_{A\bar{O}}}^{OPT} \leq \min_{e \in S} p_e^{OPT} \text{ where } S \text{ is the backup (detour) path for } e_{A\bar{O}} \text{ in } OPT \tag{1}$$

$$\leq \min_{e \in B_{e_{A\bar{O}}}^{OPT}} p_e^{OPT} \tag{2}$$

$$\leq p_{e'}^{OPT}, \text{ for any } e' \in B_{e_{A\bar{O}}}^{OPT}. \tag{3}$$

The first inequality follows from the definition of S. The second inequality holds because T_{OPT} is a maximum spanning tree of OPT based on the protection capacities p_e^{OPT}.

Lemma 4. *For all links $e_{\bar{A}O} \in E_{\bar{A}O}$, $p_{e_{\bar{A}O}}^{OPT} \leq min_{e \in B_{e_{\bar{A}O}}^A} p_e^A$.*

Proof. Note that $e_{\bar{A}O}$ is in the OPT tree T_{OPT}, but not included in T_A. Therefore the maximum protection possible on $e_{\bar{A}O}$ in any solution is at most the minimum protection assigned by Algorithm A to the links on the unique path in T_A connecting the endpoints of link $e_{\bar{A}O}$.

Lemma 5. *For any links $e_{A\bar{O}} \in E_{A\bar{O}}$, and any links $e_{\bar{A}O} \in E_{\bar{A}O}$ where $e_{\bar{A}O}$ lies on the unique path in T_{OPT} connecting the endpoints of link $e_{A\bar{O}}$, we have $p_{e_{\bar{A}O}}^{L1} \leq u_{e_{\bar{A}O}} - w_{e_{A\bar{O}}}^{OPT}$.*

Proof.

$$p_{e_{\bar{A}O}}^{L1} = \max(0, u_{e_{\bar{A}O}} - min_{e \in B_{e_{\bar{A}O}}^A} p_e^A) \qquad (4)$$

$$\leq u_{e_{\bar{A}O}} - p_{e_{\bar{A}O}}^{OPT} \qquad (5)$$

$$\leq u_{e_{\bar{A}O}} - w_{e_{A\bar{O}}}^{OPT} \qquad (6)$$

The first equality follows from the definition of $p_{e_{\bar{A}O}}^{L1}$. The second inequality holds by Lemma 4 and the fact that $u_{e_{\bar{A}O}} \geq p_{e_{\bar{A}O}}^{OPT}$. The third inequality holds by Lemma 3, with $e' = e_{\bar{A}O}$.

Lemma 6. *For any links $e_{A\bar{O}} \in E_{A\bar{O}}$, and any links $e_{\bar{A}O} \in E_{\bar{A}O}$ where $e_{\bar{A}O}$ lies on the unique path in T_{OPT} connecting the endpoints of link $e_{A\bar{O}}$, we have $p_{e_{A\bar{O}}}^A + W_{e_{A\bar{O}}} + p_{e_{\bar{A}O}}^{L1} \leq p_{e_{A\bar{O}}}^{OPT} + 2p_{e_{\bar{A}O}}^{OPT}$.*

Proof.

$$p_{e_{A\bar{O}}}^A + W_{e_{A\bar{O}}} + p_{e_{\bar{A}O}}^{L1} \leq u_{e_{A\bar{O}}} + (u_{e_{\bar{A}O}} - w_{e_{A\bar{O}}}^{OPT}) \qquad (7)$$

$$= p_{e_{A\bar{O}}}^{OPT} + u_{e_{\bar{A}O}} \qquad (8)$$

$$\leq p_{e_{A\bar{O}}}^{OPT} + 2p_{e_{\bar{A}O}}^{OPT} \qquad (9)$$

The first inequality follows from Lemma 5. The second equality follows from the definitions of protection and working traffic. The third inequality follows from the fact that p_e^{OPT} must be at least $u_e/2$ for any link $e \in T_{OPT}$. The proof is along the same lines as the proof of Case 1 in Lemma 2.

Lemma 7. $p_{E_{A\bar{O}}}^A + W_{E_{A\bar{O}}} + p_{E_{\bar{A}O}}^{L1} \leq p_{E_{A\bar{O}}}^{OPT} + 2p_{E_{\bar{A}O}}^{OPT}$.

Proof. For each link $e_{A\bar{O}} \in E_{A\bar{O}}$, we will show how to pair it up with a unique link $e_{\bar{A}O} \in E_{\bar{A}O}$, where $e_{\bar{A}O}$ lies on the unique path in T_{OPT} connecting the endpoints of link $e_{A\bar{O}}$. Summing Lemma 6 over all such pairs, the lemma follows.

Consider a bipartite graph $(E_{A\bar{O}} \cup E_{\bar{A}O}, E_m)$, where E_m contains edge $(e_{A\bar{O}}, e_{\bar{A}O})$ if $e_{\bar{A}O}$ lies on the unique path in T_{OPT} connecting the endpoints of link $e_{A\bar{O}}$. A one to one pairing of edges in $E_{A\bar{O}}$ with edges in $E_{\bar{A}O}$ can be done by finding a perfect matching on the bipartite graph. We now show a perfect matching must exists, using Hall's Theorem [21]. Given any subset S of $E_{A\bar{O}}$, find a forest F which is the union of all the unique paths in T_{OPT} connecting the endpoints of links in $E_{A\bar{O}}$. Note that all links in F are in T_{OPT}. It is easy to see that at least $|S|$ links in F are not in T_A. Otherwise we can create a cycle in T_A involving a link in S and links of F which are in T_A. Therefore links in S may be paired up with at least $|S|$ links in $E_{\bar{A}O}$, and the condition in Hall's Theorem is satisfied.

Theorem 2. *A is a polynomial time 2-approximation algorithm, i.e., $p_{E_{AO}}^A + p_{E_{A\bar{O}}}^A + p_{E_{\bar{A}O}}^{L1} + p_{E_{A\bar{O}}}^{L1} + W_{E_{AO}} + W_{E_{A\bar{O}}} \leq 2p_E^{OPT}$.*

Proof. From Lemma 7 we have $p^A_{E_{A\bar{O}}} + W_{E_{A\bar{O}}} + p^{L1}_{E_{A\bar{O}}} \leq p^{OPT}_{E_{A\bar{O}}} + 2p^{OPT}_{E_{A\bar{O}}}$. Also $p^A_{E_{AO}} + W_{E_{AO}} = u_{E_{AO}} \leq 2p^{OPT}_{E_{AO}}$ since OPT has to reserve at least $\frac{1}{2}$ the capacity of each link in E_{AO} for protection (see Lemma 6). Also $p^{L1}_{E_{A\bar{O}}} \leq p^{OPT}_{E_{A\bar{O}}}$ because there is no solution in which the working capacity of a cross link $e_{\bar{A}\bar{O}}$ can exceed $u_{e_{\bar{A}\bar{O}}} - p^{L1}_{e_{\bar{A}\bar{O}}}$. The proof follows.

4.1 Worst Case Example

We give a worse case example to show that the analysis is almost tight. Consider a ring of size n with unit capacity, the optimal solution is to put 0.5 units of protection on each link. This gives a total protection of $\frac{n}{2}$ units. Our algorithm would return a tree of size $n - 1$ and put full protection to each link, which gives a total protection of $n - 1$ units. Therefore, the approximation factor is close to 2.

5 LP-Based Approach

In this section we study a natural relaxation to our problem which can be modeled as a linear program. In this relaxation on the failure of a link its traffic can be arbitrarily split and rerouted over multiple detours. That is, in this relaxation we require that the graph $G - e$ must support at least w_e flow between nodes u and v, for every link $e = (u, v)$, using only the protection capacities on the remaining links. The LP is shown below.

For the LP, we transform the undirected graph into a directed graph by replacing each link into two directed links. We use the notation $pair(e)$ to denote the link in the opposite direction of link e. We assume that both the directed links fail together. Note that in this relaxation on the failure of link e, at most w_e traffic on it is rerouted over multiple detour paths P_e using only the protection capacities on the links on the paths P_e. We use $flow(e, i)$ to denote the amount of flow rerouted on link i on the failure of link e. The relaxed LP is

Minimize $\sum_{e \in E}(p_e/2)$
Subject to the following constraints

$$\sum_{i \in \delta_{in}(v)} flow(e, i) = \sum_{i \in \delta_{out}(v)} flow(e, i) \qquad \forall e \in E, v \in V \qquad (10)$$

$$flow(e, pair(e)) = w_e \qquad \forall e \in E \qquad (11)$$

$$flow(e, e) = 0 \qquad \forall e \in E \qquad (12)$$

$$flow(e, i) \leq p_i \qquad \forall e, i \in E, e \neq pair(i) \qquad (13)$$

$$w_e \geq W_e \qquad \forall e \in E \qquad (14)$$

$$p_e + w_e = u_e \qquad \forall e \in E \qquad (15)$$

$$w_e = w_{pair(e)} \qquad \forall e \in E \qquad (16)$$

Note that since the protection capacity of each (undirected) edge is counted twice, once for each direction, we divide the total protection capacity by 2 in the objective function.

Constraint 10 takes care of the flow conservation at each node (including the end nodes of e). For link $e = (u, v)$, a flow of w_e from node u to node v is guaranteed by sending a flow of w_e from v to u (constraint 11) and using the flow conservation equations at nodes u and v. Constraint 12 guarantees that this flow conservation is not achieved using the link (u, v). Constraint 13 guarantees that the flow does not exceed the protection capacity of any link. Constraint 14 guarantees that the working capacity of a link is at least the existing traffic. Constraint 15 guarantees that sum of protection and working capacity does not exceed the total capacity. Constraint 16 is for guaranteeing a solution for the undirected graph.

The results of [1] imply that this LP has an integrality gap of 2 when the existing working traffic W_e on all links is zero. We show however that when W_e is not restricted then this gap is $\Omega(n)$. The proof of the following lemma appears in the full paper [2].

Lemma 8. *The integrality gap of the LP or the ratio of the optimal integral solution to the optimal relaxed solution is $\Omega(n)$.*

6   NP-Hardness

In this section we show using a reduction from 3-SAT that the decision version of the problem to pre-provision existing networks to support fast restoration with minimum over-build (PREPROVISION) is NP-hard.

Let $\{X_1, X_2, \ldots, X_n\}$ be the set of variables and $\{C_1, C_2, \ldots, C_m\}$ be the set of clauses in a given instance C of 3-SAT. We construct an instance of PREPROVISION such that it has a feasible solution of total protection capacity at most $4.5n + 24m$ if and only if C is satisfiable. As shown in Figure 1, for each variable X_i we create a ring consisting of 6 unit-capacity links (arranged like a hexagon) without any existing traffic. We have a unit-capacity link splitting the ring into two equal halves. This link, with existing traffic of 1 unit, is represented by a thick segment in Figure 1. Three of the links, on the left part of the ring, correspond to literal X_i, and the other three links, on the right part of the ring, correspond to literal $\bar{X}_i$. For each clause C_j we create an octopus-like structure. In the center we have a unit-capacity link l_{C_j} with existing traffic of 1 unit (represented by a thick segment in Figure 1). Three pairs of feet, where each pair corresponds to a literal in the clause, are attached to the two endpoints of link l_{C_j}. Each foot consists of 6 serially-connected unit-capacity links without any existing traffic (represented by a dash segment). If literal X_i appears in clause C_j, one foot connects the upper endpoint of link l_{C_j} to the node between the first and second link of the half-ring corresponding to X_i. The other foot of this pair of feet connects the lower endpoint of link l_{C_j} to the node between the second and third link of the half-ring corresponding to X_i. Therefore, we have $6n + 32m$ nodes and $7n + 37m$ links in the network.

Note that if each foot in the octopus consists of only two links (rather than 6 links), we can still prove that the 3-SAT instance is satisfiable if and only if there exists a feasible solution to the PREPROVISION instance with total protection capacity of $4.5n + 8m$ (with $6n + 8m$ nodes and $7n + 13m$ links). However the analysis is much more complicated and thus we present the proof with 6 links.

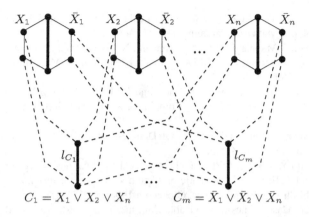

Fig. 1. Gadgets

Lemma 9. *If the 3-SAT instance is satisfiable, then there exists a feasible solution to the PREPROVISION instance with total protection capacity of* $4.5n + 24m$.

Proof. If a literal $X_i(\bar{X}_i)$ is true in the satisfying assignment, we put protection of 1 unit on each of the three links in the half-ring corresponding to $X_i(\bar{X}_i)$. Otherwise we put protection of half units on each of these three links. Therefore a total of 4.5 units of protection is reserved for each variable. In each clause, one of its literals must be true. We put protection of 1 unit on each of the 12 links between the clause and one of the true literals. We put protection of half units on each link in the remaining two pairs of feet (each pair of feet consists of 12 links). Therefore a total of 24 units of protection is reserved for each clause. For links with an existing traffic of 1 unit (the thick segments), we have no choice but to put 0 units of protection. The total protection capacity reserved is therefore $4.5n + 24m$.

It is easy to see that with this reservation all links are protected by a detour path. The link which splits a ring in a variable is protected by the 3-link half-ring corresponding to the true literal for the variable. Since there is at least one true literal in each clause, the link l_{C_j} in the middle of the gadget of a clause is protected too. The working traffic in the remaining links is either zero or half, and therefore they can be protected because all the remaining links have protection capacity of half units.

Due to lack of space, we give the proof of the other direction in the full paper [2].

Theorem 3. *The problem of pre-provisioning existing networks with minimum over-build is NP-hard.*

References

1. M. Alicherry and R. Bhatia. Pre-Provisioning Networks to Support Fast Restoration with Minimum Over-Build. Proc. INFOCOM 2004.
2. M. Alicherry, R. Bhatia, and Y. C. Wan. Designing Networks with Existing Traffic to Support Fast Restoration. Technical Report CS-TR-4598, University of Maryland, Jun 2004.

3. D. Bienstock and G. Muratore. Strong Inequalities for Capacitated Survivable Network Design Problems. Math. Programming, 89:127–147, 2001.
4. A. Balkrishnan, T. Magnanti, and P. Mirchandani. Network Design Annotated Bibliographies in Combinatorial Optimization, M.Dell'Amcio, F. Maffioli, S. Martello (eds.), John Wiley and Sons, New York, 311–334, 1997.
5. C. Chekuri, A. Gupta, A. Kumar, J. Naor, and D. Raz. Building Edge-Failure Resilient Networks. Integer Programming and Combinatorial Optimization (IPCO) 2002, pp. 439–456.
6. H. Choi, S. Subramaniam, and H. Choi. On Double-Link Failure Recovery in WDM Optical Networks. Proc. INFOCOM 2002.
7. B. S. Davie and Y. Rekhter. MPLS: Technology and Applications. Morgan Kaufmann, 2000.
8. G. Ellinas and T. E. Stern. Automatic protection switching for link failures in optical networks with bi-directional links. Proc. Globecom 1996, pp. 152–156.
9. W. D. Grover. Case studies of survivable ring, mesh and mesh-arc hybrid networks. Proc. Globecom 1992, pp. 633–638.
10. W. D. Grover and D. Stamatelakis. Cycle-oriented distributed preconfiguration: Ring-like speed with mesh-like capacity for self-planning network reconfiguration. ICC 1998, pp. 537–543.
11. W. J. Goralski. SONET, 2nd ed. McGraw-Hill Professional, New York, 2000
12. D. Haskin and R. Krishnan. A Method for Setting an Alternative Label Switched Path to Handle Fast Reroute. Internet Draft, draft-haskin-mpls-fast-reroute-05.txt, Nov 2000.
13. F. Jaeger. A survey of the double cycle cover conjecture. Cycles in Graphs, Annals of Discrete Mathematics 115: North-Holland, 1985.
14. M. Medard, S. G. Finn, and R. A. Barry. WDM Loop-back Recovery in Mesh Networks. Proc. INFOCOM 1999, vol. 2, pp. 752–759.
15. M. Medard, R. A. Barry, S. G. Finn, W. He, and S. Lumetta. Generalized Loop-back Recovery in Optical Mesh Networks. IEEE/ACM Transactions on Networking, Volume 10, Issue 1, pp. 153–164, Feb 2002.
16. P. Pan et al. Fast Reroute Techniques in RSVP-TE. Internet Draft, draft-ietf-mpls-rsvp-lsp-fastreroute-02.txt, Feb 2003.
17. J. Shi and J. P. Fonseka. Hierarchical self-healing rings. IEEE/ACM Tran. on Networking, vol. 3, pp. 690–697, Dec 1995.
18. T. E. Stern and K. Bala. Multiwavelength Optical Networks: A Layered Approach Addison-Wesley, Reading, Mass., 1999.
19. C. Thomassen. On the Complexity of Finding a Minimum Cycle Cover of a Graph. SIAM Journal of Computing, Volume 26, Number 3, pp. 675–677, June 1997.
20. J. P. Vasseur et al. Traffic Engineering Fast Reroute: Bypass Tunnel Path Computation for Bandwidth Protection. Internet Draft, draft-vasseur-mpls-backup-computation-02.txt, Aug 2003.
21. D. B. West. Introduction to Graph Theory, 2nd ed. Prentice Hall, 2001

Simultaneous Source Location

Konstantin Andreev[1,*], Charles Garrod[1,*],
Bruce Maggs[1], and Adam Meyerson[2,*]

[1] Carnegie Mellon University, Pittsburgh, PA 15213
konst@cmu.edu, {charlie,bmm}@cs.cmu.edu
[2] UCLA, 4732 Boelter Hall, Los Angeles, CA 90095
awm@cs.ucla.edu

Abstract. We consider the problem of Simultaneous Source Location –
selecting locations for sources in a capacitated graph such that a given
set of demands can be satisfied. We give an exact algorithm for trees and
show how this can be combined with a result of Räcke to give a solution
that exceeds edge capacities by at most $O(\log^2 n \log \log n)$, where n is
the number of nodes. On graphs of bounded treewidth, we show the
problem is still NP-Hard, but we are able to give a PTAS with at most
$O(1+\epsilon)$ violation of the capacities, or a $(k+1)$-approximation with exact
capacities, where k is the treewidth and ϵ can be made arbitrarily small.

1 Introduction

Suppose we are given a capacitated network and we have various demands for
service within this network. We would like to select locations for servers in order
to satisfy this demand. Such problems arise naturally in a variety of scenarios.
One example of this would be the placement of web caches in the Internet, or
file servers in an intranet. Another example would be choosing the locations of
warehouses in a distribution network.

What does it mean for a server to "serve" a demand? Most previous work
has assumed that each server can service each demand for some cost (typically
this cost is linear in some underlying distance between server and demand) [20,
16, 3]. In some cases the servers have been considered to be capacitated (each
one can provide for only some number of demands) [17]. Still, the primary goal
can be considered as minimizing the aggregate distance of demands to servers.

In many natural applications, there is no meaningful notion of distance. Consider serving a web page across the internet. The latency (travel time of a single
small packet) under low-congestion conditions tends not to be noticeable to the
end-users. The real difficulty here is the underlying capacity of the network. If
links become congested, then latency will increase and throughput will suffer. In
the case of a distribution network, there may be some relation (typically nonlinear) of costs to distance traveled. But we will definitely have to consider the

* The authors want to thank the Aladdin Center. This work was in part supported by
the NSF under grants CCR-0085982 and CCR-0122581.

K. Jansen et al. (Eds.): APPROX and RANDOM 2004, LNCS 3122, pp. 13–26, 2004.
© Springer-Verlag Berlin Heidelberg 2004

available throughput! The transportation network has capacities as well as costs, and previous work (assuming costs only) tended to ignore this constraint except at the warehouses themselves.

We consider the problem of Simultaneous Source Location (SSL) – selecting locations for sources in a capacitated network in order to satisfy given demands. Our goal is to minimize the number of sources used. Arata et al. previously gave an exact algorithm for the *Source Location* problem, in the scenario where the sources must be able to satisfy *any single* demand [2]. They also show that the Source Location problem is NP-hard with arbitrary vertex costs. In our problem we must satisfy all demands simultaneously (thus the name Simultaneous Source Location). This is a better model of the natural problems described above. Simultaneous Source Location is easier than the Source Location problem with arbitrary vertex costs explored by Arata et al. – our version of the problem can be reduced to theirs using a high-cost super-sink that is satisfied if and only if all other demands are met. However, we show that our version of the problem is NP-Hard as well and describe various approximations for it.

Our results take several forms. We describe techniques for solving Simultaneous Source Location on a variety of simple graphs. We give an exact algorithm on trees and an approximation for graphs of bounded treewidth. We observe that, in contrast to many other NP-Hard problems (for example vertex cover), Simultaneous Source Location is still NP-Hard even on graphs of treewidth two. Using our algorithm for trees combined with a result of Räcke et al. [18, 6] and later Harrelson et al. [11], we show how to solve source location on a general undirected graph while overflowing the capacity by an $O(\log^2 n \log \log n)$ factor. Combining this with a hardness result for directed graphs allows us to show that no tree decomposition similar to Räcke's decomposition can be found in the directed case. We show Simultaneous Source Location is at least hard on general undirected graphs, but there remains a significant gap between the best approximation algorithm and the known lower bound.

2　Problem Statement

An instance of Simultaneous Source Location (SSL) consists of a graph $G = (V, E)$ along with a capacity function $u : E \to R^+$ on the edges and a demand function $d : V \to R^+$ on the vertices. We must select some subset of the vertices $S \subseteq V$ to act as sources. A source set is considered to be feasible if there exists a flow originating from the nodes S that simultaneously supplies demand d_v to each node $v \in V$ without violating the capacity constraints. This is single-commodity flow – we can imagine adding a single "super-source" that is connected to each node of S with an infinite capacity edge, and a single "super-sink" that is connected to each node of V with an edge of capacity d_v, and asking whether the maximum flow equals the sum of the demands. Our goal is to select such a source set S of the smallest size.

From the above description, it is clear that the problem is in NP. A source set S can be checked for feasibility by simply solving a flow problem. Unfortunately, finding the set S of minimum size is NP-Hard.

At times we consider various generalizations of this problem. For example, we might have certain nodes in V that are not permitted to act as sources, or have costs associated with making various nodes sources and seek to find a set S of minimum cost. We will also consider simplifications of the problem that place restrictions on the graph G (for example by bounding the treewidth).

3   Solving Simultaneous Source Location on Trees

Suppose our graph $G = (V, E)$ is a tree. This special case allows us to solve SSL exactly by dynamic programming. For each vertex v and number of sources i, we define $f(v, i)$ to be the amount of flow that must be sent to v by its parent in order to satisfy all demands in the subtree of v, assuming this subtree contains i sources. Our algorithm assumes that the tree is rooted and binary; in general we can create an equivalent binary tree by creating virtual nodes and connecting them by edges of infinite capacity, as shown in Figure 2. For convenience, for each node v we define $u(v)$ to be the capacity of the edge from v to its parent.

Our algorithm is described in Figure 1.

Algorithm BinaryTree(G, d)

1. Initialize $f(v, i) = \infty$ for all $v \in V$ and $0 \le i \le |V|$
2. For each leaf vertex $v \in V$:
    (a) Set $f(v, 0) = d_v$.
    (b) Set $f(v, i) = -u(v)$ for all $i \ge 1$.
3. Consider any vertex v with children v_1, v_2 for whom f has been computed:
    (a) Loop over all values of i_1 and i_2 with $0 \le i_1, i_2 \le |V|$.
    (b) If $f(v_1, i_1) \le u(v_1)$ and $f(v_2, i_2) \le u(v_2)$ then:
        i. Set $f(v, i_1 + i_2) = \min(f(v, i_1 + i_2), max(f(v_1, i_1) + f(v_2, i_2) + d_v, -u(v)))$.
        ii. Also set $f(v, i_1 + i_2 + 1) = -u(v)$.
4. Continue until f is defined at all vertices.
5. Return the minimum k such that $f(r, k) \le 0$ where r is the root.

Fig. 1. Algorithm for SSL on a Binary Tree

Assuming that the above algorithm computes $f(v, i)$ correctly for each vertex, the correctness of the result is immediate. We need to show that $f(v, i)$ correctly represents the minimum amount of flow that must be sent from the parent of v provided the number of sources in the subtree is i.

Theorem 1. *The algorithm described produces an exact solution to the source location problem on a binary tree.*

Proof. The proof will be by induction. Our base case is at the leaves. Either a leaf is a source or it is not. If the leaf v is a source, then it requires no flow from its parent, and can send at most $u(v)$ flow upwards along its edge. This yields $f(v, 1) = -u(v)$. On the other hand, if the leaf is not a source it requires flow d_v

(its demand) from the parent, so $f(v, 0) = d_v$. Of course, it might not be feasible to send this amount of demand if $d_v > u(v)$.

We now consider any node v. Suppose we have correctly computed $f(v_1, i_1)$ and $f(v_2, i_2)$ for all values i_1, i_2 for the children of node v. Suppose we would like to compute $f(v, i)$. There are i sources to be placed in this subtree. If v is not a source itself, then all the sources are in the child trees. The total demand sent into v will have to be enough to satisfy the demand d_v and additionally to satisfy any residual demand on the children. This means $f(v, i) = \min(\max(d_v + f(v_1, i_1) + f(v_2, i - i_1), -u(v)))$ where the minimum is over choices of i_1 and the "max" term ensures that a child node cannot provide more flow to its parent than the capacity of their connecting edge. This is exactly what will be computed in the algorithm, in step 3b(i). Notice that if satisfying either child tree in this way would require overflowing a capacity $u(v_1)$ or $u(v_2)$ then this allocation of sources to subtrees is not feasible and so should not be considered; this is resolved in step 3b. However, it is also possible that v is itself a source. In this case, provided there is some choice of $i_1 \leq i - 1$ such that $f(v_1, i_1) \leq u(v_1)$ and $f(v_2, i - i_1 - 1) \leq u(v_2)$, we will be able to produce a solution. This solution can send $u(v)$ upwards since v itself is a source. This is dealt with in step 3b(ii). It follows that the algorithm computes exactly the correct values $f(v, i)$ and the algorithm solves SSL.

Theorem 2. *We can solve SSL in time $O(n^3)$ on a tree, even if some nodes are disallowed as sources.*

If our tree is non-binary we can replace any node with more than two children by multiple nodes as shown in Figure 2. This increases the number of nodes by at most a constant factor.

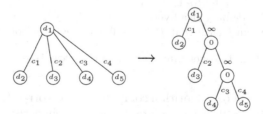

Fig. 2. By adding additional nodes linked by infinite capacity, we can convert any non-binary tree into an equivalent binary tree by only doubling the number of nodes in the tree

We can also modify our algorithm to disallow certain nodes as sources. If a leaf v cannot be a source, then we ignore step 2b and instead set $f(v, i) = d_v$ for all i. If some higher node cannot be a source, then we remove step 3b(ii) for that node (this step considers the case where v is a source). The correctness proof is the same as before.

4   Using Räcke's Result

In recent work, Harold Räcke showed that for any undirected graph, it is possible to construct a tree which approximately captures the flow properties of the original graph[18]. Later work [6, 11] improved the result and described a polynomial-time algorithm to construct such a tree. The most recent result is stated more precisely in the following theorem:

Theorem 3. *Given any capacitated, undirected graph* $G = (V, E, u : E \to R^+)$, *there exists a capacitated tree* $T = (V_T, E_T, u : E_T \to R^+)$ *with the following properties:*

1. *The vertices of* V *are the leaves of* T.
2. *For any multicommodity flow* F *which is feasible on* G, *there exists a flow of equal value between the corresponding leaves on* T.
3. *For any flow* F_T *feasible between the leaves of* T, *there is a feasible* $\frac{1}{\rho} F_T$ *flow on* G *for some* $\rho = O(\log^2 n \log \log n)$.

This gives us an approximation for the source location problem. We first construct a tree T as described above. We then solve SSL on the tree, permitting only the leaves to act as sources, using the algorithm of Section 3. We consider using the sources we have obtained on the original graph. We know there exists a flow F_T on the tree from our selected source nodes which satisfies all the demands. It follows that $\frac{1}{\rho} F_T$ is feasible on the graph. We conclude that if we violate the capacities by a factor of $\rho = O(\log^2 n \log \log n)$, then we have a feasible solution to source location. On the other hand, any feasible solution on the graph must also be feasible on the tree. This allows us to produce an exact solution (in terms of the number of sources) while increasing the capacities by $O(\log^2 n \log \log n)$.

Theorem 4. *We can produce an optimum (in terms of number of sources) solution to SSL in polynomial time, if we permit* $O(\log^2 n \log \log n)$ *stretch on the capacities.*

Our results also have interesting implications for directed graphs. Consider the possibility of a Räcke-like representation of a directed graph by a tree, where the capacity on an edge when routing "upwards" might differ from the "downwards" capacity. Suppose such a thing existed, and could be computed in polynomial time, for some factor ρ. This would enable us to solve SSL exactly on directed graphs, while exceeding capacities by a factor of ρ. But this is NP-Hard, as will be shown in Section 6.2. Thus we have the following:

Theorem 5. *No Räcke-like decomposition of a directed graph into an* ρ-*approximately flow-conserving tree can be computed in polynomial time, for any value of* ρ *polynomial in* n.

Note that our hardness result is a computational hardness result and assumes a tree-like structure decomposition. Azar et al. [5] have found an existential hardness result which is true for any decomposition, but their bound is weaker: $O(\sqrt{n})$.

5   Simultaneous Source Location with Bounded Treewidth

5.1   Defining Treewidth

The notion of treewidth was introduced by Robertson and Seymour [19]. Many problems that are in general intractable become polynomial-time solvable when restricted to graphs of bounded treewidth. Furthermore, many graphs arising from natural applications have bounded treewidth. A good survey on the topic is given by Bodlaender [7]. Here is one of the many equivalent definitions of treewidth:

Definition 1. *A graph $G = (V, E)$ has treewidth k if there exists a tree $\tau = (V_\tau, E_\tau)$ along with a mapping $f : V_\tau \to 2^V$ with the following properties:*

1. *For all $\alpha \in V_\tau$, $|f(\alpha)| \leq k + 1$.*
2. *For any $(u, v) \in E$ there exists some $\alpha \in V_\tau$ such that $u, v \in f(\alpha)$.*
3. *For any $\alpha, \beta, \gamma \in V_\tau$ where β lies along the path from α to γ, if for some $x \in V$ we have $x \in f(\alpha)$ and $x \in f(\gamma)$, then we must also have $x \in f(\beta)$.*

The above conditions essentially state that each tree vertex represents some subset of at most k graph vertices, each edge in the graph has its endpoints represented together in at least one of the tree vertices, and the set of tree vertices which represent a single graph vertex must form a contiguous subtree.

We observe that it is possible to produce such a tree decomposition for a graph of treewidth k in time linear in the number of nodes and vertices (but exponential in k). Assuming k is constant, we are able to produce a tree decomposition – and thereby implicitly detect graphs of constant treewidth – in polynomial time. In general, however, computing the treewidth of an arbitrary graph is NP-Hard.

5.2   Nice Decompositions

Bodlaender [8] also introduced the notion of a nice decomposition and proved that any tree decomposition of a graph can be transformed into a nice decomposition still of polynomial size. In a nice decomposition, each node $\alpha \in V_\tau$ has one of the following types:

- A *leaf* node α has no children, and $|f(\alpha)| = 1$
- An *add* node α has one child β with $f(\alpha) = f(\beta) \bigcup \{v\}$ for some node $v \in V$
- A *subtract* node α has one child β with $f(\alpha) = f(\beta) - \{v\}$ for some node $v \in f(\beta)$
- A *merge* node α has two children β, γ with $f(\alpha) = f(\beta) = f(\gamma)$

In addition, the nice decomposition has a root node ρ (which is a *subtract* node) with $f(\rho)$ empty.

5.3   Approximation for Graphs of Bounded Treewidth

Suppose we are given a graph $G = (V, E)$ with treewidth k, for which we would like to approximate the SSL problem. Our algorithm takes in some set of current sources S along with a graph (V, E) and returns a set S' of sources which are feasible for the given graph. Our algorithm for this problem appears in Figure 3.

Algorithm SL(S, V, E)

1. Check whether the source set S is feasible for (V, E); if so return S
2. If not, find sets X and B_X that have the following properties:
   (a) $|B_X| \le k + 1$
   (b) For all $(x, y) \in E$ with $x \in X$ and $y \in V - X$, we have $x \in B_X$
   (c) $S \bigcup (V - X)$ is not a feasible source set
   (d) $S \bigcup (V - X) \bigcup B_X$ is a feasible source set
3. Recursively solve $SL(S \bigcup B_X, (V - X) \bigcup B_X, E)$

Fig. 3. Algorithm for SSL with treewidth k

We claim that the set S^R of returned sources has $|S^R| \le (k + 1)|S^*|$ where $|S^*|$ is the smallest feasible set of sources for (V, E) which includes the given set S as a subset.

Lemma 1. *Assuming we are always able to find sets X and B_X with the properties described, algorithm SL is a $(k+1)$-approximation for the source location problem.*

The prove is done by induction on the number of sources required by the optimum solution.

Of course, we still need to prove that we can always find the sets X, B_X with the required properties in polynomial time. In a general graph, such a pair of sets might not even exist, but we will use the assumption of treewidth k to prove existence and the ability to find the sets in polynomial time.

Lemma 2. *If the current set of sources S is not feasible for the graph $G = (V, E)$ with treewidth k, then there exists a pair of sets X, B_X with the required properties; furthermore, such a pair of sets can be found in polynomial time.*

Proof. We produce a tree decomposition (τ, f) of G. For each tree node α, we define τ_α to be the subtree of τ rooted at α. We define $f(\tau_\alpha) = \bigcup_{\beta \in \tau_\alpha} f(\beta)$. For each node α we will test whether $S \bigcup (V - f(\tau_\alpha))$ is a feasible set of sources. We find node α such that $S \bigcup (V - f(\tau_\alpha))$ is infeasible, but such that $S \bigcup (V - f(\tau_\beta))$ is feasible for each child β of α. Note that such an α must exist; we simply travel upwards from the each leaf of the tree until we find one. We now consider returning $X = f(\tau_\alpha)$ and $B_X = f(\alpha)$. We will show that these sets satisfy the required properties.

Since the graph has treewidth k we know $|B_X| = |f(\alpha)| \le k+1$. Consider any $(x, y) \in E$ with $x \in X$ and $y \in V - X$. From the definition of treewidth, there must exist some node β with $x, y \in f(\beta)$. Since y is not in $f(\tau_\alpha)$ we conclude that β is not in τ_α. On the other hand, there is some node $\gamma \in \tau_\alpha$ such that $x \in f(\gamma)$ (this follows from $x \in X$). The path from γ to β must pass through α, so the treewidth definition implies $x \in f(\alpha)$ and therefore $x \in B_X$ as desired. The selection of α guarantees that $S \bigcup (V - X)$ is not a feasible source set. This leaves only the fourth condition for us to prove.

We consider the children of α. A pair of them γ_1, γ_2 must have $f(\gamma_1) \bigcap f(\gamma_2) \in f(\alpha)$ because of the contiguity property of tree decomposition. Thus the sets of nodes represented by the subtrees of the children can intersect only in the nodes of $B_X = f(\alpha)$. We know that for each child γ, the set of nodes $S \bigcup (V - f(\tau_\gamma))$ would be feasible. It follows that we can cover all the demand of $f(\tau_\gamma)$ using nodes of S and nodes external to the set. Since the children sets are disjoint except for nodes which we have declared to be sources, the flows to satisfy each child subtree are edge-disjoint; any flow from external nodes must also pass through B_X, and we conclude that we can cover $f(\tau_\gamma)$ using $S \bigcup B_X$. It follows that we can cover all of X using the sources $S \bigcup B_X$, making $S \bigcup B_X \bigcup (V - X)$ a feasible source set for the entire graph.

Theorem 6. *Algorithm SL produces a $k + 1$-approximation to the SSL problem on a graph of treewidth k.*

5.4   Bounded Treewidth with Capacity Stretch

We will describe an exact algorithm for the SSL problem on graphs of bounded treewidth. The running time of this algorithm will be exponential in the treewidth, and also depends polynomially on the maximum edge capacity. In the general case where the capacities may be large, we will use the technique of Appendix A to obtain a solution with $1 + \epsilon$ stretch on the edge capacities.

Suppose we have found a nice tree decomposition (τ, f). We will construct a set of vectors of dimension $k + 3$. Each vector has the following form:

$$(\alpha, i, f_1, f_2, ..., f_{k+1})$$

Here $\alpha \in V_\tau$, $0 \le i \le |V|$, and the f_i are feasible flow values (we assume these are from a polynomially-bounded range of integers). Let $S_\alpha = f(\tau_\alpha) - f(\alpha)$ represent the nodes represented by the subtree rooted at α minus the nodes of its boundary (the nodes of $f(\alpha)$ itself). A feasible vector represents the excess flow needed to be sent directly from the nodes of $f(\alpha)$ to S_α to satisfy all the demand in S_α if S_α contains i sources.

We observe that if k is a constant and flow is polynomially-bounded, then the number of possible vectors is polynomial in size. We will determine which such vectors are feasible in polynomial time, then use this to solve SSL. Our algorithm is as follows:

1. Start with the empty set of feasible vectors.
2. Consider each node α from the bottom up:
   – If α is a leaf, then add vector $(\alpha, 0, 0)$.
   – If α is an add node with child β, then for each feasible vector for β, copy that vector for α, placing flow 0 in the new position corresponding to the additional node in $f(\alpha) - f(\beta)$.
   – If α is a merge node with children β, γ then for each pair of feasible vectors x_β, x_γ for the children, create a vector for α: $x_\alpha = x_\beta + x_\gamma$ (adding the number of sources and the flows while changing the choice of node from V_τ to α).
   – If α is a subtract node with child β, then consider each feasible vector x_β for the child. Let the subtracted node be $b \in f(\beta) - f(\alpha)$. This node requires some flow r_b which is the sum of the demand d_b and the flow value for b in x_β. We consider all feasible allocations of flow $F(a)$ to nodes $a \in f(\alpha)$ such that $|F(a)| \leq u(b, a)$ and $\sum_{a \in f(\alpha)} F(a) \geq r_b$. For each such allocation we construct a vector x_α whose number of sources is equal to x_β and with flow value at a equal to the flow value in x_β plus $F(a)$. This corresponds to refusing to make b a source. We now consider making b a source. This corresponds to creating a vector x_α with one more source than the vector x_β. We set the flow value for a node $a \in f(\alpha)$ to be the flow value in x_β minus $u(b, a)$.
3. Now consider all vectors for the root node ρ. These are simply pairs (ρ, i) since $f(\rho)$ is empty. We return the minimum value of i such that (ρ, i) is in the feasible set.

Theorem 7. *If there are F possible flow values, the above BTW algorithm runs in time $O(kn^2 N F^{2k+2})$ where $n = |V|$ and $N = |V_\tau|$ and k is the treewidth.*

This running time is polynomial assuming that F is polynomial and k is a constant. If the number of possible flow values is not polynomial, we can use the result of Appendix A to effectively reduce the number of flow values. This will cause us to exceed the graph capacities by a factor of $1 + \epsilon$.

5.5   Lower Bound for Treewidth 2 Graphs

We show that the SSL problem is NP-Hard even on graphs with treewidth two.

Theorem 8. *SSL is NP-Hard even on graphs with treewidth two.*

Proof. The proof is by reduction from subset sum. We are given a set of numbers $\{x_1, x_2, \ldots, x_n\}$, and would like to determine whether some subset of the given inputs sums to A. Suppose the sum of all the numbers is S. We construct an SSL instance with $2n + 2$ nodes. Each number will be represented by a pair of nodes of demand S with an edge of capacity S between them. Our example is a four level graph. On the first level we have a node of demand A which connects to one side of every pair of nodes that represent a number. The capacity on the edge between the A node and the number x_i node is exactly x_i. The number nodes

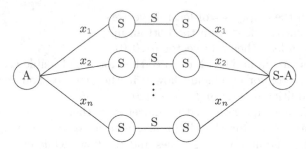

Fig. 4. Simultaneous Source Location is NP-Hard even on graphs of treewidth two. To satisfy each demand in this graph using only n sources we must find a partition of $\{x_1, x_2, \ldots, x_n\}$ whose sums are A and $S - A$

from the second level are paired with the nodes from level 3. All nodes from level 3 are connected to a single node at level 4 with an edge of capacity corresponding to their number. The node at level 4 has demand $S - A$. This graph is shown in Figure 4. If there exists a subset of the numbers summing to A, then we can place sources on the lefthand node for each of the numbers in that subset and the righthand node for all the other numbers; it is straightforward to see that this is a feasible SSL making use of n sources. On the other hand, consider a SSL solution. We must select one of the two nodes for each of the numbers (otherwise there is not enough capacity to satisfy them). It follows that the SSL uses at least n sources. If exactly n sources are used, then the result corresponds to a subset sum solution. It follows that solving source location exactly on this graph will solve subset sum. The graph given has treewidth two; we can see this because if we remove the node of demand A, the remaining graph is a tree. We take the (treewidth one) tree decomposition and add the node of demand A to the subset $f(\alpha)$ for all α. This is a tree decomposition of width two.

6 Simultaneous Source Location on Directed Graphs

6.1 Greedy $O(\log n)$ Approximation

We are given a directed graph $G = (V, E)$ for which we would like to approximate the SSL problem. We propose a simple greedy algorithm. We start with no sources and no demand satisfied. We add the source which maximizes the increase in the total satisfied demand. We repeat this until all demand is satisfied.

Theorem 9. *The greedy algorithm gives an $O(\log n)$ approximation on the number of sources with no violation of the capacities.*

Proof. Suppose that the optimum solution uses t sources. At some point in time, suppose our current set of sources can cover demand d and the total demand in the graph is D. Consider the residual graph after adding the flows from our sources to satisfy demand d. The residual demand is $D - d$, and if we were to add all the optimum sources we would be able to satisfy the full residual demand (note that this is essentially single commodity flow since sources are equivalent). It follows that we can add one source to the residual graph to satisfy demand $\frac{D-d}{t}$. We apply the standard greedy analysis for problems like $SETCOVER$ to show that the full demand will be covered in $O(\log D)$ steps. Assuming that the maximum and minimum demand are within a polynomial factor of one another, we are done. Otherwise, we can apply scaling arguments of Appendix A to give the desired $O(\log n)$ factor.

6.2   Lower Bound for Directed Graphs

We show that $O(\log n)$ is the best approximation factor we can expect to obtain for the directed SSL problem in polynomial time, due to a reduction from set cover.

Theorem 10. *We cannot approximate directed SSL to better than $O(\log n)$ in the worst case, unless $NP \subset DTIME(n^{O(\log \log n)})$.*

Proof. Suppose we would like to solve a set cover instance. We construct a graph with one node for each set, one node for each element, and one additional node. The additional node is connected by a directed edge of capacity one to each of the sets. Each set is connected by a directed edge of capacity N, where N is more than the sum of the number of sets and elements, to each of its elements. The additional node has demand one, each set has demand one, each element has demand N. We solve the SSL problem to some approximation factor ρ on this graph. We first observe that no element should be selected as a source; otherwise we could simply select one of the sets containing that element as a source instead. Second, we observe that the additional node will be selected as a source. We have a solution consisting of ρt nodes, where t is the optimum. Consider the set nodes that we selected as sources. Every element node must be connected to one of these set nodes in order for its demand to be satisfied (note N is greater than the number of sets). It follows that we have a set cover of size $\rho t - 1$. Similarly, observe that any set cover, plus the additional node, forms a feasible SSL solution. So we have obtained a $\frac{\rho t-1}{t-1} \geq \rho$ approximation to set cover. This is unlikely for ρ smaller than $\log n$ due to the results of Feige [10].

We observe that even if we are allowed to violate edge capacities by some large factor (say less than N), the reduction from set cover still holds.

7   Lower Bound for Undirected Graphs

We show that SSL does not have a polynomial-time approximation scheme via an approximation-preserving reduction from vertex cover.

Assume we can approximate SSL on any graph to a constant α. Now for an instance of Vertex Cover on a graph G=(V,E), we will setup an instance of SSL.

For every vertex the demand is equal to its degree and all edge capacities are unit.

The main observation is that a vertex cover set and a source location set on this graph are equivalent. It is easy to see that a feasible vertex cover is a feasible source location: for every edge at least one end is in the vertex cover and therefore a source. Thus every vertex will have its demand satisfied. On the other hand a source location set can't have a two hop path over a non source vertex, because this vertex will be starved. Hence every edge has at least one end in the source location set, i.e. the source location set is a feasible vertex cover.

Using the assumption we can approximate the SSL problem on G. Therefore we find a set $S \subset V$ which covers all edges in E and is within α of the optimal Vertex Cover. However as Hastad [12] showed Vertex Cover is inapproximate if $\alpha < 7/6$ (although an approximation factor better than 2 would be a surprising improvement on existing vertex cover results).

Theorem 11. *Simultaneous Source Location is* $1.36067 - \epsilon$ *hard on general undirected graphs if edge capacities are not violated.*

The proof follows from the reduction above and the recent hardness results by Dinur and Safra [9].

8   Conclusions

We define the Simultaneous Source Location problem and solve the problem exactly on trees. We present a $(1+\epsilon)$ violation of the capacities PTAS for graphs of bounded treewidth. On general graphs we find a solution with exact number of sources which can exceed the capacities by at most a factor of $O(\log^2 n \log \log n)$. We show a $O(\log n)$ factor approximation on the number of sources with no violation of the capacities for general directed graphs. We believe that many interesting applications of this problem involve graphs of low treewidth; many of the connectivity graphs of real networks have been observed to have low tree width [8].

The main open problem is the approximability of SSL on undirected graphs of large treewidth. No constant approximation on the number of sources is known, even if we allow constant violation of the capacities. The only lower bound on approximability with exact capacities is $1.36067 - \epsilon$. An approximation factor asymptotically better than 2 would be a surprising improvement on existing vertex cover results [15]. One can also consider adding costs on the edges and/or the vertices.

References

1. W. A. Aiello, F. T. Leighton, B. M. Maggs, M. Newman. Fast algorithms for bit-serial routing on a hypercube *Mathematical Systems Theory*, 1991
2. K. Arata, S. Iwata, K. Makino, and S. Fujishige. Locating sources to meet flow demands in undirected networks. *Journal of Algorithms, 42*, 2002.

3. V. Arya, N. Garg, R. Khandekar, V. Pandit, A. Meyerson, and K. Munagala. Local search heuristics for k-median and facility location problems. *Proceedings of the 33rd ACM Symposium on Theory of Computing*, 2001.
4. Y. Aumann, Y. Rabani. An $O(\log k)$ approximate mincut max-flow theorem and approximation algorithms *SIAM Journal of Computing, 27(1)*, 1998
5. Y. Azar, E. Cohen, A. Fiat, H. Kaplan, and H. Räcke. Optimal oblivious routing in polynomial time. *Proc. 35th Annual ACM Symposium on Theory of Computing*, 2003.
6. M. Bienkowski, M. Korzeniowski, and H. Räcke. A practical algorithm for constructing oblivious routing schemes. *Fifteenth ACM Symposium on Parallelism in Algorithms and Architectures*, 2003.
7. H. L. Bodlaender. Treewidth: Algorithmic Techniques and Results. *Proceedings 22nd International Symposium on Mathematical Foundations of Computer Science*, 1997
8. H. L. Bodlaender. A partial k-arboretum of graphs with bounded treewidth. *Theoretical Compututer Science, 209(1)*, 1998.
9. I. Dinur and S. Safra. On the importance of being biased. *Proceedings of the 34rd ACM Symposium on Theory of Computing*, 2002.
10. U. Feige. A threshold of $\ln n$ for approximating set cover. *Journal of the ACM*, 1995.
11. C. Harrelson, K. Hildrum, and S. Rao. A polynomial-time tree decomposition to minimize congestion. *Fifteenth ACM Symposium on Parallelism in Algorithms and Architectures*, 2003.
12. J. Hastad. Some optimal inapproximability results. *Proceedings 29th Ann. ACM Symp. on Theory of Computing, ACM, 1-10*, 1997
13. P. Klein, S. A. Plotkin, S. Rao. Excluded minors, network decomposition, and multicommodity flow. *Proceedings of the 25th ACM Symp. on Theory of Computing*, 1993
14. T. Leighton, S. Rao. An approximate max-flow min-cut theorem for uniform multicommodity flow problems with applications to approximation algorithms *Proceedings of the 29th Symp. on Foundations of Computer Science*, 1988
15. S. Khot and O. Regev. Vertex cover might be hard to approximate within $2 - \epsilon$. *Proceedings of the 17th IEEE Conference on Computational Complexity*, 2002.
16. M. Mahdian, Y. Ye, and J. Zhang. Improved approximation algorithms for metric facility location problems. *APPROX*, 2002.
17. M. Pál, É Tardos, and T. Wexler. Facility location with nonuniform hard capacities. *Proceedings of the 42nd IEEE Symposium on the Foundations of Computer Science*, 2001.
18. H. Räcke. Minimizing congestion in general networks. *IEEE Symposium on Foundations of Computer Science*, 2002.
19. N. Robertson, P. D. Seymour. Graph Minors II. Algorithmic aspects of tree width *Journal of Algorithms 7*, 1986.
20. D. Shmoys, É Tardos, and K. Aardal. Approximation algorithms for facility location problems. *Proceedings of the 29th ACM Symposium on Theory of Computing*, 1997.

A   Dealing with Super-Polynomial Capacities

Some of our analyses assume that there are only a polynomial number of capacities. Here we present a method to extend those analyses to when the capacities are super-polynomial, allowing us to satisfy demands while exceeding capacities by a $1 + \epsilon$ factor in those cases.

When capacities are super-polynomial, we express the flow and capacity of each edge as αF^i for some value of i (which might be different from edge to edge) and some value of α which is between F and F^2. This can be done by rounding every flow up to the nearest value. Once this is done, we might no longer have a feasible flow. The new capacities (which might be larger than the old capacities by a $1 + \frac{1}{F}$ factor) will not be exceeded. However, flow conservation may no longer hold. Consider any node. If the original inflow was f then the outflow was f also. But it is possible that after this rounding upwards, the inflow is still f and the outflow has increased to $f(1+\frac{1}{F})$. We consider such a flow to be feasible. This means that a node might effectively "generate" some amount of flow, but this amount is at most $\frac{1}{F}$ of the inflow.

Now consider any graph with a feasible flow under the representation described above. We would like to transform this into a real flow. If we consider splitting the "generated" flow equally among outgoing edges, we will see that as flow passes through a node the percentage of "real" flow might decrease by a factor of $\frac{F}{F+1}$. In the worst case, some edge might have a fraction of "real" flow equal to $(\frac{F}{F+1})^n \geq 1 - \frac{n}{F}$.

We let $F = \frac{n}{\epsilon}$. It follows that we can satisfy at least $1 - \epsilon$ of each demand while exceeding capacities by a factor of $1 + \frac{\epsilon}{n}$. Since we can scale the flows and capacities, this means we can satisfy the demands while exceeding capacities by $1 + \epsilon$ factor.

This is useful in bounding the running time and approximation factor of various techniques for approximating SSL.

Computationally-Feasible Truthful Auctions
for Convex Bundles

Moshe Babaioff and Liad Blumrosen[*]

School of Engineering and Computer Science
The Hebrew University of Jerusalem, Jerusalem 91904, Israel
{mosheb,liad}@cs.huji.ac.il

Abstract. In many economic settings, convex figures on the plane are for sale. For example, one might want to sell advertising space on a newspaper page. Selfish agents must be motivated to report their true values for the figures as well as to report the true figures. Moreover, an approximation algorithm should be used for guaranteeing a reasonable solution for the underlying NP-complete problem. We present truthful mechanisms that guarantee a certain fraction of the social welfare, as a function of a measure on the geometric diversity of the shapes. We give the first approximation algorithm for packing arbitrary weighted compact convex figures. We use this algorithm, and variants of existing algorithms, to create polynomial-time truthful mechanisms that approximate the social welfare. We show that each mechanism achieves the best approximation over all the mechanisms of its kind. We also study different models of information and a discrete model, where players bid for sets of predefined building blocks.

1 Introduction

The intersection between Micro-Economic theory and Computer-Science theory raises many new questions. These questions were studied recently by researchers from both disciplines (see, e.g., the surveys in [13,6]). A leading example for a problem in this intersection is the *Combinatorial Auction* problem. In a combinatorial auction, a finite set of heterogenous items is for sale, and each selfish agent has a valuation for every subset of these items. As the auction designers, we try to find an allocation of the items among the agents that maximizes the "social welfare" (i.e., a set of disjoint packages that maximizes the sum of valuations) or at least to find a good approximation.

In this paper, we study a variant of combinatorial auctions: e.g., a newspaper wants to sell an advertising space on a newspaper page. Agents might have different preferences about their desired space: the size of the ad, its location on the page, whether its figure is rectangular, square, or elliptic etc. Each agent submits a bid for her favorite figure, and we try to find the allocation that maximizes the social welfare. The underlying packing problem is known to be NP-hard, even for the very simple case of packing 2×2 squares ([15]). Thus, we settle for a computationally-efficient mechanisms that approximate the social welfare.

---

[*] The authors are grateful to Noam Nisan for many helpful discussions, and to Ron Lavi for his comments on an earlier draft.

K. Jansen et al. (Eds.): APPROX and RANDOM 2004, LNCS 3122, pp. 27–38, 2004.
© Springer-Verlag Berlin Heidelberg 2004

In our model, the plane ($\Re^2$) is for sale. Let N be a finite set of agents ($|N| = n$). Each agent has a private non-negative valuation $v_i \in \Re^+$ for a single compact convex[1] figure s_i and for any figure that contains it (other figures have a valuation of 0). Every agent submits a bid for her desired figure (e.g., an advertiser might want the lower half of the first page in a newspaper). After receiving all bids, the auctioneer determines a set of winning agents with disjoint figures and the payment that each agent should pay. Note that the agents demand figures in fixed locations in the plane, and the auctioneer cannot translate or rotate them. Bidding for convex figures is common in many real-life scenarios that involve "geometric" bids, e.g., selling real-estate lots, newspaper ads and spectrum licenses in different locations. In most existing real-estate or advertising auctions, agents are forced to bid on predefined disjoint figures. This might result in inefficient allocations that can be avoided by allowing the agents to bid for arbitrary figures (which in turn makes the computational problem harder).

Note that the problem addressed in this paper is more than just finding an algorithm with a good approximation ratio. The agents in our model are selfish, and they may report untruthful information if this is beneficial for them. We want to design *incentive-compatible* mechanisms, in which each agent uses a strategy that is best for her own selfish interest (a *dominant strategy*), and yet, a certain approximation for the social welfare is guaranteed. A general scheme for achieving a welfare-maximizing incentive-compatible mechanism is the family of Vickrey-Clarke-Groves (VCG) mechanisms (see [11] for a review). However, for implementing such mechanisms we must allocate the goods optimally (otherwise it will not be truthful [14]). Thus, since finding the optimal allocation is an *NP*-hard problem, we must find incentive-compatible mechanisms that are non-VCG. Almost all the non-VCG mechanisms currently known are for models where the agents hold a single secret value (*single-parameter* models)[2].

In light of these impossibilities, we assume that each agent is interested in a single figure. Lehmann et al. [10] initiated the study of the *Single-Minded Bidders model* for combinatorial auctions. Our model is unique since the bids have some common geometric properties (e.g., convexity), and also because we actually auction an infinite (even uncountable) number of goods (the points in the plane).

We differentiate between two models of information (similar to the differentiation done in [12, 3]) . In the first model, the auctioneer knows which figure each agent wants, but does not know how much this agent is willing to pay for this figure. This model is the *Known Single-Minded (KSM) model*, and this is indeed a "single-parameter" model. In the second model, called the *Unknown Single-Minded (USM) model*, both the figures and the values are unknown. In the KSM model we should motivate the agents to truthfully declare their true values, where in the USM model the agents might submit untruthful bids both for their desired figures and their values[3].

---

[1] Actually we prove our results for a more general model that allows non-convex bids as well, with some more general restrictions.

[2] Recent results ([9]) show that in many reasonable settings, essentially no IC mechanisms exists for multi-parameter models, except the family of weighted VCG mechanisms.

[3] Note that the USM model does not fit into the "single-parameter" definition, since the agents have both their values and their figures as their secret data.

Another differentiation we make is between a continuous and a discrete model. In the *continuous model*, each agent is interested in an arbitrary compact convex figure in $\Re^2$. For example, if a piece of land is for sale, each agent can draw an arbitrary figure on the map. In the *discrete model*, the plane contains predefined atomic building blocks (or tiles), and each agent is restricted to bids for a set of building blocks which are contained in some convex figure. For example, if we wanted to resell the island of Manhattan, the basic building blocks would be street blocks bounded between two consecutive streets and two consecutive avenues. These blocks are typically convex, though not necessarily rectangular (e.g., because of the diagonal Broadway Avenue).

Related Work:
Our research relates to a sub-field of Micro-Economics called *Mechanism Design*, which studies ways to design mechanisms that encourage agents to behave in a way that results in some desired global properties (see, e.g., [11]). Nisan and Ronen [14] introduced this concept to CS by the name *Algorithmic Mechanism Design*.

Weighted packing of rectangles in the plane was studied in several papers. Hochbaum and Maass [7] proposed a *shifting strategy* for a special case of square packing, and generalizations for arbitrary squares appear in, e.g., [5, 4]. Khanna et al. [8] used similar methods in a model where axis-parallel rectangles lie in a $n \times n$ grid. They presented an algorithm that runs in polynomial time and achieves an $O(log(n))$-approximation for the optimal welfare. However, it is an open question whether a better approximation (and in particular, a constant approximation) exists for this problem.

Our Contribution:
We measure the quality of the approximations achieved by mechanisms in our model according to an *aspect ratio R*, which measures how diverse are the dimensions of the figures demanded by the agents. R is defined as the ratio between the maximal *diameter* of a figure and the minimal *width* of a figure (formally defined in Section 2). For different families of figures, we construct IC mechanisms, either for the KSM or for the USM models. This mechanisms are also *individually-rational*, i.e., agents will not pay more than they value the figure they receive (if any). Therefore, our approximation improves as the dimensions of the figures become closer[4].

We study three different families of figures: compact convex figures, rectangles, and axis-parallel rectangles. For convex figures in the USM model, we achieve an $O(R^{\frac{4}{3}})$-approximation to the social welfare. If the bids are restricted to rectangles (not necessarily axis-parallel), we achieve a better approximation of $O(R)$.

If the agents bid for axis-parallel rectangles, we can use a slight modification of the algorithm due to Khanna et al. [8] to design an IC mechanism that achieves an approximation ratio of $O(log(R))$ (the best known approximation ratio for this problem).

We also present a novel allocation algorithm that achieves an $O(R)$-approximation for packing arbitrary compact convex figures, and as far as we know this is the first approximation algorithm for this problem. We use this algorithm for constructing an IC mechanism, with the same approximation ratio, for the KSM model.

---

[4] For instance, when all figures are disks with the same radius up to a constant, our mechanisms achieve a constant approximation.

The incentive-compatible mechanisms we present for the USM model are based on a family of greedy algorithms presented by Lehmann et al. [10]. For standard combinatorial auctions, Lehmann et al. normalized the values by $|S|^\alpha$, where $|S|$ is the number of items in the bundle S and α is some real constant, and then run a simple greedy allocation algorithm on the normalized values. They showed that choosing $\alpha = \frac{1}{2}$ guarantees the best polynomial approximation ratio (unless $NP = ZPP$).

We present mechanisms, called α-greedy mechanisms, that normalize the values using the geometric area of the figures. That is, for $\alpha \in \Re$ we assign a normalized value of $\frac{v}{q^\alpha}$ to a bid with a value v for a figure with a geometric area q. We show that, somewhat surprisingly, for compact convex figures the optimal value for α is $\frac{1}{3}$, resulting an $O(R^{\frac{4}{3}})$-approximation. The difference between the results of Lehmann et al. and ours derives from the different divisibility properties of packages in the two models. In their model, a finite set of goods is traded, and for a package to intersect "many" disjoint packages, its size must be "large". However, in our continuous model, a small package can intersect many disjoint packages.

For our discrete model, we present a mechanism that achieves an $O(R^{\frac{4}{3}})$ approximation. However, if the ratio between the minimal width of a figure and the sizes of the building blocks (we denote by Q) is smaller than the aspect ratio R, we can achieve a better approximation of $O(R \cdot Q^{\alpha^*})$, by running the α-greedy algorithm with $\alpha^* = \frac{log(R)}{2log(R)+log(Q)}$ [5].

The paper's organization: Section 2 describes our model. Section 3 describes our results for the USM model, both for the continuous case and the discrete case. Section 4 presents the results for the KSM model, and Section 5 concludes with a discussion of future work. All proofs are given in the full version of the paper ([2]).

2   Model

Let B denote the family (set) of bids (figure-value pairs) of the agents[6], that is $B = \{(s_i, v_i)|i \in N\}$. Let F denote the family of agents figures, that is $F = \{s_i|i \in N\}$. Given a family of bids B, we aim to maximize the social welfare, i.e., find a collection of non-conflicting bids (bids for disjoint figures) that maximizes the sum of valuations. For a subset $C \subseteq N$ of agents with disjoint figures, denote the value of C by $V(C) = \sum_{i \in C} v_i$. We denote the set of disjoint figures that achieves the maximal welfare by OPT (to simplify the notation we assume that there are no ties, so there is a single optimal solution), i.e.,

$$V(OPT) = \max_{C \subseteq N| \ \forall i,j \in C \ s_i \cap s_j = \emptyset} V(C)$$

---

[5] Thus, if one can embed the goods of a traditional combinatorial auction as building-blocks in the plane, such that each agent bids for building-blocks contained in some convex figure, then our approximation scheme improves the approximation ratio achieved in [10].

[6] Since all the mechanisms we consider are truthful, we use the same notation for the secret information and the declared information (bid), except of the IC proofs.

Definition 1. *A mechanism consists of a pair of functions (G,P) where:*

- *G is an allocation scheme (rule) that assigns a figure in $\mathcal{T}$ (where $\mathcal{T}$ is the set of compact convex figures in the plane) to every agent such that the figures are disjoint, i.e. $G(B) \in \mathcal{T}^N$ and for every $i \neq j$ in N, $G_j(B) \cap G_i(B) = \emptyset$ (where we denote the figure received by agent i by $G_i(B)$).*
- *P is a payment scheme, i.e. for any B, $P(B) \in \mathfrak{R}^n$. Denote the payment paid by agent i by $P_i(B)$.*

All allocation rules we present in the paper, allocate to an agent either her requested figure or the empty figure. All the payment rules we consider are *normalized*, that is, a losing agent pays zero. Additionally, each agent pays a non-negative payment. We assume quasi-linear utilities and that the agents have no externalities (the utility for each agent does not depend on the packages received by the other agents), i.e., the utility of each agent i is $u_i(B) = v_i(G_i(B)) - P_i(B)$. The agents are rational, so each agent chooses a bid that maximizes her own utility.

A mechanism is *incentive-compatible (IC)* if declaring their true secret information is a dominant strategy for all the agents. In the KSM model, it means that for any set of values reported by the other agents, each agent cannot achieve a higher utility by reporting an untruthful value, i.e., $\forall i \ \forall B_{-i} \ \forall v'_i \ u_i((s_i, v_i), B_{-i}) \geq u_i((s_i, v'_i), B_{-i})$, where B_{-i} denote the family of all bids except i's bid . In the USM model, IC means that each agent's best strategy is to report *both* her figure and her value truthfully, regardless of the other agents' reports, i.e., $\forall i \ \forall B_{-i} \ \forall v'_i, s'_i \ u_i((s_i, v_i), B_{-i}) \geq u_i((s'_i, v'_i), B_{-i})$.

An incentive-compatible mechanism is also *individually rational (IR)* if for any agent i, bidding truthfully ensures him a non-negative utility. That is, $\forall i \ \forall B_{-i} \ u_i((s_i, v_i), B_{-i}) \geq 0$.

Geometric Definitions:
We state our approximation bounds as functions of few geometric properties of the family of figures the agents bid for. We use standard definitions of *diameter* and *width* of compact figures in $\mathfrak{R}^2$:

The *diameter d_z* of a compact set z is the maximal distance between any two points in the set, i.e. $d_z = \max_{p_1, p_2 \in z} ||p_1 - p_2||_2$ ($||p_1 - p_2||_2$ is the Euclidean distance between p_1 and p_2). The *width w_z* of a compact set z is the minimal distance between the closest pair of parallel lines such that the convex set z lies between them.

Definition 2. *Given a family of figures F in $\mathfrak{R}^2$, the maximal diameter L is the maximal diameter of a figure in F, and the minimal width W is the minimal width of a figure in F. The aspect ratio R is the ratio between the maximal diameter and the minimal width. That is, $L = \max_{z \in F} d_z$, $W = \min_{z \in F} w_z$, $R = \frac{L}{W}$.*

The aspect ratio describes how diverse is the family of figures with respect to the figures' diameter and width[7]. The approximations our mechanisms achieve are asymptotic functions of the aspect ratio R.

---

[7] For example, if all the figures are disks with the same radius, then $R = 1$. If we have disks of diameter 10 and 5×2 rectangles, then $R = \frac{10}{2} = 5$.

Denote the *geometric area* of a compact figure z by $q(z)$. We assume that the diameter, width and area of any agent's figure are polynomial-time computable. We also assume that given any two agent's figures, we can decide if the two figures are disjoint in polynomial time[8].

The Discrete Model:
In the discrete model, there is a set of atomic building blocks (we call *tiles*) embedded in the plane. Each agent desires a bundle of tiles that are exactly the ones that are fully contained in some compact convex figure, and she reports this figure and her value for the set of tiles contained in it. We assume that all tiles have similar dimensions (specifically, each tile contains a disk of some positive diameter W_0 and its diameter is at most $2W_0$). Two agents are non conflicting if there is no tile which is *fully* contained in the two figures they report.

For a given family of bids, we define the *width-ratio* $Q = \frac{W}{W_0}$. The width-ratio gives an upper bound on the width of any figure, with respect to the size of the tiles. Clearly, we can assume that $Q \geq 1$.

The Greedy Mechanism:
Lehmann et al. [10] presented the following family of *greedy mechanisms* for combinatorial auctions:

Given a family of bids and some function f on the figures, such that f assigns a positive real value to any non empty figure (w.l.o.g. all figures are non empty), the *greedy allocation algorithm* picks an allocation *ALG*, by the following scheme:
Create a list of the bids sorted from high to low according to their values normalized by f (i.e., $\frac{v_1}{f(s_1)} \geq \frac{v_2}{f(s_2)} \geq \ldots \geq \frac{v_n}{f(s_n)}$).

While the list is not empty, choose a figure s_i for which the normalized value is highest in the remaining list (with a consistent tie breaking). Add i to the allocation ALG and update the list by removing all bids for figures that intersect s_i.

The specific algorithm is determined by the choice of the function (or norm) f. Lehmann et al. suggested using the norm $\frac{v_s}{|s|^\alpha}$ for combinatorial auctions, where $|s|$ is the size of the package s and α is some non-negative constant. We generalize this method for compact figures in $\Re^2$ and define the *α-greedy algorithm* to use the norm $\frac{v_z}{q(z)^\alpha}$, where $q(z)$ is the area of figure z [9].

Definition 3. *The α-greedy mechanism is a mechanism which uses the α-greedy algorithm as its allocation scheme, where a winning agent i pays according to the following payment scheme:*

Let j be the first agent to win, among all the agents whose figures intersects agent i's figure, when the greedy algorithm runs without i. If such j exists, i pays $\frac{q(s_i)^\alpha \cdot v_j}{q(s_j)^\alpha}$, otherwise i pays 0. Losing agents pay 0.

The properties of the α-greedy mechanisms, proved in [10], also hold in our model:

---

[8] Note that for polygons the above assumptions hold, and that any compact convex figure can be approximated (as good as one wants) by a polygon.

[9] For example, the 0-greedy algorithm sorts the figures according to their values and the 1-greedy algorithm sorts the figures according to their value per unit of area.

Theorem 4. *(essentially due to [10]) For every α, the α-greedy mechanism is polynomial time, individually rational and incentive compatible for agents bidding for compact figures in the USM model.*

3   The Unknown Single-Minded Model

This section considers the problem of designing a polynomial-time, individually-rational and incentive-compatible mechanisms, which guarantee some fraction of the social efficiency for the USM model. We study three families of figures: convex figures and rectangles in the continuous model, and convex figures in the discrete model. We use the α-greedy mechanism to create mechanisms with the desired properties for the three families. For each family, we find the value of α that optimizes the asymptotic approximation for the social welfare, over all α-greedy mechanisms. For convex figures in the continuous model, we show that $\alpha = \frac{1}{3}$ achieves an $O(R^{\frac{4}{3}})$-approximation for the social welfare. For rectangles in the continuous model, we improve the above result and show that $\alpha = \frac{1}{2}$ achieves an $O(R)$-approximation for the social welfare. Finally, for convex figures in the discrete model, we show that a careful choise of α as a function of R and Q, gives an approximation ratio between $O(R^{\frac{4}{3}})$ and $O(R)$. The proofs of all above results are based on a single general result presented in the full version [2].

3.1   Convex Figures and Rectangles in the Continuous Model

The following lemma presents a lower bound on the approximation ratio that can be achieved by α-greedy mechanisms, by presenting two constructions that are in a sense the hardest inputs for these mechanisms.

Lemma 5. *The α-greedy mechanism for compact convex figures achieves an $\Omega(R^{2(1-\alpha)})$ approximation for any $\alpha < 1$, and an $\Omega(R^{1+\alpha})$-approximation for any $\alpha > 0$. Therefore, the α-greedy mechanism for compact convex figures achieves an $\Omega(R^{\frac{4}{3}})$-approximation.*

Proof sketch: Each of the lower bounds is achieved by a construction that can be built for any R large enough. The left part of Figure 1 illustrates the construction used to prove the $\Omega(R^{2(1-\alpha)})$ bound and the right part is used to prove the $\Omega(R^{1+\alpha})$ bound. In the left example, a large rectangle contains $\Theta(R^2)$ small disjoint squares with a side of length W. On the right example, a small disk intersects $\Theta(R)$ disjoint triangles, with equal area of $\Theta(WL)$. In both constructions, there is one figure z (filled by small vertical lines) that is chosen by the greedy mechanism, while the socially optimal mechanism chooses a family of disjoint figures (small horizontal lines), each intersects z. The value of z is chosen such that its normalized value $\frac{v_z}{q(z)^\alpha}$ is a bit greater than 1, and the rest of the figures have a normalized value of 1. The value for α that minimizes the worst case approximation is therefore $\frac{1}{3}$, yielding an $\Omega(R^{\frac{4}{3}})$ lower bound. □

Next, we show that the $\frac{1}{3}$-greedy mechanism achieves an $O(R^{\frac{4}{3}})$-approximation (which, by the above lemma, is the best over all the α-greedy mechanisms). To prove this result, we use few elementary geometric properties of convex figures. First, for any

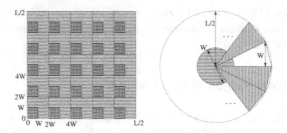

Fig. 1. Approximation bounds for the α-greedy mechanism

compact convex figure z, $q(z) = \Theta(d_z w_z)$. Additionally, the perimiter of z (denoted by p_z) is contiguous, and $p_z = \Theta(d_z)$ (the constants in both cases are independent of the figure z). These properties are sufficient for the approximation to hold.

Theorem 6. *When agents bid for compact convex figures in the plane with an aspect ratio R, the $\frac{1}{3}$ - greedy mechanism achieves an $O(R^{\frac{4}{3}})$-approximation, and this is the best asymptotic approximation achievable by an α-greedy mechanism (for any α). This mechanism is individually rational and incentive compatible for the USM model, and it runs in polynomial time.*

Proof sketch: By Theorem 4 the mechanism is IR, IC and it runs in polynomial time. Next, we present the idea behind the proof of the $O(R^{\frac{4}{3}})$-approximation ratio. By the definition of the α-greedy algorithm, any figure is either a winner or intersects a winner. It is the hardest to prove the approximation bound if the set of winners in the optimal solution *OPT* is disjoint to the set of winners *ALG* picked by the α-greedy algorithm. We map each agent $x \in OPT$ to a winner $z \in ALG$ that intersects x. We then bound the sum of values of any disjoint set of agents' figures that intersects z, by partitioning them to figures that are contained in z and to figures that are not. We use Hölder inequality and some simple geometric properties of convex figures to show that the upper bounds for $\alpha = \frac{1}{3}$, for both the contained figures and the rest of the intersecting figures, match the lower bounds presented in Lemma 5. □

If agents are only interested in rectangles (not necessarily axis parallel), than we can derive a stronger result of an $O(R)$-approximation for the social welfare. While the construction of a large rectangle containing $\Omega(R^2)$ small rectangles (as presented in Lemma 5) is still possible, the second construction is not. For rectangles, it is impossible for a small rectangle to hit (intersect but not contain) many disjoint rectangles.

Theorem 7. *When agents bid for rectangles in the plane with an aspect ratio R, the $\frac{1}{2}$ - greedy mechanism achieves an $O(R)$-approximation for the social welfare, and this is the best asymptotic approximation achievable by an α-greedy mechanism (for any α)[10]. This mechanism is individually rational and incentive compatible for the USM model, and it runs in polynomial time.*

---

[10] We actually prove a stronger statement. We show that $\Omega(R)$-approximation is the best over all the greedy mechanisms that sort the bids according to some function of the value and the area only (specifically, this includes the function $\frac{v_z}{q(z)^\alpha}$).

3.2   Convex Figures in the Discrete Model

We now turn to look at the discrete model. We first define the mechanism we use for the discrete model, we then present the mechanism properties[11].

Definition 8. *The Discrete Model Greedy Mechanism is a mechanism that given bids for compact convex figures in $\Re^2$ in the discrete model, does the following: If $Q \geq R$ then it runs the $\frac{1}{3}$-greedy mechanism, and if $Q \leq R$ then it runs the α^* - greedy mechanism for $\alpha^* = \frac{\log(R)}{2\log(R)+\log(Q)}$.*

Theorem 9. *Consider that the agents bid for compact convex figures in $\Re^2$ in the discrete model, with an aspect-ratio R and a width-ratio Q. Then, the* Discrete Model Greedy Mechanism *achieves an $O(R^{\frac{4}{3}})$-approximation for the social welfare. Moreover, when $Q \leq R$ it achieves a better approximation of $O(R \cdot Q^{\alpha^*})$. This mechanism achieves the best asymptotic approximation among all the mechanisms that choose α as a function of R and Q, and in particular it is asymptotically better than the α-greedy mechanism for any α. Additionally, the mechanism is IR and IC for the USM model, and it runs in polynomial time.*

4   The Known Single-Minded Model

In this section, we present polynomial-time mechanisms for different families of figures in the Known Single-Minded (KSM) model (where the auctioneer knows the desired figure of each agent, but does not know her value for the figure). We start by presenting an auction for general compact convex figures. We achieve an $O(R)$-approximation for the social welfare, which is better than the $O(R^{\frac{4}{3}})$-approximation that we proved for the USM model. Next, we present a mechanism (based on an algorithm from [8]), which gives an $O(\log(R))$-approximation for axis-parallel rectangles[12].

4.1   Mechanisms for Convex Figures

Consider the mechanism called the *"Classes-by-Area 1-Greedy Mechanism" (CBA-1G mechanism)* presented in Figure 2. This mechanism divides the bids of the agents to classes according to the figures' geometric area, runs a 1-greedy algorithm in each class, and allocates the figures to agents in the class that achieved the highest result. From the algorithmic aspect, this is the first algorithm for packing weighted convex figures that we know of, and it achieves an $O(R)$-approximation. We use this algorithm to construct a polynomial-time and IC mechanism with the same approximation ratio for the social welfare. The payments are exactly the "critical-values" for the agents, i.e., the minimal declaration for which they still win the auction.

---

[11] Note that an agent can manipulate the values of α by affecting R and Q. Therefore, for incentive compatibility, the mechanism is assumed to know the true values of R and Q.

[12] This approximation is exponentially better than the approximation ratio we achieve for this problem in the USM model and than the ratio we achieve for general convex figures in the KSM model.

The Classes-by-Area 1-Greedy (CBA-1G) Mechanism:

Allocation:

   Step 1: Divide the given input to $m = 2log(R)$ classes according to their area.
   A figure s belongs to class c if $q(s) \in [W^2 \cdot 2^c, W^2 \cdot 2^{c+1})$ (for $c \in \{0,\ldots,m-1\}$).
   Step 2: Perform the 1-greedy algorithm per each class. Denote the welfare achieved
   by class c by V^c.
   Step 3: Output the allocation in the class c for which the 1-greedy algorithm achieved
   the highest welfare, i.e., $c \in argmax_{\tilde{c} \in \{0,\ldots,m-1\}} V^{\tilde{c}}$.

Payments:

   Denote the winning class as class 1, and the class with the second-highest welfare as
   class 2. Let $V^1_{-i} = V^1 - v_i$, and let j be the first figure that intersects figure i and wins,
   when we run the greedy algorithm where agent i is removed. Let z_i be $\frac{v_j q(i)}{q(j)}$ if
   such j exists, and 0 otherwise.
   A winning agent i pays: $\mathbf{P(i)} = \mathbf{max}\{\mathbf{V^2 - V^1_{-i}}, \mathbf{z_i}\}$, and any losing agent pays 0.

Fig. 2. A mechanism for selling arbitrary convex figures. This mechanism is incentive compatible
in the KSM model and achieves an $O(R)$-approximation for the social welfare

The payments in the CBA-1G mechanism are chosen as follows: to win the auction,
each agent should be both a winner in her class and her class should beat all other
classes. Bidding above the value z_i in the mechanism's description, is a necessary and
sufficient condition for agent i to win in her class. However, if agent i bids below $V^2 - V^1_{-i}$ and still wins in her class, her class will definitely lose.

Theorem 10. *When the agents bid for compact convex figures in $\Re^2$ with an aspect
ratio R, the CBA-1G mechanism achieves an $O(R)$-approximation. This mechanism is
IR and IC for the KSM model[13], and runs in polynomial time.*

Proof sketch: We show that the approximation ratio achieved in each class is $O(R_c)$,
where R_c is the aspect ratio[14] in class c. Due to a general proposition we prove, the approximation ratio achieved by choosing the best class is $O(\sum_c R_c)$. Finally, we show that
$\sum_c R_c = O(R)$, by dividing this sum to two geometric series. For proving IC, we show
that the given payments are indeed the critical values for the agents, i.e. the smallest
declarations for which they still win. □

4.2 Mechanisms for Axis-Aligned Rectangles

In the full version of this paper ([2]) we present an allocation algorithm, called the
Shifting Algorithm, which is based on an algorithm by Khanna et al.([8]) with some minor changes. They studied a model where axis-aligned rectangles lie in an $n \times n$ array,
and they proved an $O(log(n))$-approximation for the weighted packing problem. This
approximation is the best approximation currently known for weighted packing of axis-
parallel rectangles. Our algorithm gives an $O(log(R))$-approximation in a slightly more
general model where the rectangles can lie in any axis-parallel location in the plane. By

---

[13] We observe that this mechanism is not IC in the USM model.

[14] I.e. the ratio between the maximal diameter and the minimal width of figures in this class.

carefully defining a payment scheme, we use this allocation rule for designing an IC polynomial-time mechanism achieving an $O(log(R))$-approximation for the social welfare. We call this mechanism the *Shifting Mechanism* and we summarize its properties in the following theorem[15]:

Theorem 11. *When the agents bid for axis-parallel rectangles in $\Re^2$ with an aspect ratio R, the* Shifting Mechanism *achieves an $O(log(R))$-approximation. This mechanism is individually rational and incentive compatible for the KSM model, and runs in polynomial time.*

5 Conclusion and Further Research

In this paper, we study auctions in which the agents bid for convex figures in the plane. We present mechanisms that run in polynomial time, in which the selfish players' best strategy is to send their true private data to the auctioneer. We suggest using the aspect ratio R, which measures how diverse are the dimensions of the figures, for analyzing the economic efficiency of the mechanisms.

In the KSM model, we were able to achieve the best approximation currently known for weighted axis-parallel rectangle packing $(log(R))$ in an IC mechanism. Lehmann et al. [10] showed that the best polynomial-time approximation for combinatorial auctions (for single minded bidders) can be achieved with an IC mechanism. On the other hand, recent results showed settings in which the optimal algorithmic approximation ratio cannot be achieved by IC mechanisms (see, e.g., [1,9]). Whether such gap exists in our model is an interesting open question:

Open Problem: *Can the best polynomial-time approximation schemes for packing convex figures (general figures, rectangles, or axis-parallel rectangles) be implemented by incentive-compatible mechanisms?*

The mechanism for combinatorial auctions presented in [10] achieves the same approximation both for the USM model and the KSM model. Our results might indicate that, in our model, a gap exists between the approximation achievable in both information models. For general convex figures, the approximation we achieve in the KSM and the USM models are $O(R)$ and $O(R^{\frac{4}{3}})$, respectively. For axis-parallel rectangles, the gap in our results is even exponential.

Open Problem: *In settings where the agents are "single minded", is there a gap between the best approximation achievable in the KSM and in the USM models?*

We present some novel algorithmic results regarding packing of convex figures and arbitrary rectangles. We have not been able to show that these results are tight.

Open Problem: *Is there an $o(R)$-approximation scheme for packing general convex figures, or even for packing rectangles (not necessarily axis-parallel)?*

Our results may also be useful in deriving approximation results for the problem of packing weighted convex bodies in dimensions higher than two[16].

---

[15] An easy observation is that the Shifting Mechanism is not IC in the USM model. We also note that the Shifting Mechanism achieves an $\Omega(R)$-approximation for general rectangles (not necessarily axis-parallel).

[16] However, the economic interpretation of such auctions is not always clear.

References

1. A. Archer and E. Tardos. Truthful mechanisms for one-parameter agents. In *symposium on foundations of computer science*, pages 482–491, 2001.
2. M. Babaioff and L. Blumrosen. Computationally-feasible truthful auctions for convex bundles. Full version. Available from www.cs.huji.ac.il/~liad or ~mosheb.
3. M. Babaioff and W. E. Walsh. Incentive-compatible, budget-balanced, yet highly efficient auctions for supply chain formation. In *ACM Conference on Electronic Commerce*, pages 64–75, 2003. Extended version to appear in DSS, 2004.
4. T. M. Chan. Polynomial-time approximation schemes for packing and piercing fat objects. In *J. Algorithms*, volume 46, pages 209–218, 2003.
5. T. Erlebacj, K. Jansen, and E. Seidel. Polynomial-time approximation schemes for geometric graphs. In *SODA 2001*, pages 671–679.
6. Papadimitriou C. H. Algorithms, games, and the internet. In *proceedings of the 33rd Annual ACM Symposium on Theory of Computing*, pages 749–753, 2001.
7. D. Hochbaum and W. Maass. Approximation schemes for covering and packing problems in image processing and vlsi. In *Journal of the ACM*, pages 130–136, 1985.
8. S. Khanna, S. Muthukrishnan, and M. Paterson. On approximating rectangle tiling and packing. In *Symposium on Discrete Algorithms*, pages 384–393, 1998.
9. Ron Lavi, Ahuva Mua'lem, and Noam Nisan. Towards a characterization of truthful combinatorial auctions. In *FOCS 03*, pages 574–583, 2003.
10. D. Lehmann, L. I. O'Callaghan, and Y. Shoham. Truth revelation in approximately efficient combinatorial auctions. *Journal of the ACM*, 49(5):1–26, 2002.
11. Andreu Mas-Colell, Michael D. Whinston, and Jerry R. Green. *Microeconomic Theory*. Oxford University Press, New York, 1995.
12. Ahuva Mua'lem and Noam Nisan. Truthful approximation mechanisms for restricted combinatorial auctions. In *AAAI-02*, 2002.
13. Noam Nisan. Algorithms for selfish agents. In *16th Symposium on Theoretical Aspects of Computer Science*, 1999.
14. Noam Nisan and Amir Ronen. Algorithmic mechanism design. *Games and Economic Behavior*, 35(1/2):166–196, April/May 2001.
15. M. H. Rothkopf, A. Pekeč, and R. M. Harstad. Computationally manageable combinatorial auctions. *Management Science*, 44:1131–1147, 1998.

Randomized Approximation Algorithms for Set Multicover Problems with Applications to Reverse Engineering of Protein and Gene Networks

Piotr Berman[1], Bhaskar DasGupta[2], and Eduardo Sontag[3]

[1] Department of Computer Science and Engineering
Pennsylvania State University, University Park, PA 16802
berman@cse.psu.edu

[2] Department of Computer Science, University of Illinois at Chicago
Chicago, IL 60607-7053
dasgupta@cs.uic.edu

[3] Department of Mathematics, Rutgers University, New Brunswick, NJ 08903
sontag@hilbert.rutgers.edu

Abstract. In this paper we investigate the computational complexities of a combinatorial problem that arises in the reverse engineering of protein and gene networks. Our contributions are as follows:

- We abstract a combinatorial version of the problem and observe that this is "equivalent" to the set multicover problem when the "coverage" factor k is a function of the number of elements n of the universe. An important special case for our application is the case in which $k = n - 1$.
- We observe that the standard greedy algorithm produces an approximation ratio of $\Omega(\log n)$ even if k is "large" *i.e.* $k = n - c$ for some constant $c > 0$.
- Let $1 < a < n$ denotes the maximum number of elements in any given set in our set multicover problem. Then, we show that a nontrivial analysis of a simple randomized polynomial-time approximation algorithm for this problem yields an expected approximation ratio $\mathbf{E}[r(a, k)]$ that is an increasing function of a/k. The behavior of $\mathbf{E}[r(a, k)]$ is "roughly" as follows: it is about $\ln(a/k)$ when a/k is at least about $\mathbf{e}^2 \approx 7.39$, and for smaller values of a/k it decreases towards 2 exponentially with increasing k with $\lim_{a/k \to 0} \mathbf{E}[r(a, k)] \leq 2$. Our randomized algorithm is a cascade of a deterministic and a randomized rounding step parameterized by a quantity β followed by a greedy solution for the remaining problem.

1 Introduction

Let $[x, y]$ is the set $\{x, x+1, x+2, \ldots, y\}$ for integers x and y. The set multicover problem is a well-known combinatorial problem that can be defined as follows.

K. Jansen et al. (Eds.): APPROX and RANDOM 2004, LNCS 3122, pp. 39–50, 2004.

Problem Name: SC_k.
Instance $<n, m, k>$: An universe $U = [1, n]$, sets $S_1, S_2, \ldots, S_m \subseteq U$
with $\cup_{j=1}^m S_j = U$ and a "coverage factor" (positive integer) k.
Valid Solutions: A subset of indices $I \subseteq [1, m]$ such that, for every
element $x \in U$, $|j \in I : x \in S_j| \geq k$.
Objective: *Minimize* $|I|$.

SC_1 is simply called the Set Cover problem and denoted by SC; we will
denote an instance of SC simply by $<n, m>$ instead of $<n, m, 1>$.

Both SC and SC_k are already well-known in the realm of design and analysis
of combinatorial algorithms (*e.g.*, see [14]). Let $3 \leq a < n$ denote the maximum
number of elements in any set, *i.e.*, $a = \max_{i \in [1, m]}\{|S_i|\}$. We summarize some of
the known relevant results for them below.

Fact 1
(a) [4] *Assuming* $NP \nsubseteq DTIME(n^{\log \log n})$, *instances* $< n, m >$ *of the* SC
problem cannot be approximated to within a factor of $(1-\varepsilon) \ln n$ *for any constant*
$0 < \varepsilon < 1$ *in polynomial time.*
(b) [14] *An instance* $<n, m, k>$ *of the* SC_k *problem can be*
$(1+\ln a)$-*approximated in* $O(nmk)$ *time by a simple greedy heuristic that, at every*
step, selects a new set that covers the maximum number of those elements that
has not been covered at least k times yet. It is also possible to design randomized
approximation algorithms with similar expected approximation ratios.

1.1 Summary of Results

The combinatorial problems investigated in this paper that arise out of reverse
engineering of gene and protein networks can be shown to be equivalent to SC_k
when k is a function of n. One case that is of significant interest is when k is
"large", *i.e.*, $k = n - c$ for some constant $c > 0$, but the case of non-constant c is
also interesting (cf. Questions **(Q1)** and **(Q2)** in Section 2). Our contributions
in this paper are as follows:

- In Section 2 we discuss the combinatorial problems (Questions **(Q1)** and
  (Q2)) with their biological motivations that are of relevance to the reverse
  engineering of protein and gene networks. We then observe, in Section 2.3,
  using a standard duality that these problems are indeed equivalent to SC_k
  for appropriate values of k.
- In Lemma 1 in Section 3.1, we observe that the standard greedy algorithm
  SC_k produces an approximation ratio of $\Omega(\log n)$ even if k is "large", *i.e.*
  $k = n - c$ for some constant $c > 0$.
- Let $1 < a < n$ denotes the maximum number of elements in any given set
  in our set multicover problem. In Theorem 2 in Section 3.2, we show that a
  non-trivial analysis of a simple randomized polynomial-time approximation
  algorithm for this problem yields an expected approximation ratio $\mathbf{E}[r(a, k)]$
  that is an increasing function of a/k. The behavior of $\mathbf{E}[r(a, k)]$ is "roughly"
  as follows: it is about $\ln(a/k)$ when a/k is at least about $\mathbf{e}^2 \approx 7.39$, and for

smaller values of a/k it decreases towards 2 exponentially with increasing k with $\lim_{a/k \to 0} \mathbf{E}[r(a,k)] \leq 2$. More precisely, $\mathbf{E}[r(a,k)]$ is at most

$$1 + \ln a, \qquad \qquad \text{if } k = 1$$

$$\left(1 + e^{-(k-1)/5}\right) \ln(a/(k-1)), \qquad \text{if } a/(k-1) \geq e^2 \text{ and } k > 1$$

$$\min\{\, 2 + 2 \cdot e^{-(k-1)/5}, \, 2 + \left(e^{-2} + e^{-9/8}\right) \cdot \tfrac{a}{k} \,\}$$
$$\approx \min\{\, 2 + 2 \cdot e^{-(k-1)/5}, \, 2 + 0.46 \cdot \tfrac{a}{k} \,\} \qquad \text{if } a/(k-1) < e^2 \text{ and } k > 1$$

Some proofs are omitted due to lack of space.

1.2 Summary of Analysis Techniques

- To prove Lemma 1, we generalize the approach in Johnson's paper [6]. A straightforward replication of the sets will not work because of the dependence of k on n, but allowing the "misleading" sets to be somewhat larger than the "correct" sets allows a similar approach to go through at the expense of a diminished constant.
- Our randomized algorithm in Theorem 2 is a cascade of a deterministic and a randomized rounding step parameterized by a quantity β followed by a greedy solution for the remaining problem.
- Our analysis of the randomized algorithm in Theorem 2 uses an amortized analysis of the interaction between the deterministic and randomized rounding steps with the greedy step. For tight analysis, we found that the standard Chernoff bounds such as in [1, 2, 10, 14] were not always sufficient and hence we had to devise more appropriate bounds for certain parameter ranges.

2 Motivations

In this section is to define a computational problem that arises in the context of experimental design for reverse engineering of protein and gene networks. We will first pose the problem in linear algebra terms, and then recast it as a combinatorial question. After that, we will discuss its motivations from systems biology. Finally, we will provide a precise definition of the combinatorial problems and point out its equivalence to the set multicover problem via a standard duality.

Our problem is described in terms of two matrices $A \in \mathbb{R}^{n \times n}$ and $B \in \mathbb{R}^{n \times m}$ such that:

- A is *unknown*;
- B is *initially unknown*, but each of its columns, denoted as $B_1, B_2, \ldots, B_m$, can be retrieved with a *unit-cost query*;
- the columns of B are in *general position*, i.e., each subset of $k \leq n$ columns of B is *linearly independent*;
- the *zero structure* of the matrix $C = AB = (c_{ij})$ is known, i.e., a binary matrix $C^0 = \left(c_{ij}^0\right) \in \{0,1\}^{n \times m}$ is given, and it is known that $c_{ij} = 0$ for each i,j for which $c_{ij}^0 = 0$.

The objective, "roughly speaking", is to obtain as much information as possible about A (which, in the motivating application, describes regulatory interactions among genes and/or proteins), while performing "few" queries (each of which may represent the measuring of a complete pattern of gene expression, done under a different set of experimental conditions).

Notice that there are intrinsic limits to what can be accomplished: if we multiply each row of A by some nonzero number, then the zero structure of C is unchanged. Thus, the best that we can hope for is to identify the rows of A up to scalings (in abstract mathematical terms, as elements of the projective space $\mathbb{P}^{n-1}$). To better understand these geometric constraints, let us reformulate the problem as follows. Let A_i denote the ith row of A. Then the specification of C^0 amounts to the specification of *orthogonality relations* $A_i \cdot B_j = 0$ for each pair i, j for which $c_{ij}^0 = 0$. Suppose that we decide to query the columns of B indexed by $J = \{j_1, \ldots, j_\ell\}$. Then, the information obtained about A may be summarized as $A_i \in \mathcal{H}_{J,i}^\perp$, where "$\perp$" indicates *orthogonal complement*, and

$$\mathcal{H}_{J,i} = \text{span} \left\{ B_j, j \in J_i \right\},$$

$$J_i = \left\{ j \mid j \in J \text{ and } c_{ij}^0 = 0 \right\}. \tag{1}$$

Suppose now that the set of indices of selected queries J has the property:

$$\text{each set } J_i, \ i = 1, \ldots, n, \ \text{ has cardinality } \geq n - k, \tag{2}$$

for some given integer k. Then, because of the general position assumption, the space $\mathcal{H}_{J,i}$ has dimension $\geq n - k$, and hence the space $\mathcal{H}_{J,i}^\perp$ has dimension at most k.

The most desirable special case is that in which $k = 1$. Then $\dim \mathcal{H}_{J,i}^\perp \leq 1$, hence each A_i is uniquely determined up to a scalar multiple, which is the best that could be theoretically achieved. Often, in fact, finding the sign pattern (such as "$(+, +, -, 0, 0, -, \ldots)$") for each row of A is the main experimental goal (this would correspond, in our motivating application, to determining if the regulatory interactions affecting each given gene or protein are *inhibitory* or *catalytic*). Assuming that the degenerate case $\mathcal{H}_{J,i}^\perp = \{0\}$ does not hold (which would determine $A_i = 0$), once that an arbitrary nonzero element v in the line $\mathcal{H}_{J,i}^\perp$ has been picked, there are only two sign patterns possible for A_i (the pattern of v and that of $-v$). If, in addition, one knows at least one nonzero sign in A_i, then the sign structure of the whole row has been *uniquely* determined (in the motivating biological question, typically one such sign is indeed known; for example, the diagonal elements a_{ii}, i.e. the ith element of each A_i, is known to be negative, as it represents a degradation rate). Thus, we will be interested in this question:

find J of minimal cardinality such that $|J_i| \geq n - 1$, $i = 1, \ldots, n$. **(Q1)**

If queries have variable unit costs (different experiments have a different associated cost), this problem must be modified to that of minimizing a suitable linear combination of costs, instead of the number of queries.

More generally, suppose that the queries that we performed satisfy (2), with $k > 1$ but small k. It is not true anymore that there are only two possible sign patterns for any given A_i, but the number of possibilities is still very small. For simplicity, let us assume that we know that no entry of A_i is zero (if this is not the case, the number of possibilities may increase, but the argument is very similar). We wish to prove that the possible number of signs is much smaller than 2^n. Indeed, suppose that the queries have been performed, and that we then calculate, based on the obtained B_j's, a basis $\{v_1, \ldots, v_k\}$ of $\mathcal{H}_{\bar{J},i}^\perp$ (assume $\dim \mathcal{H}_{\bar{J},i}^\perp = k$; otherwise pick a smaller k). Thus, the vector A_i is known to have the form $\sum_{r=1}^{k} \lambda_r v_r$ for some (unknown) real numbers $\lambda_1, \ldots, \lambda_k$. We may assume that $\lambda_1 \neq 0$ (since, if $A_i = \sum_{r=2}^{k} \lambda_r v_r$, the vector $\varepsilon v_1 + \sum_{r=2}^{k} \lambda_r v_r$, with small enough ε, has the same sign pattern as A_i, and we are counting the possible sign patterns). If $\lambda_1 > 0$, we may divide by λ_1 and simply count how many sign patterns there are when $\lambda_1 = 1$; we then double this estimate to include the case $\lambda_1 < 0$. Let $v_r = \mathrm{col}\,(v_{1r}, \ldots, v_{nr})$, for each $r = 1, \ldots, k$. Since no coordinate of A_i is zero, we know that A_i belongs to the set $\mathcal{C} = \mathbb{R}^{k-1} \setminus \left(L_1 \bigcup \ldots \bigcup L_n \right)$ where, for each $1 \leq s \leq n$, L_s is the hyperplane in $\mathbb{R}^{k-1}$ consisting of all those vectors $(\lambda_2, \ldots, \lambda_k)$ such that $\sum_{r=2}^{k} \lambda_r v_{sr} = -v_{s1}$. On each connected component of $\mathcal{C}$, signs patterns are constant. Thus the possible number of sign patterns is upper bounded by the maximum possible number of connected regions determined by n hyperplanes in dimension $k-1$. A result of L. Schläfli (see [3, 11], and also [12] for a discussion, proof, and relations to Vapnik-Chervonenkis dimension) states that this number is bounded above by $\Phi(n, k-1)$, provided that $k-1 \leq n$, where $\Phi(n, d)$ is the number of possible subsets of an n-element set with at most d elements, that is, $\Phi(n, d) = \sum_{i=0}^{d} \binom{n}{i} \leq 2\frac{n^d}{d!} \leq \left(\frac{en}{d}\right)^d$. Doubling the estimate to include $\lambda_1 < 0$, we have the upper bound $2\Phi(n, k-1)$. For example, $\Phi(n, 0) = 1$, $\Phi(n, 1) = n+1$, and $\Phi(n, 2) = \frac{1}{2}(n^2+n+2)$. Thus we have an estimate of 2 sign patterns when $k = 1$ (as obtained earlier), $2n + 2$ when $k = 2$, $n^2 + n + 2$ when $k = 3$, and so forth. In general, the number grows only polynomially in n (for fixed k).

These considerations lead us to formulating the generalized problem, for each fixed k: *find J of minimal cardinality such that $|J_i| \geq n - k$ for all $i = 1, \ldots, n$.* Recalling the definition (1) of J_i, we see that $J_i = J \bigcap T_i$, where $T_i = \{j \mid c_{ij}^0 = 0\}$. Thus, we can reformulate our question purely combinatorially, as a more general version of Question **(Q1)** as follows. Given sets

$$T_i \subseteq \{1, \ldots, m\}, \quad i = 1, \ldots, n.$$

and an integer $k < n$, the problem is:

find $J \subseteq \{1, \ldots, m\}$ of minimal cardinality such that $|J \bigcap T_i| \geq n - k$,
$1 \leq i \leq n$. **(Q2)**

For example, suppose that $k = 1$, and pick the matrix $C^0 \in \{0,1\}^{n \times n}$ in such a way that the columns of C^0 are the binary vectors representing all the $(n-1)$-element subsets of $\{1, \ldots, n\}$ (so $m = n$); in this case, the set J must equal $\{1, \ldots, m\}$ and hence has cardinality n. On the other hand, also with $k = 1$, if we pick the matrix C^0 in such a way that the columns of C^0 are the binary vectors representing all the 2-element subsets of $\{1, \ldots, n\}$ (so $m = n(n-1)/2$), then J must again be the set of all columns (because, since there are only two zeros in each column, there can only be a total of 2ℓ zeros, $\ell = |J|$, in the submatrix indexed by J, but we also have that $2\ell \geq n(n-1)$, since each of the n rows must have $\geq n-1$ zeros); thus in this case the minimal cardinality is $n(n-1)/2$.

2.1   Motivations from Systems Biology

This problem was motivated by the setup for reverse-engineering of protein and gene networks described in [8, 9] and reviewed in [13]. We assume that the time evolution of a vector of state variables $x(t) = (x_1(t), \ldots, x_n(t))$ is described by a system of differential equations:

$$\dot{x}_1 = f_1(x_1, \ldots, x_n, p_1, \ldots, p_m)$$
$$\dot{x}_2 = f_2(x_1, \ldots, x_n, p_1, \ldots, p_m)$$
$$\vdots$$
$$\dot{x}_n = f_n(x_1, \ldots, x_n, p_1, \ldots, p_m)$$

(in vector form, "$\dot{x} = f(x, p)$"), where $p = (p_1, \ldots, p_m)$ is a vector of parameters, representing for instance the concentrations of certain enzymes which are maintained at a constant value during a particular experiment. There is a reference value $\bar{p}$ of p, which represents "wild type" (that is, normal) conditions, and a corresponding steady state $\bar{x}$. That is, $f(\bar{x}, \bar{p}) = 0$. We are interested in obtaining information about the Jacobian of the vector field f evaluated at $(\bar{x}, \bar{p})$, or at least about the signs of the derivatives $\partial f_i / \partial x_j(\bar{x}, \bar{p})$. For example, if $\partial f_i / \partial x_j > 0$, this means that x_j has a positive (catalytic) effect upon the rate of formation of x_i. The critical assumption, indeed the main point of [8, 9], is that, while we do not know the form of f, we do know that *certain parameters p_j do not directly affect certain variables x_i*. This amounts to *a priori* biological knowledge of specificity of enzymes and similar data. This knowledge will be summarized by the binary matrix $C^0 = (c_{ij}^0) \in \{0,1\}^{n \times m}$, where "$c_{ij}^0 = 0$" means that p_j does not appear in the equation for $\dot{x}_i$, that is, $\partial f_i / \partial p_j \equiv 0$.

The experimental protocol allows us to make small perturbations in which we change one of the parameters, say the kth one, while leaving the remaining ones constant. (A generalization would allow for the simultaneous perturbation of more than one parameter.) For the perturbed vector $p \approx \bar{p}$, we measure the resulting steady state vector $x = \xi(p)$. (Mathematically, we suppose that for each vector of parameters p in a neighborhood of $\bar{p}$ there is a unique steady state $\xi(p)$ of the system, where ξ is a differentiable function. In practice, each such perturbation experiment involves letting the system relax to steady state, and the use of some biological reporting mechanism, such as microarrays, in order to measure the expression profile of the variables x_i.) For each of the possible m experiments, in which a given p_j is perturbed, we may estimate the n "sensitivities"

$$b_{ij} = \frac{\partial \xi_i}{\partial p_j}(\bar{p}) \approx \frac{1}{\bar{p}_j - p_j} \left(\xi_i(\bar{p} + p_j e_j) - \xi_i(\bar{p}) \right) , \quad i = 1, \ldots, n$$

(where $e_j \in \mathbb{R}^m$ is the jth canonical basis vector). We let B denote the matrix consisting of the b_{ij}'s. (See [8, 9] for a discussion of the fact that division by $\bar{p}_j - p_j$, which is undesirable numerically, is not in fact necessary.) Finally, we let A be the Jacobian matrix $\partial f / \partial x$ and let C be the negative of the Jacobian matrix $\partial f / \partial p$. From $f(\xi(p), p) \equiv 0$, taking derivatives with respect to p, and using the chain rule, we get that $A = BC$. This brings us to the problem stated in this paper. (The general position assumption is reasonable, since we are dealing with experimental data.)

2.2   Combinatorial Formulation of Questions (Q1) and (Q2)

Problem Name: CP_k (the k-Covering problem that captures Question **(Q1)** and **(Q2)**)[1]
Instance $< m, n, k >$: $U = [1, m]$ and sets $T_1, T_2, \ldots, T_n \subseteq U$ with $\cup_{i=1}^n T_i = U$.
Valid Solutions: A subset $U' \subseteq U$ such that $|U' \cap T_i| \geq n - k$ for each $i \in [1, n]$.
Objective: *Minimize* $|U'|$.

2.3   Equivalence of CP_k and SC_{n-k}

We can establish a 1-1 correspondence between an instance $< m, n, k >$ of CP_k and an instance $< n, m, n - k >$ of SC_{n-k} by defining $S_i = \{ j \mid i \in T_j \}$ for each $i \in [1, m]$. It is easy to verify that U' is a solution to the instance of CP_k if and only if the collection of sets S_u for each $u \in U'$ is a solution to the instance of SC_{n-k}.

3   Approximation Algorithms for SC_k

An ε-approximate solution (or simply an ε-approximation) of a minimization problem is defined to be a solution with an objective value no larger than ε times the value of the optimum. It is not difficult to see that SC_k is NP-complete even when $k = n - c$ for some constant $c > 0$.

---

[1] CP_{n-1} is known as the hitting set problem [5, p. 222].

3.1   Analysis of Greedy Heuristic for $\mathbf{SC}_k$ for Large k

Johnson [6] provides an example in which the greedy heuristic for some instance of **SC** over n elements has an approximation ratio of at least $\log_2 n$. This approach can be generalized to show the following result.

Lemma 1. *For any fixed $c > 0$, the greedy heuristic (as described in Fact 1(b)) has an approximation ratio of at least $\left(\frac{1}{2} - o(1)\right)\left(\frac{n-c}{8n-2}\right)\log_2 n = \Omega(\log n)$ for some instance $<n, m, n - c>$ of $\mathbf{SC}_{n-c}$.*

3.2   Randomized Approximation Algorithm for $\mathbf{SC}_k$

As stated before, an instance $<n, m, k>$ of $\mathbf{SC}_k$ can be $(1 + \ln a)$-approximated in $O(mnk)$ time for any k where $a = \max_{S \in \mathcal{S}}\{|S|\}$. In this section, we provide a randomized algorithm with an expected performance ratio better than $(1 + \ln a)$ for larger k. Let $\mathcal{S} = \{S_1, S_2, \ldots, S_m\}$.

Our algorithm presented below as well as our subsequent discussions and proofs are formulated with the help of the following vector notations:

- All our vectors have m coordinates with the ith coordinate indexed with the ith set S_i of $\mathcal{S}$.
- if $V \subset \mathcal{S}$, then $v \in \{0, 1\}^m$ is the characteristic vector of V, *i.e.*, $v_{S_i} = \begin{cases} 1 \text{ if } S_i \in V \\ 0 \text{ if } S_i \notin V \end{cases}$
- $\mathbf{1}$ is the vector of all 1's, *i.e.* $\mathbf{1} = s$;
- $S^i = \{A \in \mathcal{S} : i \in A\}$ denotes the sets in $\mathcal{S}$ that contains a specific element i.

Consider the standard integer programming (IP) formulation of an instance $< n, m, k>$ of $\mathbf{SC}_k$ [14]:

$$\textit{minimize } \mathbf{1}x \textit{ subject to } \begin{array}{l} s^i x \geq k \qquad \text{for each } i \in U \\ x_A \in \{0, 1\} \text{ for each } A \in \mathcal{S} \end{array}$$

A linear programming (LP) relaxation of the above formulation is obtained by replacing each constraint $x_A \in \{0, 1\}$ by $0 \leq x_A \leq 1$. The following randomized approximation algorithm for $\mathbf{SC}_k$ can then be designed:

---

1. Select an appropriate positive constant $\beta > 1$ in the following manner:
$$\beta = \begin{cases} \ln a & \text{if } k = 1 \\ \ln(a/(k-1)) & \text{if } a/(k-1) \geq \mathbf{e}^2 \text{ and } k > 1 \\ 2 & \text{otherwise} \end{cases}$$
2. Find a solution x to the LP relaxation via any polynomial-time algorithm for solving linear programs (e.g. [7]).
3. (**deterministic rounding**) Form a family of sets $\mathcal{C}^0 = \{A \in \mathcal{S} : \beta x_A \geq 1\}$.
4. (**randomized rounding**) Form a family of sets $\mathcal{C}^1 \subset \mathcal{S} - \mathcal{C}^0$ by independent random choices such that $\mathbf{Pr}[A \in \mathcal{C}^1] = \beta x_A$.
5. (**greedy selection**) Form a family of sets $\mathcal{C}^2$ as:
while $s^i(c^0 + c^1 + c^2) < k$ for some $i \in U$, insert to $\mathcal{C}^2$ any $A \in S^i - \left(\cup_{i=0}^2 \mathcal{C}^i\right)$.
6. Return $\mathcal{C} = \mathcal{C}^0 \cup \mathcal{C}^1 \cup \mathcal{C}^2$ as our solution.

Let $r(a, k)$ denote the performance ratio of the above algorithm.

Theorem 2.[2] $\mathbf{E}[r(a, k)]$ *is at most*

$$1 + \ln a, \qquad\qquad\qquad\qquad\qquad\qquad \text{if } k = 1$$

$$\left(1 + e^{-(k-1)/5}\right) \ln(a/(k-1)), \qquad\qquad \text{if } a/(k-1) \geq e^2 \text{ and } k > 1$$

$$\min\left\{2 + 2 \cdot e^{-(k-1)/5},\ 2 + \left(e^{-2} + e^{-9/8}\right) \cdot \tfrac{a}{k}\right\}$$
$$\approx \min\left\{2 + 2 \cdot e^{-(k-1)/5},\ 2 + 0.46 \cdot \tfrac{a}{k}\right\} \qquad \text{if } a/(k-1) < e^2 \text{ and } k > 1$$

Let OPT denote the minimum number of sets used by an optimal solution. Obviously, OPT$\geq 1x$ and OPT$\geq \frac{nk}{a}$. A proof of Theorem 2 follows by showing the following upper bounds on $\mathbf{E}[r(a, k)]$ and taking the best of these bounds for each value of a/k:

$$
\begin{cases}
1 + \ln a, & \text{if } k = 1 \\
\left(1 + e^{-(k-1)/5}\right) \ln(a/(k-1)), & \text{if } a/(k-1) \geq e^2 \text{ and } k > 1 \\
2 + 2 \cdot e^{-(k-1)/5}, & \text{if } a/(k-1) < e^2 \text{ and } k > 1 \\
2 + \left(e^{-2} + e^{-9/8}\right) \cdot \tfrac{a}{k}, & \text{if } a/(k-1) < e^2 \text{ and } k > 1
\end{cases}
$$

3.2.1 Proof of $\mathbf{E}[r(a, k)] \leq 1 + \ln a$ if $k = 1$, $\mathbf{E}[r(a, k)] \leq \left(1 + e^{-(k-1)/5}\right) \ln(a/(k-1))$ if $a/(k-1) \geq e^2$ and $k > 1$, and $\mathbf{E}[r(a, k)] \leq 2 + 2 \cdot e^{-(k-1)/5}$ if $a/(k-1) < e^2$ and $k > 1$

For our analysis, we first define two following two vector notations:

$$x_A^0 = \begin{cases} x_A & \text{if } \beta x_A \geq 1 \\ 0 & \text{otherwise} \end{cases} \qquad x_A^1 = \begin{cases} 0 & \text{if } \beta x_A \geq 1 \\ x_A & \text{otherwise} \end{cases}$$

Note that $c_A^0 = \lceil x_A^0 \rceil \leq \beta x_A^0$. Thus $1x^0 \leq 1c^0 \leq \beta 1 x^0$. Define $bonus = \beta 1 x^0 - 1c^0$. It is easy to see that $\mathbf{E}[1(c^0 + c^1)] = \beta 1 x - bonus$.

The contribution of set A to $bonus$ is $\beta x_A^0 - c_A^0$. This contribution to $bonus$ can be distributed equally to the elements if A. Since $|A| \leq a$, an element $i \in [1, n]$ receives a total of *at least* b^i/a of $bonus$, where $b^i = s^i(\beta x^0 - c^0)$ The random process that forms set $\mathcal{C}^1$ has the following goal from the point of view of element i: pick at least g^i sets that contain i, where $g^i = k - s^i c^0$ These sets are obtained as successes in Poisson trials whose probabilities of success add to at least $p^i = \beta(k - s^i x^0)$. Let y^i be random function denoting the number that element i contributes to the size of $\mathcal{C}^2$; thus, if in the random trials in Step 4 we found h sets from S^i then $y^i = \max\{0, k - h\}$. Thus, $\mathbf{E}[r(a, k)] = \mathbf{E}[1(c^0 + c^1 + c^2)] \leq \beta 1 x + \sum_{i=1}^{n} \mathbf{E}[y^i - \tfrac{b^i}{a}]$ Let $q^i = \frac{\beta}{\beta-1} s^i(c^0 - x^0)$. We can parameterize the random process that forms the set $\mathcal{C}^2$ from the point of view of element i as follows:

---
[2] The case of $k = 1$ was known before and included for the sake of completeness only.

- g^i is the *goal* for the number of sets to be picked;
- $p^i = \beta(k - s^i x^0) = \beta g^i + (\beta - 1)q^i$ is the sum of probabilities with which sets are picked;
- b^i/a is the *bonus* of i, where $b^i = s^i(\beta x^0 - c^0) \geq (\beta - 1)(k - g^i - q^i)$;
- $q^i \geq 0$, $g^i \geq 0$ and $g^i + q^i \leq k$;
- y^i measures how much the goal is *missed*;
- to bound $\mathbf{E}[r(a, k)]$ we need to bound $\mathbf{E}[y^i - \frac{b^i}{a}]$.

3.2.1.1 g-Shortage Functions

In this section we prove some inequalities needed to estimate $\mathbf{E}[y^i - \frac{b^i}{a}]$ tightly. Assume that we have a random function X that is a sum of N independent 0-1 random variables X_i. Let $\mathbf{E}[X] = \sum_i \mathbf{Pr}[X_i = 1] = \mu$ and $g < \mu$ be a positive integer. We define g-*shortage function* as $Y_g^\mu = \max\{g - X, 0\}$. Our goal is to estimate $\mathbf{E}[Y_g^\mu]$.

Lemma 2. $\mathbf{E}[Y_g^\mu] < \mathrm{e}^{-\mu} \sum_{i=0}^{g-1} \frac{g-i}{i!} \mu^i$.

From now on we will assume the worst-case distribution of Y_g^μ, so we will assume that the above inequality in Lemma 2 is actually an equality (as it becomes so in the limit), *i.e.*, we assume $\mathbf{E}[Y_g^\mu] = \mathrm{e}^{-\mu} \sum_{i=0}^{g-1} \frac{g-i}{i!} \mu^i$. For a fixed β, we will need to estimate the growth of $\mathbf{E}[Y_g^{g\beta}]$ as a function of g. Let $\rho_g(\beta) = \mathrm{e}^{g\beta} \mathbf{E}[Y_g^{g\beta}]$.

Lemma 3. $\rho_g(1) = \sum_{i=0}^{g-1} \frac{g-i}{i!} g^i = \frac{g^g}{(g-1)!}$

Lemma 4. For $\beta > 1$, $\frac{\rho_{g+1}(\beta)}{\beta \rho_g(\beta)}$ is a decreasing function of β.

Lemma 5. If $g > 1$ and $\beta > 1$ then $\frac{\mathbf{E}[Y_g^{g\beta}]}{\mathbf{E}[Y_{g-1}^{(g-1)\beta}]} \leq \mathrm{e}^{-\beta} \left(\frac{g}{g-1}\right)^g$

Lemma 6. $\frac{\mathbf{E}[Y_g^{g\beta+q}]}{\mathbf{E}[Y_g^{g\beta}]} < \mathrm{e}^{-q(1-1/\beta)}$

3.2.1.2 Putting All the Pieces Together

In this section we put all the pieces together from the previous two subsections to prove our claim on $\mathbf{E}[r(a, k)]$. We assume that $\beta \geq 2$ if $k > 1$. Because we perform analysis from the point of view of a fixed element i, we will skip i as a superscript as appropriate. As we observed in Section 3.2.1, we need to estimate $\mathbf{E}[y - \frac{b}{a}]$ and $b \geq (\beta - 1)(k - g - q)$. We will also use the notations p and q as defined there.

Obviously if $g = 0$ then $y = 0$. We omit the case of $k = 1$ and assume that $k > 1$ for the rest of this section. We first consider the "base" case of $g = 1$ and $q = 0$. Since $q = 0$, $c^0 = x^0$. Thus, $b = s^i(\beta c^0 - c^0) = (\beta - 1)s^i c^0 = (\beta - 1)(k - 1)$. Next, we compute $\mathbf{E}[y]$. Since $p = \beta g = \beta$, $\mathbf{E}[y] = \mathbf{E}[Y_1^\beta] = \mathrm{e}^{-\beta}$.

We postulate that

$$\mathbf{E}[y - \frac{b}{a}] \leq 0 \equiv \mathbf{e}^{-\beta} \leq \frac{(\beta - 1)(k - 1)}{a}$$

$$\equiv \frac{\mathbf{e}^{-\beta}}{\beta - 1} \leq \frac{k - 1}{a}$$

$$\equiv \mathbf{e}^{\beta}(\beta - 1) \geq \frac{a}{k - 1}$$

$$\equiv \beta + \ln(\beta - 1) \geq \ln \frac{a}{k - 1} \tag{3}$$

It is easy to see that, for the base case, $\mathbf{E}[\mathbf{1}(c^0 + c^1 + c^2)] \leq \beta 1 x \leq \ln(a/(k - 1))\mathrm{OPT}$.

Now we consider the "non-base" case when either $g > 1$ or $q > 0$. Compared to the base case, in a non-base case we have bonus $\frac{b}{a}$ decreased by at least $(\beta - 1)(g + q - 1)/a$. Also, $\mathbf{E}[y] = \mathbf{E}[Y_g^p] = \mathbf{E}[Y_g^{\beta g + (\beta - 1)q}]$.

Lemma 7. $\frac{\mathbf{E}[Y_y^{\beta g + (\beta - 1)q}]}{\mathbf{E}[Y_1^\beta]} \leq \mathbf{e}^{-(g + q - 1)/5}$.

Summarizing, when bonus is decreased by at most $(\beta - 1)(g + q - 1)/a = (\beta - 1)t/a$, we decrease the estimate of $\mathbf{E}[y]$ by multiplying it with at least $\mathbf{e}^{-t/5}$. As a function of $t = y + q - 1$ we have

$$\mathbf{E}[y] - b/a \leq \mathbf{e}^{-\beta - t/5} - \frac{\beta - 1}{a}(k - 1 - t) = \frac{(\beta - 1)(k - 1)}{a}\left(\mathbf{e}^{-t/5} - 1 + \frac{t}{k - 1}\right)$$

This is a convex function of t, so its maximal value must occur at one of the ends of its range. When $t = 0$ we have 0, and when $t = k - 1$ we have $\frac{(\beta - 1)(k - 1)}{a}\mathbf{e}^{-(k-1)/5}$. As a result, our expected performance ratio for $k > 1$ is given by

$$\mathbf{E}[r(a, k)] \leq \beta 1 x + \sum_{i=1}^{n} \mathbf{E}[y^i - \frac{b^i}{a}]$$

$$\leq \beta \mathrm{OPT} + \frac{\beta n k}{a} \mathbf{e}^{-(k-1)/5}$$

$$\leq \beta(1 + \mathbf{e}^{-(k-1)/5})\mathrm{OPT}$$

$$\leq \begin{cases} (1 + \mathbf{e}^{-(k-1)/5}) \ln(a/(k - 1)) \, \mathrm{OPT} & \text{if } a/(k - 1) \geq \mathbf{e}^2 \\ 2 \cdot (1 + \mathbf{e}^{-(k-1)/5}) \, \mathrm{OPT} & \text{if } a/(k - 1) < \mathbf{e}^2 \end{cases}$$

Acknowledgements

Berman was supported by NSF grant CCR-O208821, DasGupta was supported in part by NSF grants CCR-0296041, CCR-0206795, CCR-0208749 and a CAREER grant IIS-0346973, and Sontag was supported in part by NSF grant CCR-0206789.

References

1. N. Alon and J. Spencer, *The Probabilistic Method*, Wiley Interscience, New York, 1992.
2. H. Chernoff. *A measure of asymptotic efficiency of tests of a hypothesis based on the sum of observations*, Annals of Mathematical Statistics, 23: 493–509, 1952.
3. T. Cover, Geometrical and statistical properties of systems of linear inequalities with applications in pattern recognition, *IEEE Trans. Electronic Computers* EC-14, pp. 326–334, 1965. Reprinted in *Artificial Neural Networks: Concepts and Theory*, IEEE Computer Society Press, Los Alamitos, Calif., 1992, P. Mehra and B. Wah, eds.
4. U. Feige. *A threshold for approximating set cover*, JACM, Vol. 45, 1998, pp. 634-652.
5. M. R. Garey and D. S. Johnson. *Computers and Intractability - A Guide to the Theory of NP-Completeness*, W. H. Freeman & Co., 1979.
6. D. S. Johnson. *Approximation Algorithms for Combinatorial Problems*, Journal of Computer and Systems Sciences, Vol. 9, 1974, pp. 256-278.
7. N. Karmarkar. *A new polynomial-time algorithm for linear programming*, Combinatorica, 4: 373–395, 1984.
8. B. N. Kholodenko, A. Kiyatkin, F. Bruggeman, E.D. Sontag, H. Westerhoff, and J. Hoek, *Untangling the wires: a novel strategy to trace functional interactions in signaling and gene networks*, Proceedings of the National Academy of Sciences USA 99, pp. 12841-12846, 2002.
9. B. N. Kholodenko and E.D. Sontag, *Determination of functional network structure from local parameter dependence data*, arXiv physics/0205003, May 2002.
10. R. Motwani and P. Raghavan, *Randomized Algorithms*, Cambridge University Press, New York, NY, 1995.
11. L. Schläfli, *Theorie der vielfachen Kontinuitat (1852)*, in *Gesammelte Mathematische Abhandlungen*, volume 1, pp. 177–392, Birkhäuser, Basel, 1950.
12. E. D. Sontag, *VC dimension of neural networks*, in *Neural Networks and Machine Learning* (C.M. Bishop, ed.), Springer-Verlag, Berlin, pp. 69-95, 1998.
13. J. Stark, R. Callard and M. Hubank, *From the top down: towards a predictive biology of signaling networks*, Trends Biotechnol. 21, pp. 290-293, 2003.
14. V. Vazirani. *Approximation Algorithms*, Springer-Verlag, July 2001.

On the Crossing Spanning Tree Problem

Vittorio Bilò[1], Vineet Goyal[2,*], R. Ravi[2,*], and Mohit Singh[2,*]

[1] Dipartimento di Informatica Università di L'Aquila
Via Vetoio, Coppito 67100 L'Aquila, Italy
bilo@di.univaq.it
[2] Tepper School of Business, Carnegie Mellon University, Pittsburgh PA 15213
{vgoyal,ravi,mohit}@andrew.cmu.edu

Abstract. Given an undirected n-node graph and a set $\mathcal{C}$ of m cuts, the *minimum crossing spanning tree* is a spanning tree which minimizes the maximum crossing of any cut in $\mathcal{C}$, where the crossing of a cut is the number of edges in the intersection of this cut and the tree. This problem finds applications in fields as diverse as Computational Biology and IP Routing Table Minimization.
We show that a greedy algorithm gives an $O(r \log n)$ approximation for the problem where any edge occurs in at most r cuts. We then demonstrate that the problem remains NP-hard even when G is complete. For the latter case, we design a randomized algorithm that gives a tree T with crossing $O((\log m + \log n) \cdot (\text{OPT} + \log n))$ w.h.p., where OPT is the minimum crossing of any tree.
Our greedy analysis extends the traditional one used for set cover. The randomized algorithm rounds a LP relaxation of a corresponding subproblem in stages.

1 Introduction

Given a graph $G = (V, E)$ with n nodes and a family of cuts $\mathcal{C} = \{C_1, \ldots, C_m\}$, *the minimum crossing tree* is a spanning tree T, which minimizes the maximum crossing of any cut, where the crossing of a cut C_i is defined as $|E(T) \cap C_i|$. If the family of cuts is $\mathcal{C} = \{(v, V \setminus v) : v \in V\}$, then the MCST problem reduces to finding the minimum degree spanning tree problem which has been widely studied [8]. Hence, NP-completeness of the minimum degree spanning tree problem [7] shows that MCST problem is NP-hard.

In this paper, we show approximation guarantees for the greedy algorithm for the MCST problem.

Theorem 1. *Given a graph $G = (V, E)$ and a family of m cuts $\mathcal{C}=\{C_1, \ldots, C_m\}$, a greedy algorithm for MCST problem gives a spanning tree T which crosses any cut in $\mathcal{C}$ at most $O(r \cdot \log n)$ times the maximum crossing of an optimal tree.*

Although the minimum degree spanning tree problem is trivial on complete graphs, surprisingly, the MCST problem remains difficult even for this special case. We show that the decision version of even this version of the MCST problem is NP-complete.

Theorem 2. *Given a complete graph G, set of cuts $\mathcal{C}$ and a positive integer k, the problem of determining whether there exists a spanning tree of G which crosses any cut in $\mathcal{C}$ at most k times, is NP-complete.*

---

* Supported in part by NSF grant CCR-0105548 and ITR grant CCR-0122581 (The ALADDIN project).

K. Jansen et al. (Eds.): APPROX and RANDOM 2004, LNCS 3122, pp. 51–60, 2004.

A proof of the above theorem appears in the Appendix. The particular case of complete graphs finds application in fields as varied as IP routing and computational biology. We give improved algorithm for the MCST problem on complete graph which gives better performance guarantees.

Theorem 3. *There is a randomized LP rounding based algorithm, which given a complete graph G and a family of cuts $\mathcal{C}=\{C_1, \ldots, C_m\}$, outputs a spanning tree T such that crossing for any cut $C_i \in \mathcal{C}$ is $O((\log m + \log n) \cdot (OPT + \log n))$, where OPT is the maximum crossing of an optimal tree.*

1.1   Motivation: Chimerism in Physical Mapping

The MCST problem finds important applications in computational biology. The *physical mapping* problem of the human genome project is to reconstruct the relative position of fragments of DNA along the genome from information on their pairwise overlap. One has a collection of clones and a set of short genomic inserts (called *probes*). A probe defines a single location where a given subset of clones coincide. For each probe/clone pair, it can be determined whether the clone contains the probe as a subsequence using biological techniques. The problem is to construct the order in which the probes would occur along the original chromosome that is consistent with the given the probe/clone incidence matrix. This can be done efficiently if there is no *chimerism*. *Chimerism* is the result of concatenating two or more clone from different parts of the genome, producing a chimeric clone -one that is no longer a simple substring of the chromosome. More formally, the problem is as follows: Given a probe-clone incidence matrix A, with rows indexed by probes and columns by clones, and the entry a_{ij} is 1 iff probe i occurs in clone j. If there is no chimerism, then the problem is reduced to finding a permutation of rows so that ones in each column are consecutive (called as 1-C1P) and this can be solved efficiently in polynomial time [1]. However, in the presence of chimerism, the problem is more difficult. Then, we need to find a permutation π of rows, such that each column has at most k blocks of consecutive ones (called as k-consecutive ones property or k-C1P), if the chimeric clones are a concatenation of at most k clones. The decision version of this problem i.e *"Does a given 0-1 matrix have the k-consecutive ones property?"* has been proven to be NP-complete for $k \geq 2$ in [5].

1.2   k-C1P and Vector TSPs

A classical way to solve the k-C1P problem is to reduce it to a particular multidimensional TSP problem called the Vector TSP (vTSP). This problem is defined on a complete graph $G = (V, E)$, where each edge $e \in E$ is assigned an m-dimensional cost $c : E \rightarrow \{0,1\}^m$. The cost of a tour T in G is the m-dimensional vector $c(T) = \sum_{e \in E(T)} c(e)$ and the objective is to minimize $\|c(T)\|_\infty$.

The reduction from k-C1P to vTSP is straightforward. Each row of A becomes a node in G and the cost assigned to edge $e = (i, j)$ is set to the XOR-vector between the two rows a_i and a_j. Now, let π be the permutation induced by a solution T of vTSP, and let $b(A^\pi)$ be the maximum number of blocks of consecutive ones in A^π. Then, we have that $b(A^\pi) = \frac{\|c(T)\|_\infty}{2}$. Solving the vTSP problem is NP-hard by this reduction from the

2-C1P problem. However, since the Hamming distance obeys the triangle inequality, it is possible to use the standard Euler Tour short-cutting technique in order to compute a $2r$-approximate solution, given an r-approximation to the related Vector MST problem (vMST).

The vMST can be formulated as the *minimum crossing spanning tree* problem on a complete graph G. Any column j of A can be seen as a cut $C_j = (V_j, V \setminus V_j)$ defined on G by setting $V_j = \{v_i \in V | a_{ij} = 0\}$. The cost of edge $e = (i, j)$ is as before the XOR-vector between a_i and a_j i.e. $c(e)$ is a 0-1 vector, where the lth entry corresponding to a cut C_l is 1 iff the edge (i, j) crosses C_l. Here, the terminology that an edge e crosses a cut C is used interchangeably with $e \in C$. For any tree T, let $c(T) = \sum_{e \in E(T)} c(e)$. The ith entry of the vector $c(T)$ is exactly the number of edges of T crossing the cut C_i. Thus, the *minimum crossing spanning tree* minimizes $\|c(T)\|_\infty$.

1.3   Motivation: IP Routing

Another useful application of the MCST problem can be found in [2] where it is shown that the an efficient solution for the min-k-C1P can be used to obtain an good approximation for the Interval Routing problem: given a set of IP routing tables sharing the same host space, the problem is to reassign the IP addresses to the hosts in order to minimize the maximum size of any IP routing table.

This IP routing table minimization problem, MIN-IP for short, can be formalized as follows. We are given a set $R = \{r_1, \dots, r_n\}$ of n routers and a set $H = \{h_1, \dots, h_m\}$ of m destination hosts. Each router $r_j \in R$ has a degree δ_j, that is δ_j out-edges, and a routing table specifying which of the out-edges to take for every host. The problem is to choose the IP addresses of the m hosts and construct the n IP routing tables so as to minimize the maximum size of a table, that is the maximum number of used entries in a table.

In [2] it is shown that, given any r-approximation algorithm for the problem of determining a row permutation that minimizes the maximum number of blocks (of ones) in a boolean matrix A, an efficient $2r \log m$-approximation algorithm exists for MIN-IP, which exploits a matrix representation of the instances of the problem.

Similar applications can be found also in designing interval routing schemes as proposed in [3,4].

1.4   Related Work

As observed earlier, the minimum degree spanning tree problem is a special case of the MCST problem. The best result for the minimum degree spanning tree problem are due to Furer and Raghavachari [8]. They construct a spanning tree with maximum degree at most $\Delta^* + 1$ where Δ^* is the maximum degree of the optimal tree. The $vMST$ problem has been considered by Greenberg and Istrail [6]. They give solution of cost $O(s(A) \cdot OPT + \log n)$. Here $s(A) = max_{1 \le n} \sum_{j=1}^{n} a_{ij}$. Note that r in Theorem 1 is different from $s(A)$ in [6]: r is the maximum number of cuts a given edge e can cross, where the cuts are defined by columns of A; $s(A)$ is the sparsity of A i.e. the maximum number of 1's in any row in A. Observe that $r \le 2 \cdot s(A)$, but $s(A)$ can be as bad as m. Hence, our algorithm gives comparable or better performance guarantee than the algorithm in [6].

The paper is organized as follows. In Section 2, we describe a greedy algorithm for the MCST problem and prove Theorem 1. In Section 3, we give a randomized algorithm for the special case and prove the guarantees of Theorem 3. In the Appendix, we show that the MCST problem is NP-hard even for complete graphs.

2   Greedy Algorithm for the General Case

In this section, we show that the greedy algorithm gives an $O(r \cdot \log n)$ approximation for the MCST problem where r is defined as $\max_{e \in G} |\{C \in \mathcal{C}: e \in C\}|$. Given any subgraph H, the maximum number of times H crosses any cut in $\mathcal{C}$ is denoted by $Cross(H, \mathcal{C})$.

Greedy Algorithm:
$\quad F \leftarrow \phi$
$\quad$ while F is not a tree
$\quad$ do

$\qquad\qquad$ Let e' be an edge which minimizes $Cross(F \cup e, \mathcal{C})$
$\qquad\qquad$ over all edges $e \in G$ which join two components of F.

$\qquad\qquad F \leftarrow F \cup e'$
$\quad$ od

First, we give a lower bound for the MCST problem.

Lemma 1. *Given any $S \subset \mathcal{C}$, let k be the number of components formed after removing the edges from G of all cuts in S. Then*

$$opt \geq \frac{k-1}{|S|}$$

Proof. Any spanning tree of G must choose at least $k-1$ edges to join the k components formed after removing the edges of cuts in S. Each of these $k-1$ edges crosses at least one of the cuts in S. Hence, the average crossing of such a cut in S is at least $\frac{k-1}{|S|}$.

The Proof of Theorem 1. Let the solution returned by the greedy algorithm be T_g and let $l = Cross(T_g, \mathcal{C})$. We can divide the running of the greedy algorithm in l phases. The ith phase of the algorithm is the period when $Cross(F, \mathcal{C}) = i$. Let k_i denote the number of components in F when the ith phase ends. Let M_i be the cuts which are crossed by i edges at the end of ith phase and $m_i = |M_i|$.

Consider the running of the algorithm in the ith phase. The crossing number of at least m_i cuts increases by 1 in the ith phase. Each edge can increase the crossing number of at most r cuts. Hence, in the ith phase we must include at least $\lceil \frac{m_i}{r} \rceil$ edges in F. Every edge, when included in F, reduces the number of components in F by exactly one. Therefore, we have the following inequality

$$k_i \leq k_{i-1} - \frac{m_i}{r} \tag{1}$$

When the ith phase ends, every edge joining two components of F must cross at least one of the cuts in M_i, else the greedy algorithm would choose such an edge in the ith phase. Applying Lemma 1, we get the for each i,

$$opt \geq \frac{k_i - 1}{m_i} \qquad (2)$$

Using (1) and (2), we have that for each i,

$$k_{i-1} - k_i \geq \frac{k_i - 1}{r * opt} \qquad (3)$$

Using $k_i \geq 2$ for each $i \leq l - 1$ and $k_{l-1} > k_l$, we have for each i,

$$\begin{aligned}
k_{i-1} - k_i &\geq \tfrac{k_i}{2r*opt} \\
\Rightarrow k_{i-1} \quad &\geq k_i(1 + \tfrac{1}{2r*opt}) \\
\Rightarrow k_0 \quad &\geq k_l(1 + \tfrac{1}{2r*opt})^l
\end{aligned}$$

As, $k_0 = n$ and $k_l = 1$, we get that

$$\begin{aligned}
n \quad &> (1 + \tfrac{1}{2r*opt})^l \\
\Rightarrow \log n &\geq l \log(1 + \tfrac{1}{2r*opt})
\end{aligned}$$

Using, $\log(1 + x) \geq x - \frac{x^2}{2}$ and $r * opt \geq 1$ we get

$$\begin{aligned}
\log n &\geq l(\tfrac{1}{2r*opt}(1 - \tfrac{1}{4r*opt})) \geq l\tfrac{1}{4r*opt} \\
\Rightarrow \quad l &\leq 4r \log n * opt
\end{aligned}$$

Hence, the greedy algorithm is a $O(r \log n)$ approximation. □

3 A Randomized Algorithm for the Case of Complete Graphs

In this section, we describe a randomized algorithm for MCST for complete graphs and prove that it gives a tree with maximum crossing $O((\log m + \log n) \cdot (OPT + \log n))$ with high probability, where n is the number of vertices in G and m is the number of cuts in $\mathcal{C}$.

The idea is the following : Start with each vertex as a different component and merge components in phases until a connected subgraph is obtained. In a phase, each component is represented by an arbitrarily chosen vertex of the component. We carefully select some edges between the representative vertices by solving a multicommodity flow problem in each phase, so that the cuts in $\mathcal{C}$ are not crossed "too much". We ensure that at least one edge is chosen out of each representative in every phase. Hence, the number of components reduces by at least a factor of two and thus a connected subgraph is obtained in at most $\log_2 n$ phases.

In phase p, we solve the following multicommodity flow problem on a graph G' constructed from a complete graph G_p (on the representative vertices in this phase) as follows. Let $V(G_p) = \{v_1, v_2, \ldots, v_{n_p}\}$.

- For each undirected edge $e = (u, v)$, add two directed edges $e_f = (u, v)$ and $e_r = (v, u)$ in G',
- For each vertex $v_i \in V(G_p)$ introduce a new vertex s_{v_i} in $V(G')$ and
- $\forall v_j \in V(G_p), j \neq i$, add the directed edge (v_j, s_{v_i}) in G'.

Now, the flow problem on G' is the following. Each vertex $v_i \in V(G')$ is required to send a unit flow of commodity i to s_{v_i}. Let $f_1, f_2, \ldots, f_n$ be the flows associated with each of the n commodities. Let $f_i(v)$ denote the net flow of ith commodity into the vertex v. The following integer program accomplishes our goal.

$$
\begin{aligned}
\min \quad & z \\
\text{s.t.} \quad z \quad & \geq \textstyle\sum_{e \in E(G) \cap C} X_e \qquad \forall C \in \mathcal{C} \\
X_e \quad & = \textstyle\sum_{i=1,\ldots,n_p} f_i(e) \\
\forall i = 1, \ldots, n_p & \\
f_i(v) \quad & = \textstyle\sum_{(v,u) \in E(G')} f_i(v, u) - \sum_{(u,v) \in E(G')} f_i(u, v) \qquad \forall v \in V(G') \\
f_i(v) \quad & = 0 \qquad \forall v \in V(G') \setminus \{v_i, s_{v_i}\} \\
f_i(v_i) \quad & = 1 \\
f_i(s_{v_i}) \quad & = -1 \\
f_i(e) \quad & \in \{0, 1\} \quad \forall e \in E(G')
\end{aligned}
$$

We now describe the algorithm for the MCST problem. We will construct a connected subgraph H with a low *maximum crossing*.

1. Initialize $V(H) \leftarrow V(G)$, $E(H) \leftarrow \phi$, $G_0 \leftarrow G$, $R_0 \leftarrow V(G)$, $p \leftarrow 0$.
2. While H is not connected
   (a) Construct G' from G_p. Solve the LP-relaxation of the corresponding integer program for phase p and obtain an integral solution $\hat{X}$ by *randomized rounding* of the optimum LP solution [10].
   (b) Let $E' = \{e \in G_p : \hat{X}_e > 0\}$. $E(H) \leftarrow E(H) \cup E'$.
   (c) $p \leftarrow p + 1$. Let R_p be the set of representative vertices (chosen arbitrarily one for each connected component of H), G_p is the complete graph on the vertices of R_p.

Let T^* be a optimal tree for the MCST problem and let OPT be the maximum crossing of any cut in T^*.

Proposition 1. *Let z_p^* be the optimum to the LP-relaxation in phase p. Then $z_p^* \leq$ 2OPT.*

Proof. We can construct a feasible solution of the LP from the optimum tree T^* of value at most 2OPT. Let $R_i = \{v_1, \ldots, v_{n_p}\}$ be the set of representatives in phase i. From the Tree Pairing Lemma in [9], there exists a matching M between vertices of R_i such that the paths in T^* between the matched pairs of vertices are edge disjoint. We can use this matching to construct a feasible solution to the LP. Send a unit flow of commodity i on the directed path $P_{v_i, v_j} \cup (v_j, s_{v_i})$ and of commodity j on the path $P_{v_j, v_i} \cup (v_i, s_{v_j})$, where $P(u, v)$ is the unique path in tree T^* between matched pairs u and v. The above flow is a feasible flow as it satisfies all the flow constraints of the LP. Every edge of T^* carries at most two units of flow. Hence, the objective value z for this feasible flow, is at most 2OPT. Therefore, $z_p^* \leq$ 2OPT.

Proposition 2. *If an edge $e = (u, v)$ crosses a cut C, then any other path between u and v also crosses the cut C at least once.*

Proof. If we remove all the edges in C from G, then u and v would be disconnected. Thus, every path from u to v contains an edge of C.

We will use Proposition 2, to obtain and work with a special kind of optimum solution such that each flow path uses only two edges. Consider the flow decomposition for commodity i in the optimum solution of the LP-relaxation and consider a flow path $P = < v_i, v_{i_1}, v_{i_2}, \ldots, v_{i_k}, s_{v_i} >$. We can replace P by the path $P' = < v_i, v_{i_k}, s_{v_i} >$ without increasing the maximum crossing. From Observation 2, we know that any cut that the edge (v_i, v_{i_k}) crosses will be crossed at least once by the path P. Therefore, P' only reduces the number of crossings for the cuts in $\mathcal{C}$ and so we can replace P by P'. Thus, we can obtain a fractional optimum solution S^* such that each flow path uses only two edges. This step greatly simplifies the subsequent analysis of the randomized rounding since every cut crosses every flow path at most once after this preprocessing.

3.1   Rounding S^* to an Integral Solution

Let us describe the rounding of the fractional multicommodity flow obtained by solving the LP relaxation corresponding to phase p. The flow corresponding to each commodity is rounded independently of others. For each commodity $i = 1, \ldots, n_p$, choose an edge $e = (v_i, v_j)$ with probability $f_i(v_i, v_j)$. The corresponding flow is routed through the path $< v_i, v_j, s_{v_i} >$ and the edge (v_i, v_j) is included in the subgraph H. This is repeated for every commodity independently.

In phase p, let the fractional optimum flow be f^* and the optimum LP solution be z^*. Let $z(C)$ denote the number of edges crossing a cut $C \in \mathcal{C}$. Consider Y_j, a 0-1 variable associated with the jth commodity, where

$$Y_j = \begin{cases} 1 \text{ if the integral flow crosses C} \\ 0 \text{ otherwise} \end{cases}$$

Therefore,

$$Pr(Y_j = 1) = \sum_{e \in E(G_p) \cap C} \tilde{f}_j(e)$$

$$z(C) = \sum_{j=1}^{n_p} Y_j$$

$$\begin{aligned} E[z(C)] &= \sum_{j=1}^{n_p} \sum_{e \in E(G_i) \cap C} \tilde{f}_j(e) \\ &= \sum_{e \in E(G_i) \cap C} \sum_{j=1}^{n_p} \tilde{f}_j(e) \\ &= \sum_{e \in E(G_i) \cap C} X_e \\ &\leq \tilde{z} \quad \leq \quad 2 \cdot \text{OPT} \end{aligned}$$

$z(C)$ is the sum of independent Bernoulli trials. Thus, we can use Chernoff bounds to bound the tail probability

$$Pr(|z(C) - E[z(C)]| > k\beta) \quad \leq \quad exp(-\frac{k^2 \beta^2}{2E[z(C)]})$$

Let $\beta = E[z(C)] + \log n$ and $k = \log_n m + 2$. Therefore,

$$Pr(|z(C) - E[z(C)]| > k\beta) \le exp(-\frac{k^2(E[z(C)]+\log n)}{2})$$
$$< exp(-\frac{(2\log_n m+4)\log n}{2})$$
$$= \frac{1}{mn^2}$$

Since $E[z(C)] \le 2\text{OPT}$, we have that $Pr(z(C) > (2(k+1)\text{OPT} + k\log n)) < \frac{1}{mn^2}$ or $Pr(z(C) > 2(\log_n m + 3) \cdot OPT + (\log_n m + 2) \cdot \log n) < \frac{1}{mn^2}$ for any cut $C \in \mathcal{C}$ in any phase p. We say that a "bad" event occurs in a phase p if some cut $C \in \mathcal{C}$ has a high crossing in that phase. Thus, from the union bound we have $Pr(\text{bad event occurs in phase p}) < \frac{1}{n^2}$. The algorithm has at most $\log_2 n$ phases. Thus,

$$Pr(\text{"bad" event occurs in any phase}) < \frac{\log n}{n^2} \tag{4}$$

Thus, we have shown that in every phase the crossing of every cut is $O((\log_n m + 3)\text{OPT} + (\log_n m + 2) \cdot \log n)$ with high probability. Hence, we obtain a solution of maximum crossing $O((\log_2 m + \log_2 n) \cdot (\text{OPT} + \log_2 n))$ with probability at least $(1 - \frac{\log n}{n^2})$. $\square$

Remark: For $m \ge n$, setting $k = \sqrt{(\log_n m + 2)}$ gives a slightly better solution with maximum crossing $O(\sqrt{\log m \log n}(OPT + \log n))$.

4   Future Work

We believe that better performance ratios can be obtained particularly for the MCST problem on complete graphs. Furthermore, more sophisticated methods than a simple greedy approach should be able to remove the factor of r in the general case.

References

1. K. Booth and G. Luker. Testing for the consecutive ones property, interval graphs and graph planarity using pq-tree algorithms. *Journal of Computer and System Sciences 13: 335-379,1976.*
2. Vittorio Bilo and Michele Flammini. On the IP routing tables minimization with addresses reassignments. *In Proc. of the 18th International Parallel and Distributed Processing Symposium(IPDPS), IEEE Press, 2004.*
3. Michele Flammini, Giorgio Gambosi and Stefano Salomone. Interval Routing schemes. *Algorithmica 16(6): 549-568, 1996.*
4. Michele Flammini, Alberto Marchetti-Spaccamela and Jan van Leeuwen. The Complexity of Interval Routing on Random Graphs. *The Computer Journal 41(1): 16-25, 1998.*
5. Paul W. Goldberg, C. Golumbic, Martin, Haim Kaplan, and Ron Shamir. Four Strikes against Physical Mapping. *Journal of Computational Biology, 2(1): 139-152, 1995.*
6. David S. Greenberg and Sorin Istrail.Physical mapping by STS Hybridization: Algorithmic strategies and the challenges of software evaluation. *Journal of Computational Biology, 2(2): 219-273, 1995.*
7. Michael R. Garey and David S. Johnson *Computers and Intractibility: A guide to the Theory of NP-completeness.* W. H. Freeman and Company, New York, 1979.

8. M. Furer and B. Raghavachari. Approximating the minimum degree spanning tree to within one from the optimal degree. *In Proceedings of the Third Annual ACM–SIAM Symposium on Discrete Algorithms (SODA '92), 317-324, 1992.*
9. Philip N. Klein and R. Ravi. A Nearly Best-Possible Approximation Algorithm for Node-Weighted Steiner Trees. *J. Algorithms 19(1): 104-115, 1995.*
10. P. Raghavan and C. Thompson. Randomized Rounding. *Combinatorica, Volume 7, 365-374, 1987.*

Appendix: MCST for Complete Graphs Is NP-Hard

In this section, we consider the MCST problem for complete graphs. We show that the problem is NP-hard even for this special case. In fact, we show that the decision version of the problem is NP-complete.

Clearly, the decision problem is in NP. We reduce the 2-consecutive ones problem, 2-C1P, to MCST. Given a $n \times m$ matrix A, 2-C1P is the problem of determining whether there exists a permutation of rows such that in each column all ones occur in at most 2 consecutive blocks. This problem has been shown to be NP-complete in [6].

Given any arbitrary $n \times m$ matrix A, make a complete graph G over $n+1$ vertices, with one vertex corresponding to each row and a new dummy vertex s. For each column in A, include a cut in C naturally defined by the column: vertices with rows with 1 form one side of the cut. The dummy vertex s is always on the 0-side of each cut. Also include in C singleton cuts, $C_v = (\{v\}, V \setminus \{v\})$ for every vertex in G. For each pair of vertices u and v, include in C the cut $C_{uv} = (\{u,v\}, V \setminus \{u,v\})$. Finally, let $k = 4$.

We first show that if there exists a permutation of rows, π, such that it has 2-C1 property, then there exists a spanning tree which crosses each cut in C at most four times. Consider the Hamiltonian path H which starts at s and then traverses the vertices in the order corresponding to permutation π. Each cut corresponding to a column is crossed by the Hamiltonian path H exactly when the row permutation π switches from a row with 0 with a row with 1 or vice versa. As all the ones are in 2 consecutive blocks, each cut can be crossed at most four times. Introducing the dummy node corresponds to introducing a row with all zeros as the first row which clearly does not change 2-C1 property. Also, a Hamiltonian path crosses each singleton cut at most two times and cut C_{uv} at most two times for any $u, v \in V$. Hence, there exists a spanning tree which crosses every cut in C at most four times.

Now, for the other direction we show that if there exists a spanning tree T which crosses every cut in C at most 4 times then there exists that a permutation π which has the 2-C1P property. We claim that any such tree must be a Hamiltonian path. As each singleton vertex is a cut in C, hence degree of each vertex is at most four. Suppose there exists a vertex u with degree four. For $n > 5$, there exists a vertex v which is not a neighbor of u. But, then the cut C_{uv} is crossed at least five times. Hence, all vertices have degree at most three. Suppose, for the sake of contradiction there exists a vertex u such that $deg_T(u) = 3$. Consider any vertex v which is not a neighbor of u. As T crosses C_{uv} at most four times, so $deg_T(v) = 1$. This implies that the total sum of degrees of nodes in T is at most $3 * 4 + (n - 3)$. Hence, $2n - 2 \leq n + 9$ or equivalently, $n \leq 11$ which is a contradiction assuming larger n. Hence, every vertex must have degree at most two in T showing that T is a Hamiltonian path.

Let the hamiltonian path be $(v_1, \ldots, v_k, s, v_{k+1}, v_n)$. Consider the following permutation of rows $(r_{k+1}, \ldots, r_n, r_1, \ldots, r_k)$ where v_i corresponds to row r_i in the transformation. We claim that in each column, there cannot be more than two blocks of ones. Suppose for the sake of contradiction, there exists a column c_i which has three blocks of ones. Thus, the cut corresponding to the Hamiltonian cycle formed by joining v_n and v_1 must cross the cut corresponding to column c_i at least five times. But any cycle crosses any cut even number of times. Hence, it must cross the cut at least six times, but then the hamiltonian path must cross the cut at least five times, a contradiction. Hence, there exists a permutation which satisfies the 2-C1 property. This reduction shows that decision version of MCST problem for complete graphs is NP-complete.   □

A 3/4-Approximation Algorithm
for Maximum ATSP with Weights Zero and One

Markus Bläser

Institut für Theoretische Informatik, ETH Zürich
CH-8092 Zürich, Switzerland
mblaeser@inf.ethz.ch

Abstract. We present a polynomial time 3/4-approximation algorithm
for the maximum asymmetric TSP with weights zero and one.
As applications, we get a 5/4-approximation algorithm for the (minimum) asymmetric TSP with weights one and two and a 3/4-approxi-
mation algorithm for the Maximum Directed Path Packing Problem.

1 Introduction

Traveling salesperson problems with weights one and two have been studied
for many years. They are an important special case of traveling salesperson
problems with triangle inequality. Papadimitriou and Yannakakis [8] showed that
the undirected minimization problem is MaxSNP-complete. On the other hand,
they presented a 7/6-approximation algorithm with polynomial running time.
Vishwanathan [9] considered the corresponding asymmetric problem $\texttt{ATSP}(1,2)$
and gave a 17/12-approximation algorithm.

Let $\texttt{MaxATSP}(0,1)$ be the following problem: Given a directed complete loopless graph with edge weights zero and one, compute a TSP tour of *maximum*
weight. $\texttt{MaxATSP}(0,1)$ is a generalization of $\texttt{ATSP}(1,2)$ in the following sense:
Vishwanathan [9] showed that any $(1 - \alpha)$-approximation algorithm for the former problem transforms into an $(1 + \alpha)$-algorithm for the latter when replacing
weight two with weight zero. (The other direction is not known to be true.)

By computing a matching of maximum weight and patching the edges together arbitrarily, one easily obtains a polynomial time 1/2-approximation algorithm for $\texttt{MaxATSP}(0,1)$. (Note that each edge has weight at least zero, thus
we cannot loose any weight during the patching process.) Vishwanathan [9] was
the first to improve on this by designing a 7/12-approximation algorithm with
polynomial running time. In 1994, Kosaraju, Park, and Stein [6] gave a 48/63-
approximation algorithm with polynomial time that also worked for maximum
ATSP with arbitrary nonnegative weights. In their work, they also formulated
the so-called path coloring lemma, which will be crucial for our algorithm.
Bläser and Siebert [3] obtained a 4/3-approximation algorithm with running
time $O(n^{5/2})$ for $\texttt{ATSP}(1,2)$. This algorithm can also be modified to give a 2/3-
approximation algorithm for $\texttt{MaxATSP}(0,1)$ with the same running time [4]. Finally, Kaplan et al. [5] generalize this result by designing a 2/3-approximation

K. Jansen et al. (Eds.): APPROX and RANDOM 2004, LNCS 3122, pp. 61–71, 2004.

algorithm that works for maximum ATSP with arbitrary nonnegative weights but has a worse running time.

Closely related to MaxATSP(0, 1) is the Directed Path Packing Problem DPP. Here we are given a directed graph $G = (V, E)$. The aim is to find a subset P of node-disjoint paths of G such that the number of edges in P is maximized. By giving edges in G weight one and "non-edges" weight zero, any path packing transforms into a TSP tour by patching the paths arbitrarily together. On the other hand, any TSP tour yields a path packing by discarding all edges of weight zero. The only exception is the case where an optimum TSP tour has weight n. Here one weight one edge has to be discarded.

Our main result is a 3/4-approximation algorithm for MaxATSP(0, 1) with polynomial running time. As corollaries, we get a 5/4-approximation algorithm for ATSP(1, 2) and a 3/4-approximation algorithm for DPP.

1.1　Notations and Conventions

For a set of nodes V, $K(V)$ denotes the set of all edges (u, v) with $u \neq v$. In other words, $(V, K(V))$ is the complete loopless graph with node set V.

For a multigraph H and an edge e of H, $m_H(e) \in \mathbb{N}$ denotes the multiplicity of e in H, that is, the number of copies of e that are present in H. If H is clear from the context, we will also omit the subscript H.

A multigraph is called *2-path-colorable* if its edges can be colored with two colors such that each color class is a collection of node-disjoint paths. (Double edges are not allowed in such a collection of paths.)

1.2　Outline of Our Algorithm

Kosaraju, Park, and Stein [6] formulate the so-called path coloring lemma. It states that if each node of a multigraph H has indegree and outdegree at most two and total degree at most three and H does not contain any 2-cycles (that is, a cycle with exactly two edges) or triple edges, then H is 2-path colorable. Kosaraju, Park, and Stein proceed with computing a cycle cover and a matching. (A cycle cover of a graph is a collection of node-disjoint directed cycles such that each node belongs to exactly one cycle.) If the matching is carefully chosen, then the union of the cycle cover and the matching fulfills the premises of the path coloring lemma and henceforth, is 2-path-colorable. (One also has to deal with the 2-cycles in the cycle cover separately, the interested reader is referred to the original work.) If one now takes the color class with the larger weight and patches the paths arbitrarily together, one gets a TSP tour that has at least half the weight of the combined weight of the cycle cover and the matching. The weight of an optimum cycle cover is at least the weight of an optimum TSP tour and the weight of an optimum matching is at least half the weight of an optimum TSP tour. Thus in the ideal case, this would yield an 3/4-approximation. However, Kosaraju, Park, and Stein have to deal with 2-cycles and have to avoid triple edges. Therefore, they only get a 48/63-approximation. This approach is refined in subsequent works [2, 7].

In this work, we directly compute a maximum weight multigraph that fulfills the premises of the path coloring lemma. This is done via an LP approach (Section 3). The fractional solution H^* is then rounded to an integer one via an iterated decomposition scheme (Section 4). Finally the integer solution is transformed into a TSP tour (or Path Packing) via the path coloring lemma.

2 Admissible Multigraphs

A directed multigraph is called *admissible*, if

1. the indegree and outdegree of each node is at most two,
2. the total degree of each node is at most three,
3. between each pair of nodes, there are at most two edges (counted with multiplicities).

Let $\omega : K(V) \to \{0, 1\}$ be a weight function. (This will be the weight function of the input of our algorithm for MaxATSP$(0, 1)$.) Our goal is to find an admissible multigraph H of maximum weight where the edges are weighted according to ω, that is, each edge contributes weight $m_H(e) \cdot \omega(e)$.

Lewenstein and Sviridenko [7], by reduction to the path coloring lemma of Kosaraju, Park, and Stein [6] (see Bläser [2] for a proof), show the following variant of the path coloring lemma:

Lemma 1 (Path coloring lemma). *If there are no 2-cycles on a cycle in an admissible multigraph G, then G is 2-path-colorable.*

Above, a 2-cycle on a cycle is a directed cycle $v_1, \ldots, v_k, v_1$ with $k \geq 3$ such that (v_i, v_{i-1}) (if $i = 1$, then $i - 1$ is k) is also an edge for some i.

We first compute an admissible multigraph of maximum weight. Then we have to remove 2-cycles on a cycle. In the case of weights zero and one, we are able to remove these 2-cycles without any loss of weight.

3 LP for Maximum Weight Admissible Multigraphs

For each edge $e \in K(V)$, there is a variable x_e. If $e = (u, v)$, we also write x_{uv} instead of x_e. The following integral LP solves the problem of finding a maximum weight admissible multigraph:

$$\text{Maximize } \sum_{e \in K(V)} \omega(e) x_e \text{ subject to}$$

$$\sum_{u \neq v} x_{uv} \leq 2 \qquad \text{for all } v \in V \qquad \qquad \text{(indegree)}$$

$$\sum_{v \neq u} x_{uv} \leq 2 \qquad \text{for all } u \in V \qquad \qquad \text{(outdegree)}$$

$$\sum_{v \neq u} x_{uv} + \sum_{w \neq u} x_{wu} \leq 3 \quad \text{for all } u \in V \qquad \qquad \text{(total degree)}$$

$$x_{uv} + x_{vu} \leq 2 \qquad \text{for all } u, v \in V, \ u \neq v \qquad \text{(triple edge)}$$

$$x_{uv} \in \{0, 1, 2\} \qquad \text{for all } u, v \in V, \ u \neq v \qquad \text{(multiplicity)}$$

Constraints (outdegree) and (indegree) assure that the outdegree and indegree of each node are at most two. By (total degree), each node has total degree at most three. Constraint (triple edge) forbids more than two edges (counted with multiplicities) between each pair of nodes. The last condition ensures that each edge has integral multiplicity. By (triple edge), each edge can have multiplicity at most two. We now relax the integrality constraint:

$$0 \leq x_{uv} \leq 2 \quad \text{for all } u, v \in V,\, u \neq v \qquad \text{(multiplicity')}$$

(Note that only the lower bound of (multiplicity') is actually needed, the upper bound also follows from (triple edge).) Let x^*_{uv} be an optimal solution of the relaxed LP. Let ω^* be its weight. We may assume that the x^*_{uv} are rational numbers. Their smallest common denominator D can be represented with $\text{poly}(n)$ bits. We define the multigraph H^* as follows. For all u and v with $u \neq v$, there are $x_{uv} \cdot D$ copies of the edge (u, v) in H^*. Furthermore, we define the bipartite multigraph B^* with bipartition $V \cup V'$ (V' is a copy of V) as follows: If there are d copies of the edge (u, v) in H^*, then there are d copies of the edge (u, v') in B^*. The graph B^* will be helpful in the decomposition of H^*. By construction, H^* and B^* fulfill the following properties:

(P1) The outdegree of each node v in H^* is at most $2D$. The degree of each node $v \in V$ in B^* is at most $2D$.

(P2) The indegree of each node v in H^* is at most $2D$. The degree of each node $v' \in V'$ in B^* is at most $2D$.

(P3) The total degree of each node v in H^* is at most $3D$. The sum of the degrees of v and v' in B^* is at most $3D$.

(P4) Between each pair of distinct nodes u and v there are at most $2D$ edges in H^*. For any pair of distinct nodes u and v, the sum of the number of edges between u and v' and between v and u' is at most $2D$.

4   A Rounding Procedure for H^*

We first assume that $D = 2^\delta$ is a power of two. We will deal with the general case later. We now decompose H^* into two subgraphs H^*_1 and H^*_2 such that both H^*_1 and H^*_2 fulfill the properties (P1)–(P4) of the preceding section with D replaced by $D/2$. If we now replace H^* with the heavier of H^*_1 and H^*_2 and proceed recursively, we will end up with an optimum admissible multigraph after $\log D = \text{poly}(n)$ such decomposition steps.

In the following, H denotes the graph of the current iteration of the process outlined above and B is the corresponding bipartite graph. Our rounding procedure uses the fact that the edge weights are either zero or one.

Alon [1] gives a similar procedure for edge-coloring bipartite multigraphs. This procedure is then used by Kaplan et al. [5] for decomposing a fractional solution of an LP for computing cycle covers.

4.1 Normalization of H

We first ensure that H fulfills also the following property (P5) in addition to (P1)–(P4).

(P5) For all u and v: If $m(u,v) + m(v,u) = 2D$, then $m(u,v)$ and $m(v,u)$ are both even.

Algorithm 1 takes a multigraph H that fulfills (P1)–(P4) and transforms it into a multigraph H' with the same weight that fulfills (P1)–(P5).

The algorithm basically works as follows: Let u and v be two nodes. W.l.o.g. $m(u,v) \geq m(v,u)$. Furthermore, assume that $m(u,v) + m(v,u) = 2D$ and both $m(u,v)$ and $m(v,u)$ are odd. (As D is even, if one of $m(u,v)$ and $m(v,u)$ is odd, then both are.) Since D is a power of two, $m(u,v) > D$.

If $\omega(u,v) = 0$, then we can simply remove all copies of (u,v). This does not change the weight of H. Furthermore, this does not affect (P1)–(P4), too. If $\omega(v,u) = 0$, then we remove all copies of (v,u). The interesting case is $\omega(v,u) = \omega(u,v) = 1$. Here we remove one copy of (u,v) and add one copy of (v,u). This does not change the weight. Since $m(u,v) > D$, $m(u,v) \geq D$ thereafter. In particular, the indegree of u and the outdegree of v are both still at most $2D$. Therefore, the resulting graph H' fulfills (P1)–(P4), too.

By construction, H' fulfills (P5). This shows the following lemma.

Lemma 2. *Given a multigraph H fulfilling (P1)–(P4) for some $D = 2^\delta$, Algorithm 1 computes a multigraph H' that has the same weight as H and fulfills (P1)–(P5) for D.*

4.2 Decomposition of H

Next we describe how to decompose a graph H fulfilling (P1)–(P5) for D into two graphs H_1 and H_2 fulfilling (P1)–(P4) for $D/2$. Let H_1 be the heavier of H_1 and H_2. We thereafter normalize H_1 and proceed recursively.

In the first for loop of Algorithm 2, we run through all edges (u,v). If the multiplicity of (u,v) is even in H, then we simply divide all the copies evenly between H_1 and H_2. If the multiplicity is odd, then we divide all but one copy evenly between H_1 and H_2. Thereafter, all edges in H have multiplicity zero or one.

If the indegree and outdegree of a node u in H are both odd, then we add the edge (u, u') to B. Then we take two nodes of odd degree in a connected component of B. (If one such node exists, then there must exist another one.) We compute a path P connecting two such nodes and add the edges of P in an alternating way to H_1 and H_2. If there are not any nodes of odd degree, each connected component of B is Eulerian. We compute Eulerian tours and distribute the edges in the same way as in the case of a path P.

We claim that H_1 and H_2 fulfill (P1)–(P4) with parameter $D/2$. For (P1) note that we always remove the edges in pairs. If for instance an edge (u,v) is

---

Algorithm 1 Normalization of H

Input: Multigraph H fulfilling (P1)–(P4) for $D = 2^\delta$
Output: Multigraph $H^\square$ fulfilling $\omega(H) = \omega(H^\square)$ and (P1)–(P5) for D
    for all unordered pairs of nodes $u, v \in V$, $u \neq v$ **do**
        Let m be the multiplicity of (u, v) in H and $m^\square$ be the multiplicity of (v, u) in H.
        W.l.o.g. $m \geq m^\square$.
        if $m + m^\square < 2D$ or m or $m^\square$ is even **then**
            insert (u, v) and (v, u) with multiplicities m and $m^\square$ into $H^\square$, respectively.
        else
            if $\omega(u, v) = 0$ and $\omega(v, u) = 1$ **then**
                insert $m^\square$ copies of (v, u) into $H^\square$.
            end if
            if $\omega(u, v) = 1$ and $\omega(v, u) = 0$ **then**
                insert m copies of (u, v) into $H^\square$.
            end if
            if $\omega(u, v) = \omega(v, u) = 1$ **then**
                insert $m - 1$ copies of (u, v) insert $m^\square + 1$ copies of (v, u) into $H^\square$
            end if
        end if
    end for

---

removed from H and added to H_1, then also an edge (u, w) is remove from H and added to H_2. In other words, the outdegree of u in H_1 and H_2 is always the same. The only exception is the case when the original outdegree of u is odd. Then one more edge is added to H_1, say, than to H_2. However, since the outdegree in H was odd, it was at most $D - 1$. Therefore, the degree in H_1 is at most D and in H_2, it is at most $D - 1$. The same argument works if the roles of H_1 and H_2 are interchanged. In the same way, we can show that H_1 and H_2 fulfill (P2) with $D/2$.

For (P3), we distinguish four cases, depending on the parity of the indegree i and outdegree o of a node u in H. If both i and u are even, then the indegree and outdegree of u are $i/2$ and $o/2$, respectively, in both H_1 and H_2 by construction. If both i and u are odd, then the indegree and outdegree are $\lfloor i/2 \rfloor$ and $\lceil o/2 \rceil$, respectively, in one of H_1 and H_2, and $\lceil i/2 \rceil$ and $\lfloor o/2 \rfloor$ in the other of H_1 and H_2. This is due to the fact that we added the edge (u, u') to B. In both cases, the indegree and outdegree sums up to $(i + o)/2 \leq 3D/2$. Finally, we consider the case where either i or o is odd. We assume that i is odd, the other case is treated symmetrically. Then the indegree of u in H_1, say, is $\lceil i/2 \rceil$ and in H_2, it is $\lfloor i/2 \rfloor$. In both subgraphs, the outdegree of u is $o/2$. The total degree of u in H is however at most $3D - 1$, since $3D$ is even. Therefore, $i + o \leq 3D - 1$ and $\lceil i/2 \rceil + o/2 \leq 3D/2$.

It remains to show that (P4) holds: Let m be the multiplicity of (u, v) in H and m' be the multiplicity of (v, u) in H. If both m and m' are even, then half of the copies of both (u, v) and (v, u) is added to H_1 and the other half are added to H_2 in the first for loop. Thus the number of edges between u and v is $(m + m')/2 \leq 2D/2$ in both H_1 and H_2. If m is odd and m' is even or vice

---

Algorithm 2 Decomposition of H

Input: Multigraph H fulfilling (P1)–(P5) for some $D = 2^\delta$
Output: Multigraphs H_1 and H_2 fulfilling (P1)–(P4) for $D/2$
   for all (ordered) pairs of nodes u and v, $u \neq v$ **do**
      Let m be the multiplicity of (u, v).
      Let $h = \lfloor m/2 \rfloor$.
      Remove $2h$ copies of (u, v) from H.
      Add h copies to H_1 and h copies to H_2.
      Update B accordingly.
   end for
   for all nodes u **do**
      if the indegree and outdegree of u in H are both odd **then**
         add $(u, u^\square)$ to B
      end if
   end for
   while B contains nodes of odd degree **do**
      Choose two nodes $a, b \in V \cup V^\square$ of odd degree in B that lie in the same connected
      component.
      Compute a path P from a to b.
      Remove the edges of P from H and add them to H_1 and H_2 in an alternating way.
      Skip all edges of the form $(u, u^\square)$. Update B.
   end while
   for each connected component C of B **do**
      Compute a Eulerian tour of C.
      Remove the edges of C from H and add them to H_1 and H_2 in an alternating way,
      again skipping all edges of the form $(u, u^\square)$.
   end for

---

versa, then there are $\lfloor (m + m')/2 \rfloor$ between u and v in both H_1 and H_2 after the first for loop. Then one further copy is added to either H_1 and H_2. But since $2D$ is even, $\lfloor (m + m')/2 \rfloor < 2D/2$ and thus $\lfloor (m+m')/2 \rfloor + 1 \leq 2D/2$. The last case is the one where both m and m' are odd. Then $\lfloor m/2 \rfloor + \lfloor m'/2 \rfloor$ copies are added to H_1 and H_2, respectively, in the first for loop. In the remaining steps of Algorithm 2, two further copies are added to H_1 or H_2. In the worst case, they both go to one graph, say H_1. Since H fulfills (P5), $m + m' < 2D$. Thus $\lfloor m/2 \rfloor + \lfloor m'/2 \rfloor < D - 1$. Therefore, $\lfloor m/2 \rfloor + \lfloor m'/2 \rfloor + 2 \leq D = 2D/2$. Thus H_1 and H_2 also fulfill (P4) for $D/2$. Thus we obtain the next result.

Lemma 3. *Given a multigraph H that fulfills (P1)–(P5) for some $D = 2^\delta$, Algorithm 2 computes two multigraphs H_1 and H_2 such that both H_1 and H_2 fulfill (P1)–(P4) and for each edge e, $m_H(e) = m_{H_1}(e) + m_{H_2}(e)$.*

4.3 An Algorithm for Maximum Weight Admissible Multigraphs

Algorithm 3 repeatedly takes the multigraph H, normalizes it via Algorithm 1, decomposes it via Algorithm 2, and proceeds iteratively with the heavier of H_1 and H_2. Lemmas 2 and 3 immediately prove the following result.

Algorithm 3 Maximum weight admissible subgraph

Input: Multigraph H fulfilling (P1)–(P4) for some even $D = 2^\delta$
Output: Maximum weight admissible multigraph S
   for $i = 1, \ldots, \delta$ do
      Normalize H via Algorithm 1.
      Compute multigraphs H_1 and H_2 from H via Algorithm 2.
      Let w.lo.g. H_1 be the heavier of the two computed multigraphs.
      Set $H = H_1$.
   end for
   Set $S = H$.

Lemma 4. *Given a multigraph H that fulfills (P1)–(P4) for some $D = 2^\delta$ and $\omega(H) = \omega^*$, Algorithm 3 computes a maximum weight admissible multigraph S.*

4.4 Making D a Power of Two

If D is not a power of two, then we use the following standard trick and replace it by $\hat{D} = 2^{\hat{\delta}}$ where $\hat{\delta}$ is the smallest natural number such that $2n^2 D \le 2^{\hat{\delta}}$. Each value of the optimal solution x_{uv}^* is rounded down to the next multiple of $2^{-\hat{\delta}}$. Let $\hat{x}_{uv}$ be the values obtained. We have

$$\hat{x}_{uv} \ge x_{uv}^* - 2^{-\hat{\delta}} \tag{1}$$

Let $\hat{H}$ be the multigraph that has $\hat{x}_{uv} \cdot \hat{D}$ copies of each edge (u, v).

Since we round each value down, $\hat{H}$ fulfills (P1)–(P4). It remains to estimate the loss of weight. By (1),

$$\begin{aligned}
w(\hat{H}) &= \sum_{(u,v) \in K(V)} \omega(u, v) \hat{x}_{uv} 2^{\hat{\delta}} \\
&\ge \omega^* 2^{\hat{\delta}} - \sum_{(u,v) \in K(V)} \omega(u, v) 2^{-\hat{\delta}} \cdot 2^{\hat{\delta}} \\
&\ge \omega^* 2^{\hat{\delta}} - n^2,
\end{aligned}$$

because $\omega(u, v) \in \{0, 1\}$. If we now run Algorithm 3 on the graph $\hat{H}$ for $\hat{\delta}$ iterations, then we end up with an admissible multigraph S of weight

$$\omega(S) \ge (\omega^* 2^{\hat{\delta}} - n^2)/2^{\hat{\delta}} \ge \omega^* - 1/(2D).$$

Therefore, $D\omega(S) \ge D\omega^* - 1/2$. Since both quantities $D\omega(S)$ and $D\omega^*$ are integers, we have that even $D\omega(S) \ge D\omega^*$.

Therefore, we have a polynomial time solution for computing maximum weight admissible multigraphs.

Theorem 1. *On input $\hat{H}$, Algorithm 3 computes a maximum weight admissible multigraph in polynomial time.*

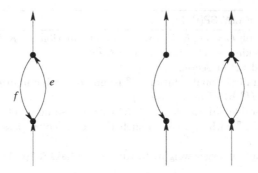

Fig. 1. On the lefthand side: A potential 2-cycle on a cycle. On the righthand side: The two ways how a potential 2-cycle on a cycle is treated

5   Coloring Admissible Multigraphs

We show that for any admissible multigraph G, there is another admissible multigraph G' of the same weight that is even 2-path colorable. Given G, G' can be found in polynomial time. Our aim is to exploit the path coloring result by Lewenstein and Sviridenko. To do so, we have to deal with 2-cycles on a cycle.

A 2-cycle c on a cycle locally looks as depicted on the lefthand side of Figure 1. If c has weight zero, then we can simply discard e. This does not change the weight. The resulting graph is still admissible and c is no longer a 2-cycle on a cycle. If e has weight one, then we remove f and add another copy of e. This can only increase the weight, since the weight of f is either zero or one. The resulting graph is still admissible and c is no longer a 2-cycle on a cycle. (Note that the procedure of Lewenstein and Sviridenko can deal with double edges.)

We now consider all (unordered) pairs of nodes and deal with them as described above. This shows the following result.

Lemma 5. *Given an admissible multigraph G and a weight function $\omega : K(V) \rightarrow \{0,1\}$, there is a 2-path-colorable admissible multigraph G' with $\omega(G) \leq \omega(G')$. Given G, G' can be found in polynomial time.*

6   Applications

Algorithm 4 now computes a maximum path packing or a maximum TSP tour, respectively. We first solve the fractional LP, round the optimum solution as in Section 4.2, make it 2-path-colorable, and finally take the color class of larger weight as a path packing, in the case of DPP, or patch the paths of this class together to form a TSP tour, in the case of MaxATSP$(0,1)$.

Theorem 2. *Algorithm 4 is a polynomial time 3/4-approximation algorithm for* DPP *and* MaxATSP$(0,1)$.

Algorithm 4 DPP, MaxATSP$(0, 1)$

Input: Directed Graph $G = (V, K(V))$ with weight function $\omega : K(V) \to \{0, 1\}$
Output: A path packing or TSP tour, respectively
Solve the fractional LP in Section 3.
Round the optimum fractional solution $H^\square$ to an admissible multigraph S via Algorithm 3. (Replace $H^\square$ by $\hat{H}$, if necessary.)
Find a 2-path colorable admissible graph $S^\square$ with the same weight as S.
Color the edges of $S^\square$ with two colors such that each color class is a collection of node-disjoint paths.
Return the collection of larger weight. In the case of MaxATSP$(0, 1)$, patch these path together arbitrarily.

Proof. By construction, the output of the algorithm is a feasible solution and it is computed in polynomial time. It remains to estimate the approximation performance.

If G has a path packing P with ℓ edges, then there is an admissible multigraph of weight $\frac{3}{2}\ell$. To see this, we decompose P into two matchings M_1 and M_2 by placing the edges of each path in P into M_1 and M_2 in an alternating fashion. Let w.lo.g. be M_1 the matching with more edges. Then $P \cup M_1$ is an admissible multigraph of weight $\frac{3}{2}\ell$.

In the same way we see that if there is a TSP tour of weight ℓ, there is an admissible multigraph of weight $\frac{3}{2}\ell$. (Strictly speaking, this is only true if the number of nodes is even. We can make it even by either adding a dummy node that is connected with weight zero edges to all other nodes. This gives a multiplicative loss of $(1 - 1/n)$ in the approximation factor. Or we can guess two consecutive edges and contract them into one. This increases the running time by a quadratic factor, but does not affect the approximation performance.)

The optimum admissible multigraph is divided into two color classes. Therefore, the heavier of the two classes has weight at least $\frac{3}{4}\ell$. □

Corollary 1. *There is a 5/4-approximation algorithm with polynomial running time for* ATSP$(1, 2)$.

Proof. Vishwanathan [9] showed that any $(1 - \alpha)$-approximation algorithm for MaxATSP$(0, 1)$ yields an $(1 + \alpha)$-approximation for ATSP$(1, 2)$, too. □

References

1. N. Alon. A simple algorithm for edge-coloring bipartite multigraphs. *Inform. Processing Letters* 85:301–302, 2003.
2. M. Bläser. An 8/13-approximation algorithm for the maximum asymmetric TSP. *J. Algorithms*, 50(1): 23–48, 2004.
3. M. Bläser and B. Siebert. Computing cycle covers without short cycles. In *Proc. 9th Ann. European Symp. on Algorithms (ESA)*, Lecture Notes in Comput. Sci. 2161, 369–379, Springer, 2001.

4. M. Bläser and B. Manthey. Approximating maximum weight cycle covers in directed graphs with edge weights zero and one. Manuscript, 2003.
5. H. Kaplan, M. Lewenstein, N. Shafrir, and M. Sviridenko. Approximation algorithms for asymmetric TSP by decomposing directed regular multigraphs. In *Proc. 44th Ann. IEEE Symp. on Foundations of Comput. Sci. (FOCS)*, pages 56–65, 2003.
6. S. R. Kosaraju, J. K. Park, and C. Stein. Long tours and short superstrings. In *Proc. 35th Ann. IEEE Symp. on Foundations of Comput. Sci. (FOCS)*, pages 166–177, 1994.
7. M. Lewenstein and M. Sviridenko. A 5/8 approximation algorithm for the maximum asymmetric TSP. *SIAM J. Discrete Math.*, 17:237–248, 2003.
8. C. H. Papadimitriou and M. Yannakakis. The traveling salesman problem with distances one and two. *Math. Oper. Res.*, 18:1–11, 1993.
9. S. Vishwanathan. An approximation algorithm for the asymmetric travelling salesman problem with distances one and two. *Inform. Proc. Letters*, 44:297–302, 1992.

Maximum Coverage Problem with Group Budget Constraints and Applications

Chandra Chekuri[1] and Amit Kumar[2]

[1] Bell Labs, 600 Mountain Avenue
Murray Hill, NJ 07974
chekuri@research.bell-labs.com
[2] Department of Computer Science and Engineering
IIT Delhi, India - 110016
amitk@cse.iitd.ernet.in

Abstract. We study a variant of the maximum coverage problem which we label the maximum coverage problem with group budget constraints (MCG). We are given a collection of sets $S = \{S_1, S_2, \ldots, S_m\}$ where each set S_i is a subset of a given ground set X. In the maximum coverage problem the goal is to pick k sets from S to maximize the cardinality of their union. In the MCG problem S is partitioned into *groups* $G_1, G_2, \ldots, G_\ell$. The goal is to pick k sets from S to maximize the cardinality of their union but with the additional restriction that at most *one* set be picked from each group. We motivate the study of MCG by pointing out a variety of applications. We show that the greedy algorithm gives a 2-approximation algorithm for this problem which is tight in the *oracle* model. We also obtain a constant factor approximation algorithm for the cost version of the problem. We then use MCG to obtain the first constant factor approximation algorithms for the following problems: (i) multiple depot k-traveling repairmen problem with covering constraints and (ii) orienteering problem with time windows when the number of time windows is a constant.

1 Introduction

In this paper we are interested in a variant of the *set cover* problem and its maximization version, the *maximum coverage* problem. The set cover problem is the following: we are given a ground set X of n elements and a set of subsets $S = \{S_1, S_2, \ldots, S_m\}$ such that for $1 \leq i \leq m$, $S_i \subseteq X$. The objective is to find a minimum number of sets from S such that their union is X. In the cost version of the set cover problem each set S_i has a cost $c(S_i)$ and we seek a minimum cost collection of sets to cover all the elements of X. The maximum coverage problem is a close relative of the set cover problem. We are given a ground set X and $S = \{S_1, S_2, \ldots, S_m\}$ of subsets of X. In addition we are given an integer k and the goal is to pick at most k sets from S such that the size of their union is maximized. In the profit version of the maximum coverage problem the items in X have profits and the goal is to maximize the profit of items picked.

K. Jansen et al. (Eds.): APPROX and RANDOM 2004, LNCS 3122, pp. 72–83, 2004.

Set cover and maximum coverage problems are fundamental algorithmic problems that arise frequently in a variety of settings. Their importance is partly due to the fact that many *covering* problems can be reduced to these problems. The *greedy* algorithm that iteratively picks the set that covers the maximum number of uncovered elements is a $(\ln n + 1)$ approximation for the set cover problem and an $\frac{e}{e-1}$ approximation[1] for the maximum coverage problem [15]. Feige [12] showed that these ratios are optimal unless NP is contained in quasi-polynomial time. In a number of applications the set system is implicitly defined and the number of sets is exponential in the number of elements. However, the greedy algorithm can still be applied if a polynomial time *oracle* that returns a set with good properties is available.

In this paper, motivated by several applications, we study a variant of the maximum coverage problem that we call the maximum coverage problem with group budget constraints (MCG). We start with a simple motivating example, the multiple knapsack problem (MKP), and the analysis of the greedy algorithm for that problem in [9]. MKP is a generalization of the classical knapsack problem to several knapsacks: we are given n items, where item i has a size s_i and a profit p_i; we are also given m knapsacks potentially of different capacities $b_1, b_2, \ldots, b_m$. The objective is to find a maximum profit subset of items that can be feasibly packed into the given set of knapsacks. We first consider the case where all the knapsacks have the same capacity b. For simplicity, in the following discussion, we assume that we have an exact algorithm to solve the single knapsack problem. We can apply the following greedy algorithm: pick an unused knapsack and use the single knapsack algorithm to pack it optimally with a subset of items from the remaining items. It is easy to show, via standard set cover style arguments, that this algorithm gives an $\frac{e}{e-1}$ approximation [9]. That the greedy algorithm gives this ratio can also be seen via a reduction to the maximum coverage problem as follows. Let $\mathcal{S}$ be the set of all distinct subsets of the items that can be feasibly packed into a knapsack of size b. The MKP problem can be rephrased as a maximum coverage problem on this implicit exponential sized set system and we are required to pick m sets. The greedy algorithm that we described for the MKP can be seen to be the greedy algorithm for maximum coverage problem on the implicit set system above where the oracle is the optimal single knapsack algorithm.

Now we consider instances of MKP in which the knapsack sizes could be different. In this case too we can define a greedy algorithm, however since the knapsacks are not identical, it is not a priori clear in which order to consider them. In [9] it is shown that *irrespective* of the ordering, the greedy algorithm results in a 2-approximation. Once again it is instructive to understand the implicit set system. Let $\mathcal{S}_i$ be set of distinct subsets of items that can be feasibly packed into a knapsack of size b_i and let $\mathcal{S} = \cup_i \mathcal{S}_i$. It is clear that if the knapsacks are not identical, we no longer have a maximum coverage problem although the greedy algorithm still gives a constant factor approximation algorithm. The

---

[1] In this paper approximation ratios for both minimization and maximization problems will be greater than or equal to 1.

problem we have is the following: we are required to choose at most *one* set from each of the S_i and cover as many elements as possible by the union of sets picked.

Motivated by MKP and other problems that we consider later in this paper, we define a variant of the maximum coverage problem that provides an abstraction for covering problems that have *groups* of sets.

Maximum Coverage with Group Budgets (MCG): We are given subsets $S_1, S_2, \ldots, S_m$ of a ground set X. We are also given sets $G_1, \ldots, G_\ell$, each G_i being a subset of $\{S_1, \ldots, S_m\}$. We call G_i a *group*. By making copies of sets, if necessary, we can assume that the groups G_i are disjoint from each other. We define two versions of the problem, the cardinality version and the cost version.

In the cardinality version, we are given an integer k, and an integer bound k_i for each group G_i. A solution is a subset $H \subseteq \{S_1, \ldots, S_m\}$ such that $|H| \leq k$ and $|H \cap G_i| \leq k_i$ for $1 \leq i \leq \ell$. The objective is to find a solution such that the number of elements of X covered by the sets in H is maximized. In fact, we can assume without loss of generality that all k_i are equal to 1. Otherwise, we can make k_i copies of each group G_i.

In the cost version, we associate a cost $c(S_j)$ with each set S_j. Further, we are given a budget B_i for group G_i, $1 \leq i \leq \ell$, and an overall budget B. A solution is a subset $H \subseteq \{S_1, S_2, \ldots, S_m\}$ such that the total cost of the sets in H is at most B. Further for any group G_i, the total cost of the sets in $H \cap G_i$ can be at most B_i. The objective is to find such a subset H to maximize the size of the union of sets in H.

In many applications, m, the number of given subsets of X, is exponential in n and the sets are defined implicitly. In such cases we require a polynomial time *oracle* with some desirable properties. We now make this more precise. In the cardinality version, we assume there exists an oracle $\mathcal{A}$ that takes as input, a subset X' of X, and an index i. $\mathcal{A}(X', i)$ outputs a set $S_j \in G_i$ such that $|S_j \cap X'|$ is maximized over all sets in G_i. We also work with approximate oracles. $\mathcal{A}$ is an α-approximate oracle if $\mathcal{A}(X', i)$ outputs a set $S_j \in G_i$ such that $|S_j \cap X'| \geq \frac{1}{\alpha} \max_{D \in G_i} |D \cap X'|$. For the cost version, we shall assume we are given an oracle $\mathcal{B}$ that takes as input a subset X' of X and a group index i. $\mathcal{B}(X', i)$ outputs a set $S_j \in G_i$ such that $\frac{|S_j \cap X'|}{c(S_j)}$ is maximized – we shall assume that all sets in G_i have cost at most B_i. As with the cardinality case we also work with approximate oracles.

We note that the maximum coverage problem is a special case of the cardinality case of MCG and the budgeted maximum coverage problem [17] is a special case of the cost version of MCG.

In this paper we show that a simple greedy algorithm gives a constant factor approximation for both the cardinality and the cost versions of MCG. The analysis differs from the usual analysis for the maximum coverage problem and is based on the analysis for MKP in [9]. For the cardinality version we show that greedy gives a 2-approximation and our analysis is tight. For the cost version we obtain a 12-approximation. The greedy algorithm works in the oracle model as well and in fact this is the main thrust of the paper and leads to the applications that we mention below.

We note that for the cardinality version, an $\frac{e}{e-1}$-approximation is achievable if the input is given in an explicit form. This ratio is achieved by rounding an LP relaxation either by the pipage rounding technique of Ageev and Sviridenko [1] or the probabilistic rounding technique of Srinivasan [18].

Set Cover with Group Budgets (SCG): We define the set cover problem with group constraints. We only consider the cardinality version in this paper, it is easy to extend it to the cost case. We are given a ground set X of n items and a set $S = \{S_1, S_2, \ldots, S_m\}$ of subsets of X. The set S is partitioned into groups $G_1, G_2, \ldots, G_\ell$. The objective is to find a subset H of S such that all elements of X are covered by sets in H and $\max_{i=1}^\ell |H \cap G_i|$ is minimized. Note that if we have a single group containing all sets then the problem is the same as the set cover problem. Elkin and Kortsarz [10] seem to be the first to consider this problem for its applications and they call it the multiple set cover problem. They present an $O(\log n)$ approximation using a randomized rounding of a natural linear programming relaxation. Kortsarz [16] asked if there is a combinatorial algorithm for this problem. From the 2-approximation bound on Greedy we obtain a simple combinatorial $(\log n + 1)$ algorithm for SCG. It also has the advantage of working in the oracle model.

We consider two problems for which we design the first constant factor approximation algorithms by using reductions to instances of MCG in the oracle model. We describe them next.

Multiple Depot k-Traveling Repairmen Problem: The k-traveling repairmen problem was considered by Fakcharoenphol, Harrelson, and Rao [11]. Their problem is the following. We are given a finite metric space on a set of nodes V induced by an edge weighted undirected graph G, and k not necessarily distinct vertices $s_1, s_2, \ldots, s_k$ from V. A feasible solution to the problem is a set of k tours $T_1, \ldots, T_k$, with tour T_i starting at s_i such that every vertex in V is visited by one of the tours. Given a feasible solution, we define the *latency* of a vertex as the time at which it gets visited by one of these tours. The objective is to minimize the sum of the latencies of the vertices in V. The problem models the case where k-repairmen are available at the k locations (depots) and we need to visit all the sites that have repairs. The goal is to minimize the average waiting time of the sites. If $k = 1$, the problem is the same as the the minimum latency problem for which a constant factor approximation was first given by Blum et al. [5]. The current best approximation ratio for the minimum latency problem is 3.59 [8]. In [11] a constant factor approximation algorithm is presented for the k-traveling repairmen problem when all the repairmen start at the same vertex s, that is $s_1 = s_2 = \ldots = s_k = s$. In the same paper [11], the generalization of the k-traveling repairmen problem to the case with multiple sources (depots) is left as an open problem. We obtain a constant factor approximation for this problem. We also obtain constant factor approximation algorithms for several generalizations as well.

Orienteering (or TSP) with Time Windows: The orienteering problem is defined as follows. We are given a metric space on a set of vertices V, a starting vertex s and a budget B. A feasible solution is a tour starting at s and having length at most B. The objective is to maximize the number of vertices covered by this tour. Blum et al. [7] gave the first constant factor approximation algorithm for this problem in general graphs, they obtained a ratio of 5 which has recently been improved to 3 in [6]. Previously, a $2 + \epsilon$-approximation was known for the case when the metric was induced by points in the Euclidean plane [2].

In this paper we consider the more general problem where we associate a window $[r_v, d_v]$ with each vertex v. A vertex v can be visited only in the time interval $[r_v, d_v]$. We shall say that r_v is the *release time* of v and d_v is the *deadline* of v. The objective, again, is to find a path that starts at s and maximizes the number of vertices visited, however we can only visit a vertex within its time window. This problem models a variety of situations when technicians or robots have to visit a number of different locations in a time period. The problem is referred to by different names in the literature including *prize collecting traveling salesman problem with time windows* and *TSP with deadlines*. Tsitsikilis [19] showed that this problem is strongly NP-complete even when the metric space is a line. Bar-Yehuda et al. [4] gave an $O(\log n)$ approximation when the vertices lie on a line. Recently Bansal et al. [6] gave an $O(\log^2 n)$ for the general problem. In this paper we consider the case when the number of *distinct* time windows, k, is a fixed constant independent of the input. We give a constant factor approximation algorithm for this problem using a reduction to MCG and using the algorithm in [7, 6] as an oracle.

The thrust of this paper is the definition of MCG and its applications. In this extended abstract we have not attempted to optimize the constants that can be achieved for the problems we consider. We defer this to the final version of the paper.

2   Greedy Algorithm for MCG

In this section we show that simple greedy algorithms give constant factor approximation ratios for MCG. First we consider the cardinality version of the problem.

2.1   Cardinality Version

We can assume without loss of generality that the number of groups $\ell \geq k$. We work in the oracle model and assume that we have an α-approximate oracle. The greedy algorithm we consider is a natural generalization of the greedy algorithm for the maximum coverage problem. It iteratively picks sets that cover the maximum number of uncovered elements, however it considers sets only from those groups that have not already had a set picked from them. The precise algorithm is stated below.

Algorithm **Greedy**
$H \leftarrow \emptyset$, $X' \leftarrow X$.
For $j = 1, 2, \ldots, k$ do
  For $i = 1, \ldots, \ell$ do
    If a set from G_i has not been added to H then $A_i \leftarrow \mathcal{A}(G_i, X')$
    Else $A_i \leftarrow \emptyset$
  EndFor
  $r \leftarrow \mathrm{argmax}_i |A_i|$
  $H \leftarrow H \cup \{A_r\}$, $X' \leftarrow X' - A_r$
EndFor
Output H.

By renumbering the groups, we can assume that Greedy picks a set from group G_j in the jth iteration. Let OPT denote some fixed optimal solution and let $i_1 < i_2 < \ldots < i_k$ be the indices of the groups that OPT picks sets from. We set up a bijection π from $\{1, 2, \ldots, k\}$ to $\{i_1, i_2, \ldots, i_k\}$ as follows. For $1 \le h \le k$, if $h \in \{i_1, i_2, \ldots, i_k\}$ then we require that $\pi(h) = h$. We choose π to be any bijection that respects this constraint.

Let C_j be the set that Greedy picks from G_j, and let O_j be the set that OPT picks from $G_{\pi(j)}$. We let $A'_j = A_j - \cup_{h=1}^{j-1} A_h$ denote the set of new elements that Greedy adds in the jth iteration. Let $C = \cup_j A_j$ and $O = \cup_j O_j$ denote the number of elements that Greedy and OPT pick.

Lemma 1. *For* $1 \le j \le k$, $|A'_j| \ge \frac{1}{\alpha}|O_j - C|$.

Proof. If $O_j - C = \emptyset$ there is nothing to prove. When Greedy picked A_j, the set O_j was available to be picked. Greedy did not pick O_j because $|A'_j|$ was at least $\frac{1}{\alpha}|O_j - \cup_{h=1}^{j-1} A_h|$. Since $\cup_{h=1}^{j-1} A_h \subseteq C$, the lemma follows. $\qquad\square$

Theorem 1. *Greedy is an* $(\alpha + 1)$-*approximation algorithm for the cardinality MCG with an* α-*approximate oracle.*

Proof. From Lemma 1, we have that

$$|C| = \sum_j |A'_j| \ge \sum_j \frac{1}{\alpha}|O_j - C| \ge \frac{1}{\alpha}(|\cup_j O_j| - |C|) \ge \frac{1}{\alpha}(|O| - |C|).$$

Hence $|C| \ge \frac{1}{\alpha+1}|O|$. $\qquad\square$

Corollary 1. *If* $k = \ell$, *Greedy is an* $(\alpha + 1)$-*approximation algorithm even if it is forced to pick sets from an adversarially chosen ordering of the groups.*

Proof. If $k = \ell$, the permutation π is the identity permutation. In this case Lemma 1 holds again. $\qquad\square$

Easy examples show that our analysis of the Greedy algorithm is tight. In fact, in the oracle model, the ratio of 2 cannot be improved. When the set system is available as part of the input, the problem is hard to approximate to within a factor of $\frac{e}{e-1}$ via a reduction from the maximum coverage problem. As we mentioned earlier, a matching ratio can be obtained via linear programming [1, 18].

A $\log n + 1$ Approximation for SCG: We observe that the 2-approximation bound for MCG can be used to obtain a $\log n + 1$ approximation for SCG as follows. We simply guess the optimal value λ^*, that is there is an optimal cover H^* such that $\max_i |H^* \cap G_i| \leq \lambda^*$. We then create an instance of MCG by having a budget of λ^* on each group G_i. Greedy covers at least $1/2$ the elements in X. Iterating Greedy $\log n + 1$ times results in a solution that covers all elements. In each iteration the number of sets added from any given group is upper bounded by λ^*. Hence, when all elements are covered, the number of sets added from any group is at most $(\log n + 1)\lambda^*$.

2.2   Cost Version

We now consider the cost version of MCG. We give a greedy algorithm for this problem which is similar in spirit to the one for the cardinality case but differs in some technical details. The algorithm that we describe below may violate the cost bounds for the groups or the overall cost bound B. We will later show how to modify the output to respect these bounds. We work in the oracle model again. Recall that the oracle $\mathcal{A}$, given a set of elements X' and an index i returns a set $S \in G_i$ that approximately minimizes the ratio $\max_{D \in G_i} \frac{|D \cap X'|}{c(D)}$. The algorithm is described in more detail below. We assume without loss of generality that $B \leq \sum_{i=1}^{\ell} B_i$.

Algorithm **CostGreedy**
$H \leftarrow \emptyset, X' \leftarrow X$.
Repeat
  For $i = 1, 2, \ldots, l$ do
    If $c(H \cap G_i) < B_i$ then $A_i \leftarrow \mathcal{A}(X', G_i)$.
    Else $A_i \leftarrow \emptyset$.
  EndFor
  $r \leftarrow \text{argmax}_i \frac{|A_i|}{c(A_i)}$
  $H \leftarrow H \cup \{A_r\}, X' \leftarrow X' - A_r$.
Until $(c(H) \geq B)$.
Output H.

Note that H need not obey the budget requirements. Define $H_i = H \cap G_i$. For a set S chosen by the algorithm, define $X(S)$ as the extra set of elements in X that are covered at the time S is added to H. Define $X(H_i) = \cup_{S \in H_i} X(S)$. Similarly, $X(H)$ is the set of elements covered by the algorithm. Let OPT be some fixed optimal solution to the given problem instance. Let O be the set of sets chosen by OPT. Define $Y_i = O \cap G_i$. We call an index i good if $c(H \cap G_i) \leq B_i$, that is the algorithm did not exceed the budget for G_i. Otherwise we call i bad. We omit proofs of the following lemmas in this version.

Lemma 2. *If i is bad, then $|X(H_i)| \geq \frac{1}{\alpha} \sum_{A \in Y_i} |A - X(H)|$.*

Corollary 2. $|X(H)| \geq \frac{1}{\alpha+1} |\cup_{i:i} \text{ bad } \cup_{A \in Y_i} A|$.

Lemma 3. $|X(H)| \geq \frac{1}{\alpha+1} |\cup_{i: i} \text{ good } \cup_{A \in Y_i} A|$.

From Corollary 2 and Lemma 3, it follows that $X(H) \geq \frac{1}{2(\alpha+1)} \text{OPT}$. But H does not respect all the budget constraints. We partition H into three subsets H_1, H_2, H_3. H_3 is the last set picked by our algorithm. H_2 contains those sets S which when added to H caused the budget of some group G_i to be violated – however we do not include the set in H_3. H_1 contains all the remaining sets in H. It is easy to see that H_1, H_2 and H_3 do not violate the budget constraints. Further, one of these three sets must be covering at least $1/3$ the number of elements covered by H. Thus we get the following theorem.

Theorem 2. *The algorithm* **CostGreedy** *is a $6(\alpha + 1)$-approximation algorithm for the cost version of MCG.*

3   Applications of MCG

3.1   The k-Traveling Repairmen Problem

Recall from Section 1 that in the k-traveling repairmen problem we are given a metric space on a set of nodes V induced by an edge weighted undirected graph G. We are given a set of k *source* vertices in V, call them $s_1, s_2, \ldots, s_k$. A feasible solution to the problem is a set of k tours, one tour starting at each source s_i, such that every vertex in V is visited by one of the tours. Given a feasible solution, we define the *latency* of a vertex as the time at which it gets visited by one of these tours. The objective is to minimize the sum of the latencies of the vertices in V.

We give the first constant factor approximation algorithm for the multiple depot k-traveling repairmen problem. We define a related problem, which we call the *budgeted cover* problem, as follows. The input to the problem is a subset V' of V and a positive integer B. A solution is a set of k tours, one tour starting at each source s_i, such that no tour has length more than B. The objective is to maximize the number of vertices of V' covered by these tours. We can view this problem in the framework of MCG as follows. The ground set is the set of vertices in V'. For each source s_i, we have a collection of sets $S_1^i, \ldots, S_{\ell_i}^i$: each set corresponds to a distinct tour of length at most B beginning at s_i. There are k groups, one group for each source vertex s_i. The group G_i corresponding to s_i is $\{S_1^i, \ldots, S_{\ell_i}^i\}$. Clearly, the cardinality version of MCG for this set system is the same as the budgeted cover problem for the original graph. From Section 2 we will get a constant factor approximation algorithm for the budgeted cover problem provided we have the following oracle $\mathcal{A}$.

The oracle $\mathcal{A}$ should be able to solve the *budget-MST* problem. In this problem, we are given a graph $G = (V, E)$, a source $s \in V$ and a budget B. A solution is a tour starting at s and having cost at most B. The objective is to maximize the number of vertices covered by the tour. Unfortunately the *budget-MST* problem is NP-hard. However, by using the algorithm for the i-MST problem [13, 3], we can obtain a polynomial time algorithm, $\mathcal{A}$, which covers OPT vertices and costs at most $\beta \cdot B$. Here $\beta > 1$ is the approximation ratio for the i-MST problem. Hence, if we are willing to violate the budget constraint by a factor of β we can cover as many vertices as the optimal solution.

It follows from Theorem 1 that we can get a polynomial time algorithm for the budgeted cover problem which covers at least half as many vertices as the optimum, and constructs k tours, each tour of length at most βB. We call this algorithm $\mathcal{C}$.

We can now describe our algorithm for the traveling k-traveling repairmen problem. Our algorithm works in phases. We assume without loss of generality that all distances are at least 1. Let V_j be the set of *uncovered* vertices at the beginning of phase j (so $V_0 = V$). In phase j, we cover as many vertices as possible so that the budget of each tour is about 2^j. More precisely, we do the following

Algorithm **Visit**(j) :
For $p = 1, 2$ do
    Run $\mathcal{C}$ on the budget cover problem instance with inputs V_j and 2^j.
    Remove from V_j the covered vertices.

This describes phase j. We invoke the subroutine **Visit**(j) with increasing values of j until all vertices are covered. Given a source s_i, we have constructed several tours starting from s_i. We just stitch them together starting in the order these tours were found by the algorithm. Clearly, our algorithm produces a feasible solution. It remains to prove that it is a constant factor approximation algorithm.

We begin with some notation. Fix an optimal solution OPT. Let O_j denote the set of nodes in OPT's solution which have latency at most 2^j. Let C_j be the set of nodes visited by our algorithm by the end of phase j.

Lemma 4. Visit(j) *covers at least* $\frac{3}{4}|O_j - C_{j-1}|$ *vertices.*

Proof. Let R_j denote $O_j - C_{j-1}$. Let A_j be the set of nodes covered by **Visit**(j) when $p = 1$. Theorem 1 implies that $|A_j| \geq \frac{1}{2}|R_j|$. One more application of this theorem when $p = 2$ gives the desired result. □

The rest of the proof goes along the same lines as in [11]. Let n_j be the set of nodes in OPT whose latency is more than 2^j. Let n'_j be the set of nodes in our tour which do not get visited by the end of phase j.

Lemma 5. $n'_j \leq \frac{1}{4}n'_{j-1} + \frac{3}{4}n_j.$

Proof. From Lemma 4 it is easy to see that $n'_j \leq n'_{j-1} - 3/4|O_j - V_{j-1}|$. Clearly, $|O_j| = n - n_j$ and $|V_{j-1}| = n - n'_{j-1}$. Combining these proves the lemma. □

The total latency of the tours obtained by our algorithm is upper bounded by $\sum_j 4\beta 2^j n'_j$ and that produced by the tours in the optimal solution is lower bounded by $\sum_j 2^{j-1} n_j$. From Lemma 5 we obtain that

$$\sum_j 2^j n'_j \leq \frac{1}{2} \sum_j 2^{j-1} n'_{j-1} + \frac{3}{4} \sum_j 2^j n_j$$

which implies that

$$\sum_j 2^j n'_j \leq 3 \sum_j 2^{j-1} n_j.$$

This proves that our algorithm yields a 12β approximation. We can improve the ratio by using ideas from [14, 11, 8]; we defer the details to the final version. We can also obtain constant factor approximations for each of the following generalizations: (i) each vertex v can be serviced only by a given subset S_v of repairmen, (ii) each vertex v has a service time p_v that the repairmen needs to spend at the vertex, and (iii) each vertex v has a weight w_v and the objective is to minimize the sum of the weighted latencies.

3.2   The Orienteering Problem with Time Windows

We now consider the orienteering problem with time windows. We assume that the number of *distinct* time windows is some fixed constant k. We use, as a subroutine, the algorithm of Bansal et al. [6] which provides a 3-approximation for the case when there is a single time window $[0, D]$ for all vertices and the tour is required to start at a vertex s and end at a vertex t. In the rest of the section we use β to denote the approximation ratio for the single deadline case. All our ratios will be expressed as functions of β. Let Δ be the maximum distance in the metric space. We begin by describing approximation algorithms for two special cases: (1) when all release times are zero, and (2) when all deadlines are the same.

Release Times Are Zero: We consider the special case when $r_v = 0$ for all nodes v. Let $d_1 < d_2 < \ldots < d_k$ be the k distinct deadlines. Let V_i denote the set of vertices whose deadline is d_i. Let P^* be the tour constructed by some optimal solution. Define v_0^* as the source vertex s. For $1 \leq i \leq k$, let v_i^* as the last vertex in the tour P^* which is visited by the deadline d_i. It is possible that $v_i^* = v_{i'}^*$ for two distinct indices i and i'. Suppose v_i^* is visited at time t_i^*, then it follows that that $t_i^* \leq d_i$.

Our algorithm first *guesses* the vertices $v_1^*, v_2^*, \ldots, v_k^*$ and the time instances $t_1^*, t_2^*, \ldots, t_k^*$. Note that $t_i^* \leq n\Delta$. Hence the total number of guesses is $O(n^{2k}\Delta^k)$. Since Δ need not be polynomially bounded, the number of guesses is not poly-bounded. We omit details on how to use a polynomial number of guesses. Now, we define k groups $G_1, \ldots, G_k$ as follows. G_i is the set of all paths on the vertex set $V_i \cup V_{i+1} \cup \cdots \cup V_k$ which originate at v_{i-1}^* and end at v_i^* with the additional constraint that the length of the path is at most $t_i^* - t_{i-1}^*$.

Lemma 6. *Consider the instance of the MCG with groups defined as above where we need to pick exactly one set from each group. A γ-approximation algorithm for this problem instance implies the same for the corresponding orienteering problem.*

Proof. Suppose we are given a solution to the MCG which picks paths $P_1, \ldots, P_k$ from the corresponding groups. If we stitch these tours sequentially, it is easy to see that we get a path which satisfies the deadlines of the vertices visited by the individual paths. Therefore the number of vertices covered by this tour is $|P_1 \cup P_2 \cup \cdots \cup P_k|$. Further, if we consider the tour P^*, we can get a solution to the MCG which covers $|P^*|$ vertices. This proves the lemma. $\qquad\square$

Thus, it is enough to approximate the MCG induced by the guess of $v_1^*, \ldots, v_k^*$ and $t_1^*, \ldots, t_k^*$. The oracle needed for this instance is an algorithm for the orienteering problem where the time windows for all the nodes are of the form $[0, D]$, hence we can use the algorithm of Blum et al. [7] or Bansal et al. [6]. From Corollary 1 we obtain a a $(\beta + 1)$-approximation algorithm for the case of k deadlines. The running time of the algorithm is $O(n^{2k} \Delta^k T)$ where T is the running time of the approximation algorithm for the single deadline case.

Single Deadline: We now consider the special case when all deadlines are the same, say D but the release dates can be different. Consider any feasible tour P which starts at s and ends at a vertex u. Let the length of P be $\ell(P)$. Suppose we reverse the tour, i.e., we view the tour as a path P^r starting at u and ending at s. If P visits a vertex v at time t, then P^r visits v at time $\ell(P) - t$. So $r_v \le t \le \ell(P)$ implies that $0 \le \ell(P) - t \le \ell(P) - r_v$. Thus, we can view this tour as one in which the release time of a vertex is 0 and the deadline is $\ell(P) - r_v$. Therefore, if we could guess the length of the optimal path P^* and the last vertex u^* in this path, then we could just use the algorithm mentioned in the previous section. Thus, we can get a $(\beta + 1)$-approximation algorithm for this problem as well. The running time of the algorithm increases by a factor of Δ from the algorithm for single release date.

k Time Windows: Now we address the case where there are both release times and deadlines. Let $r_1 < r_2 < \ldots < r_k$ be the k distinct release time and let $d_1 < d_2 < \ldots < d_k$ be the k distinct deadlines. Let P^* be the tour constructed by the optimal solution. As before let v_i^* be the last vertex in P^* to be visited before d_i and let t_i^* be the time at which v_i^* is visited. Recall that V_i is the set of vertices with deadline d_i. We define group G_i to be the set of tours that start at v_{i-1}^* at t_{i-1}^* and end at v_i^* by t_i^*. The vertices that a tour in G_i can visit are constrained to be in $V_i \cup V_{i+1} \cup \ldots \cup V_k$. Lemma 6 trivially generalizes to this setting to yield a γ approximation provided we have an oracle for the MCG instance. Consider a group G_i. The vertices all have a deadline at least as large as t_i^*, hence we have a single deadline. The vertices might have different release times, however there are at most k distinct release times. Hence the oracle needed for G_i can be obtained from the algorithm described above for this case that has an approximation ratio of $\beta + 1$. Thus, once again applying Theorem 1 we can obtain a $(\beta + 2)$-approximation algorithm for the case of k time windows.

Theorem 3. *Given a β-approximation algorithm for the orienteering problem with a single time window for all vertices, there is a $(\beta + 2)$-approximation algorithm for the orienteering problem with at most k distinct time windows that runs in time polynomial in $(n\Delta)^k$.*

Acknowledgments

We thank Moses Charikar for suggesting that we write a paper about MCG. We thank Chris Harrelson for pointing out a mistake.

References

1. A. Ageev and M. Sviridenko. Pipage Rounding: a New Method of Constructing Algorithms with Proven Performance Guarantee. To appear in *J. of Combinatorial Optimization.*
2. E. Arkin, J. Mitchell, and G. Narasimhan. Resource-constrained geometric network optimization. In *Proceedings of SoCG*, 1998.
3. S. Arora and G. Karakostas. A $2 + \epsilon$ approximation for the k-MST problem. In *Proceedings of SODA*, 2000.
4. R. Bar-Yehuda, G. Even, and S. Sahar. On Approximating a Geometric Prize-Collecting Traveling Salesman Problem with Time Windows. *Proc. of ESA*, 2003.
5. A. Blum, P. Chalasani, D. Coppersmith, B. Pulleyblank, P. Raghavan, and M. Sudan. The minimum latency problem. In *Proceedings of STOC*, 1994.
6. N. Bansal, A. Blum, S. Chawla, and A. Meyerson. Approximation Algorithms for Deadline-TSP and Vehicle Routing with Time-Windows. *Proc. of STOC*, 2004.
7. A. Blum, S. Chawla, D. Karger, T. Lane, A. Meyerson, and Maria Minkoff. Approximation Algorithms for Orienteering and Discounted-Reward TSP. *Proc. of FOCS*, 2003.
8. K. Chaudhuri, B. Godfrey, S. Rao, and K. Talwar. Paths, trees and minimum latency tours. *Proc. of FOCS*, 2003.
9. C. Chekuri and S. Khanna. A PTAS for the Multiple Knapsack Problem. *Proc. of SODA*, 2000.
10. M. Elkin and G. Kortsarz. Approximation Algorithm for the Directed Telephone Multicast Problem. *Proc. of ICALP*, 2003.
11. J. Fakcharoenphol, C. Harrelson, and S. Rao. The k-Traveling Repairmen Problem. In *Proceedings of SODA*, 2003.
12. U. Feige. A Threshold of ln n for Approximating Set Cover. *Journal of the ACM*, 45(4), 634–652, July 1998.
13. N. Garg. A 3-approximation for the minimum tree spanning k vertices. In *Proceedings of FOCS*, 1996.
14. M. Goemans and J. Kleinberg. An improved approximation ratio for the minimum latency problem. In *Proceedings of SODA*, 1996.
15. Approximation Algorithms for NP-Hard Problems. Edited by D. Hochbaum. PWS Publishing Company, Boston, 1996.
16. G. Kortsarz. Personal communication, July 2003.
17. S. Khuller, A. Moss, and J. Naor. The Budgeted Maximum Coverage Problem. *Information Processing Letters*, Vol. 70(1), pp. 39–45, (1999)
18. A. Srinivasan. Distributions on level-sets with Applications to Approximation Algorithms. *Proc. of FOCS*, 2001.
19. J. Tsitsiklis. Special Cases of Traveling Salesman Problem and Repairmen Problems with Time Windows. *Networks*, 22:263-28, 1992.

The Greedy Algorithm for the Minimum Common String Partition Problem

Marek Chrobak[1], Petr Kolman[2,1], and Jiří Sgall[3]

[1] Department of Computer Science, University of California, Riverside, CA 92521
marek@cs.ucr.edu
[2] Institute for Theoretical Computer Science, Charles University
Malostranské nám. 25, 118 00 Praha 1, Czech Republic
kolman@kam.mff.cuni.cz
[3] Mathematical Institute, AS CR, Žitná 25, CZ-11567 Praha 1, Czech Republic
sgall@math.cas.cz

Abstract. In the Minimum Common String Partition problem (M CSP) we are given two strings on input, and we wish to partition them into the same collection of substrings, minimimizing the number of the substrings in the partition. Even a special case, denoted 2-M CSP, where each letter occurs at most twice in each input string, is NP-hard. We study a greedy algorithm for M CSP that at each step extracts a longest common substring from the given strings. We show that the approximation ratio of this algorithm is between $\Omega(n^{0.43})$ and $O(n^{0.69})$. In case of 2-M CSP, we show that the approximation ratio is equal to 3. For 4-M CSP, we give a lower bound of $\Omega(\log n)$.

1 Introduction

By a *partition* of a string A we mean a sequence $\mathcal{P} = (P_1, P_2, \ldots, P_m)$ of strings whose concatenation is equal to A, that is $P_1 P_2 \ldots P_m = A$. The strings P_i are called the *blocks* of $\mathcal{P}$. If $\mathcal{P}$ is a partition of A and $\mathcal{Q}$ is a partition of B, then the pair $\pi = \langle \mathcal{P}, \mathcal{Q} \rangle$ is called a *common partition* of A, B, if $\mathcal{Q}$ is a permutation of $\mathcal{P}$. For example, $\pi = \langle (ab, bccad, cab), (bccad, cab, ab) \rangle$ is a common partition of strings $A = abbccadcab$ and $B = bccadcabab$.

The *minimum common string partition* problem (MCSP) is defined as follows: given two strings A, B, find a common partition of A, B with the minimal number of blocks, or report that no common partition exists. By k-MCSP we denote the version of MCSP where each letter occurs at most k times in each input string.

The necessary and sufficient condition for A, B to have a common partition is that each letter has the same number of occurrences in A and B. Strings with this property are called *related*. Verifying whether two strings are related can be done easily in linear time, and for the rest of the paper we assume, without loss of generality, that the input strings are related.

In this article, we deal with approximations of MCSP. In particular, we study the greedy algorithm for MCSP that constructs a common partition by iteratively extracting the longest common substring of the input strings. More precisely, the algorithm can be described in pseudo-code as follows:

K. Jansen et al. (Eds.): APPROX and RANDOM 2004, LNCS 3122, pp. 84–95, 2004.
© Springer-Verlag Berlin Heidelberg 2004

Algorithm GREEDY

Let A and B be two related input strings

while there are symbols in A or B outside marked blocks **do**

$S \leftarrow$ longest common substring of A, B that does not
overlap previously marked blocks

mark one occurrence of S in each of A and B as blocks

$(\mathcal{P}, \mathcal{Q}) \leftarrow$ sequence of consecutive blocks in A and B, respectively

For example, if $A = cdabcdabceab$, $B = abceabcdabcd$, then GREEDY first marks substring $abcdabc$, then ab, and then three single-letter substrings c, d, e, so the resulting partition is $\langle(c, d, abcdabc, e, ab), (ab, c, e, abcdabc, d)\rangle$, while the optimal partition is $\langle(cdabcd, abceab), (abceab, cdabcd)\rangle$. As illustrated by the above example, the common partition computed by GREEDY is not necessarily optimal. The question we study is what is the approximation ratio of GREEDY on MCSP and its variants. We prove the following results (by n we denote the length of the input strings):

Theorem 1.1. (a) *The approximation ratio of* GREEDY *for* MCSP *is between* $\Omega(n^{0.43})$ *and* $O(n^{0.69})$.
(b) *For* 4-MCSP, *the approximation ratio of* GREEDY *is at least* $\Omega(\log n)$.
(c) *For* 2-MCSP, *the approximation ratio of* GREEDY *is equal to* 3.

Our results can be extended to the variation of the minimum common partition problem, where letters have *signs* associated with them (cf. [1]). A substring P may now appear in A either as P or as its reversal P^R, where in the reversal the signs of all letters are reversed as well. As in MCSP, we want to find a minimum common partition of A and B into such substrings.

Related Work. The minimum common string partition problem was introduced by Chen *et al.* [1]. They pointed out that MCSP is closely related to the well-known problem of sorting by reversals and they use MCSP for comparison of two DNA sequences. In this application, the letters in the alphabet represent different genes in the DNA sequences, and the cardinality of the minimum common partition measures the similarity of these sequences. The restricted case of k-MCSP is of particular interest here. Goldstein *et al.* [3] proved that 2-MCSP is NP-hard.

The size of the minimum partition of A and B can be thought of as a distance between A and B. The classical edit-distance of two strings is defined as the smallest number of insertions, deletions, and substitutions required to change one string into another [5]. Kruskal and Sankoff [4], and Tichy [8] started to consider block operations in string comparison, in addition to the character operations. Lopresti and Tomkins [6] investigated several different distance measures; one of them is identical to the MCSP measure.

Shapira and Storer [7] study the problem of *edit distance with moves* in which the allowed string operations are the following: insert a character, delete a character, move a substring. They observe that if the input strings A, B are related, then the minimum number of the above listed operations needed to convert A into B is within a constant factor of the minimum number of only substring movements needed to convert A into B; the later is within a constant factor of the minimum common partition size. Shapira and Storer also considered a greedy algorithm nearly identical to ours and claimed an $O(\log n)$ upper bound on its approximation ratio; as it turns out, however, the analysis is flawed.

Cormode and Muthukrishnan [2] describe an $O(\log n \log^* n)$-approximation algorithm for the problem of edit distance with moves. As explained above, this result yields an $O(\log n \log^* n)$-approximation for MCSP. Better bounds for MCSP are known for some special cases. A 1.5-approximation algorithm for 2-MCSP was given by Chen *et al.* [1]; a 1.1037-approximation algorithm for 2-MCSP and a 4-approximation algorithm for 3-MCSP were given by Goldstein *et al.* [3]. All these algorithms are considerably more complicated than GREEDY. Due to its simplicity and ease of implementation, GREEDY is a likely choice for solving MCSP in many practical situations, and thus its analysis is of its own independent interest.

2   Preliminaries

By $A = a_1 a_2 \ldots a_n$ and $B = b_1 b_2 \ldots b_n$ we denote the two arbitrary, but fixed, input strings of GREEDY. W.l.o.g., we assume that A and B are related. If π is a common partition of A, B, then we use notation $\#blocks(\pi)$ for the number of blocks in π, that we refer to as the *size* of π. The size of a minimum partition of A, B is denoted by $dist(A, B)$.

We typically deal with occurrences of letters in strings, rather than with letters themselves. By a "substring" we mean (unless stated otherwise) a specific occurrence of one string in another. Thus we identify a substring $S = a_p a_{p+1} \ldots a_{p+s}$ of A with the set of indices $\{p, p+1, \ldots, p+s\}$ and we write $S = \{p, p+1, \ldots, p+s\}$, where $|S| = s + 1$ is the *length* of S. Of course, the same convention applies to substrings of B. If S is a *common* substring of A, B, we use notations S^A and S^B to distinguish between the occurrences of S in A and B.

Partitions as Functions. Suppose that we are given a bijection $\xi : [n] \to [n]$ (where $[n] = \{1, 2, \ldots, n\}$) that *preserves letters* of A and B, that is, $b_{\xi(i)} = a_i$ for all $i \in [n]$. A pair of consecutive positions $i, i + 1 \in [n]$ is called a *break* of ξ if $\xi(i + 1) \neq \xi(i) + 1$. Let $\#breaks(\xi)$ denote the number of breaks in ξ. For a common substring S of A, B, say $S = a_p a_{p+1} \ldots a_{p+s} = b_q b_{q+1} \ldots b_{q+s}$, we say that ξ *respects* S if it maps consecutive letters of S^A onto consecutive letters in S^B, that is, $\xi(i) = i + q - p$ for $i \in S^A$.

A letter-preserving bijection ξ induces a common partition (also denoted ξ, for simplicity) whose blocks are the maximum length substrings of A that do not contain breaks of ξ. The partition obtained in this way does not have any "unnecessary" blocks, i.e., $\#blocks(\xi) = \#breaks(\xi) + 1$. And vice versa, if $\pi = \langle \mathcal{P}, \mathcal{Q} \rangle$ is a common partition of A, B, we can think of π as a letter-preserving bijection $\pi : [n] \to [n]$ that respects each block of the partition. Obviously, we then have $\#blocks(\pi) \geq \#breaks(\pi) + 1$. We use this relationship throughout the paper, identifying common partitions with their corresponding functions.

Reference Partitions. Let π be a minimum common partition of A and B. (This partition may not be unique, but for all A, B, we choose one minimum common partition in some arbitrary way.) In the first step, GREEDY is guaranteed to find a substring S_1 of length at least the maximum length of a block in π. For the analysis of GREEDY, we would like to have a similar estimate for all later steps, too. However, already in the second step there is no guarantee that GREEDY finds a substring as long as the second longest block in π, since this block might overlap S_1 and it may be now partially marked (in A or B). To get a lower estimate on $|S_t|$, for $t > 1$, the idea is introduce a corresponding *reference common partition* of A, B that respects all the blocks $S_1, \ldots, S_{t-1}$ selected by GREEDY in steps 1 to $t - 1$. This partition may gradually "deteriorate" (in comparison to the minimum partition of A and B), that is, it may include more blocks and its blocks may get shorter. Furthermore, it may not be a minimum common partition of the unmarked segments. It is only used to estimate the "damage" caused by GREEDY when it makes wrong choices (that is, when it marks strings which are not in the optimum partition).

Denote by g the number of steps of GREEDY on A, B. For $t = 0, 1, \ldots, g$, the *reference common partition* ρ_t is defined inductively as follows. Initially, $\rho_0 = \pi$. Consider any $t = 1, \ldots, g$. Suppose that $S_t^A = \{p, p + 1, \ldots, p + s\}$ and $S_t^B = \{q, q + 1, \ldots, q + s\}$. Define function $\delta : S_t^A \to S_t^B$ such that $\delta(i) = i + q - p$ for $j \in S_t^A$. Then ρ_t is defined by

$$\rho_t(i) = \begin{cases} \delta(i) & \text{for } i \in S_t^A \\ \rho_{t-1}(\delta^{-1}\rho_{t-1})^{\ell(i)}(i) & \text{for } i \in [n] - S_t^A \end{cases} \tag{1}$$

where $\ell(i) = \min\left\{l \geq 0 : \rho_{t-1}(\delta^{-1}\rho_{t-1})^l(i) \notin S_t^B\right\}$. We now show that each ρ_t is well-defined and bound the increase of the number of breaks from ρ_{t-1} to ρ_t:

Lemma 2.1. *For each $t = 1, \ldots, g$, (a) ρ_t is a common partition of A, B, (b) ρ_t respects $S_1, \ldots, S_t$, (c) $\#breaks(\rho_t) \leq \#breaks(\rho_{t-1}) + 4$.*

Proof. The proof of the lemma is by induction. For $t = 0$, (a) and (b) are trivially true. Suppose that $t > 0$ and that the lemma holds for $t - 1$. To simplify notation let $S = S_t$, $\rho = \rho_{t-1}$ and $\rho' = \rho_t$.

Consider a bipartite graph $G \subseteq [n] \times [n]$, with edges $(i, \rho(i))$, for $i \in [n]$, and $(i, \delta(i))$, for $i \in S^A$. These two types of edges are called ρ-*edges* and δ-*edges*, respectively.

Let $\bar{S}^A = [n] - S^A$ and $\bar{S}^B = [n] - S^B$. In this proof, to avoid introducing additional notation, we think of S^A and $\bar{S}^A$ as the sets of nodes on the "left-hand" side of G and S^B and $\bar{S}^B$ as the nodes on the "right-hand" side. Then, any node in $\bar{S}^A$ or $\bar{S}^B$ is incident to one ρ-edge, and each node in S^A or S^B is incident to one ρ-edge and one δ-edge. Thus, G is a collection of vertex disjoint paths and cycles whose edges alternate between ρ-edges and δ-edges. We call them G-*paths* and G-*cycles*. All G-cycles have even length and contain only nodes from S^A and S^B. All maximal G-paths have odd lengths, start in $\bar{S}^A$, end in $\bar{S}^B$, and their interior vertices are in S^A or S^B. The G-path starting at i has the form $i, \rho(i), \delta^{-1}\rho(i), \rho\delta^{-1}\rho(i), \ldots, \rho(\delta^{-1}\rho)^{\ell(i)}(i)$. Thus, for $i \in \bar{S}^A$, $\rho'(i)$

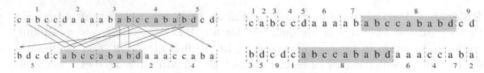

Fig. 1. An example illustrating the construction of $\rho^\square$. The left part shows ρ and some G-paths. The right part shows $\rho^\square$. The strings in the partitions are numbered, and the common substring $S_t = abccababd$ is shaded.

is simply the other endpoint of the G-path that starts at i. This implies that ρ is 1-1 and letter-preserving, so it is indeed a common partition. Condition (b) follows immediately from the inductive assumption and the definition of ρ'. It remains to prove (c).

Lemma 2.2. *Suppose that $i, i+1$ is a break of ρ'. Then there is l such that $l \leq \min\{\ell(i), \ell(i+1)\}$ and one of the following conditions holds:*
(B0) *Exactly one of $i, i+1$ is in S^A.*
(B1) *$i, i+1 \in \bar{S}^A$ and $(\delta^{-1}\rho)^l(i)$, $(\delta^{-1}\rho)^l(i+1)$ is a break of ρ.*
(B2) *$i, i+1 \in \bar{S}^A$ and exactly one of $\rho(\delta^{-1}\rho)^l(i)$, $\rho(\delta^{-1}\rho)^l(i+1)$ belongs to S^B.*

We refer to breaks (B0), (B1), (B2), respectively, as breaks induced by the endpoints of S^A, breaks induced by the breaks inside S^A (only if $i, i+1$ is a new break), and breaks induced by the endpoints of S^B.

Proof. If exactly one of i, $i+1$ is in S^A, the case (B0) holds. Since i, $i+1$ is never a break in ρ' if both i and $i+1$ are in S^A, we assume that $i, i+1 \in \bar{S}^A$ for the rest of the proof.

Consider the largest integer $l \leq \min\{\ell(i), \ell(i+1)\}$ for which $(\delta^{-1}\rho)^l(i)$, $(\delta^{-1}\rho)^l(i+1)$ are consecutive in S^A, that is $(\delta^{-1}\rho)^l(i+1) = (\delta^{-1}\rho)^l(i) + 1$. Let $j = (\delta^{-1}\rho)^l(i)$. We have two sub-cases. If $\rho(j+1) \neq \rho(j) + 1$, then $j, j+1$ is a break of ρ, so condition (B1) is satisfied. If $\rho(j+1) = \rho(j) + 1$, then at least one of $\rho(j)$, $\rho(j+1)$ must be in S^B, for otherwise $i, i+1$ would not be a break of ρ'. But we also cannot have both $\rho(j), \rho(j+1) \in S^B$, since then $(\delta^{-1}\rho)^{l+1}(i)$, $(\delta^{-1}\rho)^{l+1}(i+1)$ would be consecutive in S^A, violating the choice of l. Therefore case (B2) holds. □

We now complete the proof of part (c) of Lemma 2.1. There are no breaks of ρ' inside S^A, and we have at most two breaks of type (B0) corresponding to the endpoints of S^A. By the disjointness of G-paths and cycles, there are at most two breaks of ρ' of type (B2), each corresponding to one endpoint of S^B. Similarly, each break of ρ (inside or outside S^A) induces at most one break of ρ' of type (B1). This implies (c), and the proof of the lemma is now complete. □

Note that we did not use the fact that S has maximum length. So our construction of ρ_t can be used to convert any common partition π into another partition π' that respects a given common substring S, and has at most four more breaks than π.

Lemma 2.1 implies that in every step t, every block in ρ_t is either completely marked or unmarked.

3   Upper Bound for MCSP

In this section we show that GREEDY's approximation ratio is $O(n^{0.69})$. The proof uses reference common partitions introduced in Section 2 to keep track of the length of the common substrings selected by GREEDY.

For $p \geq q \geq 1$, we define $H(p, q)$ to be the smallest number h with the following property: for any input strings A, B, if at some step t of GREEDY there are at most p unmarked symbols in A and at most q unmarked blocks in the current reference partition ρ_t, then GREEDY makes at most h more steps until it stops (so its final common partition has at most $t+h$ blocks.) For convenience, we allow non-integral p and q in the definition. Note that $H(p, q)$ is non-decreasing in both variables.

Before proving the $O(n^{0.69})$ upper bound, note that Lemma 2.1 immediately gives a recurrence $H(p, q) \leq H(p(1 - 1/q), q + 3) + 1$ (whenever both values of H are defined), as in one step of GREEDY, the longest common substring has at least $\frac{p}{q}$ letters (which will be marked in the next partition), and the number of unmarked blocks in the reference partition increases by at most 3. With some calculation, this gives a slightly weaker but simpler bound of $O(n^{0.75})$.

The idea of the proof of the improved bound is to consider, instead of one step of GREEDY, a number of steps proportional to the number of blocks in the original optimal partition, and show that during these steps GREEDY marks a constant fraction of the input string. This yields an improved recurrence for $H(p, q)$.

Lemma 3.1. *For all p, q satisfying $p \geq 9q/5 + 3$, we have*
$H(p, q) \leq H(5p/6, (3q + 5)/2) + (q + 5)/6$.

Proof. Consider a computation of GREEDY on A, B, where, at some time t (i.e., after t blocks have already been marked), there are p unmarked symbols in A, and q unmarked reference blocks of ρ_t. We denote these blocks by $R_1, R_2, \ldots, R_q$, in the order of non-increasing length, that is $|R_z| \geq |R_{z+1}|$, for $z = 1, \ldots, q-1$. We analyze the computation of GREEDY starting at time t. Let g be the number of additional steps that GREEDY makes. Our goal is to show that $g \leq H(5p/6, (3q + 5)/2) + (q + 5)/6$. (Since the bound is monotone in p and q, we do not need to consider the case of fewer than q unmarked blocks or fewer than p unmarked symbols.) If $g \leq (q + 5)/6$, this trivially holds, so in the rest of the proof we assume that $g > (q + 5)/6$.

Let $T_i = S_{t+i}$ be the common substring selected by GREEDY in step $t + i$. We say that GREEDY *hits* R_z in step $t + i$ if T_i overlaps R_z, either in A or in B, that is, if either $T_i^A \cap R_z^A \neq \emptyset$ or $T_i^B \cap R_z^B \neq \emptyset$.

Claim A. For all $j = 1, \ldots, g$, the total length of the blocks $R_1, \ldots, R_q$ that are hit by GREEDY in A in steps $t + 1, \ldots, t + j$ is at most $6 \sum_{i=1}^{j} |T_i|$.

We estimate the total length of the blocks R_z that are hit at step $t + i$ in A but have not been hit in steps $t + 1, \ldots, t + i - 1$. The total length of the blocks that are contained in S_i^A and S_i^B is at most $2|T_i|$. There are up to four blocks that are hit partially, but by the greedy choice of T_i, each has length at most $|T_i|$, and the claim follows.   □

Claim B. $6 \sum_{i=1}^{\lfloor (q+5)/6 \rfloor} |T_i| \geq p$.

Let ℓ be the minimum integer such that $6 \sum_{i=1}^{\ell} |T_i| \geq p$. Since $\sum_{i=1}^{g} |T_i| = p$, ℓ is well defined and $\ell \leq g$. For $j = 1, \ldots, \ell$, define λ_j as the maximal index for which $\sum_{x=1}^{\lambda_j} |R_x| \leq 6 \sum_{i=1}^{j-1} |T_i|$. Since $6 \sum_{i=1}^{\ell-1} |T_i| < p = \sum_{x=1}^{q} |R_x|$, all λ_j are well defined, and $\lambda_\ell < q$. We also note that $\lambda_1 = 0$. For each $j = 1, \ldots, \ell$, Claim A implies that one of the blocks $R_1, \ldots, R_{\lambda_j + 1}$ is not hit by any of the blocks $T_1, \ldots, T_{j-1}$ and thus, by the definition of GREEDY and the ordering of the blocks R_z, $|T_j| \geq |R_{\lambda_j + 1}|$. Considering again the ordering of the blocks R_z, we have $6|T_j| \geq |R_{\lambda_j + 1}| + \ldots + |R_{\lambda_j + 6}|$. We conclude that $\lambda_{j+1} \geq \lambda_j + 6$, for $j = 1, \ldots, \ell - 1$. This, in turn, implies that $q \geq \lambda_\ell + 1 \geq 6\ell - 5$. So $\ell \leq (q+5)/6$, and Claim B follows, by the choice of ℓ and its integrality. $\square$

By Claim B, after exactly $\lfloor (q+5)/6 \rfloor$ steps, GREEDY marks at least $p/6$ letters, so the number of remaining unmarked letters is at most $p' = 5p/6$. By Lemma 2.1, the number of unmarked blocks increases by at most 3 in each step (since one new block is marked), so the number of unmarked blocks induced by GREEDY in these $\lfloor (q+5)/6 \rfloor$ steps is at most $3\lfloor (q+5)/6 \rfloor \leq (q+5)/2$. Thus the total number of unmarked blocks after these steps is at most $q' = q + (q+5)/2 = (3q+5)/2$. The condition in the lemma guarantees that $H(p', q')$ is defined, so, by induction, the total number of steps is at most $H(p', q') + (q+5)/6$. $\square$

Finally, we prove the upper bound in Theorem 1.1(a). We prove by induction on p that for $p \geq q$ and a sufficiently large constant C, $H(p, q) \leq Cp^\gamma (q + 5)^{1-\gamma} - \frac{1}{3}q$. We choose C so that for all $q \leq p < 9q/5 + 3$, the right-hand side is at least p and thus the inequality is valid. For $p \geq 9q/5 + 3$, by Lemma 3.1, the inductive assumption, and the choice of γ, we have $H(p, q) \leq H(\frac{5}{6}p, \frac{3q+5}{2}) + \frac{q+5}{6} \leq C(\frac{5}{6}p)^\gamma (\frac{3}{2}(q+5))^{1-\gamma} - \frac{1}{3} \cdot \frac{3q+5}{2} + \frac{q+5}{6} = Cp^\gamma(q+5)^{1-\gamma} - \frac{1}{3}q$. Let A, B be input strings of with length n and $dist(A, B) = m$. Then the number of blocks in GREEDY's partition is at most $H(n, m) = O(n^\gamma)m$, and the theorem follows.

4   Lower Bound for MCSP

We show that the approximation ratio of GREEDY is $\Omega(n^{1/\log_2 5}) = \Omega(n^{0.43})$. We first construct strings C_i, D_i, E_i, F_i as follows. Initially, $C_0 = a$ and $D_0 = b$. Suppose we already have C_i and D_i, and let Σ_i be the set of letters used in C_i, D_i. Create a new alphabet Σ_i' that for each $a \in \Sigma$ has a new letter, say, a'. We first create strings E_i and F_i by replacing all letters $a \in \Sigma$ in C_i and D_i, respectively, by their corresponding letters $a' \in \Sigma_i'$. Then, let

$$C_{i+1} = C_i D_i E_i D_i C_i, \quad \text{and} \quad D_{i+1} = D_i E_i F_i E_i D_i.$$

For each i, we consider the instance of strings $A_i = C_i D_i$ and $B_i = D_i C_i$. For example, $E_0 = c$, $F_0 = d$, $A_0 = ab$, $B_0 = ba$, $C_1 = abcba$, $D_1 = bcdcb$, $A_1 = abcbabcdcb$, and $B_1 = bcdcbabcba$, etc.

Let $n = 2 \cdot 5^i$. We have $|A_i| = |B_i| = n$ and $dist(A_i, B_i) \leq 2$. We claim that GREEDY's common partition of A_i and B_i has $2^{i+2} - 2 = \Omega(n^{1/\log_2 5})$ substrings. We assume here that GREEDY does not specify how the ties are broken, that is, whenever a longest substring can be chosen in two or more different ways, we can decide which choice GREEDY makes.

The proof is by induction. For $i = 0$, GREEDY produces two substrings, as claimed. For $i \geq 0$,

$$A_{i+1} = C_i \, D_i \, E_i \, D_i \, C_i \, D_i \, E_i \, F_i \, E_i \, D_i, \qquad B_{i+1} = D_i \, E_i \, F_i \, E_i \, D_i \, C_i \, D_i \, E_i \, D_i \, C_i.$$

There are three common substrings of length 5^{i+1}: $C_i D_i E_i D_i C_i$, $D_i E_i F_i E_i D_i$, and $E_i D_i C_i D_i E_i$, and no longer common substrings exist. To justify this, we use the fact that the alphabet of C_i, D_i is disjoint from the alphabet of E_i, F_i. Suppose that S is a common substring of length at least 5^{i+1}. To have this length, S must contain either the first or the second E_i from A_{i+1}. We now have some cases depending on which E_i is contained in S, and where it is mapped into B_{i+1} via the occurrence of S in B_{i+1}. If S contains the first E_i, then, by the assumption about the alphabets, this E_i must be mapped into either $E_i F_i E_i$ or into the last E_i in B_{i+1}. If it is mapped into $E_i F_i E_i$, then S must be $E_i D_i C_i D_i E_i$. If it is mapped into the last E_i in B_{i+1}, then S must be $C_i D_i E_i D_i C_i$. In the last case, S contains the second E_i. By the same considerations as in the first case, it is easy to show that then S must be either $D_i E_i F_i E_i D_i$ or $E_i D_i C_i D_i E_i$.

Breaking the tie, assume that GREEDY marks substring $E_i D_i C_i D_i E_i$. The modified strings are:

$$C_i \, D_i \, \overline{E_i \, D_i \, C_i \, D_i \, E_i} \, F_i \, E_i \, D_i, \qquad D_i \, E_i \, F_i \, \overline{E_i \, D_i \, C_i \, D_i \, E_i} \, D_i \, C_i,$$

where the overline indicates the marked substring. In the first string the unmarked segments are A_i, $A'_i D_i$, and in the second string the unmarked segments are B_i and $D_i B'_i$, where $A'_i = F_i E_i$ and $B'_i = E_i F_i$ are identical as A_i, B_i respectively, but with the letters renamed. The argument in the previous paragraph and the disjointness of the alphabets implies that the maximum length of a non-marked common substring is now 5^i. We break the tie again, and make GREEDY match the two D_i's in $F_i E_i D_i$ and $D_i E_i F_i$, leaving us with non-marked pairs of substrings $\{A_i, B_i\}$ and $\{A'_i, B'_i\}$. By induction, GREEDY produces $2^{i+2} - 2$ substrings from A_i, B_i, and the same number from A'_i and B'_i. So we get the total of $2(2^{i+2} - 2) + 2 = 2^{i+3} - 2$ strings.

5   Lower Bound for GREEDY on 4-MCSP

In this section we show that GREEDY's approximation ratio is $\Omega(\log n)$ even on 4-MCSP instances. To simplify the description, we allow the input instances $\mathcal{A}, \mathcal{B}$ to be *multisets* of equal number of strings, rather than single strings. It is quite easy to see that this does not significantly affect the performance of GREEDY, for we can always replace $\mathcal{A}, \mathcal{B}$ by two strings A, B, as follows: If $\mathcal{A} = \{A_1, \ldots, A_m\}$ and $\mathcal{B} = \{B_1, \ldots, B_m\}$, let $A = A_1 x_1 y_1 A_2 x_2 y_2 \ldots A_{m-1} x_{m-1} y_{m-1} A_m$ and $B = B_1 y_1 x_1 B_2 y_2 x_2 \ldots B_{m-1} y_{m-1} x_{m-1} B_m$, where $x_1, y_1, \ldots, x_{m-1}, y_{m-1}$ are new letters. Then both the optimal partition and the partition produced by GREEDY on A, B are the same as on $\mathcal{A}, \mathcal{B}$, except for the singletons $x_1, y_1, \ldots, x_{m-1}, y_{m-1}$.

Since in our construction m is a constant, it is sufficient to show a lower bound of $\Omega(\log n)$ for multisets of m strings.

For $i = 1, 2, \ldots$, we fix strings q_i, q_i', r_i, r_i' that we will refer to as *elementary strings*. Each elementary string q_i, q_i', r_i, r_i' has length 3^{i-1} and consists of 3^{i-1} distinct and unique letters (that do not appear in any other elementary string.)

We now recursively construct instances $\mathcal{A}^i$, $\mathcal{B}^i$ of 4-MCSP. The invariant of the construction is that $\mathcal{A}^i$, $\mathcal{B}^i$ have the form:

$$\mathcal{A}^i: P_1 q_i, \quad P_2 q_i r_i, \quad P_3 q_i, \quad P_4 q_i r_i', \quad P_5 q_i', \quad P_6 q_i' r_i, \quad P_7 q_i', \quad P_8 q_i' r_i'$$
$$\mathcal{B}^i: P_1 q_i r_i, \quad P_2 q_i, \quad P_3 q_i r_i', \quad P_4 q_i, \quad P_5 q_i' r_i, \quad P_6 q_i', \quad P_7 q_i' r_i', \quad P_8 q_i'$$

where $P_1, \ldots, P_8$ are some strings of length smaller than 3^{i-1} with letters distinct from q_i, q_i', r_i, r_i'.

Initially, we set all $P_1, \ldots, P_8 = \epsilon$, and construct $\mathcal{A}^1$, $\mathcal{B}^1$ as described above. In this case q_i, q_i', r_i, r_i' are unique, single letters.

To construct $\mathcal{A}^{i+1}$, $\mathcal{B}^{i+1}$, we append pairs of elementary strings to the strings from $\mathcal{A}^i$, $\mathcal{B}^i$. For convenience, we omit the subscripts for q and r, writing $q = q_i$, $\bar{q} = q_{i+1}$, etc. After rearranging the strings, the new instance is

$$\mathcal{A}^{i+1}: P_1 qr\,\bar{q}, \; P_4 qr^\square\,\bar{q}\bar{r}, P_7 q^\square r^\square\,\bar{q}, \; P_6 q^\square r\,\bar{q}\bar{r}, P_3 qr^\square\,\bar{q}^\square, \; P_2 qr\,\bar{q}\bar{r}, P_5 q^\square r\,\bar{q}^\square, \; P_8 q^\square r^\square\,\bar{q}\bar{r}^\square$$
$$\mathcal{B}^{i+1}: P_1 qr\,\bar{q}\bar{r}, P_4 qr^\square\,\bar{q}, \; P_7 q^\square r^\square\,\bar{q}\bar{r}, P_6 q^\square r\,\bar{q}, \; P_3 qr^\square\,\bar{q}^\square\bar{r}, P_2 qr\,\bar{q}^\square, \; P_5 q^\square r\,\bar{q}\bar{r}^\square, P_8 q^\square r^\square\,\bar{q}^\square$$

Note that this instance has the same structure as the previous one, since we can take $P_1' = P_1 qr$, $P_2' = P_4 qr'$, etc., thus we can continue this construction recursively. Each letter appears at most four times in $\mathcal{A}^i$ and $\mathcal{B}^i$, so this is indeed an instance of 4-MCSP; the claimed bound on the length of P_j also follows easily.

Consider now the i-th instance, $\mathcal{A}^i$ and $\mathcal{B}^i$. To estimate the optimal partition, we can match the 8 pairs of strings as aligned above, adding the shorter string from each pair to the common partition, leaving us with additional 4 strings $\bar{r}$, $\bar{r}', \bar{r}, \bar{r}'$, so $dist(\mathcal{A}^i, \mathcal{B}^i) \leq 12$.

We now show that GREEDY computes a partition with $\Theta(i) = \Theta(\log n)$ blocks. To this end, we claim that, starting from $\mathcal{A}^{i+1}, \mathcal{B}^{i+1}$, GREEDY first matches all suffixes that consist of two elementary strings as shown below ($\mathcal{A}^{i+1}$ and $\mathcal{B}^{i+1}$ are rearranged to show the matched strings aligned vertically):

$$\mathcal{A}^{i+1}: P_4 qr^\square\,\bar{q}\bar{r}, \; P_6 q^\square r\,\bar{q}\bar{r}^\square, \; P_2 qr\,\bar{q}^\square\bar{r}, P_8 q^\square r^\square\,\bar{q}^\square\bar{r}^\square, \; P_1 q\,r\,\bar{q}, P_7 q^\square r^\square\bar{q}, \; P_3 q\,r^\square\bar{q}^\square, P_5 q^\square\,r\bar{q}^\square$$
$$\mathcal{B}^{i+1}: P_1 qr\,\bar{q}\bar{r}, P_7 q^\square r^\square\,\bar{q}\bar{r}^\square, \; P_3 qr^\square\,\bar{q}^\square\bar{r}, \; P_5 q^\square r\,\bar{q}^\square\bar{r}^\square, \; P_6 q^\square\,r\bar{q}, \; P_4 q\,r^\square\bar{q}, P_8 q^\square r^\square\bar{q}^\square, \; P_2 q\,r\bar{q}^\square$$

Indeed, the instance has four common substrings of length $2 \cdot 3^i$, namely $\bar{q}\bar{r}$, $\bar{q}\bar{r}'$, $\bar{q}'\bar{r}$, $\bar{q}'\bar{r}'$, and, by the choices of the lengths of elementary strings and the bound on the lengths of the P_l's, all other common substrings are shorter. Thus GREEDY starts by removing (marking) these four suffixes. Similarly, at this step, the new instance will have four common substrings of length $3^i + 3^{i-1}$, namely $r\bar{q}$, $r'\bar{q}$, $r'\bar{q}'$, $r\bar{q}'$, and all other common substrings are shorter. So now GREEDY will remove these four suffixes. The resulting instance is simply $\mathcal{A}^i$, $\mathcal{B}^i$, so we can continue recursively, getting $\Theta(i)$ blocks. If n is the length (total number of characters) of $\mathcal{A}^i$, we have $i = \Theta(\log n)$, so the of the lower bound is complete.

6   Upper Bound for GREEDY on 2-MCSP Instances

Consider two arbitrary, but fixed, related strings $A = a_1 a_2 \cdots a_n$ and $B = b_1 b_2 \cdots b_n$ in which each letter appears at most twice. Let π be a minimum common partition of A, B, and denote by g the number of steps of GREEDY on A, B. For each $t = 0, \ldots, g$, let ρ_t be the common reference partition of A, B at step t, as defined in Section 2. In particular, $\rho_0 = \pi$, and ρ_g is the is the final partition computed by GREEDY.

Our proof that GREEDY's approximation ratio on 2-MCSP instances is at most 3 is based on amortized analysis. We show how to define a *potential* Φ_t of ρ_t that has the following three properties:

(P1) $\Phi_0 \leq 3 \cdot \#blocks(\rho_0) + 1$,
(P2) $\Phi_t \leq \Phi_{t-1}$ for $t = 1, \ldots, g$, and
(P3) $\Phi_g \geq \#blocks(\rho_g) + 1$.

If such Φ_t's exist, then, using the optimality of ρ_0 and conditions (P1), (P2), (P3), we obtain $\#blocks(\rho_g) \leq \Phi_g - 1 \leq \Phi_0 - 1 \leq 3 \cdot \#blocks(\rho_0) = 3 \cdot \#blocks(\pi) = 3 \cdot dist(A, B)$, and the 3-approximation of GREEDY follows immediately. Thus it remains to define the potential and show that it has the desired properties.

Classification of Breaks. Consider some step t. A break $i, i + 1$ of ρ_t is called *original* if it is also a break of π; otherwise we call this break *induced*. Letters inside blocks marked by GREEDY are called *marked*. For any letter a_i in A, we say that a_i is *unique* in ρ_t if a_i is not marked, and there is no other non-marked appearance of a_i in A.

Suppose that $i, i + 1$ is an original break in ρ_t. We say that this break is *left-mature* (resp. *right-mature*) if a_i (resp. a_{i+1}) is unique; otherwise it is called *left-immature* (resp. *right-immature*). If a break is both left- and right-mature, we call it *mature*. If it is neither, we call it *immature*. (The intuition behind these terms is that, if, say, a break $i, i + 1$ is left-mature, then the value of $\rho_t(i)$ does not change anymore as t grows.) We extend this terminology to the endpoints of A. For the left endpoint, if a_1 is unique in ρ_t we call this endpoint *right-mature*, otherwise it is immature. Analogous definitions apply to the right endpoint.

Potential. We first assign potentials to the breaks and endpoints of ρ_t:

(ϕ1) Each induced break has potential 1.
(ϕ2) The potential of an original break depends on the degree of maturity. If a break is immature it has potential 3. If it is left-mature and right-immature, or vice versa, it has potential 2. If it is mature, it has potential 1.
(ϕ3) The left endpoint has potential 2 or 1, depending on whether it is right-immature or right-mature, respectively. The potential of the right endpoint is defined in a symmetric fashion.

Then Φ_t is defined as the sum of the potentials of the breaks and endpoints in ρ_t. For $t = 0$, all breaks in π have potential at most 3 and the endpoints have potential at most 2, so we have $\Phi_0 \leq 3 \cdot \#breaks(\pi) + 4 = 3 \cdot \#blocks(\pi) + 1$. For $t = g$, all letters in A are marked, so the potentials of all breaks and endpoints are equal 1. Therefore $\Phi_g \geq \#breaks(\rho_g) + 2 = \#blocks(\rho_g) + 1$. Thus properties (P1) and (P3) hold.

To prove property (P2), first note that the potentials of the breaks of ρ_{t-1} that remain in ρ_t cannot increase. Thus only the new breaks (those in ρ_t but not in ρ_{t-1}) can contribute to the increase of the potential. All these new breaks have potential 1 in ρ_t. With each new break we can associate some old break (or an endpoint) of ρ_{t-1} that either disappears in ρ_t or whose potential decreases. This mapping is not necessarily one-to-one. However, the number of new breaks associated with each old break does not exceed the decrease of the potential of this old break. This needs some case analysis along the lines of Lemma 2.2; details are omitted in this abstract. Summarizing, we obtain the the upper bound of 3 for 2-MCSP, i.e., the upper bound in Theorem 1.1(c).

7   Lower Bound for GREEDY on 2-MCSP

Let l be a large even integer. Let $A' = a_1a_2 \ldots a_{l^2}$, $B' = b_1b_2 \ldots b_{l^2}$, and

$$A'' = a_{l(l-1)+2}a_{l(l-1)+3} \cdots a_{l^2} \quad \cdots \quad a_{l+1}a_{l+2} \cdots a_{2(l-1)+1} \quad a_2a_3 \ldots a_l \quad a_1$$
$$B'' = b_{l^2} \quad b_{l(l-1)+1}b_{l(l-1)+2} \cdots b_{l^2-1} \quad \cdots \quad b_lb_{l+1} \ldots b_{2(l-1)} \quad b_1b_2 \ldots b_{l-1},$$

where the l^2 letters a_i of A are all distinct, for all $i \equiv 1 \pmod{l+1}$, b_i are new letters distinct from each other and from all $a_{i'}$, and $b_i = a_i$ for all other values of i. The instance is $A = A'B''$ and $B = A''B'$. Obviously, no letter occurs more than twice, so this is indeed an instance of 2-MCSP.

A' and B' consist of the same $l-1$ substrings of length l, separated (and ended) by l distinct markers. The strings A'' and B'' are obtained from A' and B' by cutting into $l+1$ substrings of length $l-1$ and a single letter, and taking these in the reverse order. (Note that the cuts in the two strings are not aligned.)

The cuts indicated by spaces in the definitions of A'' and B'' show that $dist(A, B) \leq dist(A', A'') + dist(B', B'') \leq 2l + 4$.

A' and B' have $l-1$ common substrings of length l, namely the substrings between the marker symbols. We claim that A and B have no other common substrings of length l. Clearly, by the choice of the markers, A' and B' have no other common substrings of length l. The longest common substring of A' and A'' as well as of B' and B'' has length $l-1$, by the definition of A'' and B''. The strings A'' and B'' also have no common string of length l or larger, since the ends of their substrings of length $l-1$ are not aligned. There is no common substring that would overlap both A' and B'' or A'' and B'.

Consequently, GREEDY starts by matching the $l-1$ common substrings of A' and B' of length l. After this phase, each letter occurs in the remaining string exactly once, and thus the remaining matching is unique. The remaining letters a_i in A' with the exception of a_{l^2} are strings of length 1, and they give another $l-1$ strings in the final partition. Now it is sufficient to bound the number of breaks forced in the string B''. There is a break before and after each letter b_i, $i \equiv 1 \pmod{l+1}$, $1 < i < l^2$, as these are strings of length 1 in the remainder of B; this gives $2l - O(1)$ breaks. There is also a break before and after each letter b_i, $i \equiv 1 \pmod{l-1}$ as for such i, b_ib_{i+1} is a consecutive pair in B''

but the possibly matching pair $a_i a_{i+1}$ is not consecutive in A''; similarly $a_{i-1}a_i$ is consecutive in A'' but $b_{i-1}b_i$ is not consecutive in B''. This gives $2l - O(1)$ breaks. Finally, since $l - 1$ and $l + 1$ are relatively prime, only $O(1)$ breaks may be counted twice, by the Chinese remainder theorem. Altogether, GREEDY partitions $A = A'B''$ into $2l - 2$ blocks of A' and $4l - O(1)$ blocks of B'', for the total of $6l - O(1)$, and the lower bound of 3 on the approximation ratio follows by taking l arbitrarily large.

Final Comments. We have established that GREEDY's approximation ratio is $O(n^{0.69})$, but not better than $\Omega(n^{0.43})$. It would be interesting to determine the exact approximation of this algorithm. In particular, is it below, above, or equal to $\Theta(\sqrt{n})$? Also, we have observed a difference between the performance of GREEDY on 2-MCSP instances and 4-MCSP instances. Whereas the approximation ratio for 2-MCSP is 3, for 4-MCSP it is not better than $\Omega(\log n)$. The reason for this is, roughly, that for 2-MCSP every new cut (i.e., induced cut) is adjacent to a unique letter and, since GREEDY does not make mistakes on unique letters, these new cuts do not induce any further cuts. However, for $k > 2$, new cuts may induce yet more new cuts again. An intriguing question is whether for 4-MCSP the upper bound matches the $\Omega(\log n)$ lower bound, or whether it is higher? The question about exact approximation ratio of GREEDY for k-MCSP remains open even for $k = 3$.

Acknowledgments

This work was supported by NSF grants CCR-0208856 (Chrobak, Kolman) and ACI-0085910 (Kolman), by Inst. for Theor. Comp. Sci., Prague, project LN00A056 of MŠMT ČR (Kolman, Sgall), and by grant IAA1019401 of GA AV ČR (Sgall).

References

1. X. Chen, J. Zheng, Z. Fu, P. Nan, Y. Zhong, S. Lonardi, T. Jiang. Assignment of orthologous genes via genome rearrangement. Submitted. 2004.
2. G. Cormode, J.A. Muthukrishnan, The string edit distance matching with moves. Proc. 13th Annual Symposium on Discrete Algorithms (SODA), pp. 667-676, 2002.
3. A. Goldstein, P. Kolman, and J. Zheng: Minimum common string partitioning problem: Hardness and approximations. Manuscript. 2004.
4. J. B. Kruskal and D. Sankoff. An anthology of algorithms and concepts for sequence comparison. In *Time Warps, String Edits, and Macromolecules: The Theory and Practice of Sequence Comparison, Edited by David Sankoff and Joseph B. Kruskal, Addison-Wesley.* 1983.
5. V. I. Levenshtein. Binary codes capable of correcting deletions, insertions and reversals (in Russian). Doklady Akademii Nauk SSSR, 163(4):845–848, 1965.
6. D. Lopresti, A. Tomkins. Block edit models for approximate string matching. Theoretical Computer Science 181 (159–179) 1997.
7. D. Shapira, J.A. Storer. Edit distance with move operations. Proc. 13th Annual Symposium on Combinatorial Pattern Matching (CPM), pp. 85–98, 2002.
8. W. F. Tichy. The string-to-string correction problem with block moves. ACM Trans. Computer Systems 2 (309–321) 1984.

Approximating Additive Distortion
of Embeddings into Line Metrics

Kedar Dhamdhere[*]

School of Computer Science
Carnegie Mellon University
Pittsburgh, PA 15213, USA
kedar@cs.cmu.edu

Abstract. We consider the problem of fitting metric data on n points
to a *path* (line) metric. Our objective is to minimize the total additive
distortion of this mapping. The total additive distortion is the sum of
errors in all pairwise distances in the input data. This problem has been
shown to be NP-hard by [13]. We give an $O(\log n)$ approximation for this
problem by using Garg *et al.*'s [10] algorithm for the multi-cut problem
as a subroutine. Our algorithm also gives an $O(\log^{1/p} n)$ approximation
for the L_p norm of the additive distortion.

1   Introduction

One of the most common methods for clustering numerical data is to fit the
data to *tree metrics*. A tree metric is defined on vertices of a weighted tree.
The distance between two vertices is the sum of the weights of edges on the path
between them. Here the main problem is to find a tree metric that represents the
input numerical data as closely as possible. This problem, known as *numerical
taxonomy*, has applications in various fields of science, such as linguistics and
evolutionary biology. For example, in evolutionary biology tree metrics represent
the branching process of evolution that leads to the observed distribution of data.
Naturally, this problem has received a great deal of attention. (e.g. see [3, 14]).

The problem of fitting data to tree metrics is usually cast as the problem of
minimization of $L_p(D, T)$: the L_p norm of *additive distortion* of the output tree
T with respect to input data D. The input data is specified as an $n \times n$ matrix,
where the entry D_{ij} denotes the distance between points i and j. Let T_{ij} denote
the distance between i and j in the output tree metric T. Then the L_p norm
of additive distortion is $L_p(D, T) = (\sum_{i,j} |D_{ij} - T_{ij}|^p)^{1/p}$. Such a formulation
was first proposed by [5] in 1967. In 1977, Waterman *et al.* [15] showed that if
there is a tree metric T coinciding exactly with the input data D, then it can be
constructed in linear time. In the case when there is no tree that fits the data
perfectly, Agarwala *et al.* [1] used the framework of approximation algorithms
to give heuristics with provable guarantees for the problem. They gave a 3-
approximation to the L_∞ norm of the additive distortion for fitting the data to

---

[*] Supported by NSF ITR grants CCR-0085982 and CCR-0122581.

a tree metric. They reduced the problem to that of fitting the data to *ultrametric*, where each leaf is at the same distance from a common root. For *ultrametrics*, they used an exact polynomial-time algorithm for the L_∞ norm due to Farach *et al.* [8]. Agarwala *et al.* [1] showed that if there is a ρ-approximation algorithm for *ultrametrics* under the L_p norm, then it implies a 3ρ-approximation for tree metrics under the L_p norm. Our work is motivated by this problem.

For a special case of fitting the data to a tree metric under L_1 norm, we make a simple observation. Suppose we know the structure of the tree along with a mapping of the points to the vertices of the tree. Then we can find the edge weights that minimize the L_1 norm of additive distortion using linear programming. However, finding the topology of the tree is the hard part. Therefore, we consider a special case of the problem in which we restrict the structure of the output tree to be a path. For this special case, we give an approximation algorithm. We believe that our techniques can be extended to solve the original problem. The main result of this paper is the following theorem.

Theorem 1. *There is an $O(\log^{1/p} n)$-approximation algorithm for the problem of fitting metric data to a line metric to minimize the L_p norm of additive distortion for $p \geq 1$.*

1.1   Related Work

Note that fitting points to a path metric is equivalent to fitting points on a real line, where the distances in the real line are defined in a natural way. The special case of the problem for the L_∞ norm (i.e. with $p = \infty$) was considered by Håstad *et al.* [11]. They gave a 2-approximation for it.

For fitting points to a line, a well-known result due to Menger (see e.g. [6]) gives the following four point criterion. The four point criterion says that, if every subset of size 4 can be mapped into the real line exactly, then all the points can be mapped into the line exactly. An approximate version of Menger's result was given by Badoiu *et al.* [2]. They proved that if every subset of size 4 can be embedded into the line with the L_∞ norm of the additive distortion being at most ϵ then all the points can be embedded with the L_∞ norm of the additive distortion being at most 6ϵ.

In a related work, Dhamdhere *et al.* [7] and Badoiu *et al.* [2] independently studied average distortion of embedding a metric into path metrics. The objective was to minimize $\sum_{i,j} f_{ij}$, subject to $f_{ij} \geq D_{ij}$ for all i,j, where D_{ij} is the distance between points i and j in the input and f_{ij} is the distance between them in the output path metric. They gave a 14-approximation for the problem. While their problem has additional constraint $f_{ij} \geq D_{ij}$, their objective function is easier than the L_1 norm of the additive distortion. We do not know how to minimize the L_p (or even the L_1) norm of additive distortion under the additional constraint. However, for the special case of $p = \infty$, we would like to point out that the algorithm for *min-excess path problem* due to Blum *et al.* [4] gives a $2 + \epsilon$ approximation. The *min-excess path problem* asks for a path from a source s to a destination t that visits at least k vertices and minimizes the objective $l(path) - d(s,t)$, where $l(path)$ is the length of the path.

1.2   Techniques and Roadmap

In Section 2, we define our problem formally. In Section 3, we show how to reduce the problem to that of finding the best r-restricted mapping. An r-restricted mapping of the input points into a line is one in which the distances of all points in the line from point r are same as that in the input. We show that this problem is equivalent to a two-cost partition problem. In Section 4, we give an approximation for this problem via the multi-cut problem [10].

2   Problem Definition

Consider a set of n points, denoted by $[n] = \{1, 2, \ldots, n\}$. The input data consists of an $n \times n$ matrix $D_{n \times n}$. The entry D_{ij} denotes the distance between points i and j. We assume that all the entries of D are non-negative and that D is symmetric[1]. Furthermore, we assume that $D_{ii} = 0$ for all i.

Let $f : [n] \to \mathbb{R}$ denote a mapping of the input points to the real line. Distance between images of points i and j in the line is given by $f_{ij} = |f(i) - f(j)|$. The total additive distortion (in the L_1 norm) is given by

$$L_1(D, f) = \sum_{i,j} |D_{ij} - f_{ij}|.$$

Generalizing this, we can write the L_p norm of the additive distortion as

$$L_p(D, f) = \Big(\sum_{i,j} |D_{ij} - f_{ij}|^p \Big)^{\frac{1}{p}}.$$

The goal is to find a map f that minimizes the $L_1(D, f)$ (or more generally $L_p(D, f)$).

3   Approximation for L_p Norm

In this section, we give an approximation algorithm for minimizing the L_p norm of the additive distortion.

In Lemma 1, we will show that it is sufficient to look at r-restricted mapping of the points into the real line. The problem of finding an optimal r-restricted mapping can be cast as a kind of partition problem given the characteristics of the real line.

3.1 r-Restricted Mappings

Let r be a point in the input. A mapping f of the input points to the real line $\mathbb{R}$ is an r-restricted mapping, if distance on the line of all points from r is same as that in the input. Formally, $D_{ri} = |f(r) - f(i)|$ for all i.

---

[1] Our results hold even if the input distances in $D_{n \times n}$ do not satisfy triangle inequality, i.e. even if D is not a "metric".

We will denote an r-restricted mapping by f^r. We next show that there is always a "good" r-restricted mapping. This will enable us to focus only on r-restricted mappings which are easier to handle. Agarwala et $al.$ [1] prove a similar lemma for tree metrics. We adapt their proof for the case of line metrics.

Lemma 1. *There exists a point* $\mathbf{r}$ *among the input points such that there is an* r-*restricted mapping* f^r *that is within a factor of* 3 *of the optimal mapping for the* L_p *norm of additive distortion, for all* $p \geq 1$.

Proof. Let f^* denote an optimal mapping of the input points to the line for the L_p norm of additive distortion. We will modify the optimal solution to produce a mapping f^i for each point i in the input. To produce the restricted mapping f^i, perturb the distances in f^* so that it becomes i-restricted. In particular, if $f^*(j) \leq f^*(i)$ for some j, then set $f^i(j) = f^*(i) - D_{ij}$ and if $f^*(j) > f^*(i)$, set $f^i(j) = f^*(i) + D_{ij}$. Our mapping f^i maps point i to $f^*(i)$. It maps rest of the points according to their distance from i, while maintaining their order to the left or right of point i in the optimal mapping f^*.

Let ϵ_{jk} denote $|D_{jk} - f^*_{jk}|$. We can write the additive distortion of the optimal mapping as $L_p(D, f^*) = (\sum_{j,k} \epsilon^p_{jk})^{1/p}$. From the construction of the map f^i, it follows that $|f^*_{jk} - f^i_{jk}| \leq \epsilon_{ij} + \epsilon_{ik}$.

Now we bound the additive distortion of f^i in terms of ϵ_{jk}'s. For all j, k we have,

$$|D_{jk} - f^i_{jk}| \leq |f^*_{jk} - f^i_{jk}| + |D_{jk} - f^*_{jk}|$$
$$\leq (\epsilon_{ij} + \epsilon_{ik}) + \epsilon_{jk} \tag{1}$$

Note that $|x|^p$ is a convex function of x for $p \geq 1$. Therefore, Equation (1) gives us the following:

$$|D_{jk} - f^i_{jk}|^p \leq (\epsilon_{ij} + \epsilon_{ik} + \epsilon_{jk})^p$$
$$\leq 3^{p-1}(\epsilon^p_{ij} + \epsilon^p_{ik} + \epsilon^p_{jk}) \tag{2}$$

By an averaging argument, we can say that

$$\min_i \{L_p(D, f^i)^p\} \leq \frac{\sum_{i=1}^n L_p(D, f^i)^p}{n}$$

We use Equation (2) to bound the sum

$$\sum_{i=1}^n L_p(D, f^i)^p \leq \sum_{i=1}^n \sum_{j,k} 3^{p-1}(\epsilon^p_{ij} + \epsilon^p_{ik} + \epsilon^p_{jk})$$
$$\leq 3^p n \cdot \sum_{j,k} \epsilon^p_{jk}$$
$$= 3^p n \cdot L_p(D, f^*)^p$$

Therefore, $\min_i L_p(D, f^i) \leq 3 \cdot L_p(D, f^*)$ which proves the result.

3.2   Algorithm

The result of Lemma 1 proves that it is sufficient to consider r-restricted mappings (with a loss of 3 in the approximation factor). Next we describe the algorithm that implements this idea.

Algorithm A:

1. For each point $r = 1, 2, \ldots, n$, find (approximately) the best r-restricted mapping f^r.
2. Output a mapping that has the smallest additive distortion among these mappings.

By Lemma 1, the additive distortion of the output of Algorithm A is within a factor of 3 of the optimal additive distortion. As we point out in Section 5, finding best r-restricted mapping is NP-hard. Therefore, we approximate the optimal a-restricted mapping within a factor of $O(\log^{1/p} n)$. From the following observation it follows that the overall approximation factor of our algorithm will be $O(\log^{1/p} n)$.

Lemma 2. *If ρ is the approximation factor of the algorithm for r-restricted mapping, then the solution produced by Algorithm A will be a 3ρ approximation for the additive distortion.*

4   Approximating r-Restricted Mappings

Let f be an r-restricted mapping. Without loss of generality, we can assume that $f(r) = 0$. Let $V_1 = \{i \mid f(i) < 0\}$ and $V_2 = \{i \mid f(i) > 0\}$. Note that $[n] = V_1 \cup \{r\} \cup V_2$. Note that, the mapping f is fully characterized by the partition $V_1 \cup V_2$ of $[n] - \{r\}$. Hence, the problem of finding the best r-restricted mapping is equivalent to the problem of finding the partition of $V = [n] - \{r\}$ that has minimum additive distortion. Henceforth, we will think of the problem as that of partitioning the input set of points to minimize the *cost* of the partition, i.e. the additive distortion. For simplicity, we describe the argument for $p = 1$. The other cases ($p > 1$) are similar.

Consider a partition $V_1 \cup V_2$ induced by an r-restricted mapping f. We can write an expression for its *cost* as follows. Consider two points x and y. If they both belong to the same side of the partition, then the contribution of the pair $\{x, y\}$ to the cost of the partition is $c(x, y) = |D_{xy} - f_{xy}| = D_{xy} - |f(x) - f(y)| = |D_{xy} - |D_{rx} - D_{ry}||$. On the other hand, if x and y belong to different sides of the partition, then the contribution is $c'(x, y) = |D_{xy} - f_{xy}| = |D_{xy} - |f(x) - f(y)|| = |D_{rx} + D_{ry} - D_{xy}|$. Note that $c(x, y)$ and $c'(x, y)$ are completely determined from the input matrix $D_{n \times n}$.

Therefore, we can think of the problem as a graph partitioning problem where each edge has two costs $c(\cdot)$ and $c'(\cdot)$ associated with it. The objective function is

$$\sum_{x,y \text{ on same side}} c(x, y) + \sum_{x,y \text{ on different sides}} c'(x, y)$$

4.1 Two-Cost Partition Problem

We are given a complete graph $G = (V, E)$ with two cost functions c and c'. We want to find a partition of the vertex set $V = V_1 \cup V_2$ which minimizes $\sum_{i=1,2} \sum_{u,v \in V_i} c(u, v) + \sum_{u \in V_1, v \in V_2} c'(u, v)$.

Note that, if $c(u, v) = 0$ for all u, v, then the problem reduces to finding a minimum cut in the graph. On the other hand, if $c'(u, v) = 0$, then the problem is the well known edge deletion for graph bipartition problem (BIP) [12]. Our algorithm generalizes the algorithm for graph bipartition given by [12, 10]. The basic idea is to create two copies of each vertex to go on different sides of the partition. To ensure that they are on different sides, we designate each pair as a source-sink pair in the multi-cut subroutine.

Algorithm B:

1. Create an auxiliary graph G' from the graph G as follows.
   (a) For each vertex u in the graph G, G' has two vertices: u and u'.
   (b) For each edge (u, v) we create 4 edges in G': $(u, v), (u, v'), (u', v)$ and (u', v').
   (c) The edges in G' have weights, denoted by $l(\cdot, \cdot)$. Set $l(u, v) = l(u', v') = c(u, v)$ and $l(u, v') = l(u', v) = c'(u, v)$.
2. Use an approximation algorithm for the multi-cut problem (E.g., [10]) as a subroutine to find a multi-cut in graph G' with (u, u'), for all u, as the source-sink pairs. Let S be the set of edges in the multi-cut returned by the subroutine.
3. Construct a set of edges T as follows. If $\{u, v\}$ or $\{u', v'\}$ is chosen in S, then include both in T. Similarly, if $\{u, v'\}$ or $\{u', v\}$ is chosen, then include both in T.
4. Find a bipartition $V_1' \cup V_2'$ of vertices of G' so that T contains all the edges going across the partition[2].
5. Output the partition $V_1 \cup V_2$, where $V_i = V_i' \cap V$.

The intuition behind this algorithm is as follows. For the cut represented by T, we will show that we can get a partition of vertices in graph G' such that only one of u and u' is in one partition. From the partition of G', we get a bipartition of G. The cost of the bipartition of G is related to the cost of multi-cut obtained by above algorithm in the graph G'. We prove this in the next lemma.

Lemma 3. *Algorithm B returns a partition $V' = V_1' \cup V_2'$ of graph G', such that if $u \in V_1'$, then $u' \in V_2'$ and vice versa. Moreover, $\sum_{x \in V_1', y \in V_2'} l(x, y)$ is at most twice that of the multi-cut found after step 2 by Algorithm B separating each u from u'.*

Proof. Consider the set S of edges found by the multi-cut subroutine whose removal separates each u from u'. For each edge $(x, y) \in S$, we also include its "mirror" edge in T. i.e. if $(x, y) \in S$, then $(x', y') \in T$ from the graph. Note

---

[2] We will show how to do this in the proof of Proposition 1.

that, the cost of an edge and its "mirror" edge is same (i.e., $l(x,y) = l(x',y')$). Therefore, the cost of the edges in T is at most twice the cost of edges in S.

Now we show that removal of the edges in T breaks the graph in two parts with the desired property. Consider the graph $G'\backslash T$. Construct a graph H whose vertices represent the connected components in G' after removing the edges in T. Two vertices h_1 and h_2 in H are connected to each other if the corresponding connected components in G' have vertices x and x'.

In Proposition 1, we prove that the graph H is bipartite. Now we can use graph H to construct a partition $V' = V_1' \cup V_2'$ in graph G'. Since the vertices in graph H were connected components in graph G', there are no edges crossing the partition $V_1' \cup V_2'$ in graph G'. Moreover, bipartiteness of graph H means that each pair of vertices x and x' in graph G is split in the partition. The cost of this partition is at most 2 times the cost of the multi-cut.

Proposition 1. *The graph H defined in the proof of Lemma 3 is bipartite.*

Proof. For the sake of contradiction, assume that H has a cycle of odd length. Consider three consecutive vertices u, v and w in this odd cycle. Let v be connected to u and w.

Let x be a vertex of G' that belongs to the connected component u and defines the edge $\{u,v\}$ in graph H. Therefore, x' is the component v. Similarly, let y be a vertex in component w and y' be the corresponding vertex in component v. Since x' and y' are in the same connected component v, there is a path $x' \rightarrow y'$ that lies completely inside the component v. Since we didn't remove any of the edges on the path $x' \rightarrow y'$, all the *mirror* edges haven't been removed either. Therefore the the *mirror* path $x \rightarrow y$ connects x and y. This contradicts the fact that x and y were in different connected components.

This proves that the graph H is a bipartite graph.

Lemma 4. *The cost of the optimal multi-cut is a lower bound on the cost of partition of graph G.*

Proof. Consider a partition $V = V_1 \cup V_2$ of graph G. From this, we can construct a partition of the vertex set of G'. Let $V_1' = V_1 \cup \{x' \mid x \in V_2\}$ and $V_2' = V'\backslash V_1'$. Then, removing all the edges in G' crossing this partition ensures that no vertex x is connected to its counterpart x'. i.e. The set of edges going across the partition is a multi-cut. The cost of these edges is exactly the cost of the partition of G.

Recall that GVY algorithm for multi-cut [10] is an $O(\log k)$ approximation for k terminals. Here we have n terminals. Therefore by Lemmas 3 and 4, we get an $O(\log n)$ approximation for the best r-restricted mapping. Along with Observation 2 give us an $O(\log n)$ approximation for the L_1 norm of additive distortion.

To get an approximation algorithm for the L_p norm, we modify the costs in the two-cost partition problem as follows. Let $c(x,y) = |D_{xy} - |D_{ax} - D_{ay}||^p$ and $c'(x,y) = |D_{ax} + D_{ay} - D_{xy}|^p$. With these costs, Algorithm B gives an $O(\log n)$ approximation for $L_p(D, f^a)^p$. Therefore, for the L_p norm of additive distortion, we get an $O(\log^{1/p} n)$ algorithm.

5 Open Questions

We can show that the problem of finding the best r-restricted mapping is NP-hard by reducing the edge deletion for graph bipartition (BIP) [9] to it. Consider a graph G. Let $V(G) = n$. We construct a distance matrix D on $n + 1$ points $V(G) \cup \{a\}$. Set the diagonal entries D_{xx} to 0. Set $D_{ax} = 1/2$ for all $x \in V(G)$. For all $\{x, y\} \in E(G)$, set $D_{xy} = 1$. Set the rest of the entries to $1/2$. Consider an r-restricted mapping. Let $V(G) = V_1 \cup V_2$ be the partition induced by the r-restricted mapping. Then the cost of the r-restricted mapping is $B(V_1, V_2) +$ $(1/2)(\binom{n}{2} - |E(G)|)$, where $B(V_1, V_2)$ is the number of edges that need to be deleted to obtain V_1 and V_2 as two sides of a bipartite graph. Therefore, finding the optimal r-restricted mapping corresponds to minimizing the number of edges deleted for making the graph G bipartite. This proves that finding the best r-restricted mapping is NP-hard. However, this reduction is not approximation preserving. So it does not preclude the possibility of a PTAS for this problem. Getting even a constant factor approximation would be quite interesting.

In the proof of NP-hardness of r-restricted mapping problem, we used an input matrix D that does not satisfy the triangle inequality. For input matrix D that is a *metric* (i.e. it satisfies the triangle inequality), it might be possible to get a polynomial time algorithm for the best r-restricted mapping.

Agarwala *et al.* [1] have shown that for the problem of fitting data to tree metrics, the best r-restricted tree metric is within a factor of 3 of the optimal solution. However, the problem of approximating the additive distortion of the best r-restricted tree is still open.

Acknowledgments

We are grateful to R. Ravi and Anupam Gupta for helpful discussions and valuable suggestions.

References

1. Richa Agarwala, Vineet Bafna, Martin Farach, Babu O. Narayanan, Mike Paterson, and Mikkel Thorup. On the approximability of numerical taxonomy (fitting distances by tree metrics). In *Symposium on Discrete Algorithms*, pages 365–372, 1996.
2. Mihai Badoiu, Piotr Indyk, and Yuri Rabinovich. Approximate algorithms for embedding metrics into low-dimensional spaces. In *Unpublished manuscript*, 2003.
3. J-P. Barthélemy and A. Guénoche. *Trees and proximity representations*. Wiley, New York, 1991.
4. Avrim Blum, Shuchi Chawla, David Karger, Adam Meyerson, Maria Minkoff, and Terran Lane. Approximation algorithms for orienteering and discounted-reward tsp. In *IEEE Symposium on Foundations of Computer Science*, 2003.
5. L. Cavalli-Sforza and A. Edwards. Phylogenetic analysis models and estimation procedures. *American Journal of Human Genetics*, 19:233–257, 1967.
6. M. Deza and M. Laurent. *Geometry of Cuts and Metrics*. Springer-Verlag, Berlin, 1997.

7. K. Dhamdhere, A. Gupta, and R. Ravi. Approximating average distortion for embeddings into line. In *Symposium on Theoretical Aspects of Computer Science (STACS)*, 2004.
8. M. Farach, S. Kannan, and T. Warnow. A robust model for finding optimal evolutionary trees. *Algorithmica*, 13:155–179, 1995.
9. M. R. Garey and D. S. Johnson. *Computers and Intractability: A Guide to the Theory of NP-completeness*. W. H. Freeman, San Fransisco, USA, 1979.
10. N. Garg, V. Vazirani, and M. Yannakakis. Approximate max-flow min-(multi)cut theorems and their applications. *SIAM Journal on Computing*, 25(2):235–251, 1996.
11. Johan Håstad, Lars Ivansson, and Jens Lagergren. Fitting points on the real line and its application to RH mapping. In *European Symposium on Algorithms*, pages 465–476, 1998.
12. Philip Klein, Ajit Agarwal, R. Ravi, and Satish Rao. Approximation through multi-commodity flow. In *IEEE Symposium on Foundations of Computer Science*, pages 726–737, 1990.
13. James B. Saxe. Embeddability of graphs into k-space is strongly np-hard. In *Allerton Conference in Communication, Control and Computing*, pages 480–489, 1979.
14. P. H. A. Sneath and R. R. Sokal. *Numerical Taxonomy*. W. H. Freeman, San Fransisco, CA, 1973.
15. M. S. Waterman, T. S. Smith, M. Singh, and W. A. Beyer. Additive evolutionary trees. *Journal of Theoretical Biology*, 64:199–213, 1977.

Polylogarithmic Inapproximability
of the Radio Broadcast Problem
(Extended Abstract)

Michael Elkin[1] and Guy Kortsarz[2]

[1] Department of Computer Science, Yale University
New Haven, CT, USA, 06520-8285*
elkin@cs.yale.edu
[2] Computer Science department, Rutgers University, Camden, NJ, USA
guyk@crab.rutgers.edu

Abstract. We prove that there exists a universal constant $c > 0$ such that the Radio Broadcast problem admits *no additive* $c \cdot \log^2 n$-approximation, unless $NP \subseteq BPTIME(n^{O(\log \log n)})$. For graphs of at most logarithmic radius, an $O(\log^2 n)$ additive approximation algorithm is known, hence our lower bound is tight. To the best of our knowledge, this is the first tight additive polylogarithmic approximation result.

1 Introduction

1.1 The Radio Broadcast Problem

Definition and Motivation. Consider a synchronous network of processors that communicate by transmitting messages to their neighbors, where a processor receives a message in a given step if and only if precisely one of its neighbors transmit. The instance of the Radio Broadcast problem, called *radio network*, is a pair $(G = (V, E), s)$, $s \in V$, where G is an unweighted undirected graph, and s is a vertex, called *source*. The objective is to deliver one single message that the source s generates to all the vertices of the graph G using the smallest possible number of communication rounds. The prescription that tells each vertex when it should broadcast is called *schedule*; the *length* of the schedule is the number of rounds it uses, and it is called *admissible* if it informs all the vertices of the graph (see Section 2 for a formal definition). From practical perspective, the interest to radio networks is usually motivated by their military significance, as well as by the growing importance of cellular and wireless communication (see, e.g., [KM-98,GM95,BGI91]). The Radio broadcast is perhaps the most important communication primitive in radio networks, and it is intensively studied starting from mid-eighties [CR-03,KP02,KP-03,CGR-00,CG+02,CMS-03,I-02] [GM95,KM-98,ABLP91,BGI91,CW-87,CK-85].

---

* Part of this work was done while the first author was in School of Mathematics, Institute for Advanced Study, Princeton, NJ, USA, 08540.

K. Jansen et al. (Eds.): APPROX and RANDOM 2004, LNCS 3122, pp. 105–116, 2004.
© Springer-Verlag Berlin Heidelberg 2004

From theoretical perspective, the study of the Radio Broadcast problem provided researchers with a particularly convenient playground for the study of such broad and fundamental complexity-theoretic issues as the power and limitations of randomization, and of different models of distributed computation [BGI91,KM-98,KP-03]. In this paper we study the *additive approximation threshold* of the *Radio Broadcast* problem; we believe that our results show that this problem is of a particular interest from the stand-point of the theory of Hardness of Approximation as well.

Previous Results. The first known algorithm for the Radio Broadcast problem was devised by Chlamtac and Weinstein in 1987 [CW-87]. That algorithm, given an instance (G, s) of the problem, constructs an admissible broadcast schedule of length $O(Rad(G, s) \cdot \log^2 n)$ for this instance, where $Rad(G, s)$ stands for the *radius* of the instance (G, s), that is, the maximum distance $d_G(s, v)$ in the graph G between the source s and some vertex $v \in V$. Their algorithm is *centralized*, i.e., it accepts the entire graph as input.

Soon afterwards Bar-Yehuda et al. [BGI91] devised a distributed randomized algorithm that provides admissible schedules of length $O(Rad(G, s) \cdot \log n + \log^2 n)$. Alon et al. [ABLP91] have shown that the additive term of $\log^2 n$ in the result of [BGI91] is inevitable, and devised a construction of infinitely many instances (G, s) of constant radius that satisfy that any broadcast schedule for them requires $\Omega(\log^2 n)$ rounds. Kushilevitz and Mansour [KM-98] have shown that for *distributed* algorithms, the multiplicative logarithmic term in the result of [BGI91] is inevitable as well, and proved that for *any distributed algorithm* for the Radio Broadcast problem there exist (infinitely many) instances (G, s) on which the algorithm constructs a schedule of length $\Omega(Rad(G, s) \cdot \log(n/Rad(G, s)))$. Finally, the gap between the $\log n$ and $\log(n/Rad(G, s))$ was recently closed by Kowalski and Pelc [KP-03], and Czumaj and Rytter [CR-03].

Gaber and Mansour [GM95] have shown that *centralized* algorithms are capable of constructing much shorter schedules, and devised an algorithm that constructs admissible schedules of length $O(Rad(G, s) + \log^5 n)$. Their result is also the current state-of-the-art in terms of the *existential* upper bounds on the length of broadcast schedules for instances of large radius.

Remark: In [EK-03] we have improved the result of [GM95] by providing schedules of length $Rad(G, s) + O(\sqrt{Rad(G, s)} \cdot \log^2 n) = O(Rad(G, s) + \log^4 n)$.

Since, obviously, any schedule for an instance (G, s) requires at least $Rad(G, s)$ rounds, the algorithms for the Radio Broadcast problem [CW-87,BGI91,GM95,KP-03,CR-03] can be interpreted as *approximation algorithms* for the problem. The state-of-the-art in this context are the results by Bar-Yehuda et al. [BGI91] for instances of small radius, and the result of Gaber and Mansour [GM95] for instances of large radius. (We term such an approximation guarantee "a *semi-additive*", as opposed to "a *purely additive*" guarantee for which the multiplicative factor of the optimum should be equal to 1.)

In terms of the lower bounds, the NP-hardness of the Radio Broadcast problem was shown by Chlamtac and Kutten [CK-85] already in 1985. The authors of the current paper have shown [EK-02] that there exists a constant $c > 0$ such that the Radio Broadcast problem cannot be approximated with a (purely) additive approximation ratio of $c \log n$ unless $NP \subseteq BPTIME(n^{O(\log \log n)})$.

1.2   Our Results

In this paper we show that there exists a constant $c > 0$ such that the Radio Broadcast problem cannot be approximated with a (purely) additive approximation ratio of $c \log^2 n$, unless $NP \subseteq BPTIME(n^{O(\log \log n)})$. This result holds even for the Radio Broadcast problem restricted to graphs of constant radius. We believe that our reduction contains lower bound ideas for radio model context that should have further applications.

Observe also that our results together with [BGI91] fully determine the additive approximation threshold (up to constants and with randomization allowed) of the Radio Broadcast problem restricted to instances (G, s) of radius $Rad(G, s) = O(\log n)$.

Hence we have demonstrated that the approximation behavior of the Radio Broadcast problem is best-described in terms of an *additive polylogarithmic threshold*. It appears that there is no other problem whose approximation threshold is known to exhibit a similar behavior. We believe that demostrating the existence of a new pattern of approximation behavior enriches the theory of Hardness of Approximation, and constitutes the main contribution of this paper.

1.3   Proof Techniques

The proof of our lower bound incorporates the combinatorial construction of Alon et al. [ABLP91] in the reduction of Lund and Yannakakis [LY94]. The latter reduction reduces the Label-Cover problem to the Set Cover problem, and establishes the logarithmic hardness of approximation of the latter. However, the importance of this reduction extends far beyond this specific proof, and the reduction was used as a building block for many inapproximability results [ABSS93,EK-02,DGKR03,HK-03].

However, it seems that the reduction of [LY94] alone cannot provide greater than logarithmic for the Radio Broadcast problem. The second crucial building block of our lower bound is the result of Alon et al. [ABLP91]. Their construction provides a bipartite graph in which it requires $\Omega(\log^2 n)$ rounds to deliver the message from the left-hand side of the graph to its right-hand side.

To explain how we incorporate the construction of [ABLP91] into the reduction of [LY94], we first outline the structure of the reduction of [LY94]. For each edge $\tilde{e}$ of the input instance of the Label-Cover problem, the reduction of [LY94] constructs a bipartite graph $(\mathcal{S}_{\tilde{e}}, M_{\tilde{e}})$, where each vertex $x \in \mathcal{S}_{\tilde{e}}$ is connected to one half of the vertices of the set $M_{\tilde{e}}$. Furthermore, these connections are ingeniously arranged in such a way that the following condition is satisfied: for any $k = o(\log n)$ vertices $x_1, x_2, \ldots, x_k \in \mathcal{S}_{\tilde{e}}$, and for any choice of sets $X_1, X_2, \ldots, X_k$ such every X_i is either equal to the set N_i of the vertex x_i in $M_{\tilde{e}}$

or to its compliment $M_{\tilde{e}} \backslash N_i$, the set $\bigcup_{i=1}^{k} X_i$ does not cover the vertex set $M_{\tilde{e}}$. Such a bipartite graph is called a *set system*.

The reduction of [LY94] also provides each such *set system* $(\mathcal{S}_{\tilde{e}}, M_{\tilde{e}})$ with a simple *"trapdoor"* that enables to cover the set $M_{\tilde{e}}$ with only two sets, specifically, the set N_i and the set $M_{\tilde{e}} \backslash N_i$. These trapdoors are geared to enable to construct small solutions for the instances of the Set Cover problem whenever the original instance of the Label-Cover problem is a YES-instance, but, however, cannot be used if the original instance is a NO-instance.

The basic idea of our proof of the polylogarithmic lower bound for the Radio Broadcast problem is to replace the set systems of [LY94] by the bipartite graphs of [ABLP91]. To accomplish it, one, however, needs to devise analogous trapdoor to the construction of [ABLP91], and doing it without ruining the (rather delicate) properties of [ABLP91]. It is, therefore, not surprising that our construction is quite complex, and its analysis is technically very involved.

Comparison with [EK-02]: The reduction of [EK-02] can be seen as a (non-trivial) adaptation of the reduction of Lund and Yannakakis to the Radio Broadcast problem. The reduction in the current paper is far more complex both conceptually and technically, particularly because it incorporates the construction of [ABLP91] in the reduction of Lund and Yannakakis.

2   Preliminaries

2.1   The Radio Broadcast Problem

We start with introducing some definitions and notations.

Definition 1. *The set of neighbors of a vertex v in an unweighted undirected graph $G(V, E)$, denoted $\Gamma_G(v)$, is the set $\{u \in V \mid (v, u) \in E\}$. For a subset $X \subseteq V$, the set of neighbors of the vertex v in the subset X, denoted $\Gamma_G(v, X)$ (or $\Gamma(v, X)$ when the graph G is clear from the context), is the set $\{u \in X \mid (v, u) \in E\}$.*

Definition 2. *Let $G = (V, E)$ be an unweighted undirected graph, and $R \subseteq V$ be a subset of vertices. The set of vertices informed by R, denoted $I(R)$, is $I(R) = \{v \mid \exists! x \in R \; s.t. \; v \in \Gamma_G(x)\}$ (the notation $\exists! x$ stands for "there exists a unique x"). For a singleton set $R = \{x\}$, $I(R) = I(\{x\}) = I(x) = \Gamma_G(x)$.*

A sequence of vertex sets $\Pi = (R_1, R_2, \ldots, R_q)$, $q = 1, 2, \ldots$, is called a *radio broadcast schedule* (henceforth referred as a *schedule*) if $R_{i+1} \subseteq \bigcup_{j=1}^{i} I(R_j)$ for every $i = 1, 2, \ldots, q - 1$. The set of vertices *informed by a schedule* Π, denoted $I(\Pi)$, is $I(\Pi) = \bigcup_{R \in \Pi} I(R)$.

An instance of the *Radio broadcast problem* $\mathcal{G}$ is a pair $(\bar{G} = (\bar{V}, \bar{E}), s)$, where G is a graph, and $s \in V$ is a vertex. The goal is to compute an admissible schedule Π of minimal length. The *value* of an instance $\mathcal{G}$ of the radio broadcast problem is the length of the shortest admissible schedule Π for this instance.

2.2   The MIN-REP Problem

Definition 3. *The MIN-REP problem is defined as follows. The input consists of a bipartite graph $G = (V_1, V_2, E)$. In addition, for $j = 1, 2$, the input contains a partition $\tilde{V}_j$ of V_j into a disjoint union of subsets, $V_1 = \bigcup_{A \in \tilde{V}_1} A$, $V_2 = \bigcup_{B \in \tilde{V}_2} B$. The triple $\mathcal{M} = (G, \tilde{V}_1, \tilde{V}_2)$ is an* instance *of the MIN-REP problem. The size of the instance is $n = |V_1| + |V_2|$. An instance G as above induces a bipartite* super-graph *$\tilde{G} = (\tilde{V}_1, \tilde{V}_2, \tilde{E})$ in which the sets A and B of the partition serve as the vertices of the super-graph. The edges of the super-graph are $\tilde{E}(\mathcal{M}) = \tilde{E} = \{(A, B) \in \tilde{V}_1 \times \tilde{V}_2 \mid a \in A, b \in B, (a, b) \in E\}$. In other words, there is a (super-)edge between a pair of sets $A \in \tilde{V}_1$, $B \in \tilde{V}_2$ if and only if the graph G contains an edge between a pair of vertices $a \in A$, $b \in B$.*

Denote $\tilde{V} = \tilde{V}_1 \cup \tilde{V}_2$. A pair of vertices $x_1, x_2 \in V_1 \cup V_2$ is called a matching pair *with respect to a super-edge $\tilde{e} = (A, B) \in \tilde{E}$ (henceforth, $\tilde{e}$-m.p.) if $(x_1, x_2) \in E$ and either $x_1 \in A$ and $x_2 \in B$ or vice versa.*

A subset $C \subseteq V_1 \cup V_2$ of vertices is said to cover *a super-edge $\tilde{e} = (A, B)$ if it contains an $\tilde{e}$-m.p.. A subset $C \subseteq V_1 \cup V_2$ that satisfies $|C \cap X| = 1$ for every $X \in \tilde{V}$ is called a MAX-cover. In other words, a MAX-cover C contains exactly one vertex from each super-vertex. An instance $\mathcal{M}$ of the MIN-REP problem is called a YES-instance if there exists a MAX-cover that covers all the superedges. Such a MIN-REP solution is called a* perfect MAX-cover. *For a positive real number $t > 1$, an instance $\mathcal{M}$ of the MIN-REP problem is called a t-NO-instance if any MIN-cover C of the instance $\mathcal{M}$ covers less than $|\tilde{E}|/t$ super-edges of $\mathcal{M}$.*

We also impose several additional (somewhat less standard) restrictions on the set of instances of the MIN-REP problem.

1. All the super-vertices $X \in \tilde{V}$ are of size $|X| = n^{0.4}/2$.
2. The set of super-edges has sufficiently large cardinality, specifically, $|\tilde{E}| = \Omega(n^{0.6} \cdot \log^3 n)$. (Note that by restriction 1, just for the super-graph to be connected, $|\tilde{E}|$ has to be at least $2n^{0.6} - 1$.)
3. **The Star property:** For every super-edge $\tilde{e} = (A, B) \in \tilde{E}$, $A \subseteq V_1$ and $B \subseteq V_2$ and every vertex $b \in B$ there exists exactly one vertex $a \in A$, denoted $\tilde{e}(b)$, such that $(a, b) \in E$. This property is called the *star* property.

Theorem 1. *[R98] No deterministic polynomial time algorithm may distinguish between the YES-instances and the $\log^{10} n$-NO-instances of the MIN-REP problem, unless $NP \subseteq DTIME(n^{O(\log \log n)})$, even when the instances of the MIN-REP problem satisfy the conditions (1)-(3).*

3   Reduction

We next describe our randomized reduction from the MIN-REP problem to the Radio broadcast problem. This reduction always maps a YES-instance of the MIN-REP problem to an instance of the Radio broadcast problem of value $T + O(\log n)$, and with high probability over its own coin tosses the reduction maps a NO-instance of the MIN-REP problem to an instance of the Radio broadcast problem of value $T + \Omega(\log^2 n)$, where $T = 1, 2, \ldots$ is a parameter of the reduction.

Consider an instance $\mathcal{M} = (G, \tilde{V}_1, \tilde{V}_2)$, $G = (V_1, V_2, E)$ of the MIN-REP problem with $V_1 = \bigcup_{A \in \tilde{V}_1} A$, $V_2 = \bigcup_{B \in \tilde{V}_2} B$. The reduction constructs an instance $\mathcal{G} = \mathcal{G}(\mathcal{M}) = (\bar{G}, s)$, $\bar{G} = (\bar{V}, \bar{E})$, $s \in \bar{V}$, of the Radio broadcast problem in the following way.

Let $N = n^{0.6}$. The vertex set $\bar{V}$ of the graph consists of the source s, and the disjoint vertex sets $\mathcal{V}_1$ and $\mathcal{V}_2$ (i.e., $\bar{V} = \{s\} \cup \mathcal{V}_1 \cup \mathcal{V}_2$ with $s \notin \mathcal{V}_1 \cup \mathcal{V}_2$, and $\mathcal{V}_1 \cap \mathcal{V}_2 = \emptyset$). It is convenient to visualize the source s on the "left-hand" side of the graph, the vertex set $\bar{V}_1$ in the "middle", and the vertex set $\bar{V}_2$ on the "right-hand" side of the graph. The vertex set $\mathcal{V}_1$ contains N copies of every vertex $x \in V = V_1 \cup V_2$; these copies are denoted $cp_1(x), cp_2(x), \ldots, cp_N(x)$. The set of all copies of a vertex x, $\{cp_1(x), cp_2(x), \ldots, cp_N(x)\}$, is denoted by χ_x. For a subset $X \subseteq V$, let χ_X denote $\chi_X = \bigcup_{x \in X} \chi_x$. Let $\hat{J}$ denote the set of indices $\{0.4 \log n, 0.4 \log n + 1, \ldots, 0.6 \log n\}$. For an index $j \in \hat{J}$, and a vertex $x \in V$, let $\chi_x(j)$ denote the subset of the set χ_x that contains only the first $J = 2^j$ copies of the vertex x. I.e., $\chi_x(j) = \{cp_1(x), cp_2(x), \ldots, cp_J(x)\}$. These copies are also called the j-relevant copies of the vertex x. For a subset $X \subseteq V$, let $\chi_X(j) = \bigcup_{x \in X} \chi_x(j)$.

The vertex set $\mathcal{V}_2$ is of the form $\mathcal{V}_2 = \bigcup_{\tilde{e} \in \tilde{E}} M_{\tilde{e}}$, where the *ground sets* $M_{\tilde{e}}$ are disjoint, and all have equal size. Each ground set $M_{\tilde{e}}$ is a disjoint union of the sets $M_{\tilde{e}}(j, q)$, $j \in \hat{J}$, $q \in [L]$, with $L = 2 \cdot n^{c_0}$, and c_0 is an integer positive universal constant that will be determined later. The sets $M_{\tilde{e}}(j, q)$ are all of equal size $M = n^{c_0}$, for the same constant c_0.

The edge set $\bar{E}$ of the graph $\bar{G}$ contains edges that connect the source s to the vertices of the set $\mathcal{V}_1$, and edges between the vertices of $\mathcal{V}_1$ and $\mathcal{V}_2$. Specifically, the edge set $\bar{E}$ is formed in the following way.

1. Connect the source s to all the vertices of the set $\mathcal{V}_1$.
2. For every super-edge $\tilde{e} = (A, B) \in \tilde{E}$, for every pair of indices $j \in \hat{J}$, $q \in [L]$ do

   (a) For every vertex $a \in A$

      i. Let $H_{\tilde{e}, a}(j, q)$ be an *exact random half* of the set $M_{\tilde{e}}(j, q)$. In other words, every subset $H \subseteq M_{\tilde{e}}(j, q)$ that has cardinality $|M_{\tilde{e}}(j, q)|/2$ (for simplicity we ignore all the Integrality issues) has the same probability to be chosen to serve as $H_{\tilde{e}, a}(j, q)$.

      ii. Let $H^{(1)}_{\tilde{e}, a}(j, q)$ be an exact random half of the set $H_{\tilde{e}, a}(j, q)$. Set $H^{(2)}_{\tilde{e}, a}(j, q) = H_{\tilde{e}, a}(j, q) \backslash H^{(1)}_{\tilde{e}, a}(j, q)$.

   (b) i. For every vertex $b \in B$, set $H_{\tilde{e}, b}(j, q) = M_{\tilde{e}}(j, q) \backslash H_{\tilde{e}, \tilde{e}(b)}(j, q)$.

      ii. Let $H^{(1)}_{\tilde{e}, b}(j, q)$ be an exact random half of the set $H_{\tilde{e}, b}(j, q)$. Set $H^{(2)}_{\tilde{e}, b}(j, q) = H_{\tilde{e}, b}(j, q) \backslash H^{(1)}_{\tilde{e}, b}(j, q)$.

      (The steps 2a-2b will be referred as the *exact partition step*.)

(c) For every vertex $x \in A \cup B$ let $\chi_x^{(1)}(j)$ be a subset of $\chi_x(j)$ that is formed by picking each element of the set $\chi_x(j)$ with probability $1/2$ independently at random.

Set $\chi_x^{(2)}(j) = \chi_x(j) \backslash \chi_x^{(1)}(j)$.

(The step 2c will be referred as the *binomial partition step.*

(d) For every vertex $x \in A \cup B$, and index $i \in \{1, 2\}$ do

Let $\chi_x^{(i)}(j) = \{cp_1^{(i)}(x, j), cp_2^{(i)}(x, j), \ldots, cp_{\bar\chi}^{(i)}(x, j)\}$, with $\bar\chi = \bar\chi(x, i, j) = |\chi_x^{(i)}(j)|$.

Let $S_{\bar\chi}$ be the set of all permutations of the set $[\bar\chi]$.

Choose a permutation $\sigma \in_R S_{\bar\chi}$ uniformly at random from the set $S_{\bar\chi}$.

Partition the set $H_{\bar e, x}^{(i)}(j, q)$ arbitrarily into $\bar\chi$ disjoint subsets $P_{\bar e, x}^{(i)}(j, q, \ell)$, $\ell \in [\bar\chi]$.

For every index $\ell \in [\bar\chi]$, connect the vertex $cp_\ell^{(i)}(x, j)$ to every vertex v in the set $P_{\bar e, x}^{(i)}(j, q, \sigma(\ell))$.

(The step 2d will be referred as the *random permutation step.*)

(e) i. For every copy $cp(x) \in \chi_x^{(i)}(j)$, and every vertex $v \in H_{\bar e, x}^{(i)}(j, q)$, with probability $1/J$ insert the edge (x, v) into the edge set $\bar E$, independently at random.

ii. Do the same for every copy $cp(x) \in \chi_x^{(i)}(j)$, and every vertex $v \in M_{\bar e}(j, q) \backslash H_{\bar e, x}(j, q)$.

iii. Do the same for every copy $cp(x) \in \chi_x \backslash \chi_x(j)$, and every vertex $v \in M_{\bar e}(j, q)$.

(The step 2e will be referred as the *mixing step.*)

We need some additional notation. For a pair of vertices $x \in \mathcal{V}_1$ and $v \in \mathcal{V}_2$, let $(x \, \mathcal{C} \, v)$ denote the event $((x, v) \in \bar E)$, and let $(x \, \mathcal{N} \, v)$ denote the complimentary event $((x, v) \notin \bar E)$. For a subset $X \subseteq \mathcal{V}_1$ and a vertex $v \in \mathcal{V}_2$, let $(X \, \mathcal{AC} \, v)$ denote the event $(\forall x \in X, (x, v) \in \bar E)$, and, analogously, let $(X \, \mathcal{AN} \, v)$ denote the event $(\forall x \in X, (x, v) \notin \bar E)$. For a pair of subsets $X \subseteq \mathcal{V}_1, U \subseteq \mathcal{V}_2$, let $(X \, \mathcal{AC} \, U)$ denote the event $(\forall x \in X, v \in U, (x, v) \in \bar E)$, and, analogously, let $(X \, \mathcal{AN} \, U)$ denote the event $(\forall x \in X, v \in U, (x, v) \notin \bar E)$.

For a subset $X \subseteq \mathcal{V}_1$, a vertex $v \in \mathcal{V}_2$, and a positive integer $\ell = 1, 2, \ldots$, let $(X \, (\geq \ell \mathcal{C}) \, v)$ denote the event $(|\{x \in X \mid (x, v) \in \bar E\}| \geq \ell)$. Analogously, let $(X \, (\leq \ell \mathcal{C}) \, v)$ denote the event $(|\{x \in X \mid (x, v) \in \bar E\}| \leq \ell)$, and let $(X \, (\ell \mathcal{C}) \, v)$ denote the event $(|\{x \in X \mid (x, v) \in \bar E\}| = \ell)$. We also use the notation $(X \, \mathcal{INF} \, v)$ for the event $(X \, (1\mathcal{C}) \, v)$; in this case we say that the subset X *informs* the vertex v. For the complimentary event we use the notation $(X \, \mathcal{NINF} \, v)$, and say that the subset X *does not inform* the vertex v. For a subset $U \subseteq \mathcal{V}_2$, we use the notation $(X \, \mathcal{INF} \, U)$ for the event $(\forall v \in U, (X \, \mathcal{INF} \, v))$; in this case we say that the subset X informs the subset U. For the complimentary event we use the notation $(X \, \mathcal{NINF} \, U)$, and say that the subset X *does not inform* the subset U. For a collection of subsets $\Pi = (X_1, X_2, \ldots, X_t)$ of $\mathcal{V}_1$ and a vertex $v \in \mathcal{V}_2$ (respectively, a subset $U \subseteq \mathcal{V}_2$), we say that the collection Π *informs the vertex* v (resp., *the subset* U), and denote it by $(\Pi \, \mathcal{INF} \, v)$ (resp., $(\Pi \, \mathcal{INF} \, U)$), if *at least one subset* $X \in \Pi$ informs v (resp., U). For the complimentary event we use the notation $(\Pi \, \mathcal{NINF} \, v)$ (resp., $(\Pi \, \mathcal{NINF} \, U)$) and say that the collection Π *does not inform the vertex* v (resp., *the subset* U).

For a fixed super-edge $\tilde{e} = (A, B) \in \tilde{E}$, and a vertex $x \in A \cup B$, let $star(x)$ denote the subset of vertices of $A \cup B$ that belong to the same star as x with respect to the super-edge $\tilde{e}$. For a subset $S \subseteq A \cup B$, let $cp(S)$ denote the set of all the copies of the vertices of S, i.e., $cp(S) = \{cp_\ell(x) \mid x \in S\}$.

4   Analysis

In this section we analyze the reduction described in the previous section.

4.1   Basic Properties

Definition 4. *For a super-edge* $\tilde{e} = (A, B)$, *a subset* $\mathcal{S} \subseteq \chi_A \cup \chi_B$ *is called* $(\tilde{e}, j)$-*partial if for every* $\tilde{e}$-*m.p.* (a, b), *the set* $\mathcal{S}$ *contains at most* 2^{j-3} j-*relevant copies of the vertex* a *or* b.

Consider some pair of vertices (a, b) with $a \in A$ and $b \in B$, for some pair of super-vertices $A \in \tilde{V}_1$ and $B \in \tilde{V}_2$. Let $j \in \hat{J}$, $q \in [L]$ be a pair of indices. Consider a subset $\mathcal{S} \subseteq \chi_a(j) \cup \chi_b(j)$, and a vertex $v \in H_{\tilde{e}, b}(j, q)$. Recall that the set $\chi_b(j)$ is split into the disjoint union of the subsets $\chi_b^{(1)}(j)$ and $\chi_b^{(2)}(j)$ by a binomial random partition (see Section 3, step 2c). The set $H_{\tilde{e}, b}(j, q)$ is split into a disjoint union of the two subsets $H_{\tilde{e}, b}^{(1)}(j, q)$ and $H_{\tilde{e}, b}^{(2)}(j, q)$, and these two subsets have the same size $M/2$. Let $\mathcal{S}_1 = \mathcal{S} \cap \chi_b^{(1)}(j)$, $\mathcal{S}_2 = \mathcal{S} \cap \chi_b^{(2)}(j)$.

Let $s_1 = |\mathcal{S}_1|$, $s_2 = |\mathcal{S}_2|$, $w = |\chi_b^{(1)}(j)|$ and $u = |\mathcal{S} \cap \chi_a(j)|$. Note that s_1, s_2, and w are random variables, and u is a fixed value that does not depend on random coins.

Lemma 1. *For a super-edge* $\tilde{e} = (A, B) \in \tilde{E}$, *a pair of vertices* $a \in A$, $b \in B$, *a pair of indices* $j \in \hat{J}$, $q \in [L]$, *a subset* $\mathcal{S} \subseteq \chi_a(j) \cup \chi_b(j)$ *and a vertex* $v \in M_{\tilde{e}}(j, q)$, $\mathcal{P}(\mathcal{S} \, \mathcal{AN} \, v) \geq (1 - 1/2^j)^{|\mathcal{S}|} \cdot \sum_{w, s_1} \frac{w - s_1}{w} \cdot \mathcal{P}(|\mathcal{S}_1| = s_1, |\chi_b^{(1)}(j)| = w)$.

Consider a subset $\mathcal{S} \subseteq \chi_A(j) \cup \chi_B(j)$. Let $S = S(\mathcal{S})$ be the subset of vertices of $A \cup B$ that have copies in the set $\mathcal{S}$. Note that the set $\mathcal{S}$ decomposes into a disjoint collection of the subsets $\mathcal{S}(a) = cp(star(a)) \cap \mathcal{S}$, for different vertices $a \in S$. For a yet more general subset $\mathcal{S} \subseteq \chi_A \cup \chi_B$, we will use the notation $S_j = S_j(\mathcal{S})$ for the subset of vertices of $A \cup B$ that have j-relevant copies in the set $\mathcal{S}$.

We will use the notation x to denote an arbitrary vertex $x \in S$, both for $x \in A$ and $x \in B$. Let $PN_x(v)$ denote the subset of copies of the vertex x that have a non-zero probability of being connected to the vertex $v \in M_{\tilde{e}}(j, q)$ by *the random permutation step*. In other words, let $j(v)$ be the index j such that $v \in M_{\tilde{e}}(j, q)$. If $v \in M_{\tilde{e}}(j, q) \backslash H_{\tilde{e}, x}(j, q)$, then set $PN_x(v) = \emptyset$. Otherwise, let $i_v(x)$ be the index i such that $v \in H_{\tilde{e}, x}^{(i)}(j, q)$. Then set $PN_x(v) = \chi_x^{(i_v(x))}(j)$.

Definition 5. *Consider a fixed super-edge* $\tilde{e} = (A, B)$, *a pair of indices* $j \in \hat{J}$, $q \in [L]$, *a subset* $\mathcal{S} \subseteq \chi_A \cup \chi_B$, *and a vertex* $v \in M_{\tilde{e}}(j, q)$. *The exact partitions* of the set $M_{\tilde{e}}(j,q)$ into the unions of the subsets $H_{\tilde{e},a}(j,q)$ and $H_{\tilde{e},b}(j,q)$ for different $\tilde{e}$-m.p.s (a,b), the exact partitions of the sets $H_{\tilde{e},x}(j,q)$ into the unions of the subsets $H_{\tilde{e},x}^{(1)}(j,q)$ and $H_{\tilde{e},x}^{(2)}(j,q)$ for different vertices $x \in A \cup B$, and the binomial partitions of the sets $\chi_x(j)$ into the unions of the subsets $\chi_x^{(1)}(j)$ and $\chi_x^{(2)}(j)$ for different vertices $x \in A \cup B$ are called safe with respect to the pair $(\mathcal{S}, v)$ if for every vertex $x \in S_j(\mathcal{S})$, $|\mathcal{S} \cap PN_x(v)| \leq 2^{j-3}$.

Lemma 2. *Let $\mathcal{S}$ be an $(\tilde{e}, j)$-partial subset of $\chi_A \cup \chi_B$ (see Definition 4) of size $|\mathcal{S}| \leq 2^j \ln n$, and $v \in M_{\tilde{e}}(j,q)$ be a vertex. The probability that the exact partitions of the set $M_{\tilde{e}}(j,q)$, and the binomial partitions of the sets $\chi_x(j)$ are safe with respect to the pair $(\mathcal{S}, v)$ is at least $1/n^8$.*

Corollary 1. *With probability at least $1/n^8$, for every vertex $x \in S$, $|PN_x(v) \cap S| \leq 2^{j-3}$.*

4.2   Large Rounds and Their Effect

In this section we consider the probability space that is determined by the random permutation and mixing steps (steps 2d - 2e) of the reduction, and, in particular, the partitions of the sets $\chi_x(j)$ are fixed for all the vertices $x \in A \cup B$ (and the index $j \in \hat{J}$ is fixed as well).

Definition 6. *We say that a round R is j-large with respect to a super-edge $\tilde{e} = (A, B)$ if $R_{\tilde{e}} = R \cap (\chi_A \cup \chi_B)$ is of size greater or equal to $c \cdot 2^{j+1} \ln n$, where $c = 2c_0 + 2$ (c_0 is a universal constant mentioned in Section 3; it will be determined later). Otherwise, the round is called j-small with respect to the super-edge $\tilde{e}$.*

A schedule Π is called short *if its length is at most $\log^2 n/100$.*

Consider a short schedule Π. Let $\pi_{\tilde{e}}(j)$ (respectively, $\Pi_{\tilde{e}}(j)$) be the subschedule of Π that contains only j-small (resp. j-large) rounds with respect to the super-edge $\tilde{e}$. Note that a vertex $v \in M_{\tilde{e}}(j)$ is informed by the schedule Π if and only if either the subschedule $\pi_{\tilde{e}}(j)$ or the subschedule $\Pi_{\tilde{e}}(j)$ informs v (or if both do).

Let $V_{\tilde{e}}(\pi_{\tilde{e}}(j)) = \bigcup_{R \in \pi_{\tilde{e}}(j)} R \cap (\chi_A \cup \chi_B) = \bigcup_{R \in \pi_{\tilde{e}}(j)} R_{\tilde{e}}$, and, analogously, $V_{\tilde{e}}(\Pi_{\tilde{e}}(j)) = \bigcup_{R \in \Pi_{\tilde{e}}(j)} R_{\tilde{e}}$.

Next, we show that the probability that the subschedule $\Pi_{\tilde{e}}(j)$ informs a significant number of vertices is negligibly small.

For a fixed super-edge $\tilde{e} = (A, B) \in \hat{E}$, and a fixed index $j \in \hat{J}$, consider a subset $M' \subseteq M_{\tilde{e}}(j)$ such that for every index $q \in [L]$, $|M' \cap M_{\tilde{e}}(j,q)| \leq 1$. Such subsets are particularly convenient for the analysis, since for a pair of distinct indices $q_1, q_2 \in [L]$, and a pair vertices $v_1 \in M_{\tilde{e}}(j, q_1)$, $v_2 \in M_{\tilde{e}}(j, q_2)$, the events $(x \mathcal{C} v_1)$ and $(x \mathcal{C} v_2)$ are independent.

Lemma 3. *For a fixed short schedule Π, with probability at least $1 - \frac{1}{2^{\Omega(n \cdot \log n)}}$, there is no subset $M' \subseteq M_{\tilde{e}}(j)$ of size $N = n$ that satisfies $|M_{\tilde{e}}(j,q) \cap M'| \leq 1$ for every index $q \in [L]$, such that the subschedule $\Pi_{\tilde{e}}(j)$ informs all the vertices of the subset M'.*

4.3   Pivots

We start with describing a combinatorial lemma from [ABLP91]. We will use this lemma in the rest of our analysis.

Lemma 4. *[ABLP91] Let $G = (V_1, V_2, E)$ be a bipartite graph with $|V_1| = n$. Let $\Pi = (R_1, R_2, \ldots, R_t)$ be a collection of at most $\frac{\log^2 n}{100}$ subsets of V_1. Then there exists a subset $\mathcal{S}$ of the vertices on the left-hand side of the graph ($\mathcal{S} \subseteq V_1$), and an index $j \in \hat{J}$, such that*

1. *$|\mathcal{S}| \leq 2^j \cdot \log n$.*
2. *Let $\Pi' = (R_1', R_2', \ldots) = \Pi \backslash \mathcal{S}$. Then for every subset $R' \in \Pi'$, $|R'| \geq 2^j$.*
3. *Let f_k be the number of subsets R' with cardinality $2^{j+k} \leq |R'| < 2^{j+k+1}$. Then $\sum_{k \geq 0} \frac{f_k}{2^k} \leq \log n$.*

Fix a super-edge $\tilde{e} = (A, B)$, and consider the subgraph of the graph $\bar{G} = (\bar{V}, \bar{E})$ induced by the vertex set $(\chi_A \cup \chi_B) \cup M_{\tilde{e}}$. Let $\Pi|_{\tilde{e}}$ be the schedule Π restricted to the set $\chi_A \cup \chi_B$. Apply Lemma 4 to this subgraph and the collection $\Pi|_{\tilde{e}}$ of subsets of the set $\chi_A \cup \chi_B$, and let $\mathcal{S}$ and j be the subset and the index whose existence are guaranteed by the lemma. (Observe that $|\chi_A \cup \chi_B| = n^{0.6} \cdot |A \cup B| = n$.)

Definition 7. *(1) The index j as above is called the* pivot *of the super-edge $\tilde{e}$ with respect to the schedule Π. (2) A round R is* small *with respect to the super-edge $\tilde{e}$ if it is j-small with respect to super-edge $\tilde{e}$, and j is the pivot of $\tilde{e}$. (3) The schedule Π is called $\tilde{e}$-proper if $V_{\tilde{e}}(\pi_{\tilde{e}}(j))$ is an $(\tilde{e}, j)$-partial set, where the index $j \in \hat{J}$ is the pivot of the super-edge $\tilde{e}$ with respect to Π.*

4.4   Small Rounds

Consider a schedule $\Pi = (R_1, R_2, \ldots, R_t)$, $t \leq (\log^2 n)/100$, and a super-edge $\tilde{e} = (A, B)$. Let $\Pi|_{\tilde{e}} = (R_1|_{\tilde{e}}, R_2|_{\tilde{e}}, \ldots, R_t|_{\tilde{e}})$, $R_\ell|_{\tilde{e}} = R_\ell \cap (\chi_A \cup \chi_B)$, be the subschedule of Π restricted to the super-edge $\tilde{e}$ (for simplicity of notation we assume that for every index $\ell \in [t]$, the set $R_\ell|_{\tilde{e}}$ is not empty; otherwise the number of rounds in the restricted schedule would be smaller than in the original one).

Lemma 5. *For a fixed super-edge $\tilde{e} \in \tilde{E}$, and a fixed $\tilde{e}$-proper schedule Π, consider the L sets $M_{\tilde{e}}(j, q)$, $q \in [L]$. For each index $q \in [L]$, let X_q denote the indicator random variable of the event $(\pi_{\tilde{e}}(j) \; \mathcal{INF} \; M_{\tilde{e}}(j, q))$. Let $X = \sum_{q=1}^{L} X_q$. Then $\mathcal{P}(X \leq 2n^{c_0} - n) \geq 1 - exp(-\Omega(n^{c_0}))$.*

Lemma 6. *For a fixed super-edge $\tilde{e}$, and a fixed short schedule Π, $\mathcal{P}(\Pi \; \mathcal{INF} \; M_{\tilde{e}}(j)) \leq exp(-\Omega(n^{c_0}))$, where j is the pivot of the super-edge $\tilde{e}$ with respect to the schedule Π.*

Definition 8. *A schedule Π is called* proper *if there exists a subset $\tilde{H} \subseteq \tilde{E}$ that contains at least one half of all the super-edges, and such that the schedule Π is $\tilde{e}$-proper with respect to every super-edge $\tilde{e} \in \tilde{H}$. Otherwise, the schedule Π is called* non-proper.

Consider a *NO-instance*. Let $\hat{\Pi}$ denote the collection of all short proper schedules.

Lemma 7. *With probability at least* $1 - exp(-\Omega(n^{1.6} \log^3 n))$, *there is no admissible schedule* $\Pi \in \hat{\Pi}$.

4.5 Deriving the Results

In this section we glue together all the pieces of the analysis and derive our main results. The details are omitted from this extended abstract.

Theorem 2. *Unless* $NP \subseteq BPTIME(n^{O(\log \log n)})$, *there exists a universal constant c such that no (deterministic or randomized) polynomial time algorithm may provide an additive* $(c \cdot \log^2 n)$-*approximation guarantee for the Radio broadcast problem.*

References

[AB+92]   B. Awerbuch, B. Berger, L. Cowen, and D. Peleg. Fast network decomposition. In *Proc. 11th ACM Symp. on Principles of Distr. Comp.*, pp. 161-173, Aug. 1992.

[ABLP01] N. Alon, A. Bar-Noy, N. Linial, D. Peleg. A lower bound for radio broadcast. In *Journal of Computer and System Sciences 43*, pp. 290-298, 1991.

[ABSS93] S. Arora, L. Babai, J. Stern, and Z.Swedyk. The hardness of approximate optima in lattices, codes and linear equations. In *Proc. 34th IEEE Symp. on Foundations of Computer Science*, pp. 724-733, 1993.

[AL96]   S. Arora and K. Lund. In *Approximation Algorithms for NP-hard Problems*, D. Hochbaum, ed., PWS Publishing, 1996.

[AS92]   N. Alon and J. Spencer. *The Probabilistic Method*. Wiley, 1992.

[B01]   R. Bar-Yehuda. Private communication.

[BGI91]   R. Bar-Yehuda, R., O. Goldreich, and A. Itai. Efficient emulation of single-hop radio network with collision detection on multi-hop radio network with no collision detection. In *Distributed Computing*, 5(2):67–72, 1991.

[C-52]   H. Chernoff. A measure of asymptotic efficiency for tests of a hypothesis based on the sum of observations. *Annals of Mathematical Statistics*, 23:493–509, 1952.

[CG+02]   B. S. Chlebus, L. Gasieniec, A. Gibbons, A. Pelc, W. Rytter. Deterministic broadcasting in ad hoc radio networks. Distributed Computing 15(1): 27-38 (2002)

[CGR-00] M. Chrobak, L. Gasieniec, W. Rytter. Fast Broadcasting and Gossiping in Radio Networks. FOCS 2000: 575-581

[CK-85]   I. Chlamtac and S. Kutten. A spatial-reuse TDMA/FDMA for mobile multihop radio networks. In Proceedings IEEE INFOCOM, pages 389-94, 1985.

[CMS-03] A. F. Clementi, A. Monti, R. Silvestri. Distributed broadcast in radio networks of unknown topology. TCS 1-3(302): 337-364 (2003)

[CR-03]   A. Czumaj and W. Rytter. Broadcasting Algorithms in Radio Networks with Unknown Topology. FOCS 2003: 492-501

[CW-87]   Chlamtac and Weinstein The wave expansion approach to broadcasting in multihop radio networks, INFOCOM 1987.

[DGKR03] I. Dinur, V. Guruswami, S. Khot, and O. Regev. A New Multilayered PCP and the Hardness of Hypergraph Vertex Cover, with Irit Dinur, Venkatesan Guruswami In *Proc. of 34th Symp. on Theory of Computing*, 2003.

[EK-02]   M. Elkin and G. Kortsarz. A logarithmic lower bound for radio broadcast. J. Algorithms, to appear.

[EK-03]   M. Elkin and G. Kortsarz. Improved broadcast in radio networks Manuscript, 2004.

[FHKS-02] U. Feige, M. Halldo'rsson, G. Kortsarz and A. Srinivasan. Approximating the domatic number. Siam journal on computing 32(1), 172–195, 2002.

[FR-94]   M. Fu"rer, B. Raghavachari. Approximating the Minimum-Degree Steiner Tree to within One of Optimal. J. Algorithms 17(3): 409-423 (1994)

[GKR-00] N. Garg, G. Konjevod, R. Ravi. A Polylogarithmic Approximation Algorithm for the Group Steiner Tree Problem. J. Algorithms 37(1): 66-84 (2000)

[GM95]   I. Gaber and Y. Mansour. Broadcast in radio networks. The 6th ACM-Siam Symposium on Discrete Algorithms" 577–585, 1995.

[HK-03]   E. Halperin and R. Krauthgamer. Polylogarithmic inapproximability, *To appear in 35th Symposium on Theory of Computing (STOC'03)*

[HK+03]   E. Halperin, G. Kortsarz, R. Krauthgamer, A. Srinivasan and N. Wang, Integrality ratio for Group Steiner Trees and Directed Steiner Trees, SODA, pages 275-284, 2003

[I-02]   P. Indyk. Explicit constructions of selectors and related combinatorial structures, with applications. SODA 2002: 697-704

[K-98]   G. Kortsarz. On the hardness of approximating spanners. *The first International Workshop APPROX-98*, pages 135–146, 1998.

[KM-98]   Kushilevitz and Mansour. Computation in Noisy Radio Networks Symposium on Discrete Algorithms, 1988.

[GPM03]   R. Gandhi, S. Parthasarathy and A. Mishra. Minimizing Broadcast Latency and Redundancy in Ad Hoc Networks. To appear in the Proc. of the Fourth ACM International *Symposium on Mobile Ad Hoc Networking and Computing (MOBIHOC'03)*, Jun. 2003.

[KP02]   D. Kowalski and A. Pelc, Deterministic broadcasting time in radio networks of unknown topology, Proc. 43rd Annual IEEE Symposium on Foundations of Computer Science (FOCS 2002), 2002, 63-72.

[KP-03]   D. Kowalski and A. Pelc, Broadcasting in undirected ad hoc radio networks Symposium on Principles of distributed computing (PODC), 73 - 82, 2003

[LS-91]   N. Linial and M. Saks. Decomposing graphs into regions of small diameter. In *Proc. of the 2nd ACM-SIAM Symp. on Discr. Alg.*, 320-30, 1991.

[LY94]   C. Lund and M. Yannakakis. On the hardness of approximating minimization problems. *J. Assoc. Comput. Mach.*, 41(5):960–981, 1994.

[MR-95]   R. Motwani, P. Raghavan. Randomized Algorithms. *Cambridge University Press.*, 1995.

[R98]   R. Raz. A Parallel Repetition Theorem, *SIAM Journal of computing*, 27(3) (1998) pp. 763-803.

[RS97]   R. Raz, S. Safra. A Sub-Constant Error-Probability Low-Degree Test, and a Sub-Constant Error-Probability PCP Characterization of NP. *STOC*, pages 575-584

On Systems of Linear Equations
with Two Variables per Equation[*]

Uriel Feige and Daniel Reichman

The Weizmann Institute, Rehovot 76100, Israel
{uriel.feige,daniel.reichman}@weizmann.ac.il

Abstract. For a prime p, max-2lin$p$ is the problem of satisfying as many equations as possible from a system of linear equations modulo p, where every equation contains two variables. Hastad shows that this problem is NP-hard to approximate within a ratio of $11/12 + \epsilon$ for $p = 2$, and Andersson, Engebretsen and Hastad show the same hardness of approximation ratio for $p \geq 11$, and somewhat weaker results (such as $69/70$) for $p = 3, 5, 7$. We prove that max-2lin$p$ is easiest to approximate when $p = 2$, implying for every prime p that max-2lin$p$ is NP-hard to approximate within a ratio of $11/12 + \epsilon$. For large p, we prove stronger hardness of approximation results. Namely, we show that there is some universal constant $\delta > 0$ such that it is NP-hard to approximate max-2lin$p$ within a ratio better than $1/p^\delta$. We use our results so as to clarify some aspects of Khot's *unique games conjecture*. Namely, we show that for every $\epsilon > 0$ it is NP-hard to approximate the value of unique games within a ratio of ϵ.

1 Introduction

Systems of linear equations can be solved in polynomial time using Gaussian elimination. However, *max-lin*, the problem of finding a solution that maximizes the number of satisfied equations (in a system that is nonsatisfiable) is NP-hard. Here we consider the approximation ratios achievable for max-lin, where for a given algorithm this ratio is taken to be the number of equations satisfied by the algorithm divided by the number of equations satisfied by the best solution. In the following discussion we parameterize max-lin by two parameters, k and p, and denote the resulting problem as max-$k$lin$p$. Here, k is the number of variables per equation, and p is a prime, signifying that all computations are performed modulo p.

For every system of linear equations modulo p, a simple greedy algorithm is guaranteed to satisfy at least a $1/p$ fraction of the equations. Hence for every k and p, max-$k$lin$p$ can be approximated within a ratio of $1/p$. A celebrated result of Hastad [8] shows that for $k = 3$, this is essentially best possible. That is, for every prime p and for every $\epsilon > 0$, max-3lin$p$ is NP-hard to approximate within

---

[*] This research was supported by the Israel Science Foundation (Grant number 263/02).

a ratio of $1/p + \epsilon$. In particular, max-3lin2 is NP-hard to approximate within a ratio better than $1/2 + \epsilon$.

The situation for the case $k = 2$ is different. Here there are approximation algorithms with approximation ratios significantly better than $1/p$. Goemans and Williamson [7] gave an approximation algorithm based on semidefinite programming that approximates max-2lin2 within a ratio of roughly 0.87856. This result was extended in [2, 9] to show that for every prime p there is some ϵ that depends on p such that max-2lin$p$ can be approximated within a ratio better than $(1 + \epsilon)/p$. The value of ϵ in these proofs converges to 0 as p grows.

In terms of hardness results, Hastad [8] shows that max-2lin2 is NP-hard to approximate within a ratio of $11/12 + \epsilon$. In [2] it is shown that it is NP-hard to approximate max-2lin$p$ within a ratio of $\frac{17}{18}, \frac{39}{40}, \frac{69}{70}$ ($+\epsilon$, in all cases) for $p = 3, 5, 7$ respectively, and within a ratio of $11/12 + \epsilon$ for all $p \geq 11$. This state of affair raises the following questions.

1. The approximation ratios of known algorithms for max-2lin$p$ deteriorate and tend to 0 as p grows. The hardness results to not exclude approximation ratios as good as $11/12$ for every p. Does, indeed, the best possible approximation ratio for max-2lin$p$ tend to 0 as p grows? This question was asked explicitly in [2]. In this paper we give a positive answer to this question.
2. The hardness of approximation ratios for small values of p ($p = 3, 5, 7$) are not as strong as those for $p = 2$. Is it indeed easier to approximate max-2lin$p$ when $p = 3, 5, 7$ than it is when $p = 2$? In this paper we give a negative answer to this question.

In the rest of the introduction we discuss our results in more detail. In particular, we also show how our results are connected to the *unique games conjecture* of Khot [9].

1.1   Small Values of p

A reduction between two optimization problems is *approximation preserving* if there is some constant $c > 0$ such that the optimum value of the target problem is exactly c times the optimum value of the source problem. A polynomial time approximation preserving reduction from problem A to problem B can be used to transform any polynomial time approximation algorithm for problem B into a polynomial time approximation algorithm for problem A, with the same approximation ratio.

Theorem 1. *For every prime $p \geq 3$, there is a polynomial-time approximation-preserving reduction from max-2lin2 to max-2lin$p$.*

Theorem 1 is proved by using *local gadgets*. Every equation modulo 2 is replaced by a collection of equations modulo p according to some fixed rule, independently of all other equations. However, the analysis of why this reduction is approximation preserving is global (considering all equations simultaneously) rather than local (analyzing every gadget application separately). The global nature of the analysis of our gadgets may be the reason why despite their simplicity, they have not been discovered earlier. In particular, Trevisan, Sorkin, Sudan and Williamson [12] present a systematic way (based on linear programming) for finding the optimal local gadgets in reductions, but their approach (as presented in [12]) is limited to gadgets whose analyses are local.

Using the fact that it is NP-hard to approximate max-2lin2 within a ratio strictly of $11/12 + \epsilon$, Theorem 1 implies:

Corollary 1. *For every $\epsilon > 0$ and prime p, it is NP-hard to approximate max-2linp within a ratio of $11/12 + \epsilon$.*

The following question remains open. Let p and q be two arbitrary primes s.t. $p > q$. Observe that the discussion in the introduction implies that the approximation ratio for max-3linq is strictly better than that for max-3linp. Is the approximation ratio for max-2linq strictly better than that for max-2linp?

1.2 Large Values of p

When p is sufficiently large, the bounds implied by Theorem 1 can be greatly improved.

Theorem 2. *There is some $\delta > 0$ such that for every prime p, max-2linp is NP-hard to approximate within a ratio better than $1/p^{\delta}$.*

The proof of Theorem 2 is by a reduction from a problem that is sometimes called *label cover* [4], a notion recalled under different terminology in Theorem 3.

1.3 The Unique Games Conjecture

The unique games conjecture of Khot [9], if true, has far reaching consequences regarding approximation of optimization problems. In particular, Khot and Regev [10] show that it would imply that vertex cover is NP-hard to approximate within a ratio of $2 - \epsilon$. We recall this conjecture. The notation and terminology that we use in our presentation were chosen so as to clarify the connection between unique games and the problem of max-2lin.

We denote a (one round, two provers) game by $G = G(X, Y, A, B, Q, P)$. There are two finite sets of variables, X and Y, and two finite sets of values, A for X, and B for Y. There is a support set $Q \subset X \times Y$. In the most general form of games, there is a probability distribution π defined over Q. In the context of the current paper, we may take π to be uniform, and so as to simplify the presentation, we omit all references to π. For every pair of variables $(x, y) \in Q$, there is a polynomial time computable predicate $P_{x,y}(x, y)$. (More explicitly, the pair (x, y) plays two roles here, both to specify which predicate is being evaluated, and as variables in the predicate itself. Equivalently, we may think of there being only one predicate P over four variables, the first two being the *names* of x and y, and the last two being the *values* of x and y. The predicate is polynomial time computable in the sense that given values for x and y, it can be evaluated in time polynomial in $|X| + |Y|$.) The goal of the game is to find two functions, $P_1 : X \longrightarrow A$ and $P_2 : Y \longrightarrow B$, that satisfy the maximum number of given predicates. (That is, for every $(x, y) \in Q$, we check whether $P_{x,y}(P_1(x), P_2(y))$ evaluates to *true*.) The value of a game G, denoted by $\omega(G)$, is the maximum over the choices of P_1 and P_2 of the number of predicates satisfied, divided by the normalizing factor $|Q|$ (which is the total number of predicates). Hence $0 \leq \omega(G) \leq 1$.

A game is said to have *oneway uniqueness* if for every $(x, y) \in Q$ and every $a \in A$, there is exactly one $b \in B$ such that $P_{x,y}(a, b)$ holds. The following theorem is well known and has been used extensively in previous work on hardness of approximation.

Theorem 3. *For every $\epsilon > 0$ there are sets A and B with $|A| + |B| \leq (1/\epsilon)^{O(1)}$, such that it is NP-hard to distinguish between oneway unique games G with $\omega(G) = 1$ and oneway unique games G with $\omega(G) \leq \epsilon$.*

Proof. For completeness, we sketch the proof (which can be regarded as folklore). The starting point is the fact that it is NP-hard to distinguish between satisfiable 3CNF formulas and those that are at most δ-satisfiable, for some $\delta < 1$ [5]. From this one can design a oneway unique game G in which X is the set of clauses, Y is the set of variables, and Q contains all clause-variable pairs where the variable appears in the particular clause. Every clause has 7 satisfying assignments. Encode these assignments as numbers in $\{1, .., 7\}$ (where i encodes the ith satisfying assignment in lexicographic order). The set A is satisfying assignments to clauses encoded as explained above, and the set B is assignments to single variables. Predicate $P_{x,y}(a, b)$ (where y is a variable in clause x) holds if the triple assignment a satisfies the clause x, and a and b are consistent in their assignment to y. $\omega(G) = 1$ if the original 3CNF formula is satisfiable, and $\omega(G) = (2 + \delta)/3$ if the original formula was only δ-satisfiable. Thereafter, the parallel repetition theorem of Raz [11] gives the desired result. ∎

To avoid the possibility of misunderstanding, let us point out that in Theorem 3 (as well as elsewhere when we discuss games) we view the sets A and B as fixed for a whole class of games, whereas other parameters of the games (X, Y, Q, P) change within the class of games. Hence the notion of NP-hardness relates to algorithms whose running time is polynomial in $|X| + |Y|$.

A *unique* game has the oneway uniqueness property in both directions, namely, also for every $b \in B$ there is exactly one $a \in A$ such that $P_{x,y}(a, b)$ holds. The unique games conjecture of Khot [9] says that a weaker form of Theorem 3 holds also for unique games.

Conjecture 1. (The unique games conjecture of Khot) For every $0 < \epsilon < 1/2$ and every $0 < \delta < 1/2$, there are sets A and B such that it is NP-hard to distinguish between unique games G with $\omega(G) \geq 1 - \epsilon$ and unique games G with $\omega(G) \leq \delta$.

Equivalently, for every $\epsilon > 0$ and every $0 < \eta < 1$, there are sets A and B such that it is NP-hard to distinguish between unique games G with $\omega(G) \geq \eta$ and unique games G with $\omega(G) \leq \epsilon \cdot \eta$.

The condition $\eta < 1$ cannot be strengthened to $\eta = 1$ because there is a polynomial time algorithm that decides whether $\omega(G) = 1$ for unique games.

A weaker version of the unique games conjecture simply asks whether for every ϵ, it is NP-hard to approximate $\omega(G)$ for unique games within a ratio better than ϵ. Technically, this weaker version replaces the condition "for every $\epsilon > 0$ and every $0 < \eta < 1$" in Conjecture 1 by the weaker condition "for every $\epsilon > 0$ there is some $0 < \eta < 1$ such that". Khot observed that even this weaker form of the unique games conjecture was not known to be true, and asked whether it is true. We give a positive answer to this question. Our approach is explained in the following.

Motivated by the unique games conjecture, we prove a stronger version of Theorem 2. Call a system of equations modulo p *proper* if it has the following properties:

1. The set of variables can be partitioned into two disjoint sets, X and Y.
2. Every linear equation contains exactly one variable from X and exactly one variable from Y (with nonzero coefficients!).
3. For every pair of variables $x \in X$ and $y \in Y$, there is at most one equation containing both x and y.

Proper max-2linp is the problem of finding an assignment that maximizes the number of satisfied equations in a proper instance of 2linp. With some extra work, the proof of Theorem 2 can be extended to show that:

Theorem 4. *There is some $\delta > 0$ such that for every prime p, proper max-2linp is NP-hard to approximate within a ratio better than $1/p^\delta$.*

As a corollary, we have that unique games are hard to approximate within any $\epsilon > 0$.

Corollary 2. *For every $\epsilon > 0$, there are some constant $0 < \eta < 1$, prime p, and sets A, B with $|A| = |B| = p$, such that it is NP-hard to distinguish between unique games G with $\omega(G) \geq \eta$ and unique games G with $\omega(G) \leq \epsilon \cdot \eta$.*

Proof. A proper 2linp system S of equations is a unique game G. The sets X, Y of variables for S serve as the sets X, Y of variables for G. The set of values A, B for G corresponds to the set $\{0, 1, \ldots, p-1\}$. The support Q corresponds to all pairs of variables (x, y) that appear in some equation in S. As each pair $(x, y) \in Q$ appears in exactly one equation in S (because S is proper), this equation serves as the predicate $P_{x,y}$ for G. The uniqueness property follows from the fact that modulo p, every equation of the form $cx = d$ (with nonzero coefficient c and arbitrary d) has a unique solution, and similarly for $cy = d$.

Having established the fact that a proper 2linp system is a unique game, it remains to choose the parameters of the system. Choose p large enough such that $1/p^\delta$ in Theorem 4 is smaller than ϵ. Then for proper max-2linp there is some constant $0 < \eta < 1$ such that it is NP-hard to distinguish between instances in which an η-fraction of the equations are satisfied, and instances in which at most an $\epsilon \cdot \eta$-fraction of the equations are satisfied. As desired. ∎

(Note: an explicit value for η can in principle be derived from the proof of Theorem 4, though at the moment there does not seem to be much interest in doing so).

1.4   Equations over the Rationals

In all our reductions, if one treats the system of equations as a system over the rationals (and then we call the underlying problem max 2lin, omitting the p) or over the integers, rather than modulo p, the same hardness of approximation result follow. As the prime p increases, the approximation ratios proved hard in Theorem 2 tend to 0. One obtains:

Theorem 5. *For every $\epsilon > 0$, it is NP-hard to approximate max-2lin within a factor of ϵ.*

Using our techniques, one can also prove hardness of approximation results for max-2lin within ratios that tend to 0 as the size of the input instance increases (e.g., of the form $\epsilon = 2^{-(\log n)^{\gamma}}$ for $\gamma < 1$, where n is the number of equations in the system). However, these results are proved under the assumption that NP does not have quasi-polynomial time algorithms. It is an open question whether such results can be proved under the assumption that $P \neq NP$.

Theorem 5 and the discussion that follows should not be confused with hardness results appearing in [1, 3], as there the number of variables per equation is not limited to 2. Additional discussion on the difference between 2 versus many variables per constraint can be found in [6].

2   Proofs

We present here the proofs of our main theorems (except for those whose proof was already sketched in the introduction).

2.1   Small Values of p

We restate Theorem 1 in more detail, and provide the proof.

Theorem 6. *For prime $p > 3$ There is a polynomial time reduction transforming an instance E of max 2lin2 to an instance F of max 2linp s.t. one can satisfy k equations in E iff one can satisfy $p \cdot k$ equations in F. For $p = 3$, there is a polynomial time reduction transforming an instance E of max 2lin2 to an instance F of max 2lin3 s.t. one can satisfy k equations in E iff one can satisfy $2k$ equations in F.*

Proof. Every equation in E is of the form $x + y \equiv_2 0$ or $x + y \equiv_2 1$. Consider first the case in which $p > 3$. Transform an equation of the first form to the set of $2(p-1)$ equations

$$x \equiv_p iy$$

for $i = 1, \ldots, p - 1$ and

$$x - 1 \equiv_p i(y - 1)$$

for $i = 1, \ldots, p - 1$. Transform an equation of the second form to the equations

$$x - 1 \equiv_p iy$$

for $i = 1, \ldots, p - 1$ and

$$x \equiv_p i(y - 1)$$

for $i = 1, \ldots, p - 1$. Call such a group of equations a *block*. Assignments of $0/1$ values to x and y that satisfy an original equation will satisfy exactly p equations in the corresponding block. Assignments of $0/1$ values which do not satisfy the original equation do not satisfy any equation in the corresponding block. Non-$0/1$ assignments to either x or y satisfy at most two equations in the corresponding block.

This proves that if one can satisfy k equations in E, one can satisfy $p \cdot k$ equations in F. To see the converse look at a satisfying assignment σ of F satisfying $p \cdot k$ equations. If all the variables receive $0/1$ values the claim follows immediately (Give to all the variables in E the values given to them by σ). If not, call every variable not receiving a value from $\{0, 1\}$ *misbehaved*, and assign to all misbehaved variables independently at random 0 w.p. half and 1 w.p. half. Every block of equations containing a variable not set to values in $\{0, 1\}$ will have p of its equations satisfied with probability $1/2$ and zero of its equations satisfied with probability $1/2$. Hence the expected number of satisfied equations in such a block is $p/2$. We conclude that there is a $0/1$ assignment to the misbehaved variables such that if we have l blocks containing misbehaved variables the number of satisfied equations in those blocks will be at least $pl/2$. σ satisfied at most $2l$ equations in those l blocks. As $p/2 > 2$ (recall we assumed that $p > 3$) we get a $0/1$ assignment satisfying more equations than σ. In particular it satisfies at least $p \cdot k$ equations and we are done as we have seen that a $0/1$ assignment satisfying $p \cdot k$ equations of F implies we can satisfy k equations in E.

In case $p = 3$ transform an equation of the type $x + y \equiv_2 0$ to the 3 equations

$$x - y \equiv_3 0$$

$$x - 1 \equiv_3 2(y - 1)$$

$$x \equiv_3 2y$$

An equation of the type $x + y \equiv_3 1$ will be transformed to the 3 equations

$$x \equiv_3 2(1 - y)$$

$$x - 1 \equiv_3 y$$

$$x \equiv_3 1 - y$$

One can check that assigning the variables $0/1$ values which satisfy the original equation will satisfy two out of the three equations, assigning the variables $0/1$

values which do not satisfy the original equation will satisfy zero out of the three equations and giving to at least one of the variables value 2 will satisfy at most one equation out of the 3. An analogue argument to the one given in the last paragraph shows that if we have an assignment satisfying $2k$ equations in the new system we have such an assignment with $0/1$ values (in our random process every block containing a misbehaved variable will have an expected number of $(2+0)/2 = 1$ satisfied equations). The rest of the arguments are as in the above paragraphs. ∎

2.2   Large Values of p

Here we prove Theorem 2.

Proof. The proof is by reduction from Theorem 3, and the reader is advised to recall the relevant notation for games. For simplicity in our presentation, rather than fix p in our proof and then figure out the relevant parameters for the game G, we shall fix parameters for the game G, and based on them choose p. It is not difficult to exchange the order of these choices, and we omit the details.

Consider an arbitrary game G with oneway uniqueness and $|A| \geq |B|$ (as is indeed the case in the games appearing in the proof of Theorem 3), and let $p = \Theta(|A|/\epsilon)$, for ϵ as in Theorem 3. Let σ be an arbitrary one to one mapping from A to Z_p and τ be an arbitrary one to one mapping from B to Z_p. Thus from now on we look at elements of A and B as elements of Z_p

For every pair $(x, y) \in Q$, replace the predicate $P_{x,y}$ by a *block* of $|A|(p-1)$ equations as follows. For every $a \in A$ the block contains $p-1$ equations $x - a = i(y - b)$, where $1 \leq i \leq p - 1$, and $b \in B$ is the unique value s.t. $P_{x,y}(a, b) = 1$. Looking at a group of $p-1$ equations at a specific a, we see that an assignment that zeroes $(x - a)$ and $(y - b)$ will satisfy all $p - 1$ equations whereas any other assignment will satisfy at most one equation.

For any assignment assigning c to x and d to y where $(x, y) \in Q$ we have that if $P_{x,y}(\sigma^{-1}(c), \tau^{-1}(d)) = 1$, at least $p - 1$ equations will be satisfied out of the block corresponding to (x, y). If on the other hand, $P_{x,y}(\sigma^{-1}(c), \tau^{-1}(d)) = 0$, at most $|A|$ equations would be satisfied in the (x, y) block. We remark that in any event we cannot satisfy more that $p - 1 + |A| - 1$ equations per block.

Since we have a block of equations for every pair $(x, y) \in Q$, we get that if $\omega(G) = 1$ then we could satisfy at least $(p - 1)|Q|$ equations.

Assume now that π is an assignment that satisfies more than $((1 - \epsilon)|A| + \epsilon(p-1+|A|-1))|Q|$ equations, where ϵ is as in Theorem 3. Since π cannot satisfy more than $p - 1 + |A| - 1$ equations per block, it must hold that the fraction of blocks in which π satisfies more than $|A|$ equations is strictly larger than ϵ. Hence by the discussion above it follows that for G, the assignment $P_1(x) = \sigma^{-1}(\pi(x))$ for all $x \in X$ and $P_2(y) = \tau^{-1}(\pi(y))$ for all $y \in Y$ shows that $\omega(G) > \epsilon$. Thus we get a hardness of (recall that $|A| \leq p$)

$$\frac{((1 - \epsilon)|A| + \epsilon(p - 1 + |A| - 1))|Q|}{(p - 1)|Q|} \leq \frac{|A|}{p - 1} + 2\epsilon$$

As $p = \Theta(|A|/\epsilon)$ and $|A| = (1/\epsilon)^{O(1)}$, the proof follows. ∎

2.3   Proper Max-2lin$p$

The proof of Theorem 2 gives a system of equations arranged in *blocks*. To every pair of variables $(x, y) \in Q$ corresponds one block of exactly $|A|(p-1)$ equations. In what follows, the parameters q and m will be used to denote $q = |A|(p-1)$ and $m = 40qp^3$. In our proof that proper max-2lin$p$ is hard to approximate, we shall make use of a notion that we call *balanced coloring*.

Definition 1. *Let c be an edge coloring with q colors of the complete bipartite graph $K_{m,m}$. Call a vertex-induced subgraph $G = (V, E)$ big if $|E| \geq m^2/p^3$. We say c is* balanced *if the following two conditions hold:*

1. *Every color is assigned to at least $m^2/2q$ edges.*
2. *For every big induced subgraph $G = (V, E)$, every color is assigned to at most $\frac{3}{2}(|E|/q)$ edges in E.*

Lemma 1. *Balanced colorings as defined above, exist.*

Proof. We give an existence proof using the probabilistic method. Color all the edges of $K_{m,m}$ with q colors by assigning independently at random to each edge a color between 1 and q (each color is chosen with probability $\frac{1}{q}$). Let $G = (V, E)$ a fixed big induced subgraph of $K_{m,m}$. The probability that a given color will occur more than $\frac{3}{2}|E|/q$ times or less than $\frac{1}{2}|E|/q$ in this induced subgraph is by the Chernoff Bound at most

$$2 \cdot (1/e)^{0.38 \cdot \frac{1}{4}|E|/q} \leq 2 \cdot (1/e)^{0.09\frac{m^2}{p^3 q}}$$

The number of induced subgraphs of $K_{m,m}$ is bounded by 2^{2m}. Taking the union bound on all colors and all induced subgraphs we get that the probability some big subgraph will not have the desired property is bounded by

$$(1/e)^{0.09\frac{m^2}{p^3 q}} \cdot 2^{\log q + 2m + 1} < (1/e)^{0.09\frac{m^2}{p^3 q} - 3m}$$

As we took m to equal $40p^3q$ the exponent of the above expression is positive. Hence, with positive probability we get a balanced coloring and the result follows. ∎

In our proofs we will need balanced colorings when the parameters m and q can be regarded as constants (even though they are "large" constants), and hence a balanced coloring can be found in constant time by exhaustive search.

Let E be an instance of max 2lin$p$ with the block structure as described above. That is, every block has q equations and all the equations in a block contain the same two variables (and every equation contains an X variable and a Y variable). For this reason, this 2lin$p$ system is not proper, in the sense defined in Section 1.3. We reduce the improper (by improper we mean an instance that is not proper) instance of 2lin$p$ (that we call E) to a proper instance of 2lin$p$ (that we call E').

The set of variables for E' will be m times as large as that of E. Every variable $x \in X$ for E will be "replaced" by m variables $x_1, \ldots, x_m$ in E'. Likewise, $y \in Y$

will be replaced by $y_1, \ldots, y_m$. As to the equations, every block of q equations in E is replaced by a block of m^2 equations in E'. This is done as follows. Let x and y be the two variables occurring in a block of E. Assign to every equation in the the block a different color from $1, 2, \ldots, q$ (in arbitrary order). Color the edges of $K_{m,m}$ with a *balanced coloring* with q colors. If in this coloring the color of the edge between i and j is l, put x_i and y_j in equation l. We remark that this coloring is used simultaneously in all the blocks of equations of $|E'|$ and that the number of blocks in E' is $|Q|$, the same as the number of blocks in E. It is not hard to see that E' is indeed proper.

Denote by X' the set of all the x variables in E' and by Y', all y variables in E'. Say a set $R = S \times T \subseteq X' \times Y'$ is a *rectangle* if both S and T are subsets of the variables in the same block of E'. A rectangle is *big* if it contains at least a $1/p^3$ fraction of the equations in a block.

Let τ be an arbitrary assignment to the variables of E'. For any block of E', let $x_1, \ldots, x_m, y_1, \ldots, y_m$ be its variables. Assignment τ partitions both $x_1, \ldots, x_m$ and $y_1, \ldots, y_m$ to at most p nonempty sets S_k, T_k where x_i belongs to the S_j if $\tau(x_i)$ is the jth value from the field and similarly for T_j with the y variables. By the properties of balanced coloring, in every big rectangle $R = S_l \times T_n$ any color in R appears $t|R|/q$ times where $1/2 \leq t \leq 3/2$. ($|R| = |S_l| \times |T_n|$).

The proof of Theorem 4 now follows from the combination of Theorem 2 and the following lemma.

Lemma 2. *Let η denote the optimum[1] of the improper 2linp system E with block structure as described above. Then η', the optimum of the proper system E' resulting from the reduction described above, satisfies $\eta/2 \leq \eta' \leq 3\eta$.*

Proof. Let σ be an assignment that satisfies $\eta q|Q|$ equations in E. Define an assignment τ to the variables of E' as follows: for each tuple of m variables $z_1, \ldots, z_m$ corresponding to a variable z of the original system put $\tau(z_1) = \tau(z_2) = \ldots = \tau(z_m) = \sigma(z)$. By property 1 of balanced coloring we have that τ satisfies at least $\frac{1}{2}\eta m^2|Q|$ equations in the new system. Hence $\eta' \geq \eta/2$.

We now prove that $\eta' \leq 3\eta$, or rather, that $\eta \geq \eta'/3$. Let τ be an optimal assignment to the variables of the new system, satisfying $\eta' m^2|Q|$ equations. Index the blocks with numbers between 1 and $|Q|$. Let k be an arbitrary block of equations in E', and let x and y be the variables in the corresponding block of E. As explained above, the assignment τ partitions the block k into rectangles $S_{i,k} \times T_{j,k}$ ($0 \leq i, j \leq p-1$), where x variables are in $S_{i,k}$ if they are given the value i by the assignment and similarly for y variables. For any *big* rectangle define $c_{i,j,k}$ to be the fraction of equations of $S_{i,k} \times T_{j,k}$ satisfied by τ. The number of equations in rectangles that are not big is bounded by $p^2 \cdot \frac{1}{p^3} \cdot m^2$. Thus, the number of equations satisfied by τ in E' is at most

$$m^2|Q|/p + \sum_{k=1}^{|Q|} \sum_{S_{i,k} \times T_{j,k} \text{ is big}} (|S_{i,k}||T_{j,k}|c_{i,j,k}) \tag{1}$$

---

[1] By optimum we mean the maximum fraction of equations that can be satisfied by an assignment.

Construct an assignment σ to the variables of E as follows: choose k and l uniformly at random from $[1, ..m]$. For every $x \in X$, let $\sigma(x) = \tau(x_k)$ (where x_k is the kth variable replacing x), and for every $y \in Y$, let $\sigma(y) = \tau(y_l)$. Let Z be the following random variable: Z counts the number of equations of E satisfied by σ. Its expectation satisfies:

$$E(Z) \geq \sum_{k=1}^{|Q|} \sum_{S_{i,k} \times T_{j,k} \text{ is big}} \left(\frac{|S_{i,k}||T_{j,k}|}{m^2}\right)\left(\frac{2}{3}c_{i,j,k} \cdot q\right) \tag{2}$$

The above inequality follows from property 2 of balanced colorings. Namely, in a big rectangle R no color appears on a fraction of more than $\frac{3}{2q}$ of the edges (equations). In a block of E, the same color appears on exactly $1/q$ of the equations. Hence in E the fraction of satisfied equations shrinks by a factor no worse than $\frac{2}{3}$ compared to R.

There must be some assignment to E that achieves at least the expectation of Z. Comparing equations (1) and (2), this implies that $\eta \geq \frac{2}{3}(\eta' - 1/p)$. Moreover, $\eta \geq 1/p$, as in every system of equations modulo p one can satisfy a $1/p$ fraction of the equations. Hence $\eta \geq \eta'/3$, as desired. ∎

References

1. E. Amaldi and V. Kann. On the approximability of minimizing nonzero variables or unsatisfied relations in linear systems. *Theoretical Computer Science*, 209(1–2):237–260, 1998.
2. G. Andersson, L. Engebretsen, and J. Håstad. A new way of using semidefinite programming with applications to linear equations mod p. *Journal of Algorithms*, 39(2):162–204, 2001.
3. S. Arora, L. Babai, J. Stern, and Z. Sweedyk. The hardness of approximate optima in lattices, codes, and systems of linear equations. *Journal of Computer and System Sciences*, 54(2):317–331, 1997.
4. S. Arora, C. Lund. Hardness of Approximations. In *Approximation Algorithms for NP-hard Problems*, D. Hochbaum (editor), PWS Publishing Company, 1997.
5. S. Arora, C. Lund, R. Motwani, M. Sudan, and M. Szegedy. Proof verification and the hardness of approximation problems. *J. ACM*, 45(3):501–555, 1998.
6. L. Engebretsen and V. Guruswami. Is constraint satisfaction over two variables always easy? *Proceedings of RANDOM'02*:224–238, 2002.
7. M. Goemans and D. Williamson. Improved approximation algorithms for maximum cut and satisfiability problems using semidefinite programming,. *Journal of the ACM*, 42:1115–1145, 1995.
8. J. Hastad. Some optimal inapproximability results. *Journal of ACM*, 48:798–859, 2001.
9. S. Khot. On the power of unique 2-prover 1-round games. *STOC*:767–775, 2002.
10. S. Khot and O. Regev. Vertex cover might be hard to approximate to within $2 - \epsilon$. *Proceedings of Computational Complexity*, 2003.
11. R. Raz. A parallel repetition theorem. *SIAM Journal of Computing*, 27:763–803, 1998.
12. L. Trevisan, G. B. Sorkin, M. Sudan, and D. P. Williamson. Gadgets, approximation, and linear programming. *SIAM Journal on Computing*, 29(6):2074–2097, 2000.

An Auction-Based Market Equilibrium Algorithm for the Separable Gross Substitutability Case

Rahul Garg[1], Sanjiv Kapoor[2], and Vijay Vazirani[3]

[1] IBM India Research Lab., Block-I, IIT Campus
Hauz Khas, New Delhi, India - 110016
grahul@in.ibm.com
[2] Department of Computer Science, Illinois Institute of Technology
Chicago, IL-60616
skapoor@iit.edu
[3] Georgia Institute of Technology, Atlanta, USA
vazirani@cc.gatech.edu

Abstract. Utility functions satisfying gross substitutability have been studied extensively in the economics literature [1, 11, 12] and recently, the importance of this property has been recognized in the design of combinatorial polynomial time market equilibrium algorithms [8]. This naturally raises the following question: is it possible to design a combinatorial polynomial time algorithm for this general class of utility functions? We partially answer this question by giving an algorithm for separable, differentiable, concave utility functions satisfying gross substitutes. Our algorithm uses the auction based approach of [10].
We also outline an extension of our method to the Walrasian model.

1 Introduction

The recent papers of Papadimitriou [16] and Deng, Papadimitriou and Safra [5], which raised the issue of efficient computability of market equilibria, have resulted in considerable activity on this topic. Some of the algorithms are based on solving nonlinear convex programs [14, 18], following the classical approach of Eisenberg and Gale [9], while others are combinatorial [6, 10, 13, 7, 8, 17]. The latter algorithms are dominated by two main techniques: primal-dual-type algorithms which were initiated by Devanur, Papadimitriou, Saberi and Vazirani [6], and auction-based algorithms introduced by Garg and Kapoor [10].

Both these papers deal with linear utility functions. Moreover, both start with very low prices and monotonically raise them until equilibrium prices are reached. The usefulness of such a scheme was clarified in [8]: prices play the role of dual variables in primal-dual algorithms, and almost all such algorithms known in the areas of exact and approximation algorithms work by greedily raising dual variables - arranging anything more sophisticated is extremely difficult. Furthermore, linear utility functions satisfy gross substitutability, i.e., increasing the price of one good cannot decrease the demand for another. Hence, for such utility functions, monotonically raising prices suffices. In contrast concave,

K. Jansen et al. (Eds.): APPROX and RANDOM 2004, LNCS 3122, pp. 128–138, 2004.
© Springer-Verlag Berlin Heidelberg 2004

and even piecewise-linear and concave, utility functions do not satisfy gross substitutability, and designing market equilibrium algorithms for them remains an outstanding open problem.

Linear utility functions have been generalized to spending constraint utility functions, which also satisfy gross substitutability, in [17]. Building on the algorithm of [6], a primal-dual-type polynomial time algorithm for spending constraint step utility functions has been obtained in [17]. The importance of gross substitutability has been recognized in the economics literature, and there have been attempts at designing algorithms for this general class of utility functions [1, 11]; however, to our knowledge no such polynomial time algorithms are known. In this paper, we use the auction-based approach of [10] to give polynomial time approximation algorithms for computing equilibria for the general class of additively separable, differentiable, concave utility functions satisfying gross substitutability. Our algorithm achieves a complexity of $O((E/\epsilon)\log(1/\epsilon)\log((evv_{max})/(e_{min}v_{min}))\log m)$ where E is the number of non-zero utilities u_{ij}, $v_{ij}(x_ij)$ is the slope of u_{ij} w.r.t. x_{ij} and e_i are the endowments (in terms of money). In the complexity description, $e_{min} = \min_i e_i$, $e = \sum_{i=1}^{n} e_i$,v_{max}/v_{min} is the ratio of the largest slope to the least slope (we assume that this ratio is bounded), $v = \sum_{j=1}^{m} v_{1j}(a_k)$ and ϵ is the tolerance parameter to which equilibrium prices are computed. This algorithm is faster by a factor of $O(n)$ as compared to the auction algorithm for the Arrow-Debreu model [10] and also extends the class of utility functions for which approximate market clearing can be achieved.

Interestingly, while the algorithm's framework remains as simple as in [10], the proof of correctness and convergence are complicated by the general nature of the utility functions. We are able to, in this paper, resolve the convergence to the equilibrium prices via a monotone change in prices for separable increasing functions satisfying gross-substitution and concavity. The problem of computing equilibrium in the more general class of (non-separable) increasing concave functions satisfying gross substitutability is a challenging one.

In Section 2 we define the market model and provide a characterization of gross substitutable functions. In Section 3 we outline our algorithm and prove correctness and the complexity bounds. Finally, we outline (Section 4) an extension to the Arrow-Debreu model.

2 Market Model

Consider a market consisting of a set B of n buyers and a set A of m divisible goods. Buyer i has, initially, an amount of money equal to e_i. The amount of good j available in the market is a_j. We will assume that the utilities are additive and separable. Buyer i has a utility function, $U_i(X_i) = \Sigma_j u_{ij}(x_{ij})$ where x_{ij} is the amount of allocation of good j to buyer i, $X_i = (x_{i1}, x_{i2} \ldots x_{im})$ represents the current allocation vector of goods to buyer i and $u_{ij}(x_{ij}) : R_+ \to R_+$ is the utility function of item j to buyer i. We assume that u_{ij} is non-negative, strictly increasing, differentiable, and concave in the range $[0, a_j]$. Let v_{ij} represent the first derivative of U_i w.r.t. x_{ij} (which is well defined). Assume that the buyers have no utility for money, however they use their money to purchase the goods.

Given prices $P = \{p_1, p_2, \ldots, p_m\}$ of these m goods, a buyer uses its money to purchase goods that maximize its total utility subject to its budget constraint. Thus a buyer i will choose an allocation X_i that solves the following buyer program $B_i(P)$:

$$\text{Maximize}: \sum_{1 \leq j \leq m} u_{ij}(x_{ij}) \tag{1}$$

$$\text{Subject to}: \sum_{1 \leq j \leq m} x_{ij} p_j \leq e_i \tag{2}$$

$$\forall j: \ x_{ij} \geq 0 \tag{3}$$

This defines a family of programs, each program parameterized by a fixed price. Since u_{ij} is concave for all i and j, the theory of duality (Kuhn-Tucker conditions) can be used to give the following necessary and sufficient conditions for optimality for a given price vector P:

$$\sum_{1 \leq j \leq m} x_{ij} p_j = e_i \tag{4}$$

$$\forall j: \alpha_i p_j \geq v_{ij}(x_{ij}) \tag{5}$$

$$\forall j: x_{ij} > 0 \Rightarrow \alpha_i p_j = v_{ij}(x_{ij}) \tag{6}$$

$$\alpha_i \geq 0, \forall j: x_{ij} \geq 0 \tag{7}$$

In the above conditions, α_i is the marginal utility per unit of money, also referred to as *marginal bang per buck*, for buyer i. If $x_{ij} > 0$, then at x_{ij} the marginal bang per buck for buyer i from good j ($v_{ij}(x_{ij})/p_j$) is exactly α_i. Otherwise, $v_{ij}(0)/p_j \leq \alpha_i$. We say that the pair (X, P), $X = (X_1, X_2 \ldots X_n)$ forms a market equilibrium if (a) the vector $X_i \in R_+^n$ solves the problem $B_i(P)$ for user i and (b) there is neither a surplus or a deficiency of any good i.e.,

$$\forall j: \sum_{1 \leq i \leq n} x_{ij} = a_j \tag{8}$$

The prices P are called market clearing prices and the allocation X is called the equilibrium allocation at price P.

The equations (8) and (4) imply that all the goods are sold and all the buyers have exhausted their budget. Equations (5) and (6) imply that (a) that every buyer has the same marginal utility per unit price on the goods it gets and (b) every good that a buyer is not allocated provides less marginal utility.

2.1 Gross Substitutes

Gross substitutes is a well-studied property that has useful economic interpretations. Goods are said to be gross substitutes for a buyer iff increasing the price of a good does not decrease the buyer's demand for other goods. Similarly, goods in an economy are said to be gross substitutes iff increasing the price of a good does not decrease the total demand of other goods. Clearly, if the goods are gross substitute for every buyer, they are gross substitutes in the economy.

We now give a formal definition of gross substitute in our model. Consider the buyer maximization problem $B_i(P)$. Let $S_i(P) \subset R^m_+$ be the set of optimal solutions of the program $B_i(P)$. Consider another price vector $P' > P$. Goods are gross substitutes for buyer i if and only if for all $X_i \in S_i(P)$ there exists $X'_i \in S_i(P')$ such that $p_j = p'_j \Rightarrow x_{ij} \leq x'_{ij}$.

Note that u_{ij} is continuous, concave and differentiable for all i and j and $v_{ij}(x) = \frac{d}{dx}u_{ij}(x)$. Since u_{ij} is concave, v_{ij} is a non-increasing function. The following result characterizes the class of separable concave gross substitute utility functions.

Lemma 1. *Goods are gross substitutes for buyer i if and only if for all j, $yv_{ij}(y)$ is a non-decreasing function of the scalar y.*

Proof. To prove the converse part (which is critical to the algorithm in Section 3), assume that there are scalars y and y' such that $y' < y$ and $y'v_{ij}(y') > yv_{ij}(y)$. Choose a price P and an optimal solution X_i of $B_i(P)$ such that $x_{ij} = y$. Let α_i be the optimal dual solution of $B_i(P)$. Construct a corresponding P' and X'_i such that $x'_{ik} = x_{ik}$, $p'_k = p_k$ for all $k \neq j$, $x'_{ij} = y'$ and $p'_j = p_j v_{ij}(x'_{ij})/v_{ij}(x_{ij})$. Now,

$$x'_{ij}p'_j = x'_{ij}p_j v_{ij}(x'_{ij})/v_{ij}(x_{ij})$$
$$> p_j x_{ij} v_{ij}(x_{ij})/v_{ij}(x_{ij})$$
$$= x_{ij}p_j$$

So, the solution X'_i satisfies (11) and (10) for price P', but $\sum_{j=1}^m x'_{ij}p'_j > \sum_{j=1}^m x_{ij}p_j = e_i$. Therefore, the optimal dual solution α'_i of $B_i(P')$ will satisfy $\alpha'_i > \alpha_i$. Therefore, the optimal solution X''_i of $B_i(P)$ will have $x''_{ik} < x'_{ij}$ for all $k \neq j$, such that $x_{ik} > 0$. Hence the goods will not be gross substitute for buyer i.

We next show that $yv_{ij}(y)$ is non-decreasing then the goods satisfy gross-substitutability. Consider an optimal solution $X_i \in S_i(P)$. The dual of the program $B_i(P)$ gives the following necessary and sufficient conditions for the optimality of X_i.

$$\sum_{j=1}^m x_{ij}p_j = e_i \tag{9}$$

$$\forall j : x_{ij} > 0 \Rightarrow v_{ij}(x_{ij}) = \alpha_i p_j \tag{10}$$

$$\forall j : \alpha_i p_j \geq v_{ij}(x_{ij}) \tag{11}$$

$$\alpha_i \geq 0, x_{ij} \geq 0$$

If $x_{ij} > 0$, equation (10) gives $x_{ij}p_j = x_{ij}v_{ij}/\alpha_i$. Consider $P' > P$. For this price vector we construct a feasible solution X'_i as follows: If $p'_j = p_j$ then $x'_{ij} = x_{ij}, \forall i$. Alternately, if $v_{ij}(0) < \alpha_i p'_j$ then set x'_{ij} to zero, else choose x'_{ij} such that $v_{ij}(x'_{ij}) = \alpha_i p'_j$. By definition, the solution X'_i satisfies the complementary slackness conditions (10). Since $P' > P$, X'_i also satisfies (11). Also since v_{ij} is a non-increasing function $p'_j > p_j \Rightarrow x'_{ij} \leq x_{ij}$. Now,

$$
\begin{aligned}
x'_{ij} p'_j &= x'_{ij} v_{ij}(x'_{ij})/\alpha_i \\
&\leq x_{ij} v_{ij}(x_{ij})/\alpha_i \\
&= x_{ij} p_j
\end{aligned}
$$

The above equations give

$$\sum_{j=1}^m x'_{ij} p'_j \leq \sum_{j=1}^m x_{ij} p_j = e_i$$

Note that u_{ij} is concave for all j. Therefore, there is an optimal solution X''_i of the program $B_i(P')$ such that $X''_i \geq X'_i$. ¿From the definition of X'_i if $p_j = p'_j$ then $x'_{ij} = x_{ij}$. Therefore $p_j = p'_j \Rightarrow x''_{ij} \geq x_{ij}$ where X_i is an optimal solution of $B_i(P)$ and X''_i is a corresponding optimal solution of $B_i(P')$.

3   An Auction Algorithm for Market Clearing

We now present an ascending price algorithm for discovering the market clearing prices approximately. The algorithm starts with a low price and an initial allocation x_i for all buyers i, such that all the goods are completely allocated and optimal allocation of buyers dominate their current allocation. Now the prices of *over-demanded* items are raised slowly and the current allocation is recomputed, until no item is over-demanded. This approach has a similarity with the Hungarian method of Kuhn [15] for the assignment problem. Unlike the Hungarian method which raises the price of all the goods in a *minimal over-demanded set* by a specific amount, our algorithm raises the price of one good at a time by a fixed multiplicative factor $(1+\epsilon)$, where $\epsilon > 0$ is a small quantity suitably chosen at the beginning of the algorithm. This algorithm has an auction interpretation, where traders outbid each other to acquire goods of their choice by submitting a bid that is a factor $(1+\epsilon)$ of the current winning bid. Prior to this, auction algorithms have been proposed for maximum weight matching in bipartite graphs, network flow problems and market clearing with linear utilities [4], [3], [2], [10].

We discuss the algorithm which is formally presented in Figure 1. In procedure `initialize` (Figure 1) all the items are allocated to the first buyer. Prices are initialized such that (a) the buyer's money is exhausted and (b) the buyer's allocation is optimal at the initial prices. An initial assignment of dual variables α_i is also required. Instead of α_i we maintain α_{ij}, a separate dual variable for each item, for computational ease. Finally we will relate α_i to α_{ij}.

It is easy to verify that the prices set in procedure `initialize` exhausts the budget of the first buyer. Since $v_{1j}(a_j)$ is assumed to be strictly positive[1], $p_j > 0$ for all j. Therefore the initial value of α_{ij} is well-defined for all i and j.

---

[1] This is not a necessary assumption. It is made for simplicity of the presentation. A weaker assumption would be that every item j has a buyer i such that $v_{ij}(a_j) > 0$. The initial allocation may still be found that satisfies the desired properties.

Also note that $v_{1j}(x_{ij})/p_j = \alpha_{1j} = \alpha_1$ for all j. Hence the initial allocation maximizes the utility of the first buyer.

The goods are allocated at two prices, p_j and $p_j/(1+\epsilon)$. The allocation of good j to buyer i at price p_j is represented by h_{ij} and the allocation at price $p_{ij}/(1+\epsilon)$ is represented by y_{ij}. The total allocation of good j to buyer i is given by $x_{ij} = h_{ij} + y_{ij}$. Define the surplus of a buyer i as $r_i = \sum_{j=1}^{m}(h_{ij}p_j + y_{ij}p_j/(1+\epsilon))$. Define the total surplus in the system as $r = \sum_{i=1}^{n} r_i$.

The auction algorithm main (Figure 1) begins with a buyer i who has significant surplus (more than ϵe_i) and tries to acquire items, with utility per unit price more than the current utility per unit price. It outbids other buyers by acquiring items at a higher prices. It raises the prices by a factor $(1 + \epsilon)$ if needed. This process continues till total surplus in the economy becomes sufficiently small.

The algorithm maintains the following invariants: (I1) items are fully sold, (I2) buyers do not exceed their budget, (I3, I4) after completely exhausting its surplus a buyer's utility is close to its optimal utility at the current prices, (I5) prices do not fall and (I6) total surplus money in the economy does not increase. Figure 2 lists these invariants formally.

It is easy to check that all the invariants (I1 through I6) are satisfied after the initialization (i.e. after procedure initialize has been called).

The allocation x_{ij} is modified only in procedure outbid. However, the modifications leave the sum $\sum_{1 < i < n} x_{ij}$ unchanged. Therefore the invariant I1 is satisfied throughout the algorithm.

For invariant I2, it is sufficient to show that $r_i \geq 0$ for all i. r_i is reduced only in procedure outbid. In this procedure, the variable t_2 is chosen such that r_i does not become negative and hence I2 remains satisfied.

For invariant I6, note that the only steps that change r are in procedure outbid. In these steps, r is reduced by ϵt. Hence I6 is satisfied in the algorithm. The invariant I5 is trivially satisfied

We now show that invariants I3 and I4 are satisfied by the algorithm.

Lemma 2. *During the approximate auction algorithm the invariants I3 and I4 are always satisfied.*

Proof. The invariants are true initially for all the buyers. We first show invariant I3. Note that when α_{ij} is modified in algorithm main after calling outbid, the invariant is satisfied. Since p_j never decreases, the invariant remains satisfied whenever p_j changes. When x_{ij} is reduced, $v_{ij}(x_{ij})$ increases causing a potential violation of the invariant. In this case, the inner while loop of the algorithm will be executed. We argue that when the inner loop ends $\alpha_{ij}p_j = v_{ij}(x_{ij})$ for all i, j.

To prove this, consider the time instant z when good j was acquired by buyer i at price p_j. Let a be the quantity of good j acquired by buyer i. Now, $\alpha_{ij}p_j = v_{ij}(a)$. Assume that the amount of good j currently acquired by buyer i is $b < a$. Let the current price of j be p'_j. Choose c such that $v_{ij}(c)/p'_j = v_{ij}(a)/p_j$. It is always possible to do so since $v_{ij}(b) > \alpha_{ij}p'_j \geq \alpha_{ij}p_j = v_{ij}(a)$. Since $p'_j > p_j$, $c < a$. Now, $cp'_j = cp_j v_{ij}(c)/v_{ij}(a)$. From the assumption that goods are gross substitutes and using Lemma 1 we have $cv_{ij}(c) \leq av_{ij}(a)$. Therefore $cp'_j \leq ap_j$.

```
procedure initialize
```
$$\forall i, \forall j : h_{ij} = 0$$
$$\forall i \neq 1, \forall j : y_{ij} = 0$$
$$\forall j : y_{1j} = a_j$$
$$\forall j : \alpha_{1j} = (\sum_j a_j v_{1j}(a_j))/e_i$$
$$\forall j : p_j = v_{1j}(a_j)/\alpha_1$$
$$\forall i \neq 1 : \alpha_i = v_{ij}(0)/p_j; \ r_i = e_i$$
$$\forall i \neq 1, \forall j : \alpha_{ij} = v_{ij}(x_{ij})/p_j$$
$$r_1 = 0$$
```
end procedure initialize
```

```
algorithm main
      initialize
      while ∃i : r_i > ϵe_i
            while (r_i > 0) and (∃j : α_{ij}p_j < v_{ij}(x_{ij}))
                  if ∃k : y_{kj} > 0 then outbid(i, k, j, α_{ij})
                  else raise_price(j)
            end while
            j = arg max_l α_{il}
            if ∃k : y_{kj} > 0
                  outbid(i, k, j, α_{ij}/(1 + ϵ))
                  α_{ij} = v_{ij}(x_{ij})/p_j
            else raise_price(j)
      end while
end algorithm main
```

```
procedure raise_price(j)
```
$$\forall i : y_{ij} = h_{ij}; h_{ij} = 0;$$
$$p_j = (1 + \epsilon)p_j$$
```
end procedure raise_price
```

```
procedure outbid(i, k, j, α)
```
$$t_1 = y_{kj}$$
$$t_2 = r_i/p_j$$
$$\text{if } (v_{ij}(a_j) \geq \alpha p_j) \text{ then } t_3 = a_j$$
$$\text{else } t_3 = \min \delta : v_{ij}(x_{ij} + \delta) = \alpha p_j$$
$$t = \min(t_1, t_2, t_3)$$
$$h_{ij} = h_{ij} + t$$
$$r_i = r_i - t p_j$$
$$y_{kj} = y_{kj} - t$$
$$r_k = r_k + t p_j/(1 + \epsilon)$$
```
end procedure outbid
```

Fig. 1. The auction algorithm

Therefore the amount of money needed to be spent on j to ensure $v_{ij}(c)p'_j = \alpha_{ij}$ is $p'_j c$ which is no more than $p_j a$; the amount of money spent on j before x_{ij} was reduced. Hence, when the inner loop ends $\alpha_{ij} p'_j \geq v_{ij}(x_{ij})$ for all i and j.

I1: $\forall j$: $\sum_i x_{ij} = a_j$

I2: $\forall i$: $\sum_j x_{ij} p_j \le e_i$

I3: $\forall i$: $r_i = 0 \Rightarrow \alpha_{ij} p_j \ge v_{ij}(x_{ij})$

I4: $\forall i,j$: $x_{ij} > 0 \Rightarrow (1 + \epsilon) v_{ij}(x_{ij}) \ge \alpha_{ij} p_j$

I5: $\forall j$: p_j does not fall

I6: r does not increase

Fig. 2. The invariants in the auction algorithm

The invariant I4 is satisfied after initialization. Whenever α_{ij} is changed in main $v_{ij}(x_{ij}) = \alpha_{ij} p_j$. Therefore, if x_{ij} is reduced, I4 remains satisfied. x_{ij} may be increased by a call to outbid in the inner loop. However, the parameter α_{ij} in outbid ensures that the variable t_3 is chosen such that $\alpha_{ij} p_j \le v_{ij}(x_{ij})$. Moreover, if $x_{ij} > 0$ at the exit of the inner loop, then $\alpha_{ij} p_j = v_{ij}(x_{ij})$. So, if p_j is raised by a factor $(1 + \epsilon)$, I4 will still be satisfied. If p_j rises by more than the factor $1 + \epsilon$, x_{ij} will be set to zero and when x_{ij} is increased again $\alpha_{ij} p_j = v_{ij}(x_{ij})$. So, I4 will remain satisfied in the algorithm.

The algorithm ends when the surplus with each buyer is small, i.e. $r_i \le \epsilon e_i$. At this stage the market-clearing conditions (4) and (6) are satisfied approximately and (5) and (8) are satisfied exactly.

3.1 Convergence of the Algorithm

In order to provide efficient convergence, bidding is organized in rounds. In each round every buyer (i) is picked once and reduces his surplus to 0, i.e. $r_i = 0$. The following lemma proves a bound on the surplus reduction in each round.

Lemma 3. *In every round the total unspent money $r = \sum_{i=1}^n r_i$ decreases by a factor of $(1 + \epsilon)$.*

Proof. The value of r is decreased in procedure outbid by $t\epsilon p_j$. Buyer i bids until $r_i = 0$. WLOG assume that these bids are of amounts $t_1, t_2, \ldots, t_k$ on items $1, 2, \ldots, k$ at prices $p_1, p_2, \ldots, p_k$. Now we have:

$$\sum_{l=1}^k t_l (1 + \epsilon) p_l = r_i$$

Reduction in r is given by

$$\Delta r = \sum_{l=1}^k t_l \epsilon p_l = \frac{\epsilon}{1 + \epsilon} r_i$$

Bidding by buyer i can only increase r_k for another buyer k. Therefore the total reduction in r in one round is given by:

$$\Delta r \ge \sum_{i=1}^n \frac{\epsilon}{1 + \epsilon} r_i = \frac{\epsilon}{1 + \epsilon} r$$

The new value of unspent money r' after every round is related to its old value r as: $r' = \frac{r}{1+\epsilon}$

Let $e_{min} = \min_i e_i$ and $e = \sum_{i=1}^n e_i$. If $r < \epsilon e_{min}$ then no buyer has significant money left. Therefore the algorithm is guaranteed to terminate in k rounds where $k = \log(\frac{e}{\epsilon e_{min}})/\log(1 + \epsilon)$.

Let E be the number of non-zero utilities. We first bound the number of calls to procedures raise_price and outbid.

The price of any item is bounded by:

$$\frac{v_{1j}(a_j)}{\sum_{k=1}^m v_{1k}(a_k)} \frac{e_1}{a_j} \le p_j \le \frac{e}{a_j}$$

At every step, the price is raised by a factor $(1 + \epsilon)$, therefore the total number of calls to raise_price is bounded by:

$$\frac{1}{\epsilon} \log(\frac{ev}{e_{min} v_{min}})$$

where $v = \sum_{j=1}^m v_{1j}(a_k)$, $e_{min} = \min_i e_i$ and $v_{min} = \min_{ij} v_{1k}(a_k)$. We will assume that the ratio of v and v_{min} is bounded. If not, i.e. if v_{min} is zero then we can perturb the utility function $u_{ij}(x_{ij})$ by addition of the term ϵx_{ij} such that the derivative is at least ϵ. Furthermore, the derivative is bounded above by $v_{ik}(a_k)$.

In a call to outbid, one of the following four events can occur: (a) y_{kj} becomes zero for some k, (b) r_i becomes zero, (c) α_{ij} reduces by a factor of $(1 + \epsilon)$ and (d) $v_{ij}(x_{ij})$ reaches α_{ij} in the inner while loop of algorithm main.

Events of type (a) are charged to calls to procedure raise_price and events of type (b) and (d) are charged to the current round of bidding. Events of type (c) are charged to a reduction in α_{ij}.

Note that for every price rise of item j, the number of events of type (a) is bounded by the number of buyers having non-zero utilities on item j. Thus the total number of type (a) events is bounded by E times the maximum possible number of price rises for any given item.

It is easy to see that number of type (b) events is exactly equal to n in every round of bidding. At every type (c) event, α_{ij} is reduced by a factor of $(1+\epsilon)$. Since $\alpha_{ij} = v_{ij}(x_{ij})/p_j$, its value varies from $v_{ij}(a_j)/p_{max}$ to $v_{ij}(0)/p_{min}$, where p_{min} and p_{max} are the minimum and maximum values of p_j, respectively. In every round only one event of type (d) occurs for each buyer.

During the algorithm evaluation of $\arg\max_l \alpha_{il}$ can be done efficiently by maintaining n heaps, one for each buyer. The heap of a buyer contains the goods with non-zero utilities sorted in decreasing order of α_{il}. Updates requires $O(\log n)$ steps.

This gives us the following time complexity of the algorithm:

Theorem 1. *The auction algorithm terminates in*
$O((E/\epsilon) \log((evv_{max})/(\epsilon e_{min} v_{min})) \log n)$ *steps.*

where $v_{max} = \max_{ij} v_{ij}(0)$.

4   Extension to the Arrow-Debreu Model

The algorithm for the Fisher model described in the previous section can be extended to the general Walrasian model, also called the Arrow-Debreu model, where there is no demarcation between buyers and sellers, and all agents come to the market with an initial endowment of goods.

In order to extend to this case, we first note that under the assumption that goods satisfy gross substitutability, a version of Lemma 1 still holds, i.e., $xv_{ij}(x)$ is a non-increasing function with respect to buyer i if $x > a_{ij}$.

The algorithm is modified to compute surplus available with each of the traders based on the endowment of the trader. In each phase of the algorithm, a trader with surplus is picked. Firstly, items that do not satisfy the invariance

$$\alpha_i p_j \geq v_i j(x)$$

are acquired by being outbid at the current price, failing which the price of the chosen item is raised. Gross substitutability ensures that the invariance would be satisfied when the surplus is exhausted. Alternately, if every item satisfies the invariance w.r.t. the chosen trader with surplus, the trader bids on other items which may either be available or whose price may have to be raised. The bidding for items is performed in rounds, with each trader being considered at least once in a round and exhausting his surplus, to ensure fast convergence. A detailed analysis of the efficient convergence of this bidding process is similar to that provided in the paper of Garg and Kapoor[10] for the linear case.

5   Conclusions

It would be of considerable interest to extend the class of utility functions for which the market equilibrium problem is solvable via the auction method. In particular, can this approach be extended to non-separable functions?

Acknowledgments

V. Vazirani was supported by NSF Grants CCR 0220343 and CCR 0311541.

References

1. K. Arrow, H. Block, and L.Hurwicz. On the stability of the competitive equilibrium, II. *Econometrica*, 27(1):82–109, 1959.
2. Vipul Bansal and Rahul Garg. Simultaneous Independent Online Auctions with Discrete Bid Increments. *Electronic Commerce Research Journal: Special issue on Dynamic Pricing Policies in Electronic Commerce*, To Appear.
3. Dimitri P. Bertsekas. Auction Algorithms for Network Flow Problems: A Tutorial Introduction. *Computational Optimization and Applications*, 1:7–66, 1992.
4. Gabrielle Demange, David Gale, and Marilda Sotomayor. Multi-item Auctions. *Journal of Political Economy*, 94(4):863–872, 1986.

5. X. Deng, C. Papadimitriou, and S. Safra. On the Complexity of Equilibria. In *34th ACM Symposium on Theory of Computing (STOC 2002)*, Montreal, Quebec, Canada, May 2002.
6. N. Devanur, C. Papadimitriou, A. Saberi, and V. Vazirani. Market Equilibrium via a Primal-Dual-Type Algorithm. In *43rd Symposium on Foundations of Computer Science (FOCS 2002)*, pages 389–395, November 2002.
7. N. R. Devanur and V. Vazirani. An Improved Approximation Scheme for Computing the Arrow-Debreu Prices for the Linear Case. In *Foundations of Software Technology and Theoretical Computer Science (FSTTCS 2003)*, 2003.
8. N. R. Devanur and V.V. Vazirani. The Spending Constraint Model for Market Equilibrium: Algorithmic, Existence and Uniqueness Results. In *Proceedings of the 36th Annual ACM Symposium on the Theory of Computing*, 2004.
9. E. Eisenberg and D. Gale. Consensus of Subjective Probabilities: The Pari-Mutuel Method. *Annals of Mathematical Statistics*, 30:165–168, 1959.
10. Rahul Garg and Sanjiv Kapoor. Auction Algorithms for Market Equilibrium. In *Proceedings of the 36th Annual ACM Symposium on the Theory of Computing*, 2004.
11. J. Greenberg. An elementary proof of the existence of a competitive equilibrium with weak gross substitutes. *The Quarterly Journal of Economics*, 91:513–516, 1977.
12. F. Gul and E. Stacchetti. Walrasian equilibrium with gross substitutes. *Journal of Economic Theory*, 87:95–124, 1999.
13. K. Jain, M. Mahdian, and A. Saberi. Approximating Market Equilibrium. In *Workshop on Approximation Algorithms for Combinatorial Optimization (APPROX 2003)*, 2003.
14. Kamal Jain. A Polynomial Time Algorithm for Computing the Arrow-Debreau Market equilibrium for Linear Utilities. Preprint.
15. H. W. Kuhn. The Hungarian Method for the Assignment Problem. *Naval Research Logistics Quarterly*, 2:83–97, 1955.
16. C. Papadimtriou. Algorithms, games, and the internet. In *Proceedings of the 33rd Annual ACM Symposium on the Theory of Computing*, pages 749–753, 2001.
17. V. V. Vazirani. Market Equilibrium When Buyers Have Spending Constraints . Submitted, 2003.
18. Yinyu Ye. A Path to the Arrow-Debreu Competetive Market Equilibrium. Preprint, 2004.

Cost-Sharing Mechanisms for Network Design[*]

Anupam Gupta[1], Aravind Srinivasan[2], and Éva Tardos[3]

[1] Department of Computer Science, Carnegie Mellon University, Pittsburgh PA 15232
anupamg@cs.cmu.edu
[2] Department of Computer Science and
University of Maryland Institute for Advanced Computer Studies
University of Maryland at College Park, College Park, MD 20742
srin@cs.umd.edu
[3] Department of Computer Science, Cornell University, Ithaca, NY 14853
eva@cs.cornell.edu

Abstract. We consider a single source network design problem from a game-theoretic perspective. Gupta, Kumar and Roughgarden (*Proc. 35th Annual ACM STOC*, pages 365–372, 2003) developed a simple method for single source rent-or-buy problem that also yields the best-known approximation ratio for the problem. We show how to use a variant of this method to develop an approximately budget-balanced and group strategyproof cost-sharing method for the problem.

The novelty of our approach stems from our obtaining the cost-sharing methods for the rent-or-buy problem by carefully combining cost-shares for the simpler problem Steiner tree problem; we feel that this idea may have wider implications. Our algorithm is conceptually simpler than the previous such cost-sharing method due to Pál and Tardos (*Proc. 44th Annual FOCS*, pages 584–593, 2003), and has a much improved approximation factor of 4.6 (over the previously known factor of 15).

1   Introduction

This paper studies the problem of giving good *cost-sharing mechanisms* for a single source network design problem. Imagine a general network design problem, where the participants (or *agents*) want to build a network connecting them to a common source (a server); however, they are autonomous and behave in a selfish (but non-malicious) fashion. Informally, a cost-sharing mechanism builds a network, and allocates the cost incurred among the agents, so that no group of agents is charged too much, thus precluding the possibility of their being unhappy and trying to secede from the system.

The type of problem we consider is where we are given an undirected graph $G = (V, E)$, and a set of *demands* $D \subseteq V$ that want to connect to a common

---

[*] Research done in part during the IMA workshop on *Network Management and Design* at the University of Minnesota, April 2003. The research of the second author was supported in part by the National Science Foundation under Grant No. 0208005, and that of the third author in part by ONR grant N00014-98-1-0589, and NSF grants CCR-0311333 and CCR-0325453.

K. Jansen et al. (Eds.): APPROX and RANDOM 2004, LNCS 3122, pp. 139–150, 2004.

source r. A cost-sharing method is an algorithm that builds a cheap network, and also specifies what portion of its cost is paid by which of the participants in the network; this is done in a manner that ensures that the cost paid by any subset of the participants is fair.

Given an instance of a network design game specified by a set of demands D, a cost-sharing scheme for the problem is simply a function $\xi(D, i)$ with $\xi(D, i) = 0$ for $i \notin D$. For a set D of demands we use $\mathsf{OPT}(D)$ to denote the minimum cost network serving the users in D, while $\mathsf{Alg}(D)$ denotes the cost, or expected cost, of the network computed by a given algorithm. There are three properties that we will be concerned with:

1. (β-**approximate budget-balance**) For a set of demands D, we require that
$$\mathsf{OPT}(D) \geq \sum_{i \in D} \xi(D, i) \geq \mathsf{Alg}(D)/\beta$$
for some given parameter $\beta \geq 1$. Equivalently, by multiplying the shares up by β, we could require that the cost-shares are at least the total cost of a solution found, but do not exceed $\beta\, \mathsf{OPT}(D)$. If $\beta = 1$, we call the cost-sharing budget-balanced.

2. (**fairness**) For any $A \subseteq D$, $\sum_{i \in A} \xi(D, i) \leq \mathsf{OPT}(A)$. I.e., the cost paid by any subset of people should not exceed the optimal cost of connecting them alone and hence they have no incentive to secede.

3. (**cross-monotonicity**) For any $A \subseteq D$, and demand $i \in A$ we require that $\xi(D, i) \leq \xi(A, i)$. I.e., the cost-share of any demand should not go up due to other demands entering the system. This property is also known under the name of population-monotone.

Cross-monotonicity is a key ingredient used in solving the following type of mechanism design problems: consider the network design problem with a set of demand nodes D, with each user (or demand) d having an associated utility u_d. Since the users have limited utilities, the service provider has to now decide which subset of customers it must serve, in addition to designing the network and deciding how to share the cost between the served customers. A mechanism for solving this problem is called *group strategyproof* if no subset of users has an incentive to deviate from the protocol (e.g., by misreporting their utility) in the hope of improving the outcome for themselves (e.g., receiving the service at a cheaper cost). Moulin and Shenker [6] show that having a cross-monotone cost-sharing method for a problem naturally gives rise to a group strategyproof mechanism for the problem in the following way. We start with all the customers; if there is some customer whose cost share (computed w.r.t. the current set of customers) exceeds its utility, we drop it from the set, recompute cost shares and repeat. At the end, we are left with the desired set of customers and their cost-shares.

As an example of cross-monotonicity, let us consider the MINIMUM SPANNING TREE game on the complete graph $G = (V, E)$ with edge weights, given by a set of players D and a *root* $r \notin D$; the objective is to find the cheapest tree $\mathsf{MST}(D)$ spanning D and r. It must be emphasized that this game does not allow the use of Steiner vertices in $\mathsf{MST}(D)$, and r is not a player, and hence should have no cost-share. It is not difficult to verify that a budget-balanced and fair cost-sharing scheme for this game can be found thus: find an MST, root it at r, and set the cost-share for vertex $i \in D$ to be the cost of the edge from i to its parent. However, this scheme is not cross-monotone, and getting a cross-monotone cost-sharing scheme for the minimum spanning tree problem takes more work. Kent and Skorin-Kapov [5] and Jain and Vazirani [7] developed such budget-balanced and cross-monotone cost-sharing schemes for the spanning tree game using a directed branching game; let us denote this cost-sharing scheme by ξ_{MST}. Note that the values $\frac{1}{2}\xi_{MST}$ serve also as cross-monotone, 2-approximately budget-balanced cost-sharing for the corresponding Steiner tree game.

The Single Source Rent-or-Buy Game. In this paper we will consider the *Single Source Rent-or-Buy* Network Design game; this combines features of Steiner tree and shortest paths. The game is defined as follows: we are given a complete undirected graph $G = (V, E)$ with edge costs c_e satisfying the triangle inequality, a special *source* (or *root*) vertex r, and a parameter $M \geq 1$. There are also a set of *users* (also called *players* or *demands*) $D \subseteq V$, each of which is identified with a vertex of the graph. We assume that there is a unique player at each vertex. Our results can be easily extended to multiple players, each with a *weight* d_j indicating its amount of demand; we leave the details for the final version of the paper.

The objective is to connect each player j to the source r via some path P_j, on which one unit of bandwidth has been allocated. What makes the game interesting is that aggregating paths is beneficial, in tune with the idea of economies of scale. Hence there are two different actions that can be performed by each edge: either the edge can be *bought* at cost Mc_e, but then an arbitrary amount of bandwidth can be sent on that edge (and hence an arbitrary number of paths P_j can use it); or bandwidth on the edge can be *rented* at cost c_e per unit bandwidth (and hence if the paths for some set S of players were using the edge, then the edge would cost $c_e \times |S|$). As usual, any edge can be used for any of the users. Our main theorem is the following:

Theorem 1. *There is a cross-monotone, fair cost-sharing method for the Single Source Rent-or-Buy network design game that is also β-budget-balanced, where $\beta = 4.6$. Furthermore, these cost-shares can be computed in deterministic polynomial time.*

This improves on the results of Pál and Tardos [9], who gave a 15-approximate cross-monotone cost-sharing scheme for the problem. They use a primal-dual algorithm to build the network and obtain cost-shares.

We construct the claimed cost-shares based on expected costs in the approximation algorithm of Gupta et al. [4]. Loosely speaking, that algorithm works by randomly reducing the network design problem to computing Steiner trees over subsets of terminals; we show that if we use the function $\frac{1}{2}\xi_{MST}$ to allocate approximate cost-shares to each vertex in this resulting Steiner game, then the expected cost share of a vertex (taken over the random reduction) gives us the cross-monotone cost-shares claimed in Therorem 1. However, since the variances of the random variables involved may be large in general, computing these expected costs in polynomial time is non-trivial. To this end, we give an alternate analysis of the algorithm of [4]; this allows us to give a derandomization using a small sample space, which has some additional useful properties that help ensure cross-monotonicity.

Note: Independently of our result, Leonardi and Schafer (personal communication) have also suggested deriving cross-monotone cost-shares from SimpleCFL; however, lacking the derandomization of the algorithm, they are not able to compute these cost-shares in polynomial time.

2   The Algorithm and the Cost-Sharing

We will use an algorithm SimpleCFL suggested by Gupta et al. [4] for the equivalent problem of connected facility location (without facility costs). First, let us recall the algorithm from [4]. Here $\alpha > 0$ is a constant that will be chosen later.

S1. *Mark* each demand j independently with probability α/M, and let D' be the set of marked demands.
S2. Construct a minimum spanning tree T on $F = D' \cup \{r\}$. (This is a 2-approximate Steiner tree on F.) The elements of F are called the *open facilities*.
S3. Assign each demand $j \in D$ to its closest demand $i(j)$ in F.

The algorithm suggests a simple and intuitive idea for the cost-shares: each player pays a cost proportional to the *expected* cost incurred by it on running the above algorithm. For a given a set of coin-tosses, the players in D' will pay for buying the MST, where their shares are derived from a cross-monotone cost-sharing scheme ξ_{MST} for the MST problem given in [5, 7]. All other players (players in $D \setminus D'$) will pay for their shortest paths to F. The cost-shares defined for a particular set of coin-tosses exactly cover the cost of the solution built. From [4] we know that this algorithm is a β-approximation algorithm for some constant β. To get a β-approximate cost-shares we divide the above defined shares by a β. Formally:

$$\xi(D, j) = \tfrac{1}{\beta} \, \mathrm{E}[\, M \, \xi_{MST}(F, j) + \ell(j, F) \,], \qquad (2.1)$$

where $\ell(j, S)$ is the length of a shortest-path from j to the closest vertex of the set S, and the expectation is over the coin tosses. (Note that the set $F = D' \cup \{r\}$ depends on the coin tosses.)

Outline of the Proof: There are two parts to proving the result: we first have to show that the properties of the cost-shares claimed in Theorem 1 are indeed satisfied; i.e., they are β-budget-balanced, fair, and cross-monotone. These properties are proved in Section 3.

The technically involved part of the proof involves showing that our cost-shares can be computed in (deterministic) polynomial time. A little thought will convince the reader that this is not obvious, even with the results of Gupta et al. [4]. Indeed, we need to estimate the expectations in (2.1) for each of the players, but the random variables in (2.1) do not have small variance in general. Furthermore, it is unclear how to derandomize the proof of [4], since it relies on some severe conditioning. To take care of this problem, we give a proof of the performance guarantee of the SimpleCFL algorithm different from the one given by [4]; this is done in Section 4. Our new proof will yield a somewhat worse constant than that of [4], but allow us to derandomize the algorithm by marking the demands in Step (S1) in t-wise independent fashion for a constant t; this appears in Section 5.

Using t-wise independent random choices in Step (S1), allows us to use a polynomial-sized sample space, letting us compute expectations such as those of (2.1) in (deterministic) polynomial time, by considering all the points in the sample space. However, we need to make sure that the properties of the expectation proved in Section 3 (fair and cross-monotone) also hold for the expectation using t-wise independent random choices. Interestingly, the properties of a *particular* construction of t-wise independent random variables turns out to be crucial, as described in the proof of Theorem 4.

3   Properties of the Cost-Sharing Scheme

Recall that given an instance of the game specified by a set of demands D, a *cost-sharing scheme* for the game is simply a function $\xi(D, i)$ (with $\xi(D, i) = 0$ for $i \notin D$). We now need to show that the function defined in (2.1) is a cost-sharing scheme with the properties we care about, namely, approximate budget-balance, fairness, and cross-monotonicity. These three properties are, in fact, not independent. A cross-monotone and approximately budget balanced cost-sharing is always fair.

Lemma 1. *If a cost-sharing function ξ is approximately budget balanced and cross-monotone then it is also fair.*

Proof. Consider a subset $A \subseteq D$. By cross-monotonicity, we get $\sum_{i \in A} \xi(D, i) \le \sum_{i \in A} \xi(A, i)$. The approximate budget balance property gives us $\sum_{i \in A} \xi(A, i) \le$ OPT(A), and completes the proof of the lemma.

We will need the following facts about the cost-sharing scheme ξ_{MST} :

Theorem 2 ([5, 7]). *There exists an efficiently computable cost-sharing scheme ξ_{MST} for the Minimum Spanning Tree game, that is budget-balanced, fair and cross-monotone.*

Since the spanning tree connects D and r, we use the notation $\xi_{MST}(D \cup \{r\}, i)$ and $\xi_{MST}(D, i)$ interchangeably; however, we always ensure that $\xi_{MST}(D \cup \{r\}, r) = 0$. To prove approximate budget-balance, we need the following result bounding the performance of the algorithm SimpleCFL; its proof appears in Sections 4 and 5.

Theorem 3. *The expected cost of the algorithm* SimpleCFL *on a set of demands*
D *is at most* $\beta\,\mathrm{OPT}(D)$ *(where* $\beta = 4.6$*), even if the demands are marked in a*
t-wise independent fashion in Step (S1), for a suitably large constant t.

Note that this is a stronger form of the performance guarantee proved in [4]
(albeit with a worse constant), since it requires only constant-wise independence
of the random variables. Armed with this result, we are now in a position to prove
the properties of the function ξ defined in (2.1), and thus to prove Theorem 1.
The above theorems give us that ξ is β-approximately budget-balanced: the
details are deferred to the final version of the paper.

Lemma 2. *The function* ξ *is a* β*-approximately budget-balanced.*

Details of a Limited-Independence Marking: In order to prove that the expectation using t-wise independent random marking is cross-monotone, we need
a particular choice of random marking. Let $\mathbb{F}$ be a field of size $\geq n$, with $|\mathbb{F}|$ chosen
large enough so that $\lceil\alpha|\mathbb{F}|/M\rceil/|\mathbb{F}|$ is sufficiently close to α/M; for convenience,
we will assume that this fraction equals α/M. (This approximation impacts our
analysis in a negligible manner, as will be shown in the full version.) Let the
elements of the field be $\{a_1, a_2, ..., a_{|\mathbb{F}|}\}$, and let the vertices V of the graph be
labeled by the first n elements $\{a_1, a_2, \ldots, a_n\}$. Let S be any pre-specified sub-
set of $\mathbb{F}$ with $|S| = \lceil\alpha|\mathbb{F}|/M\rceil$. To get a sample, generate $\omega = (x_0, x_1, \ldots, x_{t-1})$
uniformly at random from $\mathbb{F}^t$, and define Y_i to be 1 if $\sum_{j=0}^{t-1} x_j a_i^j$ lies in S, and
$Y_i = 0$ otherwise. By construction, we have $\Pr[Y_i = 1] \sim \alpha/M$; furthermore, it
is well known that the Y_i generated thus are all t-wise independent. Note that
the above distribution is generating n coin tosses Y_i, while we need only $|D|$ of
them; however, we can just ignore all the Y_j for $j \notin D$.

Theorem 4. *Assuming that the random marking is done using the t-wise in-*
dependent random variables explained above, or using independent random vari-
ables, then the function ξ *is cross-monotone, i.e.,* $\xi(A, i) \leq \xi(D, i)$ *for any*
$A \subseteq D$.

Proof. The particular type of t-wise independent distribution that we use has
the following crucial property: for any choice of $\omega \in \mathbb{F}^t$, if the set of marked
demands in the run of SimpleCFL on A is $A'(\omega)$ and the set of marked demands
in the run on D is $D'(\omega)$, then $A'(\omega) \subseteq D'(\omega)$. Define the joint probability
$p(F, E)$ to be the probability of selecting an element of the sample space ω such
that the marked demands in the run of SimpleCFL on A is F and on D it is E.
Note that if $p(F, E) > 0$, then $F \subseteq E$. A joint probability with this property can
also be defined for the case when the marking is fully independent. With this
additional property, we can prove that the scheme ξ is cross-monotone. To see
this, note that

$$\xi(A, i) = \tfrac{1}{\beta} \sum_{F,E} p(F, E)[M\,\xi_{MST}(F, i) + \ell(i, F)]$$
$$\geq \tfrac{1}{\beta} \sum_{F,E} p(E, F)[M\,\xi_{MST}(E, i) + \ell(i, E)] = \xi(D, i),$$

where the inequality uses the fact that the support of $p(F, E)$ is only on pairs
with $F \subseteq E$, and the cross-monotonicity of ξ_{MST}.

4   A New Analysis of the Performance of SimpleCFL

Suppose first that the marking is done fully independently in Step (S1); we now give a way of analyzing SimpleCFL that is different from the one of [4]. We will then use this proof in Section 5 to show that our estimate of β changes negligible when we conduct a t-wise independent marking fro a high enough value of t. Recall that a solution to the Connected Facility Location Problem with demands D is given by the facilities F and the Steiner tree on them, with the cost being $\sum_{j \in D} \ell(j, i(j)) + M c(T)$. Of the two terms in the sum, the former cost is referred to as the *connection* cost, and the latter is the *Steiner* cost. Let OPT be an optimal solution with facilities F^* and tree T^*, and let C^* and $S^* = M c(T^*)$ be the connection and Steiner costs in OPT; also let $Z^* = C^* + S^*$.

The bound for the Steiner cost C is the same as in [4]: we consider the optimal solution OPT, and compute the expected cost assuming that we construct the Steiner tree in Step (S2) using the paths of the optimal solution; the proof is omitted here.

Lemma 3. *The expected cost of Step (S2) of* SimpleCFL *is at most*

$$S \doteq \mathrm{E}[\, M c(T)\,] \leq 2\,(S^* + \alpha C^*). \tag{4.2}$$

Massaging the Optimal Solution: Before we proceed to bound the connection costs, let us modify OPT to get another solution OPT′, which costs more than OPT, but which has some more structure that allows us to complete the argument. The new solution has the following properties:

P1. OPT′ opens facilities at $F' \subseteq F^*$, and these are connected by a *cycle* T' instead of a tree T^*; each demand j is now connected to the facility $i'(j)$, which is not necessarily its closest facility.

P2. The number of demands assigned to a facility in F', except perhaps the root r, is a multiple of M. (We will call this the *mod-M* constraint).

Lemma 4. *There is a solution* OPT′ *with the properties above, which has connection cost $C' \leq C^* + S^*$ and Steiner cost $S' \leq 2S^*$. Hence, the total cost of* OPT′ *is $Z' \leq C^* + 3S^*$.*

Proof. (Sketch) Modifying the assignment of demands to nodes in F^* to satisfy the mod-M constraint is fairly simple. Given the tree T^*, we process this in a two-pass fashion. The first pass goes bottom-up, starting at the leaves which must all be in F^*, and making sure that each node satisfies the mod-M property by sending some $b < M$ demand from the node to its parent in the tree. OPT has nodes that are not in F^*, and hence cannot be facilities in F'. We eliminate these nodes during a top-down pass, where we distribute their demand among their children.

Back to Bounding the Connection Cost: From now on, we shall only consider the modified solution OPT′. For simplicity, by making copies of nodes if necessary, we can assume that each node in F' has been assigned M demands. Recall

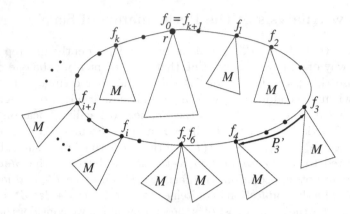

Fig. 1. The transformed instance OPT"

that T' is a cycle which contains r. (See Figure 1.) Starting from r, let us name the facilities on T' in (say) clockwise order $r = f_0, f_1, \ldots, f_k, f_{k+1} = r$. (Hence referring to f_l makes sense for all l, since we can just consider $l \bmod (k+1)$.)

Let D_l be the set of demands assigned to f_l, and hence $|D_l| = M$ for $l \neq 0$ by assumption (P2). Let us abuse notation and add r to D_0. Let P'_l be the portion of the cycle T' joining f_l and f_{l+1}, and let $c(P'_l)$ denote the length of P'_l, and hence $S' = M \sum_{l=0}^k c(P'_l)$. Let C'_l be the total connection cost of the demands in D_l in the solution OPT' (with $C' = \sum_{l=0}^k C'_l$). Our algorithm chooses, in Step (S3), the cheapest assignment of demands to nodes in F. Thus, to bound the expected connection cost, it suffices to bound the expected connection cost of an arbitrary experiment that assigns demands to nodes in F. We present one such experiment next and analyze it; an improved experiment will be presented in the final version of the paper.

The Candidate Assignment: We now consider the following experiment: if $D_l \cap F \neq \emptyset$, then we assign *all* the demands in D_l to the element in $D_l \cap F$ that is closest to f_l. Else we have to assign these demands to some other open facility (i.e., "go outside D_l"). In this case, consider the smallest t such that $D_{l+t} \cap F \neq \emptyset$; let $s = l + t$. Note that $r \in D_0$, and so $t \leq k$. We now send all the demands in D_l to the facility in D_s closest to f_s. If we assign demands in D_l to a marked node in D_{l+t}, then a path for each demand in D_l goes through $f_l, P'_l, P'_{l+1}, \ldots, P'_{l+t-1}$, and then from f_{l+t} to the element in $D_{l+t} \cap F$ closest to it. We bound the expected cost of these paths, which in turn bounds the expected connection cost of SimpleCFL.

Let X_i be the indicator variable that $D_i \cap F = \emptyset$ in our algorithm. (Note that $X_0 = 0$ with probability 1.) Let A_i be the distance from f_i and the closest element of $D_i \cap F$. (If this is empty, A_i can be set to 0.) By the above arguments, the assignment cost of the M demands in D_l is at most

$$C'_l + M \sum_{i=l}^k (X_l \cdots X_i) c(P'_i) + M \sum_{i=l}^k (X_l \cdots X_{i-1}) (1 - X_i) A_i \qquad (4.3)$$

Indeed, the first term is the distance traveled by the demands in D_l to reach f_l, the second term expresses the fact that demands use P_i (and pay $M\,c(P_i')$) if and only if $D_l, \ldots, D_i$ all do not intersect F, and the third term implies that we incur a cost of $M\,A_i$ if we assign demands in D_l to the closest member of $D_i \cap F$.

Note that X_i and X_j are independent for $i \neq j$, and that $\mathrm{E}[X_i] = (1 - \alpha/M)^M = \mathbf{q}$. To bound $M \cdot \mathrm{E}[(1 - X_i)A_i]$, let the distances from f_i to the members of D_i be $a_1 \le a_2 \le \ldots \le a_M$ with $\sum_j a_j = C_i'$. Now,

$$\mathrm{E}[(1 - X_i)A_i] = \sum_{j=1}^{M} a_j \times \alpha/M \times (1 - \alpha/M)^{j-1}.$$

Note that the coefficients of a_j decrease as j increases, and hence subject to the constraints above, the expectation is maximized when all the a_j's are equal to C_i'/M. This yields $\mathrm{E}[(1 - X_i)A_i] \le (C_i'/M)[1 - (1 - \alpha/M)^M] = (C_i'/M)(1 - \mathbf{q})$. Let C_l be the expected connection cost of the demands in D_l in our current experiment. Combining the above-seen facts with the inequality (4.3), we get that for $l > 0$,

$$C_l \le C_l' + M \sum_{i=l}^{k} c(P_i')\,\mathbf{q}^{i-l+1} + \sum_{i=l}^{k} C_i'\,(1 - \mathbf{q})\,\mathbf{q}^{i-l}. \tag{4.4}$$

Note that $C_0 = C_0'$; adding this to the sum of (4.4) over all l, the total expected connection cost is

$$C \le \sum_{l=0}^{k} C_l' + M \sum_{i=1}^{k} c(P_i') \sum_{l=1}^{i} \mathbf{q}^{i-l+1} + \sum_{i=1}^{k} C_i'\,(1 - \mathbf{q}) \sum_{l=1}^{i} \mathbf{q}^{i-l} \tag{4.5}$$

$$\le C' + M \sum_{i=1}^{k} c(P_i')\frac{\mathbf{q}}{1-\mathbf{q}} + \sum_{i=1}^{k} C_i'\,(1 - \mathbf{q})\frac{1}{1-\mathbf{q}} \le 2\,C' + \frac{\mathbf{q}}{1-\mathbf{q}}\,S' \tag{4.6}$$

However, $\mathbf{q} \le 1/e^\alpha$, and hence the second term is at most $1/(e^\alpha - 1)$. Now using Lemma 4 to replace C' by $C^* + S^*$ and S' by $2S^*$, we see that the expected connection cost of our algorithm can be bounded by $C \le 2\,C^* + \frac{2\,e^\alpha}{e^\alpha - 1}\,S^*$. Adding with (4.2), the total expected cost is at most $C + S \le 2(1 + \alpha)C^* + (2 + \frac{2e^\alpha}{e^\alpha - 1}) \cdot S^*$. Since the optimal cost is $Z^* = C^* + S^*$, we choose $\alpha \approx 1.35$ to minimize the approximation ratio, and get the following result.

Theorem 5. SimpleCFL *is a* $\beta = 4.7$-*approximation algorithm for* CFL.

We will present an improved bound of a 4.6 approximation in the final version of the paper; this improved algorithm can also be derandomized in a fashion similar to that given in Section 5. If we were to use an 1.55-approximation for Steiner tree [10] to buy edges in Step (S2), we would be getting an improved approximation ratio of 4.2 for CFL, while the analysis of [4] gets a 3.55 approximation. However, we need to use the 2-approximation algorithm of the MST, as there is a cross-monotone cost-sharing function ξ_{MST} for the MST problem.

5   Analysis of the Limited-Independence Marking

Let ϵ be an arbitrary positive constant lying in $(0, 1)$. We now prove that if the demands are marked in t-wise independent fashion where $t = a \log(1/\epsilon)$ for a suitably large constant a, then the expected approximation ratio is at most $(1 + \epsilon)$ times what it is with independent marking.

Recall that the total connection cost is a random variable that is the sum of three quantities: (i) the deterministic value $\sum_l C'_l$, which represents the cost of all demands in each D_l first getting routed to f_l; (ii) the value that corresponds to unsuccessfully traveling through some D_l, and (iii) the total cost paid in traveling from f_i to the closest marked demand in D_i, once D_i is identified as the closest cluster. We will only show that the expected total cost of (iii) gets multiplied by at most $(1+\epsilon)$ due to our $t = a \log(1/\epsilon)$-wise independent marking; the argument for the term (ii) is analogous, and, in fact, simpler. Specifically, we will show the following. Let i be the index of some arbitrary but fixed D_i. We show that the expected value of the random variable

$$\phi \doteq \sum_{j=1}^{i} M X_j X_{j+1} \cdots X_{i-1} \cdot (1 - X_i) A_i.$$

gets multiplied by at most $(1 + \epsilon)$. (Since i is fixed, we have not subscripted ϕ as ϕ_i; this remark also holds for many other definitions below.) We next present some notation.

- For each group D_j, let $Z_{j,1}, Z_{j,2}, \ldots, Z_{j,M}$ be the respective indicator random variables for the demands in D_j getting marked by our algorithm, when these demands are considered in *nondecreasing order of distance from* f_j. (This ordering will be important only in the case where $j = i$.)
- Let A be any set of ordered pairs $\{(j, k)\}$. Then, $N(A)$ is the indicator random variable for the event "for all $(j, k) \in A$, $Z_{j,k} = 0$". Also, $S(A)$ denotes $\sum_{(j,k) \in A} Z_{j,k}$. ("$N$" stands for none, and "$S$" for sum.)

The following lemma distills some results known from [1, 2, 11], to give two types of upper bounds on $\mathrm{E}[N(A)]$.

Lemma 5. *Let t_1 and t_2 be any **even** positive integers such that $t_1, t_2 \leq t$ (recall that the marking is done in a t-wise independent manner). The following hold for any set A of ordered pairs $\{(j, k)\}$.*

(i) *Let $IE(s, A)$ be the random variable denoting the inclusion-exclusion expansion of $N(A)$ truncated at the sth level; i.e.,*

$$IE(s, A) = \sum_{r=0}^{s} (-1)^r \sum_{A' \subseteq A: \, |A'|=r} \prod_{(j,k) \in A'} Z_{j,k}.$$

Then, the inequality $N(A) \leq IE(t_1, A)$ holds always; also,

$$\mathrm{E}[IE(t_1, A)] \leq (1 - 1/M)^{|A|} + \left(\frac{e|A|}{Mt_1}\right)^{t_1}. \tag{5.7}$$

(That is, $\mathrm{E}[IE(t_1, A)]$ is at most $\left(\frac{e|A|}{Mt_1}\right)^{t_1}$ more than what it would be if the marking was done completely independently.)

(ii) *Let $NCM(t_2, A)$ denote the "normalized central moment" $\frac{(S(A)-|A|/M)^{t_2}}{(|A|/M)^{t_2}}$. Then, the inequality $N(A) \leq NCM(t_2, A)$ holds always; also,*

if $t_2 \leq |A|/M$, then $\mathrm{E}[NCM(t_2, A)] \leq 8 \cdot (2t_2)^{t_2/2} \cdot (|A|/M)^{-t_2/2}$. (5.8)

Proof. (Sketch) The upper bounds that are claimed to hold always on $N(A)$ in (i) and (ii), easily follow from the fact that t_1 and t_2 are even. Bound (5.7) follows from [2]; bound (5.8) follows from [1] (see also [11] for similar bounds).

We now show how to use Lemma 5 to upper-bound $E[\phi]$. Let $a_1, a_2, \ldots, a_M$ be the distances of the demands in D_i from f_i, written in *nondecreasing* order. Then, expanding the "$(1 - X_i)A_i$" part of ϕ, we see that

$$\phi = \sum_{j=1}^{i} M X_j X_{j+1} \cdots X_{i-1} \cdot [\sum_{k=1}^{M} a_k Z_{i,k} \cdot \prod_{\ell=1}^{k-1} (1 - Z_{i,\ell})].$$

Fix k arbitrarily, and let $z_k \doteq Z_{i,k} \cdot \prod_{\ell=1}^{k-1}(1 - Z_{i,\ell})] \cdot \sum_{j=1}^{i} X_j X_{j+1} \cdots X_{i-1}$. We aim to show that $E[z_k]$ is multiplied by at most $(1+\epsilon)$ in our t-wise independent marking, as compared to the fully-independent marking. For $j = 0, 1, \ldots, i-1$, define $A_j = \{(r, s) : (i-j) \leq r \leq (i-1), 1 \leq s \leq M\}$, and $B_{i,k} = \{(i, \ell) : 1 \leq \ell \leq k-1\}$. Thus we get

$$z_k = Z_{i,k} \cdot N(B_{i,k}) \cdot \sum_{j=0}^{i-1} N(A_j).$$

Now let a_1 be a sufficiently large constant. Define a_2 to be the smallest even integer that is at least $2e \log(1/\epsilon)$, α to be $a_1 \log(1/\epsilon)$, t_1 to be the smallest even integer that is at least $2ea_1 \log(1/\epsilon)$, and t_2 to be the largest even integer that is at most $a_1 \log(1/\epsilon)/4$. Rewrite z_k as

$$Z_{i,k} \cdot N(B_{i,k}) \cdot \sum_{j \leq \alpha} N(A_j) + Z_{i,k} \cdot N(B_{i,k}) \cdot \sum_{j > \alpha} N(A_j).$$

Thus, by Lemma 5, z_k is always bounded by the sum of the following two random variables:

$$Z_{i,k} \cdot IE(a_2, B_{i,k}) \cdot \sum_{j \leq \alpha} IE(t_1, A_j), \text{ and}$$
$$Z_{i,k} \cdot IE(a_2, B_{i,k}) \cdot \sum_{j > \alpha} NCM(t_2, A_j).$$

If we expand the expectations of these two random variables using the linearity of expectation, we get terms each of which is a product of some of the random variables $Z_{\cdot,\cdot}$; importantly, the number of factors in each such term is at most $1 + a_2 + t_1$ and $1 + a_2 + t_2$ respectively. Thus, if we choose $t = 1 + a_2 + t_1$ (recall that $t_1 \geq t_2$), then the expectations of these two random variables become

$$E[Z_{i,k}] \cdot E[IE(a_2, B_{i,k})] \cdot \sum_{j \leq \alpha} E[IE(t_1, A_j)], \text{ and} \tag{5.9}$$
$$E[Z_{i,k}] \cdot E[IE(a_2, B_{i,k})] \cdot \sum_{j > \alpha} E[NCM(t_2, A_j)] \tag{5.10}$$

respectively. We next use (5.7) and (5.8) to bound these values; we will see that choosing the constant a_1 large enough results in $E[z_k]$ being at most $(1 + \epsilon)$ times what it is with independent marking. The expression (5.9) is at most

$$\frac{1}{M} \cdot \left((1 - \frac{1}{M})^{k-1} + \left(\frac{e(k-1)}{Ma_2}\right)^{a_2}\right) \cdot \sum_{j \leq \alpha} \left((1 - \frac{1}{M})^{Mj} + \left(\frac{ej}{t_1}\right)^{t_1}\right)$$
$$\leq \frac{1}{M} \cdot ((1 - \frac{1}{M})^{k-1} + \left(\frac{e}{a_2}\right)^{a_2}) \cdot \sum_{j \leq \alpha}((1 - \frac{1}{M})^{Mj} + \left(\frac{ea_1 \log(1/\epsilon)}{t_1}\right)^{t_1}).$$

Similarly, the expression (5.10) is at most $\frac{1}{M} \cdot ((1 - 1/M)^{k-1} + (e/a_2)^{a_2}) \cdot \sum_{j>\alpha}(8 \cdot (2t_2/j)^{t_2/2})$. Thus, if the demands are marked in t-wise independent fashion, then $\mathrm{E}\,[z_k]$ is at most $\frac{1}{M} \cdot ((1 - 1/M)^{k-1} + (e/a_2)^{a_2})$ times

$$\sum_{0 \le j \le \alpha} \left((1 - 1/M)^{Mj} + (ea_1 \log(1/\epsilon)/t_1)^{t_1}\right) + \sum_{j>\alpha}(8 \cdot (2t_2/j)^{t_2/2}).$$

On the other hand, under fully-independent marking, $\mathrm{E}\,[z_k] = \frac{1}{M} \cdot (1-1/M)^{k-1} \cdot \sum_{j \ge 0}(1-1/M)^{Mj}$. Recalling the definitions of a_1, a_2, t_1 and t_2, it is easy to verify that if a_1 is chosen as a sufficiently large constant, then the former value is at most $(1 + \epsilon)$ times the latter.

Further Extensions: We can improve on Theorem 5 using the following experiment: for a D_l with no open facilities, instead of going clockwise around the cycle in OPT', we find the smallest t such that D_{l+t} or D_{l-t} has an open facility, and assign all demands in D_l to this facility. This gives us a 4.6-approximation. For the model of connected facility location *with facility opening costs*, we can also get constant-approximate cross-monotone cost shares by an analysis similar to that given here. (The proofs are deferred to the final version of the paper.)

References

1. Mihir Bellare and John Rompel. Randomness-efficient oblivious sampling. In *Proc. 35th FOCS*, pp. 276–287, 1994.
2. Guy Even, Oded Goldreich, Michael Luby, Noam Nisan, and Boban Veličković. Approximations of general independent distributions. In *Proc. 24th STOC*, pp. 10–16, 1992.
3. Anupam Gupta, Amit Kumar, Jon Kleinberg, Rajeev Rastogi, and Bülent Yener. Provisioning a Virtual Private Network: A network design problem for multicommodity flow. In *Proc. 33rd STOC*, pp. 389–398, 2001.
4. Anupam Gupta, Amit Kumar, and Tim Roughgarden. Simpler and better approximation algorithms for network design. In *35th STOC*, pp. 365–372, 2003.
5. K. J. Kent and D. Skorin-Kapov. Population monotonic cost allocations on MSTs. In *Proceedings of the 6th International Conference on Operational Research (Rovinj, 1996)*, pp. 43–48. 1996.
6. H. Moulin, S. Shenker. Strategyproof sharing of submodular costs: budget balance versus efficiency. *Economic Theory* **18**:511-533, 2001.
7. Kamal Jain and Vijay Vazirani. Applications of approximation algorithms to cooperative games. In *Proc. 33rd STOC*, pp. 364–372, 2001.
8. David R. Karger and Maria Minkoff. Building Steiner trees with incomplete global knowledge. In *Proc. 41th FOCS*, pp. 613–623, 2000.
9. Martin Pál and Éva Tardos. Group Strategyproof Mechanisms via Primal-Dual Algorithms. In *Proc. 44th FOCS*, pp. 584-593, 2003.
10. Gabriel Robins and Alexander Zelikovsky. Improved Steiner tree approximation in graphs. In *Proc. 11th SODA*, pp. 770-779, 2000.
11. Jeanette P. Schmidt, Alan Siegel, and Aravind Srinivasan. *Chernoff-Hoeffding bounds for applications with limited independence*, SIAM J. Discrete Math., 8 (1995), pp. 223–250.

Approximating MAX kCSP
Using Random Restrictions

Gustav Hast

Department of Numerical Analysis and Computer Science
Royal Institute of Technology, 100 44 Stockholm, Sweden
ghast@nada.kth.se

Abstract. In this paper we study the approximability of the maximization version of constraint satisfaction problems. We provide two probabilistic approximation algorithms for MAX kCONJSAT which is the problem to satisfy as many conjunctions, each of size at most k, as possible. As observed by Trevisan, this leads to approximation algorithms with the same approximation ratio for the more general problem MAX kCSP, where instead of conjunctions arbitrary k-ary constraints are imposed. The first algorithm achieves an approximation ratio of $2^{1.40\square\,k}$. The second algorithm achieves a slightly better approximation ratio of $2^{1.54\square\,k}$, but the ratio is shown using computational evidence. These ratios should be compared with the previous best algorithm, due to Trevisan, that achieves an approximation ratio of $2^{1\square\,k}$. Both the new algorithms use a combination of random restrictions, a method which have been used in circuit complexity, and traditional semidefinite relaxation methods. A consequence of these algorithms is that some complexity classes described by probabilistical checkable proofs can be characterized as subsets of P. Our result in this paper implies that $\mathrm{PCP}_{c,s}[\log, k] \subseteq \mathrm{P}$ for any $c/s > 2^{k\square\,1.40}$, and we have computational evidence that if $c/s > 2^{k\square\,1.54}$ this inclusion still holds.

1 Introduction

In this paper we study the approximability of maximum k-constraint satisfaction problem (MAX kCSP). An instance of MAX kCSP consists of a set of Boolean variables and a set of weighted constraints, each acting over a k-tuple of the variables. The value of an assignment is the sum of the weights of the satisfied constraints. This problem was defined by Khanna et al. [9] and is a natural generalization of many well-known optimization problems. In fact it can express any MAX SNP problem.

An algorithm r-approximates a maximum optimization problem if the ratio of the value of the solution returned by the algorithm, and the optimal value of the problem, is at least r. Trevisan observed in [12] that it is sufficient to r-approximate a restricted version of MAX kCSP in order to r-approximate MAX kCSP. The restricted version is called MAX kCONJSAT and allows only constraints that are conjunctions of literals. The, to date, best approximation

K. Jansen et al. (Eds.): APPROX and RANDOM 2004, LNCS 3122, pp. 151–162, 2004.
© Springer-Verlag Berlin Heidelberg 2004

algorithm for MAX kCONJSAT, and thereby also for MAX kCSP, is based on a linear relaxation and is due to Trevisan [12]. It achieves an approximation ratio of 2^{1-k}. Since the work was first published in 1996, it has remained the best algorithm known and one could start to wonder if it, in fact, is the best possible polynomial-time approximation algorithm for MAX kCSP. In this work we provide two algorithms, with approximation ratios of $2^{1.40-k}$ and $2^{1.54-k}$, that show that this is not the case. The ratio of $2^{1.54-k}$ is shown using computational evidence rather than a proof. However, for the approximation ratio of $2^{1.40-k}$ we have a complete proof. The new approximation ratios can be considered as a moderate improvement, but are structurally interesting.

In probabilistic proof checking, the problem of maximizing the accepting probability of a verifier looking at k bits can be described as an instance of MAX kCONJSAT [12]. Thus, our result in this paper implies that $\mathrm{PCP}_{c,s}[\log, k] \subseteq \mathrm{P}$ for any $c/s > 2^{k-1.40}$, and we have computational evidence that if $c/s > 2^{k-1.54}$ this inclusion still holds.

We point out that there exist better algorithms for MAX kCSP if the instances are satisfiable. In that case, another algorithm by Trevisan [13] achieves a ratio of $2^{\log(k+1)-k}$.

In another research direction, PCP techniques have been used to show that approximating MAX kCSP within certain ratios, in polynomial-time, cannot be done unless $\mathrm{P} = \mathrm{NP}$. A long line of results led to the work of Samorodnitsky and Trevisan [11], which show that there cannot exist a polynomial-time $2^{2\sqrt{k+1}-(k+1)}$-approximation algorithm for MAX kCSP unless $\mathrm{P} = \mathrm{NP}$. More recently Engebretsen and Holmerin [2] achieved an even stronger result showing that, for $k \geq 3$, there cannot exist a polynomial-time $2^{\sqrt{2k-2}+1/2-k}$-approximation algorithm.

Semidefinite Relaxation. Let us give a brief introduction into semidefinite relaxations in connection with approximation algorithms. The work of Goemans and Williamson [5] laid ground for a large number of works providing better approximation ratios for various constraint satisfaction problems. The idea is to relax each clause into a set of semidefinite constraints. The resulting semidefinite program can be solved to any desired accuracy in polynomial-time. The solution is then rounded into a solution of the original constraint satisfaction problem. In some cases it is possible to analyze the rounding technique used and thereby obtain an approximation ratio. For MAX 2CONJSAT Goemans and Williamson [5] obtained a 0.79607-approximation algorithm. Later, Karloff and Zwick [8] introduced a methodology to create semidefinite relaxations for any constraint function and tools to analyze the rounding scheme in some cases. Using these methods Zwick [15] obtained a 0.5-approximation algorithm for MAX 3CONJSAT and Guruswami et al. [6] obtained a 0.33-approximation algorithm for MAX 4CONJSAT. We note that these algorithms outperform the random assignment algorithm with more than a factor of 3, 4 and 5, respectively. In contrast, the general MAX kCONJSAT by Trevisan, which is based on a linear relaxation instead of a semidefinite relaxation, beats the random assignment algorithm with a factor of 2. Unfortunately, applying the semidefinite relaxation method on Max kConjSAT for $k > 4$ makes the analysis of the approximation ratio problematic. An interesting question that remains open is if the algorithm achieves an approximation ratio of $2^{\log k - k}$, as it does for $k \leq 4$, but we simply do not know how to show it.

Techniques Used. The main idea of both our Max kConjSAT algorithms is to make use of the fact that semidefinite relaxation methods work well for conjunctions of small size. We shrink the size of the conjunctions by using random restrictions, an often used method in circuit complexity that was introduced by Furst et al. [4]. The random restriction is done by assigning a random value to each variable with high probability. Many of the conjunctions are unsatisfiable after this random restriction and are thrown away, but the effective sizes of the remaining conjunctions have decreased so they are tractable for traditional semidefinite relaxation methods. For the $2^{1.54-k}$ algorithm we use the above mentioned methodology of Karloff and Zwick and analyze its effect on the restricted instance. For the $2^{1.40-k}$ algorithm we instead use so-called gadgets [14] to transform the restricted instance of conjunctions into a Max 2SAT instance which is solved using the algorithm of Feige and Goemans [3].

When analyzing the approximation ratio of our $2^{1.54-k}$ algorithm we have to rely on the assumption that we have found global optima of two continuous optimization problems, one in ten variables and one in six variables. We note that this situation is quite common for many works that analyze the effectiveness of rounding procedures of semidefinite programs [3, 6–8, 15][1]. In the case of our $2^{1.40-k}$ algorithm we rely on the Feige-Goemans algorithm which has been rigorously analyzed by Zwick [16].

Probabilistic vs. Deterministic Algorithms. We note that both our algorithms are probabilistic and thus the approximation ratio is calculated using the expected value of the produced solution. We do not see any problem derandomizing them with the method of Mahajan and Ramesh [10], but we have not studied it closely. Alternatively, multiple independent executions of the algorithm guarantees that except with exponentially small probability a solution with value arbitrarily close to the one stipulated is produced. For some applications only a lower bound of the optimum value is needed and we note that both algorithms can be used to get such a bound in a deterministic way (see Sect. 6).

2   Preliminaries

We say that an instance is ρ-satisfiable if the ratio of an optimal solution and the total weight of all clauses is at least ρ. An algorithm A α-approximates a maximization problem if for all instances x of the problem $\mathrm{val}(A, x)/\mathrm{opt}(x) \geq \alpha$, where $\mathrm{val}(A, x)$ is the value of $A(x)$ and $\mathrm{opt}(x)$ is the optimal value of x.

---

[1] Some of these results have been proved in a rigorous manner since the original publication and many of the assumptions are seemingly not as strong as our is, i.e. involve less variables.

Equivalently, A is said to have an approximation ratio of α. For probabilistic algorithms $\mathrm{val}(A, x)$ is allowed to be an expected value over the random choices done by A.

An instance of the MAX kCSP problem consists of a set $\{C_1, \ldots, C_m\}$ of clauses (or constraints) with associated weights $\{w_1, \ldots, w_m\}$ and a set of Boolean variables $X = \{x_1, \ldots, x_n\}$. Each clause C_i consists of a Boolean function f_i of arity k and a size k tuple of Boolean variables $(x_{i_1}, \ldots, x_{i_k})$ where $x_{i_j} \in X$. A solution is an assignment to X and the value of the solution is the sum of the weights of the satisfied clauses. A clause $C_i = (f_i, (x_{i_1}, \ldots, x_{i_k}))$ is satisfied if $f_i(x_{i_1}, \ldots, x_{i_k})$ is true.

An instance of the MAX kCONJSAT problem is a special type of MAX kCSP problem where each clause is a conjunction of literals from X. A literal is either a variable or a negated variable.

3   A $2^{1.40-k}$-Approximation Algorithm for MAX kCONJSAT

In this section we describe how our first MAX kCONJSAT algorithm works. The algorithm runs both a random assignment algorithm and Algorithm $A1$, depicted in Fig. 1, and returns the best solution.

Input: A set of n Boolean variables $\{x_1, \ldots x_n\}$ and a set of m conjunctions $\{C_1, \ldots C_m\}$ with weights $\{w_1, \ldots w_m\}$.

1. **(Initialization)** Set the parameter $s := 2.56$.
2. **(Random restriction)** Set $I := \emptyset$. For each $i := 1, \ldots n$ do:
   – With probability s/k: put i in I, $I := I \cup \{i\}$.
   – With remaining probability: let x_i be assigned according to an unbiased coin flip.
3. **(Gadget)** Set $C_{2\mathrm{SAT}} := \emptyset$. For each $j := 1, \ldots m$ do: If C_j still is satisfiable given the already set variables and include at most five unassigned variables, apply the gadget to 2SAT as explained in Sect. 3.2. Put the 2SAT clauses in $C_{2\mathrm{SAT}}$, each with weight w_j.
4. **(Solve MAX 2SAT)** Apply the Feige-Goemans algorithm [3] on $C_{2\mathrm{SAT}}$ and let it assign the remaining variables $\{x_i\}_{i \in I}$.

Fig. 1. Algorithm $A1$: a MAX kCONJSAT algorithm.

The random assignment algorithm assigns a random Boolean value to each variable in the instance. The expected value of the solution is at least a 2^{-k}-fraction of the total weight of all clauses. Thus, if the instance is not $2^{-1.40}$-satisfiable then the random assignment algorithm is a $2^{1.40-k}$-approximation.

If the instance instead is $2^{-1.40}$-satisfiable then we can show that Algorithm $A1$ produces a $2^{1.40-k}$-approximation. The algorithm starts by applying a random restriction over the instance, thereby reducing the size of the conjunctions. The small conjunctions are then transformed, using gadgets, into a MAX 2SAT instance which is solved using the algorithm of Feige and Goemans [3]. When proving the approximation abilities of Algorithm $A1$, the main idea is to relate the optima of the MAX kCONJSAT instance and the produced MAX 2SAT instance. Theorem 1 is a consequence of the combination of Algorithm $A1$ and the random assignment algorithm. The corollary follows by using the observation made by Trevisan in [12].

Theorem 1. MAX kCONJSAT *has a polynomial-time* $2^{1.40-k}$*-approximation algorithm.*

Corollary 2. MAX kCSP *has a polynomial-time* $2^{1.40-k}$*-approximation algorithm.*

Next we describe the two main techniques used in the algorithm, random restrictions and gadgets. Then we analyze the approximation ratio of the algorithm and thereby prove Theorem 1.

3.1   Random Restrictions

In the introduction it was mentioned that semidefinite relaxation methods works well on small conjunctions. A MAX kCONJSAT instance consists of length-k conjunctions and in order to reduce the length of these conjunctions, and thereby making the instance tractable for semidefinite methods, we use a technique called random restrictions. For each variable, either assign it a random value with probability $1 - s/k$, or let it be unassigned with probability s/k. Each clause is originally a conjunction of at most k variables. Restricting a clause to the values set by the random restriction either makes the clause true (if all literals in the clause were made true by the restriction), false (if any of the literals in the clause were made false by the restriction) or it is still a conjunction but with size i (if all but i literals were set and made true by the restriction).

3.2   Gadgets Reducing CONJSAT Clauses to 2SAT Clauses

A gadget is a way to transform a constraint of one type into a set of constraints of another type. In [14] Trevisan et al. exhibit a gadget reducing a conjunction of three literals, $X_1 \wedge X_2 \wedge X_3$ into 2SAT clauses. A 2SAT clause is a disjunction of literals of size at most 2. The gadget is Y, $(\neg Y \vee X_1)$, $(\neg Y \vee X_2)$, $(\neg Y \vee X_3)$, where Y is a so-called auxiliary variable that is added. Note that if $X_1 \wedge X_2 \wedge X_3$ is true, then all four of the gadget's clauses can be made true by setting Y to true. However, if $X_1 \wedge X_2 \wedge X_3$ is false then we can never satisfy all four clauses, but always three of them by setting Y to false. For those familiar with the jargon of [14], this is a strict 4-gadget. In [14] this gadget was used to create a 0.367-approximation algorithm for MAX 3CONJSAT.

Generalizing this gadget construction we can produce a gadget reducing a conjunction of size i, $X_1 \wedge \ldots \wedge X_i$ into $i+1$ 2SAT clauses: Y, $(\neg Y \vee X_1)$, $\ldots (\neg Y \vee X_i)$. (For $i = 1$ we do not have to use a gadget because X_1 is already a 2SAT clause.) In Algorithm $A1$, after the random restriction has reduced the size of the conjunctions, we use the above gadget to transform conjunctions into 2SAT clauses. The gadget is more effective for small conjunctions, and that is why the gadget is used only on conjunctions of size five or less.

3.3   Analysis of the Approximation Ratio

If an instance is not $2^{-1.40}$-satisfiable, then a random assignment will be a $2^{1.40-k}$-approximation. Thus, left to be shown is that Algorithm $A1$ produce $2^{1.40-k}$-approximations on instances that are $2^{-1.40}$-satisfiable.

The first phase of Algorithm $A1$ is to apply a random restriction, with $s = 2.56$ (see Sect. 3.1). This yields an instance of fairly small conjunctions. On each such iCONJSAT clause containing at most five literals we apply the gadget mentioned above, which reduces the clause to i length-2 and one length-1 2SAT clauses. Thus we get a MAX 2SAT instance that can be solved using the MAX 2SAT algorithm of Feige and Goemans [3] thereby deciding the variables that remained unassigned after the random restriction.

The probability that an arbitrary conjunction of length k remains satisfiable after the random restriction and contains exactly i unassigned variables is

$$p_i = \binom{k}{i} \left(\frac{s}{k}\right)^i \left(1 - \frac{s}{k}\right)^{k-i} 2^{-(k-i)} \ .$$

Let w_{tot} be the sum of all weights in the MAX kCONJSAT instance. The expected sum of weights from length-1 clauses that an instance induce is $\alpha_1 w_{\text{tot}}$ and the corresponding value for length-2 clauses is $\alpha_2 w_{\text{tot}}$, where $\alpha_1 = \sum_{i=1}^{5} p_i$ and $\alpha_2 = \sum_{i=2}^{5} i p_i$ are the expected number of 2SAT clauses, length-1 respectively length-2, induced from a single conjunction of length k.

Let the value of an optimal assignment be $w_{\text{opt}} = \rho w_{\text{tot}}$, where $\rho \geq 2^{-1.40}$. Such an assignment, restricted to the unassigned variables, satisfy all of the length-2 clauses and an expected ρ-fraction of the weights of the length-1 clauses in the MAX 2SAT instance. Thus, the expected weight of satisfied length-2 clauses is $\alpha_2 w_{\text{tot}}$ and the expected weight of satisfied length-1 clauses is $\rho \alpha_1 w_{\text{tot}}$.

Using the MAX 2SAT algorithm of Feige and Goemans [3] we are guaranteed to obtain a solution of expected weight at least $\beta_1 \rho \alpha_1 w_{\text{tot}} + \beta_2 \alpha_2 w_{\text{tot}}$, where $\beta_1 = 0.976$ and $\beta_2 = 0.931$. The use of gadgets ensures that if we let the solution decide the value of the unassigned variables in the original MAX kCONJSAT instance, the weight of that assignment will have expected value at least $\beta_1 \rho \alpha_1 w_{\text{tot}} + \beta_2 \alpha_2 w_{\text{tot}} - \alpha_2 w_{\text{tot}} + p_0 w_{\text{tot}}$. We can now express an approximation ratio:

$$\frac{\beta_1 \rho \alpha_1 w_{\text{tot}} + \beta_2 \alpha_2 w_{\text{tot}} - \alpha_2 w_{\text{tot}} + p_0 w_{\text{tot}}}{\rho w_{\text{tot}}} = \beta_1 \alpha_1 - \rho^{-1}(1 - \beta_2)\alpha_2 + \rho^{-1} p_0 \ .$$

We see that $p_i \to (2s)^i (i!)^{-1} e^{-s} 2^{-k}$, when $k \to \infty$. Using $s = 2.56$ we get $\alpha_1 = 7.618 \cdot 2^{-k}$, $\alpha_2 = 27.401 \cdot 2^{-k}$ and $p_0 = 0.0773 \cdot 2^{-k}$ thus yielding an approximation ratio greater than $2^{1.40-k}$ for values of k large enough. To show that the ratio is valid for all values of $k \geq 5$, we have numerically checked that the ratio is higher than $2^{1.40-k}$ for all values of k from 5 up to 1000 and then obtained a lower bound of $2^{1.40-k}$ for all $k \geq 1000$ (details in the full paper).

4   A $2^{1.54-k}$-Approximation Algorithm for MAX kCONJSAT

In this section we give another approximation algorithm for MAX kCONJSAT. This algorithm also runs two different algorithms: a linear relaxation algorithm, described in Sect. 4.1, and Algorithm $A2$, depicted in Fig. 2. The linear relaxation algorithm performs well on instances that are $(1 + 2^{1-k})/2$-satisfiable and algorithm $A2$ achieves well on all other instances.

Input: A set of n Boolean variables $\{x_1, \ldots x_n\}$ and a set of m conjunctions $\{C_1, \ldots C_m\}$ with weights $\{w_1, \ldots w_m\}$.

1. **(Initialization)** Set the parameters $s := 2.8$, $P := 0.72$.
2. **(Semidefinite programming)** Generate the semidefinite programming relaxation of the input instance given in Fig. 3. Find an (almost) optimal solution $v_0, \ldots v_n$ in polynomial-time using standard techniques.
3. **(Random restriction)** Set $I := \emptyset$. For each $i := 1, \ldots n$ do:
   - With probability s/k: put i in I, $I := I \cup \{i\}$.
   - With remaining probability: let x_i be assigned according to an unbiased coin flip.
4. **(Hyperplane rounding / Random assignment)** Choose a random hyperplane normal $n_H \in S^n$. With probability P do for each $i \in I$:
   - If $sgn(n_H \cdot v_0) = sgn(n_H \cdot v_i)$ assign $x_i := 1$.
   - If $sgn(n_H \cdot v_0) \neq sgn(n_H \cdot v_i)$ assign $x_i := 0$.
   With remaining probability let x_i be assigned according to an unbiased coin flip, for all $i \in I$.

Fig. 2. Algorithm $A2$: a MAX kCONJSAT algorithm.

Algorithm $A2$ is similar in approach to Algorithm $A1$. The first step is to solve a semidefinite relaxation of the problem instance. As in the previous algorithm a random restriction is imposed on the variables (see Sect. 3.1), but instead of reducing the conjunctions of the unassigned variables into a MAX 2SAT instance we use the solution obtained from the semidefinite relaxation to assign values to the remaining variables by performing a random hyperplane rounding.

As in many previous works dealing with approximation algorithms for MAX CSP problems using semidefinite programming we only present computational evidence for the approximation ratio. The result is valid if we have found the worst possible vector configuration in two different cases. More explicitly we rely on the following technical assumption.

Assumption 3. *Let $P = 0.72$, $(v_0, v_1, \ldots v_3)$ be a vector configuration of four vectors in R^4 and z be the minimum of rel_A, where $A \subseteq \{1, 2, 3\}$, defined in Fig. 3. Then 0.493 is a lower bound on*

$$\left(P \cdot \Pr_{n_H \in R^4} \left[sgn(n_h \cdot v_0) = sgn(n_h \cdot v_1) = \ldots sgn(n_h \cdot v_3) \right] + (1 - P) \cdot 2^{-3} \right) / z \ ,$$

for all possible vector configurations of $(v_0, v_1, \ldots v_3)$.

Furthermore, let $(v_0, v_1, \ldots v_4)$ be a vector configuration of five vectors in R^5, and z be the minimum of rel_A, where $A \subseteq \{1, 2, 3, 4\}$, defined in Fig. 3. Then 0.337 is a lower bound on

$$(P \cdot \Pr_{n_H \in R^5} [sgn(n_h \cdot v_0) = sgn(n_h \cdot v_1) = \ldots sgn(n_h \cdot v_4)] + (1 - P) \cdot 2^{-4})/z \ ,$$

for all possible vector configurations of $(v_0, v_1, \ldots v_4)$.

$$\max \quad \sum_{j=1}^{m} w_j z_j$$

$$z_j \leq rel_A, \ A \subseteq C_j, |A| \leq 4$$

$$rel_\square = 1$$

$$rel_{\{i\}} = \frac{1 + v_0 \cdot v_i}{2}$$

$$rel_{\{i,j\}} = \frac{1 + v_0 \cdot v_i + v_0 \cdot v_j + v_i \cdot v_j}{4}$$

$$rel_{\{i,j,k\}} = \min_{\substack{(\sigma_0, \sigma_1, \sigma_2, \sigma_3)\square \\ permute(\{0,i,j,k\})}} \frac{1 + v_{\sigma_0} \cdot v_{\sigma_1} + v_{\sigma_0} \cdot v_{\sigma_2} + v_{\sigma_1} \cdot v_{\sigma_2}}{4}$$

$$rel_{\{i,j,k,l\}} = \min(\frac{2 + \sum_{\sigma_0, \sigma_1 \square \{0,i,j,k,l\}, \sigma_0 \square \sigma_1} v_{\sigma_0} \cdot v_{\sigma_1}}{12},$$

$$\min_{\substack{(\sigma_0, \sigma_1, \sigma_2, \sigma_3, \sigma_4)\square \\ permute(\{0,i,j,k,l\})}} \frac{2 - \sum_{t=1}^{4} v_{\sigma_0} \cdot v_{\sigma_t} + \sum_{1\square t < u\square 4} v_{\sigma_t} \cdot v_{\sigma_u}}{4})$$

$$v_i \in S^n, \ 0 \leq i \leq 2n$$

$$v_{i+n} = -v_i, \ 1 \leq i \leq n$$

Fig. 3. Semidefinite program relaxation for MAX kCONJSAT.

We have computational evidence that vector configurations of four vectors where all pairs of vectors have angle 1.438 are global minima and attain values close to the lower bound of 0.493. Vector configurations of five vectors are defined, up to rotation, by ten pair-wise angles. Let θ_{ij} be the angle between v_i and v_j. If for some a, $\theta_{ai} = 1.855$ for all $i \neq a$ and $\theta_{ij} = 1.424$ for all other pair-wise angles, then the vector configuration seems to be a global minimum and attains a value close to the lower bound of 0.337.

Due to space considerations we omit the analysis of Algorithm $A2$, which primarily consists of semidefinite relaxation techniques due to Goemans and Williamson [5] and Karloff and Zwick [8]. The complete analysis and proof of the following theorem is given in the full version of this paper. The corollary follows in the same way as in the previous section.

Theorem 4. *If Assumption 3 is true, then* MAX kCONJSAT *has a polynomial-time* $2^{1.54-k}$*-approximation algorithm.*

Corollary 5. *If Assumption 3 is true, then* MAX kCSP *has a polynomial-time* $2^{1.54-k}$*-approximation algorithm.*

¿From a practical standpoint Algorithm $A2$ seems a bit odd. First it solves a huge semidefinite program and after this only a small fraction of the variables are assigned values according to the solution. A more efficient order seems to be to do the random restriction first, as in Algorithm $A1$, and build a semidefinite program from the small fraction of conjunctions that still are satisfiable. This approach is in fact also realizable, but each weight in the semidefinite program has to be adjusted according to the length of its conjunction. Further details is given in the full version of this paper.

Next, we give a description of the linear relaxation algorithm. After that we describe the semidefinite relaxation and the rounding method used in Algorithm $A2$.

4.1 A Linear Relaxation Algorithm for MAX kCONJSAT with Threshold Rounding

The linear program relaxation is identical to the one by Trevisan [12]. Each binary variable x_i is relaxed to t_i such that $0 \leq t_i \leq 1$. The clause C_j has an associated variable z_j in the relaxation. If x_i occurs positively in C_j, then $z_j \leq t_i$ is a constraint in the relaxation, and if x_i occurs negated then $z_j \leq 1 - t_i$. It is easy to see that an assignment to MAX kCONJSAT can be transformed into a valid assignment for the linear program with the same objective value. This is done by letting $t_i = x_i$.

The difference of this algorithm compared with [12] is in how an assignment to the linear relaxed program is transformed into an assignment of the associated MAX kCONJSAT instance. In [12] the value of x_i was decided by a flip of an unbiased coin with probability $(k-1)/k$ and only in the remaining case had the value of t_i some impact on the value of x_i. We instead make a threshold rounding by setting $x_i = 1$ if $t_i > 1/2$ and $x_i = 0$ otherwise.

Theorem 6. *A* $(1+\epsilon)/2$*-satisfiable instance of* MAX kCONJSAT, *where* $\epsilon > 0$, *can be approximated in polynomial time within* $2\epsilon/(1+\epsilon)$ *of the optimal value.*

The key observation for proving the above theorem is that if the relaxed value z_j of a clause C_j is strictly larger than one half, then C_j will be satisfied by the threshold rounding. The proof is given in the full version of this paper.

4.2 Semidefinite Relaxations

Remember that a MAX CONJSAT instance consists of a set of conjunctions $\{C_1, \ldots, C_m\}$ with weights $\{w_1, \ldots, w_m\}$ and a set of variables $\{x_1, \ldots, x_n\}$. In the semidefinite relaxation each variable x_i corresponds to a vector v_i on the unit sphere S^n in R^{n+1}. A variable can appear negated in a conjunction and thus there is for each x_i a vector $v_{i+n} = -v_i$ which correspond to $\overline{x_i}$. This is done only for notational convenience and we could replace v_{i+n} by $-v_i$ everywhere. By somewhat abusing previous notation, a conjunction C_j can be described as a subset of $[2n]$ such that if x_i is part of the conjunction then $i \in C_j$ and if $\overline{x_i}$ is part of the conjunction then $i+n \in C_j$. There is also a vector v_0 which represents true. Karloff and Zwick [8] introduced the so-called canonical semidefinite relaxation which specifies how to create the strongest possible relaxations from a specific constraint function. This construction could be applied on conjunctions of k variables, but we only use the canonical semidefinite relaxation of each subset of literals of size at most four. In Fig. 3 the complete semidefinite program that we use is specified.

Hyperplane rounding is a popular method of obtaining a solution of the original problem from the solution of the semidefinite relaxation and is done as follows. Choose a random hyperplane in R^{n+1}. Set $x_i = 1$ if v_i lies on the same side of the hyperplane as v_0 does, and otherwise set $x_i = 0$. We employ the hyperplane rounding as well, but first after making a random restriction. Thus, only the variables that remain unassigned after the random restriction get their value set by the hyperplane rounding method, all others are set by unbiased coin flips.

5   Approximating MAX kCONJSAT for Small Values on k

In both Algorithm $A1$ and Algorithm $A2$ the random restriction parameter s was the same for all values of k. For specific values of k we can obtain better approximation ratios by using a larger value on s. This is the case for both algorithms. In Table 1 the obtained approximation ratios for MAX kCSP with the corresponding value on s are listed for $5 \leq k \leq 12$. For the approximation ratios of Algorithm $A2$, the parameter $P = 0.72$ was used. We note that for $k \leq 4$ there are known semidefinite relaxation algorithms achieving better ratios.

6   Relations with Probabilistical Proof Checking

We use the standard notation and definitions in connection with probabilistical checkable proofs, PCPs. For those unfamiliar with these, we refer to the work of

Table 1. Approximation ratios α_{A1} of Algorithm $A1$ (combined with a random assignment) and approximation ratios α_{A2} of Algorithm $A2$ (combined with a linear relaxation with threshold rounding) on MAX kCSP for $5 \leq k \leq 12$, with tailor-made values on s.

k	5	6	7	8	9	10	11	12
$k + \log \alpha_{A1}$	1.46	1.45	1.44	1.43	1.43	1.42	1.42	1.42
s	3.1	3.0	2.9	2.9	2.8	2.8	2.8	2.8
$k + \log \alpha_{A2}$	1.88	1.78	1.73	1.69	1.67	1.65	1.64	1.63
s	3.4	3.2	3.1	3.1	3.0	3.0	3.0	3.0

Bellare et al. [1] which also contains the history of PCPs. The complexity class $\text{PCP}_{c,s}[\log, q]$ contains all languages that have a verifier with completeness c, soundness s, which uses only a logarithmic number of random bits and asks at most q (adaptive) questions.

The following theorem due to Trevisan shows that an approximation algorithm for MAX kCONJSAT implies that certain PCP classes are contained in P.

Theorem 7 (Trevisan [12]). *If* MAX kCONJSAT *is* r-approximable for some $r \leq 1$, then $\text{PCP}_{c,s}[\log, k] \subseteq P$ for any $c/s > 1/r$.

Both algorithms presented in this work are probabilistic, thus it is not immediate that the above theorem can be applied. By inspecting the proof of Theorem 7, however, it is clear that only a lower bound of the value of an optimal solution is needed. For the algorithm described in Sect. 4 such a lower bound can be obtained deterministically from the solution of the semidefinite program in Algorithm A2 and by performing the (deterministic) linear relaxation algorithm with threshold rounding. Thus, we get the following theorem.

Theorem 8. *If Assumption 3 is true, then* $\text{PCP}_{c,s}[\log, k] \subseteq P$ for any $c/s > 2^{k-1.54}$.

We now turn to the algorithm from Sect. 3.2 and primarily Algorithm $A1$. As with Algorithm $A2$ we do not need to perform the rounding step (in the Feige-Goeman algorithm), because solving the semidefinite program is enough to get a lower bound of the value of an optimal solution. But the random restriction that is applied before the semidefinite relaxation has to be derandomized. Therefore, we construct a polynomial size family of k-wise independent random restrictions. As each conjunction of size k is analyzed separately we get the same approximation ratio if the random restriction is chosen from a family of k-wise independent random restrictions. By applying Algorithm $A1$ using each one of these random restrictions, it is guaranteed that at least one performs as well as the expected value over all different possible random restrictions. Thus, a lower bound on the value of the optimal solution, where the bound is at most a factor $2^{k-1.40}$ smaller than the optimal solution value is achieved.

Theorem 9. $\text{PCP}_{c,s}[\log, k] \subseteq P$ for any $c/s > 2^{k-1.40}$.

Acknowledgments

I wish to thank Johan Håstad for much appreciated help and ideas. I am also grateful to Gunnar Andersson for providing me with code that calculates the volume of a spherical tetrahedron.

References

1. Mihir Bellare, Oded Goldreich, and Madhu Sudan. Free bits, PCPs, and nonapproximability - towards tight results. *SIAM Journal on Computing*, 27(3):804–915, 1998.

2. Lars Engebretsen and Jonas Holmerin. More efficient queries in PCPs for NP and improved approximation hardness of maximum CSP. Unpublished manuscript, 2003.
3. Uriel Feige and Michel X. Goemans. Approximating the value of two prover proof systems, with applications to MAX 2SAT and MAX DICUT. In *Proceedings of the 3rd Symposium on Theory of Computing and Systems*, pages 182–189, 1995.
4. Merrick Furst, James Saxe, and Michael Sipser. Parity, circuits, and the polynomial-time hierarchy. *Mathematical Systems Theory*, 17(1):13–27, 1984.
5. Michel X. Goemans and David P. Williamson. Improved Approximation Algorithms for Maximum Cut and Satisfiability Problems Using Semidefinite Programming. *Journal of the ACM*, 42:1115–1145, 1995.
6. Venkatesan Guruswami, Daniel Lewin, Madhu Sudan, and Luca Trevisan. A tight characterization of NP with 3 query PCPs. In *Proceedings of the 39th Annual IEEE Symposium on Foundations of Computer Science*, pages 8–17, 1998.
7. Eran Halperin and Uri Zwick. Approximation algorithms for MAX 4-SAT and rounding procedures for semidefinite programs. *Journal of Algorithms*, 40:185–211, 2001.
8. Howard Karloff and Uri Zwick. A 7/8-approximation algorithm for MAX 3SAT? In *Proceedings of the 38th Annual IEEE Symposium on Foundations of Computer Science*, pages 406–415, 1997.
9. Sanjeev Khanna, Rajeev Motwani, Madhu Sudan, and Umesh Vazirani. On syntactic versus computational views of approximability. *SIAM Journal on Computing*, 28(1):164–191, 1999.
10. Sanjeev Mahajan and H. Ramesh. Derandomizing approximation algorithms based on semidefinite programming. *SIAM Journal on Computing*, 28(5):1641–1663, 1999.
11. Alex Samorodnitsky and Luca Trevisan. A PCP characterization of NP with optimal amortized query complexity. In *Proceedings of the 32nd ACM Symposium on Theory of Computing*, pages 191–199, 2000.
12. Luca Trevisan. Parallel approximation algorithms by positive linear programming. *Algorithmica*, 21(1):72–88, 1998.
13. Luca Trevisan. Approximating satisfiable satisfiability problems. *Algorithmica*, 28(1):145–172, 2000.
14. Luca Trevisan, Gregory B. Sorkin, Madhu Sudan, and David P. Williamson. Gadgets, approximation, and linear programming. *SIAM Journal on Computing*, 29:2074–2097, 2000.
15. Uri Zwick. Approximation algorithms for constraint satisfaction problems involving at most three variables per constraint. In *Proceedings of the 9th Annual ACM-SIAM Symposium on Discrete Algorithms*, pages 201–210, 1998.
16. Uri Zwick. Analyzing the MAX 2-SAT and MAX DI-CUT approximation algorithms of Feige and Goemans. Manuscript, 2000.

Approximation Schemes for Broadcasting in Heterogenous Networks*

Samir Khuller[1], Yoo-Ah Kim[1], and Gerhard Woeginger[2]

[1] Department of Computer Science, University of Maryland
College Park, MD 20742
{samir,ykim}@cs.umd.edu
[2] Department of Mathematics and Computer Science
Eindhoven University of Technology
Eindhoven, The Netherlands
gwoegi@igi.tu-graz.ac.at

Abstract. We study the problem of minimizing the broadcast time for a set of processors in a cluster, where processor p_i has transmission time t_i, which is the time taken to send a message to any other processor in the cluster. Previously, it was shown that the Fastest Node First method (FNF) gives a 1.5 approximate solution. In this paper we show that there is a polynomial time approximation scheme for the problems of broadcasting and multicasting in such a heterogenous cluster.

1 Introduction

Networks of Workstations (NOWs) are an extremely popular alternative to massively parallel machines and are widely used (for example the Condor project at Wisconsin [17]) and the Berkeley NOW project [16]. By simply using off-the-shelf PC's, a very powerful workstation cluster can be created, and this can provide a high amount of parallelism at relatively low cost. Since NOWs are put together over time, the machines tend to have different capabilities and this leads to a *heterogenous* collection of machines, rather than a *homogenous* collection, in which all the machines have identical capabilities.

One fundamental operation that is used in such clusters, is that of *broadcast* (this is a primitive in many message passing systems such as MPI [1, 6, 8]). In addition it is used as a primitive in many parallel algorithms. The main objective of a broadcast operation is to quickly distribute the input data to the entire network for processing. Another situation is when the system is performing a parallel search, then the successful processor needs to inform all other processors that the search has concluded successfully. Various models for heterogenous environments have been proposed in the literature. One general model is the one proposed by Bar-Noy et al [3] where the communication costs between links are not uniform. In addition, the sender may engage in another communication before the current one is complete. An approximation factor with a guarantee of $O(\log k)$ is given for the operation of performing a multicast. Other popular models

---

* Research supported by NSF ITR Award CCR-0113192.

K. Jansen et al. (Eds.): APPROX and RANDOM 2004, LNCS 3122, pp. 163–170, 2004.

in the theory literature generally assume an underlying communication graph, with the property that only adjacent nodes in this graph may communicate.

Broadcasting efficiently is an essential operation and many works are devoted to this (see [18, 9, 10, 4, 5] and references therein). In addition, for emergency notification an understanding of how to perform broadcast quickly is essential.

A simple model and algorithm was proposed by Banikazemi et al [2]. In this model, heterogeneity among processors is modeled by a non-uniform speed of the sending processor. A heterogenous cluster is defined as a collection of processors $p_1, p_2, \ldots, p_n$ in which each processor is capable of communicating with any other processor. Each processor has a transmission time which is the time required to send a message to any other processor in the cluster. Thus the time required for the communication is a function of only the sender. Each processor may send messages to other processors in order, and each processor may be receiving only one message at a time.

Thus a broadcast operation is implemented as a broadcast tree. Each node in the tree represents a processor of the cluster. The root of the tree is the source of the original message. The children of a node p_i are the processors that receive the message from p_i. The completion time of a node is the time at which it completes receiving the message from its parent. The completion time of the children of p_i is $c_i + j \cdot t_i$, where c_i is the completion time of p_i, t_i is the transmission time of p_i and j is the child number. In other words, the first child of p_i has a completion time of $c_i + t_i$ ($j = 1$), the second child has a completion time of $c_i + 2t_i$ ($j = 2$) etc. See Figure 1 for an example.

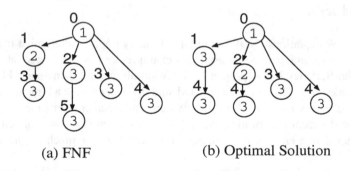

(a) FNF   (b) Optimal Solution

Fig. 1. An example that FNF does not produce an optimal solution. Transmission times of processors are inside the circles. Times at which nodes receive a message are also shown.

A commonly used method to find a broadcast tree is referred to as the "Fastest Node First" (FNF) technique [2]. This works as follows: In each iteration, the algorithm chooses a sender from the set of processors that have received the message (set S) and a receiver from the set of processors that have not yet received the message (set R). The algorithm then picks the sender from $s \in S$ so that s can finish the transmission as early as possible, and chooses the receiver $r \in R$ as the processor with the minimum transmission time in R. Then r is moved from R to S and the algorithm continues. The intuition is that sending the message to fast processors first is a more effective way to propagate the message quickly. This technique is very effective and easy to implement.

In practice it works extremely well (using simulations) and in fact frequently finds optimal solutions as well [2]. However, there are situations when this method also fails to find an optimal solution. A simple example is shown in Figure 1.

Despite several non-trivial advances in an understanding of the *fastest node first* method by Liu [13] (see also work by Liu and Sheng [15] in SPAA 2000) it was not well understood as to how this algorithm performs in the worst case. For example, can we show that in all instances the FNF heuristic will find solutions close to optimal?

Liu [13] (see also [15]) shows that if there are only two classes of processors, then FNF produces an optimal solution. In addition, if the transmission time of every slower processor is a multiple of the transmission time of every faster processor, then again the FNF heuristic produces an optimal solution. So for example, if the transmission time of the fastest processor is 1 and the transmission time of all other processors are powers of 2, then the algorithm produces an optimal solution. It immediately follows that by rounding all transmission times to powers of 2 we can obtain a solution using FNF whose cost is at most twice the cost of an optimal solution[1]. However, this still did not explain the fact that this heuristic does much much better in practice. Recently, Khuller and Kim [12] showed that the FNF heuristic actually produces an optimal solution for the problem of minimizing the sum of completion times. This property is used to show that the FNF method has a performance ratio of at most 1.5 when compared to the optimal solution for minimizing broadcast time. In addition the performance ratio of FNF is at least $\frac{25}{22}$. As a corollary of the above approximation result, it is shown that if the transmission times of the fastest $\frac{n}{2}$ processors are in the range $[1 \ldots C]$ then FNF produces a solution with makespan at most $T_{OPT} + C$. It is also shown that the problem is NP-hard, so unless $P = NP$ there is no polynomial time algorithm for finding an optimal solution.

It was conjectured in [12] that there is a polynomial time approximation scheme (PTAS) for this problem. In this paper we prove this conjecture. However, this algorithm is not practical due to its high running time, albeit polynomial.

2   Problem Definition

We are given a set of processors $(p_1, p_2, \ldots p_n)$ and there is one message to be broadcast to all the processors. Each processor p_i can send a message to another processor with transmission time t_i once it has received the message. Each processor can be either sending a message or receiving a message at any point of time. Without loss of generality, we assume that $t_1 \leq t_2 \leq \ldots \leq t_n$ and $t_1 = 1$. Also we assume that p_1 has the message at time zero.

We define the completion time of processor p_i to be the time when p_i has received the message. Our objective is to find a schedule that minimizes $C_{\max} = \max_i c_i$ where c_i is the completion time of processor p_i. In other words, we want to find a schedule that minimizes the time required to send the message to all the processors.

We recall the following definition from [12].

---

[1] One approach to obtain a PTAS might be rounding to powers of $(1+\epsilon)$. However, this does not work since their proof works only when the completion times of all processors are integers.

Definition 1. *We define the number of* fractional blocks *as the number of (fractional) messages a processor can send by the given time. In other words, given a time T, the number of fractional blocks of processor p_i is $(T - c_i)/t_i$.*

Our proof makes use of the following results from [12].

Theorem 1. *[12] The Fastest Node First algorithm maximizes the total number of fractional blocks for any value T.*

3   Approximation Scheme

We now describe a polynomial time approximation scheme for the problem of performing broadcast in the minimum possible time. Unfortunately, the algorithm has a very high running time when compared to the *fastest node first* heuristic.

We will assume that we know the broadcast time T of the optimal solution. Since $t_1 = 1$, we know that the minimum broadcast time T is between 1 and n, and we can try all possible values of the form $(1 + \epsilon)^j$ for some fixed $\epsilon > 0$ and $j = 1 \ldots \lceil \frac{\log n}{\log(1+\epsilon)} \rceil$. In this guessing process we lose a factor of $(1 + \epsilon)$.

Let $\epsilon' > 0$ be a fixed constant. We define a set of fast processors F as all processors whose transmission time is at most $\epsilon'T$. Formally, $F = \{p_j | t_j \le \epsilon'T\}$. Let S be the set of remaining (slow) processors. We partition S into collections of processors of similar transmissions speeds. For $i = 1 \ldots k$, define $S_i = \{p_j | \epsilon'T(1 + \epsilon')^{i-1} < t_j \le \epsilon'T(1 + \epsilon')^i\}$ where k is $\lceil \frac{\log(1/\epsilon')}{\log(1+\epsilon')} \rceil$.

We first send messages to processors in F using FNF. We prove that there is a schedule with broadcast time at most $(1 + O(\epsilon))T$ such that all processors in F receive the message first. We then find a schedule for slow processors based on a dynamic programming approach.

Schedule for F: We use the FNF heuristic for the set F to generate a partial broadcast schedule. Assume that the schedule for F has a broadcast time of T_{FNF}. In this schedule every processor $p_j \in F$ becomes idle at some time between $T_{FNF} - t_j$ and T_{FNF}.

We will prove that there is a schedule with broadcast time at most $(1 + O(\epsilon))T$ such that all processors in F receive the message first, and then send it to the slow processors. The following lemma relates T_{FNF} with the time taken by the optimal schedule to propagate the message to *any* $|F|$ processors.

Lemma 1. *In any schedule, we need at least $T_{FNF} - 2\epsilon'T$ time units to have $|F|$ processors receive (any portion of) the message.*

Proof. We prove this by contradiction. In any schedule let C_t be the number of processors that have completely received the message by time t. In addition, let I_t be the number of processors that have started receiving the message by time t. Suppose that at time $T' < T_{FNF} - 2\epsilon'T$, we have $C_{T'} + I_{T'} \ge |F|$. First note that $I_t \le C_t$ since each processor in I_t is getting the message from exactly one (distinct) processor in C_t. This means we should have that $C_{T'} \ge |F|/2$.

If a schedule is able to complete sending the message to at least $|F|/2$ processors by time T', then the number of fractional blocks of this schedule is at least $|F|/2$. Since FNF maximizes fractional blocks by Theorem 1, we claim that FNF also has at least $|F|/2$ fractional blocks by time T'. Let t_f be the transmission of slowest processor in F. Notice that in additional time $t_f (\leq \epsilon'T)$ all processors that had started receiving the message must have finished receiving it. In additional time $t_f (\leq \epsilon'T)$, FNF most certainly can double the number of processors that have received the message. Thus before T_{FNF}, more than $|F|$ processors would have received the message; a contradiction since T_{FNF} is the earliest time at which the fastest $|F|$ processors receive the message in FNF.

Lemma 2. *There is a schedule in which all processors in F receive the message no later than any processor in S and the makespan of the schedule is at most $(1 + 3\epsilon')T$.*

Proof. The main idea behind the proof is to show that an optimal schedule can be modified to have a certain form. Consider the set of processors of an optimal schedule that have received any portion of the message by time $T_{FNF} - 2\epsilon'T$. This consists of some fast processors, F' and some slow processors S'. Let $F'' = F \setminus F'$ and $S'' = S \setminus S'$. Note that $|F''| \geq |S'|$ since $|F'| + |S'| \leq |F|$ (Lemma 1). In the FNF schedule, by time T_{FNF} all processors in $F = F' \cup F''$ have the message. We can now have the processors in F'' send messages to all processors in S'. Since $|F''| \geq |S'|$ each processor sends only one message and this will take additional time at most $\epsilon'T$. By time $T_{FNF} + \epsilon'T$, all processors in $F' \cup S'$ certainly have the message in addition to the processors in F''. Notice that the optimal schedule now broadcasts the message to all remaining processors in additional time $T - (T_{FNF} - 2\epsilon'T)$. Thus we can also finish the broadcasting in additional time $T - (T_{FNF} - 2\epsilon'T)$. The broadcast time of this schedule is at most $T_{FNF} + \epsilon'T + T - (T_{FNF} - 2\epsilon'T) = (1 + 3\epsilon')T$.

Create All Possible Trees of S: For the processors in S, we will produce a set $\mathcal{S}$ of labeled trees $\mathcal{T}$. A tree $\mathcal{T}$ is any possible tree with broadcast time at most T consisting of a subset of processors in S. Then we label a node in the tree as i if the corresponding processor belongs to S_i ($i = 1 \ldots k$). We prove that the number of different trees is constant.

Lemma 3. *The size of $\mathcal{S}$ is constant for fixed $\epsilon' > 0$.*

Proof. First consider the size of a tree $\mathcal{T}$ (that is, the number of processors in the tree). Let us denote it as $|\mathcal{T}|$. Since the transmission time of processors in S is greater than $\epsilon'T$, we need at least $\epsilon'T$ time units to double the number of processors that received the message. It means that given a processor as a root of the tree, within time T we can have at most $2^{1/\epsilon'}$ processors receive the message. Therfore, $|\mathcal{T}| \leq 2^{1/\epsilon'}$. Now each node in the tree can have different label $i = 1 \ldots k$. To obtain an upperbound of the number of different trees, given a tree $\mathcal{T}$ we transform it to a complete binomial tree of size $2^{1/\epsilon'}$ by adding nodes labeled as 0. Then the number of different trees is at most $(k + 1)^{2^{1/\epsilon'}}$.

Attach $\mathcal{T}$ to F: Let the completion time of every processor $p_j \in F$ be c_j. Each processor p_j in F sends a message to a processor in S every t_j time unit. Therefore, a fast processor p_j can send messages to at most $X_j = \lfloor \frac{T-c_j}{t_j} \rfloor$ other processors. Let $X = \sum_{p_j \in F} X_j$. Let us consider the time x_i of each sending point in X. We sort those x_i in nondecreasing order and attach a tree from S to each point (See Figure 2). Note that we can attach at most $|X|$ trees of slow processors. Clearly $|X| \leq n$.

We check if an attachment is feasible, using dynamic programming. Recall that we partition slow processors into a collection of processors $S_1, S_2, \ldots S_k (k = \frac{\log(1/\epsilon')}{\log(1+\epsilon')})$. Let s_i denote the number of processors in set S_i. We define a state $s[j, n_1, n_2, \ldots n_k]$ $(0 \leq j \leq |X|, 0 \leq n_i \leq s_i)$ to be true if there is a set of j trees in S that we can attach to first j sending points and the corresponding schedule satifies the following two conditions: i) the schedule completes by time T and ii) exactly n_i processors in S_i appear in j trees in total. Our goal is to find out whether $s[j, s_1, s_2, \ldots s_k]$ is true for some j, which means that there is a feasible schedule with makespan at most T. The number of states is at most $O(n^{k+1})$ since we need at most n trees ($|X| \leq n$) and $s_i \leq n$.

Now we prove that each state can be computed in constant time. Given $s[j-1, \ldots]$, we compute $s[j, n_1, n_2, \ldots n_k]$ as follows. We try to attach all possible trees in S to x_j. Then $s[j, n_1, n_2, \ldots n_k]$ is true if there exists a tree T' such that the makespan of T' is at most $T - x_j$ and $s[j-1, n_1 - m_1, n_2 - m_1, \ldots n_k - m_k]$ is true where T' has m_i slow processors belonging to set S_i. It can be checked in constant time since the size of S is constant (Lemma 3).

Theorem 2. *Given a value T, if a broadcast tree with broadcast time T exists, then the above algorithm will find a broadcast tree with broadcast time at most $(1+\epsilon')(1+3\epsilon')T$.*

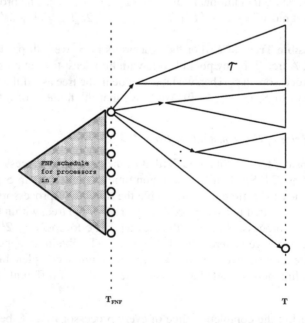

Fig. 2. Attach trees for slow processors to fast processors.

Proof. Consider the best schedule among all schedules in which processors in F receive the message first. By Lemma 2, the broadcast time of this schedule is at most $(1+3\epsilon')T$. We round up the transmission time of p_j in S_i to $\epsilon'T(1+\epsilon')^i$ where i is the smallest integer such that $t_j \leq \epsilon'(1+\epsilon')^iT$. By this rounding, we increase the broadcast time by factor of at most $1 + \epsilon'$. Therefore, the broadcast time of our schedule is at most $(1+\epsilon')(1+3\epsilon')T$.

Theorem 3. *The algorithm takes as input the transmission times of the n processors, and constants $\epsilon, \epsilon' > 0$. The algorithm finds a broadcast tree with broadcast time at most $(1+\epsilon)(1+\epsilon')(1+3\epsilon')T$ in polynomial time.*

Proof. We try the above algorithm for all possible value of the form $T = (1+\epsilon)^j$ for $j = 1 \ldots \lceil \frac{\log n}{\log(1+\epsilon)} \rceil$. This will increase the broadcast time by factor of at most $1 + \epsilon$. Therefore the broadcast time of our schedule is at most $(1+\epsilon)(1+\epsilon')(1+3\epsilon')T$.

For each given value $(1+\epsilon)^j$, we find FNF schedule for processors in F (it takes at most $O(n \log n)$) and attach trees of slow processors to processors in F, using dynamic programming. As we discussed earlier, the number of states is $O(n^{k+1})$ and each state can be checked if it is feasible in $O((k+1)^{2^{1/\epsilon'}})$ time, which is constant. Thus the running time of our algorithm is $O(\lceil \frac{\log n}{\log(1+\epsilon)} \rceil (n \log n + (k+1)^{2^{1/\epsilon'}+1} \cdot n^{k+1})$ where k is $\lceil \frac{\log(1/\epsilon')}{\log(1+\epsilon')} \rceil$.

4 Multicast

A multicast operation involves only a subset of processors. By utilizing fast processors which are not in the multicast group, we can reduce the multicasting time significantly. For example, suppose that we have m processors with transmission time t_1 and m more processors with transmission time t_2 where $t_1 < t_2$. Let we want to multicast a message to all processors with transmission time t_2. If we only use processors in the multicast group, it will take $t_2 \cdot \log m$ time. But if we utilize processors with transmission time t_1, we can finish the multicast in $t_1 \cdot (\log m + 1)$. Therefore, when $t_1 \ll t_2$, the speed-up is significant.

Theorem 4. *We have a polynomial time approximation scheme for multicasting.*

Proof. Note that if an optimal solution utilizes k processors not in the multicast group, then those processors are the k fastest ones. Therefore, if we know how many processors participate in multicasting, we can use our PTAS for broadcasting. By trying all possible k and taking the best one, we have PTAS for multicasting.

Acknowledgement

We thank Maxim Sviridenko for useful discussions.

References

1. *Message Passing Interface Forum*. March 1994.
2. M. Banikazemi, V. Moorthy and D. K. Panda. Efficient Collective Communication on Heterogeneous Networks of Workstations. *International Conference on Parallel Processing*, 1998.
3. A. Bar-Noy, S. Guha, J. Naor and B. Schieber. Multicasting in Heterogeneous Networks. *Proceedings of the 13th Annual ACM Symposium on Theory of Computing*, pp. 448-453, 1998.
4. A. Bar-Noy and S. Kipnis. Designing broadcast algorithms in the Postal Model for Message-passing Systems. *Mathematical Systems Theory*, 27(5), 1994.
5. P. Bhat, C. Raghavendra and V. Prasanna. Efficient Collective Communication in Distributed Heterogeneous Systems. *Proceedings of the International Conference on Distributed Computing Systems*, 1999.
6. J. Bruck, D. Dolev, C. Ho, M. Rosu and R. Strong, Efficient Message Passing Interface(MPI) for Parallel Computing on Clusters of Workstations. *J. Parallel Distributed Computing*, 40:19–34, 1997.
7. M.R.Garey and D.S.Johnson Computer and Intractability: A Guide to the Theory of NP-completeness, Freeman, New York, 1979.
8. W. Gropp, E. Lusk, N. Doss and A. Skjellum. A High-performance, portable Implementation of the MPI: a Message Passing Interface Standard. *Parallel Computing* 22:789–828, 1996.
9. S.M.Hedetniemi, S.T. Hedetniemi and A.L.Liestman. A Survey of Gossiping and Broadcasting in Communication Networks, *Networks* 18:129–134, 1991.
10. R. Karp, A. Sahay, E. Santos and K.E.Schauser. Optimal Broadcast and Summation in the Logp Model. *Proceedings of 5th Annual Symposium on Parallel Algorithms and Architectures*, pp. 142-153, 1993.
11. R. Kesavan, K. Bondalapati and D. Panda. Multicast on Irregular Switch-based Networks with Wormhole Routing. *Proceedings of the International Symposium on High Performance Computer Architectures*, pp. 48-57, 1997.
12. S. Khuller and Y. Kim. On Broadcasting in Heterogeneous Networks. *Proc. of 15th ACM/SIAM Symp. on Discrete Algorithms*, pp. 1004-1013, 2004.
13. P. Liu. Broadcasting Scheduling Optimization for Heterogeneous Cluster Systems. *Journal of Algorithms* 42:135-152, 2002.
14. P. Liu and D. Wang. Reduction Optimization in Heterogeneous Cluster Environments. *Proceedings of the International Parallel and Distributed Processing Symposium* pp. 477-482, 2000.
15. P. Liu and T-H. Sheng. Broadcasting Scheduling Optimization for Heterogeneous Cluster Systems. *ACM Symp. on Parallel Algorithms and Architectures (SPAA)*, pp 129–136, 2000.
16. D. A. Patterson, D. E. Culler and T. E. Anderson. A case for NOWs (Networks of Workstations). *IEEE Micro*, 15(1):54–64, 1995.
17. J. Pruyne and M. Livny. Interfacing Condor and PVM to Harness the Cycles of Workstation Clusters. *Journal on Future Generations of Computer Systems*, 12, 1996.
18. D. Richards and A. L. Liestman. Generalization of broadcasting and Gossiping. *Networks* 18:125–138, 1988.

Centralized Deterministic Broadcasting
in Undirected Multi-hop Radio Networks[*]

Dariusz R. Kowalski[1,2] and Andrzej Pelc[3]

[1] Max-Planck-Institut für Informatik
Stuhlsatzenhausweg 85, 66123 Saarbrücken, Germany
darek@mpi-sb.mpg.de
[2] Instytut Informatyki, Uniwersytet Warszawski
Banacha 2, 02-097 Warszawa, Poland
[3] Département d'informatique, Université du Québec en Outaouais
Hull, Québec J8X 3X7, Canada
Andrzej.Pelc@uqo.ca

Abstract. We consider centralized deterministic broadcasting in radio networks. The aim is to design a polynomial algorithm, which, given a graph G, produces a fast broadcasting scheme in the radio network represented by G. The problem of finding an optimal broadcasting scheme for a given graph is NP-hard, hence we can only hope for a good approximation algorithm. We give a deterministic polynomial algorithm which produces a broadcasting scheme working in time $\mathcal{O}(D \log n + \log^2 n)$, for every n-node graph of diameter D. It has been proved recently [15, 16] that a better order of magnitude of broadcasting time is impossible unless $NP \subseteq BPTIME(n^{\mathcal{O}(\log \log n)})$. In terms of approximation ratio, we have a $\mathcal{O}(\log(n/D))$-approximation algorithm for the radio broadcast problem, whenever $D = \Omega(\log n)$.

1 Introduction

A radio network is a collection of stations, equipped with capabilities of transmitting and receiving messages. Stations will be referred to as *nodes* of the network. The network is modeled as an undirected connected graph on the set of these nodes. An edge e between two nodes means that the transmitter of one end of e can reach the other end. Nodes send messages in synchronous *steps* (time slots). In every step every node acts either as a *transmitter* or as a *receiver*. A node acting as a transmitter sends a message which can potentially reach all of its neighbors. A node acting as a receiver in a given step gets a message, if and only if, exactly one of its neighbors transmits in this step. The message received in this case is the one that was transmitted by the unique neighbor. If at least

---

[*] The work of the first author was supported in part by the KBN Grant 4T11C04425. Research of the second author supported in part by NSERC grant OGP 0008136 and by the Research Chair in Distributed Computing of the Université du Québec en Outaouais. Part of this work was done during the second author's visit at the Max-Planck-Institut für Informatik.

K. Jansen et al. (Eds.): APPROX and RANDOM 2004, LNCS 3122, pp. 171–182, 2004.

two neighbors v and v' of u transmit simultaneously in a given step, none of the messages is received by u in this step. In this case we say that a *collision* occurred at u. It is assumed that the effect at node u of more than one of its neighbors transmitting is the same as that of no neighbor transmitting, i.e., a node cannot distinguish a collision from silence.

Broadcasting is one of the fundamental primitives in network communication. Its goal is to transmit a message from one node of the network, called the *source*, to all other nodes. Remote nodes get the source message via intermediate nodes, along paths in the network. A *broadcasting scheme* for a given network prescribes in which step which nodes transmit. In this paper we concentrate on one of the most important and widely studied performance parameters of a broadcasting scheme, which is its *broadcasting time*, i.e., the number of steps it uses to inform all the nodes of the network. Broadcasting time is considered as a function of two parameters of the network: the number n of nodes, and the radius D, which is the largest distance from the source to any node of the network. (For undirected graphs, the diameter is of the order of the radius.)

In this paper we study *centralized broadcasting*. All nodes of the network know its topology, i.e., they have a labeled copy of the underlying graph as input. Each node also knows its label. Labels of all nodes are distinct integers. For simplicity, we assume that these are consecutive natural numbers with source 0 but this assumption is not essential. Thus, the scheme itself, although executed in a distributed way by nodes of the network, can be considered as designed centrally, using the same sequential algorithm at each node. The input of this algorithm is the graph representing the network, and the output is a sequence of sets of nodes which are to transmit in consecutive steps. It is important to distinguish between the running time of the algorithm producing a broadcasting scheme for each graph and the broadcasting time of the output scheme, for a given graph.

Our goal is to design a deterministic algorithm, running in polynomial time, which produces a fast broadcasting scheme for any input graph. The problem of finding an optimal deterministic broadcasting scheme for any input graph (i.e. a scheme having the smallest possible broadcasting time for this graph) is NP-hard. Hence the best we can hope for is a good approximation algorithm.

1.1 Related Work

Centralized broadcasting in radio networks has been studied, e.g., in [6, 7, 18, 15–17]. In [6] it was shown that the problem of finding an optimal deterministic broadcasting scheme for any input graph is NP-hard. In [7] the authors gave a deterministic polynomial algorithm which produces a broadcasting scheme working in time $\mathcal{O}(D \log^2(n/D))$, for any n-node graph of diameter D. Hence they got a $\mathcal{O}(\log^2(n/D))$-approximation algorithm for the radio broadcast problem. In [18] the authors showed a method consisting in partitioning the underlying graph into clusters, which improves the time of broadcasting, since known broadcasting schemes can be applied in each cluster separately, and diameters of clusters are smaller than the diameter of the graph. Applied to the randomized scheme from [3], working in expected time $\mathcal{O}(D \log n + \log^2 n)$, the method from [18] produces, for any n-node graph G of diameter D, a *randomized* scheme with expected broadcasting time $\mathcal{O}(D + \log^5 n)$. Using the broadcasting scheme from [7], this method constructs (in polynomial time) a deterministic broadcasting scheme working in $\mathcal{O}(D + \log^6 n)$ steps.

Recently, the clustering method from [18] has been improved in [17]. Applied to the randomized scheme from [3], this improved method produces, for any n-node graph G of diameter D, a *randomized* scheme with expected broadcasting time $\mathcal{O}(D + \log^4 n)$. On the other hand, using the broadcasting scheme from [7] this method constructs (in polynomial time) a deterministic broadcasting scheme working in $\mathcal{O}(D + \log^5 n)$ steps. On the negative side, it has been proved in [15] that $o(\log n)$-approximation of the radio broadcast problem is impossible, unless $NP \subseteq BPTIME(n^{\mathcal{O}(\log \log n)})$. (Here $o(\log n)$ is meant as a multiplicative factor.) Under the same assumption it was also proved in [16] that there exists a constant c such that there is no polynomial-time algorithm which produces, for every n-node graph G, a broadcasting scheme with broadcasting time at most $opt(G) + c \log^2 n$, where $opt(G)$ denotes optimal broadcasting time for G.

There is also another approach to the problem of radio broadcasting, where the knowledge of the entire network topology is not assumed, and distributed broadcasting algorithms relying on limited information about the network are sought. It is not surprising that broadcasting under such scenarios is much slower than in the case of centralized broadcasting. Many authors [5, 8–10, 12, 14] studied deterministic distributed broadcasting in radio networks assuming that every node knows only its own label. In [8 10, 14, 20, 13] the model of directed graphs was used. Increasingly faster broadcasting algorithms working on arbitrary n-node (directed) radio networks were constructed, the currently fastest being the $\mathcal{O}(n \log^2 D)$-time algorithm from [13]. (Here D is the radius of the network, i.e, the longest distance from the source to any other node). On the other hand, in [12] a lower bound $\Omega(n \log D)$ on broadcasting time was proved for directed n-node networks of radius D.

Randomized broadcasting algorithms in radio networks were studied, e.g., in [3, 13, 22]. For these algorithms, no topological knowledge of the network was assumed. In [3] the authors showed a randomized broadcasting algorithm running in expected time $\mathcal{O}(D \log n + \log^2 n)$. In [21] we improved this upper bound by presenting a broadcasting algorithm with expected time $\mathcal{O}(D \log(n/D) + \log^2 n)$. (Shortly later, a similar result was obtained independently in [13].)

1.2   Our Results

Our main result is a deterministic polynomial algorithm which produces a broadcasting scheme working in time $\mathcal{O}(D \log n + \log^2 n)$, for any n-node graph of diameter D. The method from [17], applied to our deterministic algorithm rather than to the randomized algorithm from [3], gives a *deterministic* polynomial algorithm producing a scheme with broadcasting time $\mathcal{O}(D + \log^4 n)$. (Recall that the $\mathcal{O}(D + \log^4 n)$-time scheme, obtained in [17] by applying their method to the randomized algorithm from [3], was itself randomized.) The negative results from [15, 16] show that the order of magnitude of our broadcasting time cannot be improved, unless $NP \subseteq BPTIME(n^{\mathcal{O}(\log \log n)})$.

In terms of approximation ratios, we have a $\mathcal{O}(\log(n/D))$-approximation algorithm for the radio broadcast problem, whenever $D = \Omega(\log n)$. The best previous approximation ratio for the radio broadcast problem was $\mathcal{O}(\log^2(n/D))$, in [7]. It follows from the inapproximability result in [15] that we have the best possible order of approximation ratio, for $D = \Omega(\log n)$, unless $NP \subseteq BPTIME(n^{\mathcal{O}(\log \log n)})$.

1.3 Terminology and Preliminaries

The radio network is modeled as an undirected n-node graph $G = (V, E)$. The source is denoted by 0. We denote by D the radius of the graph, i.e., the largest distance from the source to any node of the network. Notice that, for undirected graphs, the diameter is of the order of the radius. Let L_j, for $j = 1, \ldots, D$, be the set of nodes of distance j from the source. L_j is called the j-th *layer*. We denote by $N_G(v)$ the set of neighbors of node v in graph G. For every node v in G, we define l_v as the index of the layer containing node v (thus, we have $v \in L_{l_v}$). A simple path from node $v \in L_j$ to node $w \in L_{j'}$ is called a *straight path*, if it intersects every layer in at most one node. A node w is called a *straight predecessor* of v in graph G if $l_w \leq l_v$ and there is a straight path from w to v in graph G. Note that v is also its own straight predecessor.

We will use the following lemma, proved in [7].

Lemma 1. *Let $H = (R, S, E)$ be an undirected bipartite graph. Suppose that all nodes in R have the source message, and no node in S has it. There exists a set $Q \subseteq R$ such that if the set of nodes transmitting in a given step is Q then at least $|S|/\ln|S|$ nodes in S get the message after this step. Moreover, the set Q can be constructed in time polynomial in the size of H.*

In [7] the authors present an algorithm, polynomial in the size of H, to construct the set Q for a given undirected bipartite graph H. Iterating this algorithm $2 \ln |S|$ times – after every step we remove all nodes in S which obtained the message – we get a scheme to inform at least $|S|/2$ nodes in S. Call this scheme CW.

We will also use another scheme. Again consider an undirected bipartite graph $H = (R, S, E)$, and suppose that all nodes in R have the source message, and no node in S has it. The simple round-robin scheme informs all nodes in S in time at most $|S|$. Call this scheme RR.

We will use log instead of ln in later definitions and calculations, where log denotes the binary logarithm.

2   The Deterministic Algorithm FDB

This section is devoted to the description of our deterministic algorithm FDB (Fast Deterministic Broadcasting) which produces a broadcasting scheme working in time $\mathcal{O}(D \log n + \log^2 n)$, for any n-node graph of diameter D. Fix such an input graph G. For simplicity of presentation we assume that n is a power of 2, in order to avoid rounding of logarithms. This assumption can be easily removed. The scheme is divided into *phases*, constructed recursively, each containing 3 stages.

In the first stage of phase 0 only the source transmits, and in remaining stages of phase 0 no action is performed. Hence after phase 0 all nodes in L_1 have the source message. For every phase $k \geq 1$ and layer number j, we define sets $S_{k,j}, R_{k,j}$ at the end of phase $k - 1$. For $k = 1$ we define (at the end of phase 0):

- $S_{1,1} = \bigcup_{j>1} L_j$ and $R_{1,1} = L_1$;
- $S_{1,j} = \emptyset$ and $R_{1,j} = \emptyset$ for layer number $j > 1$;

Suppose that we have already constructed phases $i = 0, ..., k - 1$, together with sets $S_{i+1,j}, R_{i+1,j}$ for all layer numbers j. We show how to construct phase k with corresponding sets $S_{k+1,j}, R_{k+1,j}$.

Let $H_{k,j} = (R_{k,j}, S_{k,j}, E_{k_j})$ be the bipartite graph with sets of nodes $R_{k,j}$, $S_{k,j}$, where the set of edges $E_{k,j}$ is defined as follows: $\{v, w\} \in E_{k,j}$ iff $v \in R_{k,j}$, $w \in S_{k,j}$ and there is a straight path from w to v in graph G which is one of the shortest straight paths from w to any node in $R_{k,j}$. Consider the execution of scheme CW on graph $H_{k,j}$. Since $H_{k,j}$ need not be a subgraph of G, we call the execution of scheme CW on $H_{k,j}$ a *virtual execution*. During it a node $v \in R_{k,j}$ *virtually transmits* a message (if the scheme CW schedules a transmission) and a node $w \in S_{k,j}$ *virtually receives* a message from some node in $R_{k,j}$ (if it would receive it in scheme CW). We assume that each virtually transmitted message is the same for all nodes in $R_{k,j}$, and that no node in $S_{k,j}$ has this message prior to the execution of the scheme. Actual transmitting and receiving in the underlying graph G (as opposed to their virtual counterparts) is referred to as *real* transmitting and receiving, or simply transmitting and receiving.

Phase k consists of three stages. First we present the generic scheduling scheme ($\text{GSS}(k, j)$) for phase k and layer number j, which will be used in stage 1.

Regular Substage (Lasts $2 \log n$ Steps):
Run scheme CW on graph $H_{k,j}$ to get the set $S \subseteq S_{k,j}$ of nodes that virtually receive a message in this graph. Denote by $S'_{k,j}$ the set $S_{k,j} \setminus S$.

Closing Substage (Lasts $4 \log n$ Steps):
  If $j = (k + 1) - \log n$ and $|S'_{k,j}| \leq 4 \log n$ **then**
      apply scheme RR to the subgraph of $H_{k,j}$
          induced by nodes $R_{k,j}$ and $S'_{k,j} \cap L_{j+1}$.
      $S_{k+1,j} := \emptyset$;
      $S_{k+1,j+1} := S_{k+1,j+1} \cup S_{k,j}$;
  else
      produce the sequence of empty sets of transmitters;
      $S_{k+1,j} := S_{k+1,j} \cup S'_{k,j}$;
      $S_{k+1,j+1} := S_{k+1,j+1} \cup S_{k,j} \setminus S'_{k,j}$.

Output:

List of $\mathcal{O}(\log n)$ virtual transmitter sets: concatenation of the list from the regular substage (obtained by applying scheme CW , and of the list from the closing substage (obtained by applying scheme RR or adding the list of empty sets). If the execution of scheme CW is shorter than $2 \log n$ or the execution of scheme RR is shorter than $4 \log n$, fill the remaining steps by empty sets of transmitters, to provide synchronous time in execution of phases.

Let V_k denote the set of nodes that have the source message at the end of phase $k - 1$. Note that $V_1 = N_G(0) = L_1$. Now we define phase k of the scheme.

Stage 0:

For every layer number j, initialize $S_{k+1,j} = \emptyset$, $R_{k+1,j} = \emptyset$

Stage 1:

For $r := 1$ **to 3 do**

    For all $j = r$ mod 3 **in parallel do**

        run GSS(k, j) (sets $S_{k+1,j}$ were defined in this procedure)

For every step of broadcasting, consider the union of virtual transmitter sets obtained for this step, over all layers j. This is the set of (real) transmitters to be used by the scheme in this step.

Stage 2:

Given the set V_k and the list of transmitter sets from stage 1, let V_{k+1} be the set of nodes that have the source message at the end of this stage. Construct sets $R_{k+1,j}$ as follows: for every node $w \in S_{k+1,j}$, find an arbitrary node $v \in L_j$ such that there is a straight path from w to v, and add node v to $R_{k+1,j}$.

Figure 1 shows phases 0 and 1 of the algorithm. Part (a) depicts phase 0, part (b) shows virtual transmissions during phase 1, and part (c) shows new sets $S_{2,1}, S_{2,2}$ at the end of phase 1.

Intuition. The reason of dividing Stage 1 into three phases is to avoid interference between transmitters from close layers. The intuitive meaning of sets $S_{k,j}$ is the following: every node of $S_{k,j}$ has the property that all its straight predecessors in layer L_j already have the source message after phase k. Sets $R_{k,j}$, included in layer L_j, are intended to transmit the message to sets $S_{k,j}$. As we will show in Lemma 3, messages from nodes in $R_{k,j}$ will be received by nodes in $S_{k,j} \cap L_{j+1}$ not only virtually but also in the real execution in graph G.

The idea of procedure GSS is to push at least half of the nodes from $S_{k,j}$ to $S_{k+1,j+1}$ in the regular stage, using protocol CW, and to push all these nodes in the closing stage, using protocol RR, if only fewer than $4 \log n$ nodes are remaining. Nodes pushed out to the set $S_{k+1,j+1}$, associated with the next layer $j + 1$, will be taken care of in phase $k + 1$. We will prove later that if a node v virtually receives the message then one of its straight predecessors really receives the message in graph G, hence a really informed predecessor of node v is now in layer $j + 1$, so we put node v in the set $S_{k+1,j+1}$. We will also prove that

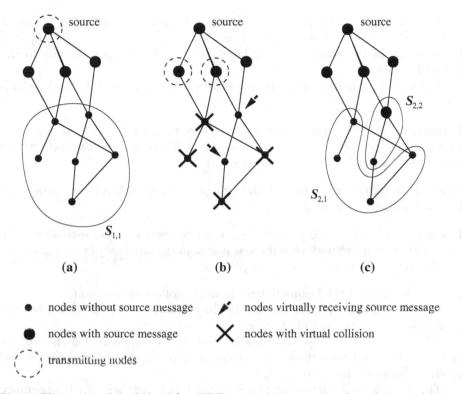

(a)   (b)   (c)

● nodes without source message   ✦ nodes virtually receiving source message

● nodes with source message   ✕ nodes with virtual collision

( ) transmitting nodes

Fig. 1. Phases 0 and 1 of algorithm FDB: transmissions, virtual receiving and corresponding sets $S_{1,1}, S_{2,1}, S_{2,2}$.

the set $S_{k,k+1-\log n}$ is small, so we can, using additional sets of transmitters in the closing substage, push all the nodes from this set to the set $S_{k+1,k+2-\log n}$ (not only half of them as in the regular substage). This is because, in the closing substage, we use protocol RR which guarantees that all nodes in set $S_{k,k+1-\log n}$ virtually receive a message. Since their informed predecessors are in the next layer now, they may be considered in the set corresponding to the next layer.

Remark. The set V_{k+1} is obtained by adding to the set V_k new nodes that receive the source message in graph G for the first time during phase k. Although sets of real transmitters used in the scheme are equal to sets of virtual transmitters output by procedure GSS, the fact of virtual receiving of a message by node $w \in S_j$, when scheme CW is executed on graph $H_{k,j}$, does not necessarily mean that node w gets the source message in the real execution of our scheme in graph G in phase k. For example, if w is not a neighbor of any node in $R_{k,j}$, it does not get the source message, and consequently it is not added to V_{k+1}.

3   Analysis of the Broadcasting Time

In this section we show that the broadcasting scheme, produced by our algorithm FDB for an n-node graph G of diameter D, has broadcasting time $\mathcal{O}(D \log n + \log^2 n)$.

Assume that $n \geq 8$ - otherwise centralized broadcasting is trivially done in constant time. Notice that scheme RR, working in time at most $|S|$, completes broadcasting in the closing stage, if $|S| \leq 4 \log n$. We will use this fact in the proof of Lemma 4

Suppose that our scheme completes broadcasting in phase t. Consider phase $k \leq t$.

Lemma 2. *Consider sets of transmitters corresponding to the same time step in Stage 1 of the scheme, for a fixed $r = 1, 2, 3$. These sets are disjoint and the sets of their neighbors in G are also disjoint.*

Proof. This follows from the fact that, for a given r, such sets are subsets of layers at distance at least 3.

Lemma 3. *If node $v \in S_{k,j}$, for $j < l_v$, virtually receives a message during the regular substage of phase k then there is a straight predecessor of v in layer $j+1$ which receives the source message in phase k.*

Proof. The proof of the lemma follows from the following invariant:

(a) if $v \in S_{k,j}$, for $j < l_v$, then a straight predecessor of v in layer j has the source message at the end of phase $k - 1$;

(b) if $v \in S_{k,j}$, for $j < l_v$, virtually receives a message during the regular substage of phase k then there is a straight predecessor of v in layer $j+1$ which receives the source message in phase k.

The proof is by induction on phase k. For $k = 1$, only set $S_{1,1}$ is nonempty among sets $S_{1,j}$. Consider node $v \in S_{1,1}$. Let $w \in L_1$ be the straight predecessor of v. Node w has the source message at the end of phase $0 = k - 1$ by definition of phase 0 (in phase 0 only the source transmits and every node in L_1 gets the source message). This proves invariant (a). The arguments for part (b) are similar as in the induction step proved below.

Suppose that the invariant holds for all positive integers $k' < k$. We prove it for k. First we prove invariant (a) for k. We may assume that $j > 1$ (otherwise, invariant (a) follows from the fact that all nodes in L_1 have the source message at the end of phase 0). Let $w \in L_j$ be a straight predecessor of node v. Since $v \in S_{k,j}$, for some phase $k' < k$ node v has been added to $S_{k'+1,j}$ during phase k'. Let w' denote a straight predecessor of v, w in layer L_{j-1} - it exists since $j - 1 < j = l_w < l_v$ and w is a straight predecessor of v. By the invariant (a) for k' we have that w' has the source message at the end of phase k'. By the invariant (b) for k' we have that w gets the source message from w' during phase k', and consequently w has the source message at the end of phase $k' \leq k - 1$.

Now we prove invariant (b) for k. Consider sets of transmitters during the regular substages in phase k. (Recall that, to transmit in graph G, we use the same sets of transmitters as in virtual transmitting.) Suppose that node $v \in S_{k,j}$ virtually receives a message from node $w \in R_{k,j}$ in step l of phase k. It follows that $w \in L_j$ is a straight predecessor of v. Let $z \in L_{j+1}$ be the second node on the straight path from w to v. Hence node z really receives the source message in step l. Indeed, its informed neighbor w (the fact that w has the source message at the end of phase $k-1$ follows from invariant (a)) transmits in step l. No node from layers $j+1$ and $j+2$ is allowed to transmit in this step, in view of the division of Stage 1 into three time periods. Also, no other neighbor of z from $R_{k,j}$ transmits, since all of them are also straight predecessors of v, and if one of them transmitted then v would not virtually receive the message in step l. This completes the proof of the invariant and of the lemma.

Lemma 4. *For every k, j, such that $j \leq (k+1) - \log n \leq n$, we have $|S_{k+1,j}| = 0$.*

Proof. The lemma follows from the following invariant, for every k, j, such that $j \leq k + 1 \leq n + \log n$:

$\text{INV}(k, j)$

(a) $\left| \bigcup_{i \leq j} S_{k+1,i} \right| \leq n 2^{-k} \cdot \sum_{i=j-\log n}^{j} \binom{k}{i}$, for $j > (k+1) - \log n$,

(b) $|S_{k+1,j}| = 0$, for $j \leq (k+1) - \log n$.

The proof of the invariant is by induction on k. It is omitted due to lack of space and will appear in the journal version.

Lemma 5. *If node $v \in L_{j_v}$ does not belong to any set $S_{k+1,j}$, for $j = 1, \ldots, j_v - 1$, at the end of phase k, then v has got the source message by the end of phase k.*

Proof. The proof is by induction on phase k. For $k = 0$, the lemma follows from the definition of sets $S_{1,j}$. Suppose that the lemma is true for phase $k - 1 \geq 0$. Consider phase k.

If node v is not in any set $S_{k,j}$, for $j = 1, \ldots, j_v - 1$, then v received the source message by the end of phase $(k-1)$, in view of the inductive hypothesis. Suppose that v is in some set S_{k,j_0}, where $1 \leq j_0 \leq j_v - 1$, but not in any $S_{k+1,j}$, for $j = 1, \ldots, j_v - 1$. By the description of the scheme, v cannot belong to any $S_{k+1,j}$, for $j > j_0 + 1$. It follows that $v \in S_{k+1,j_0+1}$ and $j_0 + 1 = j_v$. Two cases are possible:

Case 1. $j_0 \neq (k+1) - \log n$.

Since $v \in S_{k+1,j_0+1}$, it follows that v has virtually received the message during the regular substage of phase k. Consequently, using inequality $j_0 < j_v$, Lemma 3 implies that the straight predecessor of v in layer $j_0 + 1 = j_v$ has got the source message by the end of phase $k - 1$. However, the only such node is v itself. This concludes the proof in Case 1.

Case 2. $j_0 = (k+1) - \log n$.

In this case the closing substage is applied to nodes in j_0 during phase k. Note that, by Lemma 4, we have $|S_{k+1,j_0}| = 0$ and $|S'_{k,j}| \leq 4 \log n$. Hence the natural properties of round-robin scheme RR imply that node $v \in S'_{k,j_0}$ has got the source message by the end of phase k, since its predecessor does single transmission during scheme RR.

Theorem 1. *The broadcasting scheme, produced by the algorithm FDB for an n-node graph G of diameter D, has broadcasting time $\mathcal{O}(D \log n + \log^2 n)$.*

Proof. Each phase of the broadcasting scheme lasts $\mathcal{O}(\log n)$ steps. It remains to show that $\mathcal{O}(D+\log n)$ phases are enough to complete broadcasting. By Lemma 4 we get $|S_{k+1,j}| = 0$, for $k = D-1+\log n$ and all $j \leq D$. By Lemma 5 this implies that all nodes get the source message by the end of phase $D - 1 + \log n$, which completes the proof.

Corollary 1. *Algorithm FDB applied to the clustering method from [17] produces a broadcasting scheme with broadcasting time $\mathcal{O}(D + \log^4 n)$, for every n-node graph G of diameter D.*

Corollary 2. *Algorithm FDB combined with the clustering method from [17] yields a $\mathcal{O}(\log(n/D))$-approximation deterministic algorithm for the radio broadcast problem, whenever $D = \Omega(\log n)$.*

Proof. For $D = \mathcal{O}(\log^4 n)$, we have $\mathcal{O}(\log(n/D)) = \mathcal{O}(\log n)$, so our broadcasting time $\mathcal{O}(D \log n + \log^2 n)$, exceeding the lower bound D by a factor $\mathcal{O}(\log n)$, is a $\mathcal{O}(\log(n/D))$-approximation in this range. For $D = \Omega(\log^4 n)$, we have $\mathcal{O}(D+\log^4 n) = \mathcal{O}(D)$, so we have a constant approximation ratio in this range.)

Corollary 3. *The broadcasting time of the scheme produced by the algorithm FDB has optimal order of magnitude, unless $NP \subseteq BPTIME(n^{\mathcal{O}(\log\log n)})$.*

Proof. Assume that $NP \subseteq BPTIME(n^{\mathcal{O}(\log\log n)})$ does not hold. If $D = \Omega(\log n)$ then our broadcasting time is $\mathcal{O}(D \log n)$, which cannot be improved because the result from [15] implies that $o(\log n)$-approximation of the radio broadcast problem is impossible. If $D = \mathcal{O}(\log n)$ then our broadcasting time is $D = \mathcal{O}(\log^2 n)$. A better order of magnitude would give time $opt(G) + o(\log^2 n)$, precluded by the result from [16].

4   Complexity of the Algorithm FDB

We showed that the algorithm FDB constructs a broadcasting scheme with broadcasting time $\mathcal{O}(D \log n + \log^2 n)$, for every n-node graph of diameter D. In this section we estimate the running time of the algorithm itself.

It follows from [7] that the construction of scheme CW on an m-node graph can be done in time $\mathcal{O}(m^3 \log m)$. The construction of scheme RR can be done in time $\mathcal{O}(m)$. We now estimate the time to construct schemes used in one execution of procedure GSS. The regular and closing substages contain one execution of CW for $m = \mathcal{O}(n)$, and one execution of RR for $m = \mathcal{O}(\log n)$, respectively. Thus the time to construct these schemes is $\mathcal{O}(n^3 \log n)$. The rest of the procedure takes time $\mathcal{O}(n)$, hence the total time of one execution of procedure GSS is $\mathcal{O}(n^3 \log n)$.

Hence we get the following estimate of the running time of algorithm FDB:

Time of stage 0 in all D phases – $\mathcal{O}(D)$.

Time of stage 1 in all D phases – $\mathcal{O}(D)$ executions of GSS and finding the union of $\mathcal{O}(D)$ disjoint sets of transmitting nodes – the total time $\mathcal{O}(Dn^3 \log n)$.

Time of stage 2 in all D phases – For each node v in $S_{k+1,j}$, we find the shortest straight path to layer L_j. Either $v \in S_{k,j}$ and then this path is the same as it was, or $v \in S_{k,j-1}$ and the new straight path is the old one without one node in L_{j-1}. The constructed sets $R_{k,j}$ are disjoint, hence the total time is $\mathcal{O}(n)$.

Theorem 2. *Algorithm FDB runs in time $\mathcal{O}(Dn^3 \log n)$ for a give n-node graph G of diameter D.*

Remark. We stress again the distinction between the *running time* of algorithm FDB which constructs a broadcasting scheme for a given graph G, and the *broadcasting time* of the obtained broadcasting scheme for this graph.

5 Conclusion

We presented a deterministic polynomial algorithm which produces a broadcasting scheme working in time $\mathcal{O}(D \log n + \log^2 n)$, for any n-node graph of diameter D. Combined with the clustering method from [17], our algorithm yields a deterministic polynomial algorithm producing a scheme with broadcasting time $\mathcal{O}(D + \log^4 n)$. It remains open if the additive term $\log^4 n$ can be improved. More precisely: what is the fastest broadcasting scheme that can be produced by a polynomial deterministic algorithm, given an n-node graph of diameter D?

References

1. Alon, N., Bar-Noy, A., Linial, N., Peleg, D.: A lower bound for radio broadcast. Journal of Computer and System Sciences **43** (1991) 290–298
2. Awerbuch, B.: A new distributed depth-first-search algorithm. Information Processing Letters **20** (1985) 147–150
3. Bar-Yehuda, R., Goldreich, O., Itai, A.: On the time complexity of broadcast in radio networks: an exponential gap between determinism and randomization. Journal of Computer and System Sciences **45** (1992) 104–126
4. Basagni, S., Bruschi, D., Chlamtac, I.: A mobility–transparent deterministic broadcast mechanism for ad hoc networks. IEEE/ACM Transactions on Networking **7** (1999) 799–807
5. Bruschi, D., Del Pinto, M.: Lower bounds for the broadcast problem in mobile radio networks. Distributed Computing **10** (1997) 129–135
6. Chlamtac, I., Kutten, S.: On broadcasting in radio networks - problem analysis and protocol design. IEEE Transactions on Communications **33** (1985) 1240–1246
7. Chlamtac, I., Weinstein, O.: The wave expansion approach to broadcasting in multihop radio networks. IEEE Transactions on Communications **39** (1991) 426–433
8. Chlebus, B., Gąsieniec, L., Gibbons, A., Pelc, A., Rytter, W.: Deterministic broadcasting in unknown radio networks. Distributed Computing **15** (2002) 27–38
9. Chlebus, B., Gąsieniec, L., Östlin, A., Robson, J.M.: Deterministic radio broadcasting. *Proc. 27th International Colloquium on Automata, Languages and Programming* (ICALP'2000), LNCS 1853, 717–728

10. Chrobak, M., Gąsieniec, L., Rytter, W.: Fast broadcasting and gossiping in radio networks. *Proc. 41st Symposium on Foundations of Computer Science* (FOCS'2000) 575–581
11. Clementi, A., Crescenzi, P., Monti, A., Penna, P., Silvestri, R.: On computing ad-hoc selective families. *Proc. 5th International Workshop on Randomization and Approximation Techniques in Computer Science* (RANDOM'2001) 211–222
12. Clementi, A., Monti, A., Silvestri, R.: Selective families, superimposed codes, and broadcasting on unknown radio networks. *Proc. 12th Annual ACM-SIAM Symposium on Discrete Algorithms* (SODA'2001) 709–718
13. Czumaj, A., Rytter, W.: Broadcasting algorithms in radio networks with unknown topology. *Proc. 44th Symposium on Foundations of Computer Science* (FOCS'2003) 492–501
14. De Marco, G., Pelc, A.: Faster broadcasting in unknown radio networks. Information Processing Letters **79** (2001) 53–56
15. Elkin, M., Kortsarz, G.: A logarithmic lower bound for radio broadcast. Journal of Algorithms **52** (2004) 8–25
16. Elkin, M., Kortsarz, G.: Polylogarithmic additive inapproximability of the radio broadcast problem. *Proc. 7th International Workshop on Approximation Algorithms for Combinatorial Optimization Problems* (APPROX'2004), to appear
17. Elkin, M., Kortsarz, G.: Improved broadcast schedule for radio networks. Manuscript, 2004
18. Gaber, I., Mansour, Y.: Centralized broadcast in multihop radio networks. Journal of Algorithms **46** (1) (2003) 1–20
19. Kowalski, D., Pelc, A.: Deterministic broadcasting time in radio networks of unknown topology. *Proc. 43rd Symposium on Foundations of Computer Science* (FOCS'2002) 63–72
20. Kowalski, D., Pelc, A.: Faster deterministic broadcasting in ad hoc radio networks. *Proc. 20th Annual Symposium on Theoretical Aspects of Computer Science* (STACS'2003), LNCS 2607, 109–120
21. Kowalski, D., Pelc, A.: Broadcasting in undirected ad hoc radio networks. *Proc. 22nd ACM Symposium on Principles of Distributed Computing* (PODC'2003) 73–82
22. Kushilevitz, E., Mansour, Y.: An $\Omega(D \log(N/D))$ lower bound for broadcast in radio networks. SIAM Journal on Computing **27** (1998) 702–712

Convergence Issues in Competitive Games

Vahab S. Mirrokni[1] and Adrian Vetta[2]

[1] Massachusetts Institute of Technology
mirrokni@theory.lcs.mit.edu
[2] McGill University
vetta@math.mcgill.ca

Abstract. We study the speed of convergence to approximate solutions in iterative competitive games. We also investigate the value of Nash equilibria as a measure of the cost of the lack of coordination in such games. Our basic model uses the underlying best response graph induced by the selfish behavior of the players. In this model, we study the value of the social function after multiple rounds of best response behavior by the players. This work therefore deviates from other attempts to study the outcome of selfish behavior of players in non-cooperative games in that we dispense with the insistence upon only evaluating Nash equilibria. A detailed theoretical and practical justification for this approach is presented. We consider non-cooperative games with a submodular social utility function; in particular, we focus upon the class of valid-utility games introduced in [13]. Special cases include basic-utility games and market sharing games which we examine in depth. On the positive side we show that for basic-utility games we obtain extremely quick convergence. After just one round of iterative selfish behavior we are guaranteed to obtain a solution with social value at least $\frac{1}{3}$ that of optimal. For n-player valid-utility games, in general, after one round we obtain a $\frac{1}{2n}$-approximate solution. For market sharing games we prove that one round of selfish response behavior of players gives $\Omega(\frac{1}{\ln n})$-approximate solutions and this bound is almost tight. On the negative side we present an example to show that even in games in which every Nash equilibrium has high social value (at least half of optimal), iterative selfish behavior may "converge" to a set of extremely poor solutions (each being at least a factor n from optimal). In such games Nash equilibria may severely underestimate the cost of the lack of coordination in a game, and we discuss the implications of this.

1 Introduction

Traditionally, research in operation research has focused upon finding a global optimum. Computer scientists have also long studied the effects of lack of different resources, mainly the lack of computational resources, in optimization. Recently, the *lack of coordination* inherent in many problems has become an important issue in computer science. A natural response to this has been to analyze Nash equilibria in these games. Of particular interest is the *price of anarchy* in a game [8]; this is the worst case ratio between an optimal social solution and

K. Jansen et al. (Eds.): APPROX and RANDOM 2004, LNCS 3122, pp. 183–194, 2004.
© Springer-Verlag Berlin Heidelberg 2004

a Nash equilibrium. Clearly, a low price of anarchy may indicate that a system has no need for a single regulatory authority. Conversely, a high price of anarchy is indicative of a poorly functioning system in need of some regulation.

In this paper we move away from the use of Nash equilibria as the solution concept in a game. There are several reasons for this. The first reason relates to use of non-randomized (pure) and randomized (mixed) strategies. Often pure Nash equilibria may not exist, yet in many games the use of a randomized (mixed) strategy is unrealistic. This necessitates the need for an alternative in evaluating such games.

Secondly, Nash equilibria represent stable points in a system. Therefore (even if pure Nash equilibria exist), they are a more acceptable solution concept if it is likely that the system does converge to such stable points. In particular, the use of Nash equilibria seems more valid in games in which Nash equilibria arise when players iteratively engage in selfish behavior. The time it takes for *convergence* to Nash equilibria, however, may be extremely long. So, from a practical viewpoint, it is important to evaluate the speed or rate of convergence. Moreover, in many games it is not the case that repeated selfish behavior always leads to Nash equilibria. In these games, it seems that another measure of the cost of the lack of coordination would be useful.

As is clear, these issues are particularly important in games in which the use of pure strategies and repeated moves are the norm, for example, auctions. We remark that for most practical games these properties are the rule rather than the exception (this observation motivates much of the work in this paper). For these games, then, it is not sufficient to just study the value of the social function at Nash equilibria. Instead, we must also investigate the speed of convergence (or non-convergence) to an equilibrium. Towards this goal, we will not restrict our attention to Nash equilibria but rather prove that after some number of improvements or best responses the value of the social function is within a factor of the optimal social value. We tackle this by modeling the behavior of players using the underlying best response graph on the set of strategy states. We consider (best response) paths in this graph and evaluate the social function at states along these paths. The rate of convergence to high quality solutions (or Nash equilibria) can then be measured by the length of the path. As mentioned, it may the case that there is no such convergence. In fact, in Section 4.2, it is shown that instead we have the possibility of "convergence" to non-Nash equilibria with a bad social value. Clearly such a possibility has serious implications for the study of stable solutions in games.

An overview of the paper is as follows. In section 2, we describe the problem formulations and model. In section 3, we discuss other work and their relation to this paper. In section 4, we give results for valid-utility and basic-utility games. We prove that in valid-utility games we obtain a $\frac{1}{2n}$-approximate solution if each player sequentially makes one best response move. For basic-utility games we obtain a $\frac{1}{3}$-approximate solution in general, and a $\frac{1}{2}$-approximate solution if each player initially used a null strategy. We then present a valid-utility game in which every Nash equilibria is at least half-optimal and, yet, iterative selfish behavior may lead to only $O(\frac{1}{n})$-approximate solutions. In section 5, we examine market sharing games and show that we obtain $\Omega(\frac{1}{\ln n})$-approximate solutions after one best response move each. Finally, in section 6, we discuss other classes of games and present some open questions.

2   Preliminaries

In this section, we define necessary game theoretic notations to formally describe the classes of games that we study in the next sections. The game is defined as the tuple $(U, \{S_j\}, \{\alpha_j()\})$. Here U is the set of players or agents. Associated with each player j is a disjoint groundset V_i, and S_j is a collection of subsets of V_j. The elements in the a groundset correspond to acts a player may make, and hence the subsets correspond to strategies. We denote player j's strategy by $s_j \in S_j$. Finally, $\alpha_j : \Pi_j S_j \to \mathbb{R}$ is the private payoff or utility function for agent j, given the set of actions of all the players. In a non-cooperative game, we assume that each selfish agent wishes to maximize its own payoff.

Definition 1. *A function $f : 2^V \to \mathbb{R}$ is a set function on the groundset V. A set function $f : 2^V \to \mathbb{R}$ is submodular if for any two sets $A, B \subseteq V$, we have $f(A) + f(B) \geq f(A \cap B) + f(A \cup B)$. The function is non-decreasing if $f(X) \leq f(Y)$ for any $X \subseteq Y \subset V$.*

For each game we will have a social objective function $\gamma : \Pi_j S_j \to \mathbb{R}$. (We remark that γ can be viewed as a set function on the groundset $\cup V_i$.) Our goal will be to analyze the social value of solutions produced the selfish behavior of the agents. Specifically, we will focus upon the class of games called *valid-utility games*.

Definition 2. *Let $\mathcal{G}(U, \{S_j\}, \{\alpha_j\})$ be a non-cooperative game with social function γ. $\mathcal{G}$ is a valid-utility game if it satisfies the properties:*

- *γ is a submodular set function.*
- *The payoff of a player is at least equal to the difference in the social function when the player participates versus when it does not.*
- *The sum of the utility or payoff functions for any set of strategies should be less than or equal to the social function.*

This framework encompasses a wide range of games in facility location, traffic routing and auctions [13]. Here, as our main application, we consider the market sharing game which is a special case of valid-utility games (and also congestion games). We define this game formally in Section 5.

2.1   Best Response Paths

We model the selfish behavior of players using an underlying *state graph*. Each vertex in the graph represents a strategy state $S = (s_1, s_2, \ldots, s_n)$. The arcs in the graph corresponds to best response moves by the players. Formally, we have

Definition 3. *The state graph,* $\mathcal{D} = (\mathcal{V}, \mathcal{E})$, *is a directed graph. Each vertex in* $\mathcal{V}$ *corresponds to a strategy state. There is an arc from state* S *to state* S' *with label* j *if the only difference between* S *and* S' *is only in the strategy of player* j; *and player* j *plays his best response in strategy state* S *to go to* S'.

Observe that the state graph may contain loops. A best response path is a directed path in the state graph. We say that a player i plays in the best response path $\mathcal{P}$, if at least one of the edges of $\mathcal{P}$ is labelled i. Assuming that players optimize their best response function sequentially (and not in parallel), we can evaluate the social value of states on a best response path in the state graph. In particular, given a best response path starting from an arbitrary state, we will be most interested in the social value of the the last state on the path. Notice that if we do not allow every player to make a best response on a path $\mathcal{P}$ then we may not be able to bound the social value of a state with respect to the optimal solution. This follows from the fact that the actions of a single player may be very important for producing solutions of high social value. Hence, we consider the following models:

One-round path: Consider an arbitrary ordering of players $i_1, \ldots, i_n$. Path $\mathcal{P}$ is a *one-round path* if it starts from an arbitrary state and edges of P are labelled $i_1, i_2, \ldots, i_n$ in this order.

Covering path: A best response path $\mathcal{P}$ is a *covering path* if each player plays at least once on the path.

k-**Covering path:** A best response path $\mathcal{P}$ is a *k-covering path* if there are k covering paths $\mathcal{P}_1, \mathcal{P}_2, \ldots, \mathcal{P}_k$ such that $\mathcal{P} = (\mathcal{P}_1, \mathcal{P}_2, \ldots, \mathcal{P}_k)$.

Observe that a one-round path is a covering path. Note that in the one-round path we let each player play his best response exactly one time, but in a covering path we let each player play at least one time. Both of these models have justifications in extensive games with complete information. In these games, the action of each player is observed by all the other players. As stated, for a non-cooperative game $\mathcal{G}$ with a social function γ, we are interested in the social value of states (especially the final state) along one-round, covering, and k-covering paths.

A Simple Example. Here, we consider covering paths in a basic load balancing game; Even-Dar et al. [2] considered the speed of convergence to Nash equilibria in these games. There are n jobs that can be scheduled on m machines. It takes p_j units of time for job j to run on any of the machines. The social objective function is the maximum makespan over all machines. The private payoff of a job, however, is the inverse of the makespan of the machine that the job is scheduled on. Thus each job wants to be scheduled on a machine with as small a makespan as possible. It is easy to verify that the price of anarchy in this game is at most 2. It is also known that this game has pure Nash equilibria and the length of any best-response path in this game is at most n^2 [1]. In addition, from any state there is a path of length at most n to some pure Nash equilibrium [12]. It may, however, take much more than n steps to converge to a pure Nash equilibrium. Hence, our goal here is to show that the social value of any state at the end of a covering path is within a factor 2 of optimal. So take a covering path $\mathcal{P} = (S_1, S_2, \ldots, S_k)$. Let i^* be the machine with the largest makespan at state S_k and let the load this machine be L^*. Consider the last job j^* that was scheduled on machine i, and let the schedule after scheduling j^* be S_t. Ignoring job j^*, at time t the makespan of all the machines is at least $L^* - p_{j^*}$. If not, job j^* would not have been scheduled at machine i^*. Consequently, we have $\sum_{1 \leq j \leq n} p_i \geq m(L^* - p_{j^*})$. Thus, if OPT is the value of the optimal schedule, then $\text{OPT} \geq \sum_{1 \leq j \leq n} p_j/m \geq L^* - p_{j^*}$. Clearly $\text{OPT} \geq p_{j^*}$ and so $L^* = L - p_{j^*} + p_{j^*} \leq 2\text{OPT}$.

3 Related Work

Here we give a brief overview of related work in this area. The consequences of selfish behavior and the question of efficient computation of Nash equilibria have recently drawn much attention in computer science [8, 7]. Moreover, the use of the price of anarchy [8] as a measure of the cost of the lack of coordination in a game is now widespread, with a notable success in this realm being the selfish routing game [11]. Roughgarden and Tardos [10] also generalize their results on selfish routing games to non-atomic congestion games. A basic result of Rosenthal [9] defines congestion games for which pure strategy Nash equilibria exist. Congestion games belong to the class of potential games [6] for which any best-response path converges to a pure Nash equilibrium. Milchtaich [5] studied player-specific congestion games and the length of best-response paths in this set of games. Even-Dar et al. [2] considered the convergence time to Nash equilibria in variants of a load balancing game. They bound the number of required steps to reach a pure Nash equilibrium in these games. Recently, Fabrikant et al. [3] studied the complexity of finding a pure strategy Nash equilibrium in general congestion games. Their PLS-completeness results show that in some congestion games (including network congestion games) the length of a best-response path in the state graph to a pure Nash equilibrium might be exponential. Goemans et al. [4] considered market sharing games in modeling a decentralized content distribution policy in ad-hoc networks. They show that the market sharing game is a special case of valid-utility games and congestion games. In addition, they give improved bounds for the price of anarchy in some special cases, and present an algorithm to find the pure strategy Nash equilibrium in the uniform market sharing game. The results of Section 5 extend their results.

4 Basic-Utility and Valid-Utility Games

In this section we consider valid-utility games. First we present results concerning the quality of states at the end of one-round paths. Then we give negative results concerning the non-convergence of k-covering paths.

4.1   Convergence

We use the notation from [13]. In particular, a strategy state is denoted by $S = \{s_1, s_2, \ldots, s_k\} \in \mathcal{S}$. Here s_i is the strategy of player i, where $s_i \subseteq V_i$ and V_i is a groundset of elements (with each element corresponding to an action for player i); $\emptyset_i$ corresponds to a null strategy. We also let $S \oplus s_i' = \{s_1, \ldots, s_{i-1}, s_i', s_{i+1}, \ldots, s_k\}$, i.e. the strategy set obtained if agent i changes its strategy from s_i to s_i'. The social value of a state S is $\gamma(S)$, where γ is a submodular function on the groundset $\cup_i V_i$. For simplicity, in this section we will assume that γ is non-decreasing. Similar results, however, do hold in the general case.

We also denote by $\alpha_i(S)$ the private return to player i from the state S, and we let $\gamma_{s_i}'(S) = \gamma(S \cup s_i) - \gamma(S)$. Thus, formally, the second and third conditions in definition 2 are $\alpha_i(S) \geq \gamma_{s_i}'(S \oplus \emptyset_i)$ and $\sum_i \alpha_i(S) \leq \gamma(S)$, respectively. Of particular interest is the subclass of valid-utility games where we always have $\alpha_i(S) = \gamma_{s_i}'(S \oplus \emptyset_i)$; these games are called *basic-utility games* (examples of which include competitive facility location games).

Theorem 1. *In basic-utility games, the social value of a state at the end of a one-round path is at least $\frac{1}{3}$ of the optimal social value.*

Proof. Let $\Omega = \{\sigma_1, \ldots, \sigma_n\}$ denote the optimum state, and let $T = \{t_1, \ldots, t_n\}$ and $S = \{s_1, \ldots, s_n\}$ be the initial state and final states on the one-round path, respectively; we assume the agents play best response strategies in the order $1, 2, \ldots, n$. So $T^i = \{s_1, \ldots, s_{i-1}, t_i, \ldots, t_n\}$ is a state in our one-round path $\mathcal{P} = \{T = T^1, T^2, \ldots, T^{n+1} = S\}$. Thus, by basicness and the fact that the players use best response strategies, we have $\sum_i \alpha_i(T^{i+1}) = \sum_{i=1}^n \gamma_{s_i}'(T^i \oplus \emptyset_i) \geq \sum_i \gamma_{\sigma_i}'(T^i \oplus \emptyset_i)$. It follows by submodularity that $\sum_i \alpha_i(T^{i+1}) \geq \sum_i \gamma_{\sigma_i}'(S \cup T^i \oplus \emptyset_i) \geq \gamma(\Omega) - \gamma(S \cup T) \geq \gamma(\Omega) - \gamma(S) - \gamma(T)$. Moreover, by basicness, $\gamma(S) - \gamma(T) = \sum_{i=1}^n \gamma(T^{i+1}) - \gamma(T^i) = \sum_i \gamma(T^{i+1}) - \gamma(T^i \oplus \emptyset_i) - \sum_i \gamma(T^i) - \gamma(T^i \oplus \emptyset_i) = \sum_i \gamma_{s_i}'(T^i \oplus \emptyset_i) - \sum_i \gamma_{t_i}'(T^i \oplus \emptyset_i) = \sum_i \alpha_i(T^{i+1}) - \sum_i \gamma_{t_i}'(T^i \oplus \emptyset_i)$. Let $\bar{T}^i = \{\emptyset_1, \ldots, \emptyset_{i-1}, t_i, \ldots, t_n\}$. Then, by submodularity, $\gamma(S) - \gamma(T) \geq \sum_i \alpha_i(T^{i+1}) - \sum_i \gamma_{t_i}'(\bar{T}^i \oplus \emptyset_i) = \sum_i \alpha_i(T^{i+1}) - \gamma(T)$. Hence, $\gamma(S) - \gamma(T) \geq \gamma(\Omega) - \gamma(S) - 2\gamma(T)$. Since $\gamma(S) \geq \gamma(T)$, it follows that $3\gamma(S) \geq$ OPT. $\square$

We suspect this result is not tight and that a factor 2 guarantee is possible. Observe, though, that the above proof gives this guarantee for the special case in which the initial strategy state is $T = \emptyset$.

Theorem 2. *In basic-utility games, the social value of a state at the end of a one-round path beginning at $T = \emptyset$ is at least $\frac{1}{2}$ of the optimal social value and this bound is tight.* $\square$

It is known that any Nash equilibria in any valid-utility game has value within a factor 2 of optimal. So here after just one round in a basic-utility game we obtain a solution which matches this guarantee. However for non-basic-utility games, the situation can be different. We can only obtain the following guarantee, which is tight to within a constant factor.

Theorem 3. *In general valid-utility games, the social value of some state on any one-round path is at least $\frac{1}{2n}$ of the optimal social value.*

Proof. Let $\gamma(\Omega) = \text{OPT}$ and assume that $\gamma(t_1, t_2, \ldots, t_n) \leq \frac{1}{2n}\text{OPT}$. Again, agent i changes its strategy from t_i to s_i given the collection of strategies $T^i = \{s_1, \ldots, s_{i-1}, t_i, \ldots, t_n\}$. If at any state in the path $\mathcal{P} = \{T = T^1, T^2, \ldots, T^{n+1} = S\}$ we have $\alpha_i(T^{i+1}) \geq \frac{1}{2n}\text{OPT}$ then we are done. To see this note that $\alpha_i(T^{i+1}) \geq \gamma(T^{i+1}) - \gamma(T^i \oplus \emptyset_i) \geq 0$, since γ is non-decreasing. Thus $\gamma(T^{i+1}) \geq \sum_j \alpha_j(T^{i+1}) \geq \alpha_i(T^{i+1}) \geq \frac{1}{2n}\text{OPT}$. Hence we have, $\gamma(t_1 \cup s_1, \ldots, t_i \cup s_i, t_{i+1}, \ldots, t_n) - \gamma(T) = \sum_{j=1}^{i} \gamma(s_1 \cup t_1, \ldots, t_j \cup s_j, t_{j+1}, \ldots, t_n) - \gamma(s_1 \cup t_1, \ldots, s_{j-1} \cup t_{j-1}, t_j, t_{j+1}, \ldots, t_n) \leq \sum_{j=1}^{i} \gamma(T^j \cup s_j) - \gamma(T^j) \leq \sum_{j=1}^{i} \gamma(T^{j+1}) - \gamma(T^{j+1} \oplus \emptyset_j) \leq \sum_{j=1}^{i} \alpha_j(T^{j+1}) < \frac{i}{2n}\text{OPT}$. Consequently $\gamma(\sigma_1 \cup t_1 \cup s_1, \ldots, \sigma_i \cup t_i \cup s_i, \sigma_{i+1} \cup t_{i+1}, \ldots, \sigma_n \cup t_n) - \gamma(t_1 \cup s_1, \ldots, t_i \cup s_i, t_{i+1}, \ldots, t_n) \geq \text{OPT} - \gamma(t_1 \cup s_1, \ldots, t_i \cup s_i, t_{i+1}, \ldots, t_n) \geq \text{OPT} - \gamma(S) - \frac{i}{2n}\text{OPT} \geq \frac{2n-i-1}{2n}\text{OPT}$. Thus, there is a $j > i$ such that $\gamma'_{\sigma_j}(T^{i+1}) \geq \gamma'_{\sigma_j}(t_1 \cup s_1, \ldots, t_i \cup s_i, t_{i+1}, \ldots, t_n) \geq \frac{2n-2i-1}{2n(n-i)}\text{OPT} \geq \frac{1}{2n}\text{OPT}$. Therefore we must obtain $\alpha_j(T^{j+1}) \geq \frac{1}{2n}\text{OPT}$ for some $j > i$. $\square$

4.2 Cyclic Equilibria

Here we show that Theorem 3 is essentially tight, and discuss the consequences of this. Specifically, there is the possibility of convergence to low quality states in games in which every Nash equilibria is of high quality.

Theorem 4. *There are valid-utility games in which every solution on a k-covering path has social value at most $\frac{1}{n}$ of the optimal solution.*

Proof. We consider the following n-player game. The groundset of player i consists of three elements x_i, x'_i and y_i. Let $X = \cup_i x_i$ and $X' = \cup_i x'_i$. We construct a non-decreasing, submodular social utility function in the following manner. For each agent $1 \leq i \leq n$, we have

$$\gamma'_{x_i}(S) = \begin{cases} 1 & \text{if } S \cap (X \cup X') = \emptyset \\ 0 & \text{otherwise} \end{cases}$$

We define $\gamma'_{x'_i}(S)$ in an identical manner. Finally, for each agent $1 \leq i \leq n$, we let $\gamma'_{y_i}(S) = 1, \forall S$. Clearly, the social utility function γ is non-decreasing. To see that it is submodular, it suffices to consider any two sets $A \subseteq B$. If $\gamma'_{x_i}(B) = 1$ then $\gamma'_{x_i}(A) = 1$. This follows as $B \cap (X \cup X') = \emptyset$ implies that $A \cap (X \cup X') = \emptyset$. Hence $\gamma'_{x_i}(A) \geq \gamma'_{x_i}(B), \forall i, \forall A \subseteq B$. Similarly $\gamma'_{x'_i}(A) \geq \gamma'_{x'_i}(B), \forall i, \forall A \subseteq B$. Finally, $\gamma'_{y_i}(A) = \gamma'_{y_i}(B) = 1, \forall i, \forall A \subseteq B$. It is well known that a function f is submodular if and only if $A \subseteq B$ implies $f'_j(A) \geq f'_j(B), \forall j \in V - B$. Thus γ is submodular.

With this social utility function, we construct a valid utility system. To do this, we create private utility functions α_i using the following rule (except for a few cases given below), where $X_i = S \cap (x_i \cup x'_i)$.

$$\alpha_i(S) = \begin{cases} 1 + \frac{|X_i|}{|(X \cup X') \cap S|} & \text{if } y_i \in S_i \\ \frac{|X_i|}{|(X \cup X') \cap S|} & \text{if } y_i \notin S_i \end{cases}$$

In the following cases, however, we ignore the rule and use the private utilities given in the table.

s_1	s_2	s_3	$\cdots$	s_{n-1}	s_n	$\alpha_1(S)$	$\alpha_2(S)$	$\alpha_3(S)$	$\cdots$	$\alpha_{n-1}(S)$	$\alpha_n(S)$
x_1	x_2	$\emptyset_3$	$\cdots$	$\emptyset_{n-1}$	$\emptyset_n$	0	1	0	$\cdots$	0	0
x_1	x_2	x_3	$\cdots$	$\emptyset_{n-1}$	$\emptyset_n$	0	0	1	$\cdots$	0	0
				$\vdots$						$\vdots$	
x_1	x_2	x_3	$\cdots$	x_{n-1}	$\emptyset_n$	0	0	0	$\cdots$	1	0
x_1	x_2	x_3	$\cdots$	x_{n-1}	x_n	0	0	0	$\cdots$	0	1
x_1'	x_2	x_3	$\cdots$	x_{n-1}	x_n	1	0	0	$\cdots$	0	0
x_1'	x_2'	x_3	$\cdots$	x_{n-1}	x_n	0	1	0	$\cdots$	0	0
				$\vdots$						$\vdots$	
x_1'	x_2'	x_3'	$\cdots$	x_{n-1}'	x_n	0	0	0	$\cdots$	1	0
x_1'	x_2'	x_3'	$\cdots$	x_{n-1}'	x_n'	0	0	0	$\cdots$	0	1
x_1	x_2'	x_3'	$\cdots$	x_{n-1}'	x_n'	1	0	0	$\cdots$	0	0
x_1	x_2	x_3'	$\cdots$	x_{n-1}'	x_n'	0	1	0	$\cdots$	0	0
				$\vdots$						$\vdots$	
x_1	x_2	x_3	$\cdots$	x_{n-1}	x_n'	0	0	0	$\cdots$	1	0

Observe that, by construction, $\sum_i \alpha_i(S) = \gamma(S)$ for all S (including the exceptions). It remains to show that the utility system is valid. It is easy to check that $\alpha_i(S) \geq \gamma(S) - \gamma(S \oplus \emptyset_i) = \gamma_i'(S \oplus \emptyset_i)$ for the exceptions. So consider the "normal" S. If $S_i \cap (x_i \cup x_i') = \emptyset$, then $\alpha_i(S) = 1$ when $y_i \in S_i$ and $\alpha_i(S) = 0$ otherwise. In both cases $\alpha_i(S) = \gamma_i'(S \oplus \emptyset_i)$. If $S_i \cap (x_i \cup x_i') \neq \emptyset$ then

$$\gamma_i'(S \oplus \emptyset_i) = \begin{cases} 2 & \text{if } y_i \in S_i \text{ and } (S - S_i) \cap (X \cup X') = \emptyset \\ 1 & \text{if } y_i \in S_i \text{ and } (S - S_i) \cap (X \cup X') \neq \emptyset \\ 1 & \text{if } y_i \notin S_i \text{ and } (S - S_i) \cap (X \cup X') = \emptyset \\ 0 & \text{if } y_i \notin S_i \text{ and } (S - S_i) \cap (X \cup X') \neq \emptyset \end{cases}$$

Consider the first case. We have that $\alpha_i'(S) = 1 + \frac{|X_i|}{|(X \cup X') \cap S|} = 1 + \frac{|X_i|}{|(X \cup X') \cap S_i| + |(X \cup X') \cap (S - S_i)|} = 1 + \frac{|X_i|}{|X_i| + 0} = 2 = \gamma_i'(S \oplus \emptyset_i)$. It is easy to verify that in the other three cases we also have $\alpha_i(S) \geq \gamma_i'(S \oplus \emptyset_i)$. Thus our utility system is valid. It remains to choose which subsets of each players' groundset will correspond to feasible strategies in our game. We simply allow only the singleton elements (and the emptyset) to be feasible strategies. That the set of possible actions for player i are $\mathcal{A}_i = \{\emptyset_i, x_i, x_i', y_i\}$. Now it is easy to see that an optimal social solution has value n. Any set of strategies of the form $\{y_1, \ldots, y_{i-1}, z_i, y_{i+1}, \ldots, y_n\}$, where $z_i \in \{x_i, x_i', y_i\}$, $1 \leq i \leq n$ gives a social outcome of value n. However, consider the case of $n = 3$. From the start of the game, if the players behave greedily then we can obtain the sequence of strategies illustrated in Figure 1. The private payoffs given by these exceptional strategy sets mean that each arrow actually denotes a best response move by the labelled agent. However, all of the (non-trivial) strategy sets induce a social

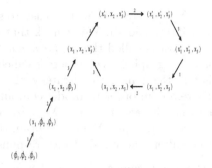

Fig. 1. Bad Cycling.

outcome of value 1, a factor 3 away from optimal. Clearly this problem generalizes to n agents. So we converge to a cycle of states all of whose outcomes are a factor n away from optimal. $\square$

So our best response path may lead to a cycle on which *every* solution is extremely bad socially, despite the fact that every Nash equilibria is very good socially (within a factor two of optimal). We call such a cycle in the state graph a *cyclic equilibria*. The presence of low quality cyclic equilibria is therefore disturbing: even if the price of anarchy is low we may get stuck in states of very poor social quality! We remark that our example is unstable in the sense that we may leave the cyclic equilibria if we permute the order in which players make there moves. We will examine in more detail the question of the stability of cyclic equilibria in a follow-up paper.

5   Market Sharing Games

In this section we consider the market sharing game. We are given a set U of n agents and a set H of m markets. The game is modelled by a bipartite graph $G = (H \cup U, E)$ where there is an edge between agent j and market i if market i is of interest to agent j (we write j is interested in market i). The value of a market $i \in H$ is represented by its query rate q_i (this is the rate at which market i is requested per unit time). The cost, to any agent, of servicing market i is C_i. In addition, agent j has a total budget B_j. It follows that a set of markets $s_j \subseteq H$ can be serviced by player j if $\sum_{i \in s_j} C_i \le B_j$; in this case we say that s_j represents a feasible action for player j. The goal of each agent is to maximise its return from the markets it services. Any agent j receives a reward (return) R_i for providing service to market i, and this reward is dependent upon the number of agents that service this market. More precisely, if the number of agents that serve market i is n_i then the reward $R_i = \frac{q_i}{n_i}$. Observe that the total reward received by all the players is equal to the total query rate of the markets being serviced (by at least one player). The resultant game is called the *market sharing game*. Observe that if $\alpha_j(S)$ is the return to agent j from state S, then the social value is $\gamma(S) = \sum_{j \in U} \alpha_j(S)$. It is then easy to show that the market sharing game is a valid-utility game [4]. We remark that the subcase in which all markets have the same cost is called the *uniform market sharing game*; the subcase in which the bipartite graph G is complete is called the *complete market sharing game*. Note that in this game, the strategy of each player is to solve a knapsack problem. Therefore, in order to model computationally constrained agents, we may assume that the agents apply λ-approximation algorithms to determine their best-response strategies. We then obtain the following theorems concerning the social value after one round of best responses moves.

Theorem 5. *In market sharing games, the social value of a state at the end of a one-round path is at least $\frac{1}{2H_n+1}$ of the optimal social value (or at least $\frac{1}{(\lambda+1)H_n+1}$ if the agents use λ-approximation algorithms).*

Proof. Let $\Omega = \{\sigma_1, \ldots, \sigma_n\}$ denote an optimum state. Here $\sigma_j \subseteq H$ is the set of markets that player j services in this optimum solution; we may also assume that each market is provided by at most one player. Let $T = \{t_1, \ldots, t_n\}$ and $S = \{s_1, \ldots, s_n\}$ be the initial state and final states on the one-round path, respectively. Again, we assume the agents play best response strategies in the order $1, 2, \ldots, n$. So in step r, using a λ-approximation algorithm, agent r changes its strategy from t_r to s_r; thus $T^r = \{s_1, \ldots, s_r, t_{r+1}, \ldots, t_n\}$ is an intermediate state in our one-round path $\mathcal{P} = \{T = T^0, T^1, \ldots, T^n = S\}$. Let $\alpha_j(S)$ be the return to agent j, then the social value of the state $S = T^n$ is $\gamma(S) = \sum_{j \in U} \alpha_j(T^n)$ So we need to show that $\sum_{j \in U} \alpha_j(T^n) \geq \frac{1}{1+(\lambda+1)H_n} \text{OPT}$. Towards this goal, we first show that $\gamma(S) = \sum_{j \in U} \alpha_j(T^n) \geq \frac{1}{H_n} \sum_{j \in U} \alpha_j(T^j)$. We know that agent j does not changes its strategy from s_r after step r. Therefore a market i has a nonzero contribution in $\gamma(S)$ if and only if market i has a nonzero contribution in the summation $\sum_{j \in U} \alpha_j(T^j)$. For any market i, if i appears in one of strategies in T^n then the contribution of i to $\gamma(S)$ is q_i. On the other hand, at most n players use market i in their strategies. Consequently, the contribution of market i in the summation $\sum_{j \in U} \alpha_j(T^j)$ is at most $(1 + \frac{1}{2} + \frac{1}{3} + \ldots + \frac{1}{n})q_i = H_n q_i$. It follows that $\sum_{j \in U} \alpha_j(T^n) \geq \frac{1}{H_n} \sum_{j \in U} \alpha_j(T^j)$, as required. We denote by $\mathcal{T}$ the summation $\sum_{j \in U} \alpha_j(T^j)$. Next consider the optimal assignment Ω, and let Y_j be the set of markets that are in serviced by agent j in σ_j but that are *not* serviced by any agent in T^n, that is, $Y_j = \sigma_j - \cup_{r \in U} s_r$. Now $\gamma(S)$ is greater than the value of all the markets in $\cup_{r \in U}(\sigma_r - Y_r)$ since these markets are a subset of markets serviced in T^n. Hence, using the notation $q(Q) = \sum_{i \in Q} q_i$ to denote the sum of query rates of a subset Q of the markets, we have $\gamma(S) \geq \sum_{r \in U} q(\sigma_r - Y_r)$. Next we will prove that $\mathcal{T} \geq \frac{1}{\lambda} \sum_{r \in U} q(Y_r)$. Let Y'_j be the markets in Y_j that are serviced in T^j, that is, $Y'_j = Y_j - (s_1 \cup \cdots \cup s_j \cup t_{j+1} \cup \cdots \cup t_n)$. Then Y'_j is a feasible strategy for agent j at step j, and thus, since player j uses a λ-approximation algorithm, we have $\lambda \alpha_j(T^j) \geq q(Y'_j)$. Therefore, $\lambda \mathcal{T} \geq \sum_{r \in U} q(Y'_r)$.

Finally, we claim that $\mathcal{T} \geq \sum_{j \in U} q(Y''_j)$. To see this, consider a any market $i \in Y''_j = Y_j - Y'_j$. Then market i is not in the strategy set of any agent in T^n, but is in the strategy set of at least one player in T^j. Therefore, somewhere on the path $\mathcal{P}$ after T^j some player must change its strategy and discontinue servicing market i. Let b_i be time step such that T^{b_i} is the first state amongst $T^{j+1}, \ldots, T^n$ that does not service market i. Let $M_j = \{i \in H | b_i = j\}$ be the set of markets for which $b_i = j$. It follows that $\cup_{r \in U} Y''(t) = \cup_{r \in U} M_r$. Notice that $M_r \subseteq t_r$ and no other agents service any market in M_r at step r. It follows that $\alpha_j(T^j) \geq q(M_j)$. Therefore, $\sum_{j \in U} q(Y_j'') = \sum_{j \in U} q(M_j) \leq \sum_{j \in U} \alpha_j(T^j) = T$. Hence we have, OPT $= \sum_{j \in U} q(\sigma_j) \leq \sum_{j \in U} q(\sigma_j - Y_j) + \sum_{j \in U} q(Y_j) \leq \gamma(S) + \sum_{j \in U} q(Y_j') + \sum_{j \in U} q(Y_j'') \leq \gamma(S) + \lambda T + T \leq (1 + (\lambda + 1)H_n)\gamma(S)$. $\square$

Theorem 6. *In market sharing games, the social value of a state at the end of a one-round path may be as bad as $\frac{1}{H_n}$ of the optimal social value. In particular, this is true even for uniform market sharing games and complete market sharing games.*

Proof. Consider the following instance of a complete market sharing game. There are $m = n$ markets, and the query rate of market i is $q_i = \frac{n}{i} - \epsilon$ for all $1 \leq i \leq n$ where ϵ is sufficiently small. The cost of market i is $C_i = 1 + (n-i)\epsilon$ for $2 \leq i \leq n$ and $C_1 = 1$. There are n players and the budget of player j is equal to $1 + (n-j)\epsilon$. Consider the ordering $1, 2, \ldots, n$ and the one-round path starting from empty set of strategies and letting each player play once in this order. The resulting assignment after this one-round path is that all players provide market number 1 and the social value of this assignment is $n - \epsilon$. However, the optimum solution is for agent j to service market j giving an optimal social value of $nH_n - n\epsilon$. Thus, the ratio between the optimum and the value of the resulting assignment is H_n at the end of a one-round path.

The bad instance for the uniform market sharing game is similar. There are n markets of cost 1 and query rates $q_i = \frac{n}{i} - \epsilon$ for all $1 \leq i \leq n$ where ϵ is sufficiently small. There are n players each with budget 1. Player j is interested in markets $j, j+1, \ldots, n$ and in market 1. It follows that the social value of the assignment after one round is $n - \epsilon$. The optimal social value covers all markets and its value is $nH_n - n\epsilon$. Thus, the ratio is $\frac{1}{H_n}$ after one round. $\square$

6   Conclusion and Open Problems

In this paper, we presented a framework for studying speed of convergence to approximate solutions in competitive games. We proved bounds on the outcome of one round of best responses of players in terms of the social objective function. More generally, one may consider longer (but polynomial-sized) best-response paths, provided the problem of cycling can be dealt with. In acyclic state graphs, such as potential games (or congestion games), the PLS-completeness results of Fabrikant et al. [3] show that there are games for which the size of the shortest best-response path from some states to any pure Nash equilibrium is exponential. This implies that in some congestion games the social value of a state after exponentially many best responses might be far from the optimal social value. However, this does not preclude the possibility that good approximate solutions are obtained when short k-covering paths are used. This provides additional motivation for the study of such paths. Here we may consider using a local optimization algorithm and evaluating the output of this algorithm after a polynomial number of local improvements.

The market sharing games are not yet well understood. In particular, it is not known whether exponentially long best-response paths may exist. Bounding the social value of a vertex at the end of a k-covering path is another open question. Goemans et al. [4] give a polynomial-time algorithm to find the pure Nash equilibrium in uniform market sharing games. Finding such an equilibrium is NP-complete for the general case, but the question of obtaining approximate Nash equilibria is open.

Acknowledgments

This work began at the Bell-Labs Algorithms Pow-Wow; we thank Bruce Shepherd for hosting us at Bell-Labs. The first author would also like to thank Michel Goemans for useful discussions.

References

1. P. Brucker, J. Hurink, and F. Werner. "Improving local search heuristics for some scheduling problems", *Discrete Applied Mathematics*, **72**, pp47-69, 1997.
2. E. Even-dar, A. Kesselman, and Y. Mansour. "Convergence time to Nash equilibria", *ICALP*, 2003.
3. A. Fabrikant, C. Papadimitriou, and K. Talwar. "On the complexity of pure equilibria", *STOC*, 2004.
4. M. Goemans, L. Li, V.S.Mirrokni, and M. Thottan. "Market sharing games applied to content distribution in ad-hoc networks", *MOBIHOC*, 2004.
5. I. Milchtaich. "Congestion games with player-specific payoff functions", *Games and Economic Behavior*, **13(1)**, pp111-124, 1996.
6. D. Monderer and L. Shapley. "Potential games", *Games and Economic Behavior*, **14(1)**, pp124-143, 1996.
7. N. Nisan and A. Ronen. "Algorithmic mechanism design", *Games and Economic Behavior*, **35(1-2)**, pp166–196, 2001.
8. C. Papadimitriou. "Algorithms, games, and the internet", *STOC*, 2001.
9. R. W. Rosenthal. "A class of games possessing pure-strategy Nash equilibria", *International Journal of Game Theory*, **2**, pp65-67, 1973.
10. T. Roughgarden and E. Tardos. "Bounding the inefficiency of equilibria in nonatomic congestion games", *Games and Economic Behavior*, **47(2)**, pp389-403, 2004.
11. T. Roughgarden and E. Tardos. "How bad is selfish routing?", *J. ACM*, **49(2)**, pp236-259, 2002.
12. P. Schuurman and T. Vredeveld. "Performance guarantees of local search for multiprocessor scheduling", *IPCO*, 2001.
13. A. Vetta. "Nash equilibria in competitive societies, with applications to facility location, traffic routing and auctions", *FOCS*, 2002.

Cuts and Orderings: On Semidefinite Relaxations for the Linear Ordering Problem

Alantha Newman*

MIT CSAIL, Cambridge, MA 02139
alantha@theory.csail.mit.edu

Abstract. The linear ordering problem is easy to state: Given a complete weighted directed graph, find an ordering of the vertices that maximizes the weight of the forward edges. Although the problem is NP-hard, it is easy to estimate the optimum to within a factor of 1/2. It is not known whether the maximum can be estimated to a better factor using a polynomial-time algorithm. Recently it was shown [NV01] that widely-studied polyhedral relaxations for this problem cannot be used to approximate the problem to within a factor better than 1/2. This was shown by demonstrating that the integrality gap of these relaxations is 2 on random graphs with uniform edge probability $p = 2^{\sqrt{\log n}}/n$. In this paper, we present a new semidefinite programming relaxation for the linear ordering problem. We then show that if we choose a random graph with uniform edge probability $p = \frac{d}{n}$, where $d = \omega(1)$, then with high probability the gap between our semidefinite relaxation and the integral optimal is at most 1.64.

1 Introduction

Vertex ordering problems comprise a fundamental class of combinatorial optimization problems that, on the whole, is not well understood. For the past thirty years, combinatorial methods and linear programming techniques have failed to yield improved approximation guarantees for many well-studied vertex ordering problems such as the linear ordering problem and the traveling salesman problem. Semidefinite programming has proved to be a powerful tool for solving a variety of cut problems, as first exhibited for the maximum cut problem [GW95]. Since then, semidefinite programming has been successfully applied to many other problems that can be categorized as cut problems such as coloring k-colorable graphs [KMS98], maximum-3-cut [GW04], maximum k-cut [FJ97], maximum bisection and maximum uncut [HZ01], and correlation clustering [CGW03], to name a few. In contrast, there is no such comparably general approach for approximating vertex ordering problems.

In this paper, we focus on a well-studied and notoriously difficult combinatorial optimization problem known as the linear ordering problem. Given a complete weighted directed graph, the goal of the linear ordering problem is to

---

* Supported in part by NSF Grant CCR0307536.

K. Jansen et al. (Eds.): APPROX and RANDOM 2004, LNCS 3122, pp. 195–206, 2004.
© Springer-Verlag Berlin Heidelberg 2004

find an ordering of the vertices that maximizes the weight of the forward edges. Although the problem is NP-hard [Kar72], it is easy to estimate the optimum to within a factor of $\frac{1}{2}$: In any ordering of the vertices, either the set of forward edges or the set of backward edges accounts for at least half of the total edge weight. It is not known whether the maximum can be estimated to a better factor using a polynomial-time algorithm. Approximating the problem to within better than $\frac{65}{66}$ is NP-hard [NV01].

The linear ordering problem is also known as the maximum acyclic subgraph problem. Given a weighted directed graph, the maximum acyclic subgraph problem is that of finding the maximum weight subgraph that contains no cycles. The forward edges in any linear ordering comprise an acyclic subgraph and a topological sort of an acyclic subgraph yields a linear ordering of the vertices in which all edges in the acyclic subgraph are forward edges.

Recently it was shown that several widely-studied polyhedral relaxations for the linear ordering problem each have an integrality gap of 2, showing that it is unlikely these relaxations can be used to approximate the problem to within a factor greater than $\frac{1}{2}$ [NV01,New00]. The graphs used to demonstrate these integrality gaps are random graphs with uniform edge probability of approximately $2^{\sqrt{\log n}}/n$, where n is the number of vertices. For sufficiently large n, such a random graph has a maximum acyclic subgraph close to half the edges with high probability. However, each of the polyhedral relaxations studied provide an upper bound for these graphs that is asymptotically close to all the edges, which is off from the optimal by a factor of 2.

In this paper, we first present a new semidefinite programming relaxation for the linear ordering problem. A vertex ordering for a graph $G = (V, E)$ with n vertices can be fully described by a series of $n + 1$ cuts. We use this simple observation to relate cuts and orderings. We derive a semidefinite program for the linear ordering problem that is related to the semidefinite program used in the Goemans-Williamson algorithm to approximate the maximum cut problem [GW95]. We note that by using different objective functions, our semidefinite programming relaxation can be used to obtain semidefinite relaxations for many other vertex ordering problems.

Second, we show that for sufficiently large n, if we choose a random directed graph on n vertices with uniform edge probability $p = \frac{d}{n}$ (i.e. every edge in the complete directed graph on n vertices is chosen with probability p), where $d = \omega(1)$, our semidefinite relaxation will have an integrality gap of no more than 1.64 with high probability. In particular, the graphs used in [NV01] to demonstrate integrality gaps of 2 for the widely-studied polyhedral relaxations fall into this category of random graphs. The main idea is that our semidefinite relaxation provides a "good" bound on the value of an optimal linear ordering for a graph if it has no small roughly balanced bisection. With high probability, a random graph with uniform edge probability contains no such small balanced bisection.

2   Relating Cuts and Orderings

Given an undirected weighted graph $G = (V, E)$, the maximum cut (maxcut) problem is to find a bipartition of the vertices that maximizes the weight of the edges crossing the partition. In 1993, Goemans and Williamson used a semidefinite programming relaxation to obtain a .87856-approximation algorithm for this fundamental graph optimization problem [GW95]. The goal of the Goemans-Williamson algorithm for the maxcut problem is to assign each vertex $i \in V$ a vector $v_i \in \{1, -1\}$ so as to maximize the weight of the edges (i, j) such that $v_i \neq v_j$.

A closely related graph optimization problem is the maximum directed cut (dicut) problem. Given a directed weighted graph $G = (V, A)$, the dicut problem is to find a bipartition of the vertices – call these disjoint sets V_1 and V_2 – that maximizes the weight of the directed edges (i, j) such that vertex i is in set V_1 and vertex j is in set V_2. Note that the edges in a directed cut form an acyclic subgraph. We can generalize the dicut problem to that of dividing the vertices into k labeled sets $V_1, V_2, \ldots, V_k$ so as to maximize the weight of the edges (i, j) such that vertex i is in set V_k and vertex j is in set V_h and $k < h$. We call this the k-acyclic dicut problem. The linear ordering problem is equivalent to the n-acyclic dicut problem.

2.1   A Relaxation for the Linear Ordering Problem

We can generalize the semidefinite programming relaxation for the dicut problem [FG95,GW95] to obtain a new semidefinite programming relaxation for the linear ordering problem. The basic idea behind this formulation is a particular description of a vertex ordering that uses $n+1$ unit vectors for each vertex. Each vertex $i \in V$ has $n + 1$ $(n = |V|)$ associated unit vectors: $v_i^0, v_i^1, v_i^2, \ldots v_i^n$. In an integral solution, we enforce that $v_i^0 = -1$, $v_i^n = 1$ and that v_i^h and v_i^{h+1} differ for only one value of h, $0 \leq h < n$. Constraint (1) enforces that in an integral solution, v_i^h and v_i^{h+1} differ for only one such value of h. This position h denotes vertex i's position in the ordering. For example, suppose we have a graph G that has four vertices, arbitrarily labeled 1 through 4. Consider the vertex ordering in which vertex i is in position i. An integral description of this vertex ordering is:

$$\{v_1^0, \ v_1^1, \ v_1^2, \ v_1^3, \ v_1^4\} \ = \ \{-1, \ \ 1, \ \ 1, \ \ 1, \ \ 1\},$$
$$\{v_2^0, \ v_2^1, \ v_2^2, \ v_2^3, \ v_2^4\} \ = \ \{-1, -1, \ \ 1, \ \ 1, \ \ 1\},$$
$$\{v_3^0, \ v_3^1, \ v_3^2, \ v_3^3, \ v_3^4\} \ = \ \{-1, -1, -1, \ \ 1, \ \ 1\},$$
$$\{v_4^0, \ v_4^1, \ v_4^2, \ v_4^3, \ v_4^4\} \ = \ \{-1, -1, -1, -1, \ \ 1\}.$$

Let $G = (V, A)$ be a directed graph. The following is an integer quadratic program for the linear ordering problem. For the sake of convenience, we assume that n is odd since this simplifies constraint (2). By $P(G)$, we denote the optimal value of the integer quadratic program P on the graph G.

(P)

$$\max \sum_{ij \in A} \sum_{1 \leq h < \ell \leq n} w_{ij}(v_i^h \cdot v_j^\ell + v_i^{h-1} \cdot v_j^{\ell-1} - v_i^h \cdot v_j^{\ell-1} - v_i^{h-1} \cdot v_j^\ell) \tag{1}$$

$$v_i^h \cdot v_j^\ell + v_i^{h-1} \cdot v_j^{\ell-1} - v_i^h \cdot v_j^{\ell-1} - v_i^{h-1} \cdot v_j^\ell \geq 0 \quad \forall i,j \in V,\ h, \ell \in [n]$$

$$v_i^h \cdot v_i^h = 1 \quad \forall i \in V,\ h \in [n]$$

$$v_i^0 \cdot v_0 = -1 \quad \forall i \in V$$

$$v_i^n \cdot v_0 = 1 \quad \forall i \in V$$

$$\sum_{i,j \in V} v_i^{\frac{n}{2}} \cdot v_j^{\frac{n}{2}} = 0 \tag{2}$$

$$v_i^h \in \{1, -1\} \quad \forall i, h \in [n]. \tag{3}$$

We obtain a semidefinite programming relaxation for the linear ordering problem by relaxing constraint (3) to: $v_i^h \in \mathcal{R}^n$, $\forall i, h$. We denote the optimal value of the relaxation of P on the graph G as $P_R(G)$.

2.2 Cuts and Uncuts

Suppose we have a directed graph $G = (V, A)$ and we are given a set of unit vectors $\{v_i\} \in \mathcal{R}^n$, $0 \leq i \leq n$. We will define the *forward* value of this set of vectors as the value obtained if we compute the value of the dicut semidefinite programming relaxation [GW95,FG95] using these vectors. Specifically, the forward value for this set of vectors is:

$$\max \sum_{ij \in A} \frac{1}{4}(1 - v_i \cdot v_j - v_0 \cdot v_i + v_0 \cdot v_j). \tag{4}$$

In an integral solution for the dicut problem, there will be edges that cross the cut in the backward direction, i.e. they are not included in the dicut. For a specified set of unit vectors, we can view the dicut semidefinite programming relaxation as having forward and *backward* value. We define the backward value of the set of vectors $\{v_i\}$ as:

$$\max \sum_{ij \in A} \frac{1}{4}(1 - v_i \cdot v_j - v_0 \cdot v_j + v_0 \cdot v_i). \tag{5}$$

The *difference* between the forward and backward value of a set of vectors $\{v_i\}$ is:

$$\sum_{ij \in A} \frac{1}{2}(v_j \cdot v_0 - v_i \cdot v_0). \tag{6}$$

Lemma 1. *If a directed graph $G = (V, A)$ has a maximum acyclic subgraph of $(\frac{1}{2} + \delta)|A|$ edges, then there is no set of vectors $\{v_i\}$ such that the difference between the forward and backward value of this set of vectors exceeds $2\delta|A|$.*

Proof. We will show that given a vector solution $\{v_i\}$ to the semidefinite program in which the objective function is (6) and all the v_i vectors are unit vectors, we can find an integral (i.e. an actual cut) solution in which the difference of the forward and backward edges crossing the cut is exactly equal to the objective value. If the difference of an actual cut exceeds $2\delta|A|$, e.g. suppose it is $(2\delta+\epsilon)|A|$, then we can find an ordering with $(\frac{1}{2} + \delta + \epsilon/2)|A|$ forward edges, which is a contradiction. This ordering is found by taking the cut that yields $(2\delta + \epsilon)|A|$ more forward than backward edges and ordering the vertices in each of the two sets greedily so as to obtain at least half of the remaining edges.

Suppose we have a set of unit vectors $\{v_i\}$ such that the value of equation (6) is at least $(2\delta + \epsilon)|A| = \beta|A|$. We will show that we can find an actual cut such that the difference between the forward and the backward edges is at least $\beta|A|$. Note that $v_0 \cdot v_i$ is a scalar quantity since v_0 is a unit vector that without loss of generality is $(1, 0, 0, \dots)$. Thus, our objective function can be written as $\sum_{ij \in A} \frac{1}{2}(z_j - z_i)$ where $1 \geq z_i \geq -1$. We transform the z_i variables into x_i variables that range between 0 and 1 by letting $z_i = 2x_i - 1$. Then we have that $\sum_{ij \in A} \frac{1}{2}(z_j - z_i) = \sum_{ij \in A}(x_j - x_i)$. This results in a linear program. We claim that an optimal solution to the following linear program is integral.

$$\sum_{ij \in A} (x_j - x_i) \tag{7}$$

$$0 \leq x_i \leq 1, \quad \forall i \in V.$$

To show this, consider rounding the variables by letting i be 1 with probability x_i and 0 otherwise. Then the expected value of the solution is exactly the objective value. However, note that the value of the solution cannot be less than the expected value, since then there must exist a solution with value greater than the expected value, which contradicts the optimality of the expected value. Thus, the integral solution obtained must have difference of forward and backward edges that is equal to the objective value (7). $\qquad \square$

We will also find a discussion of the following problem useful. Consider the problem of finding a balanced partition of the vertices of a given graph (i.e. each partition has size $\frac{n}{2}$) that maximizes the weight of the edges that do *not* cross the cut. This problem is referred to as the max-$\frac{n}{2}$-uncut problem by Halperin and Zwick [HZ01]. Below is a integer quadratic program for the max-$\frac{n}{2}$-uncut problem.

(T)

$$\max \sum_{ij \in A} \frac{1 + v_i \cdot v_j}{2}$$

$$\sum_{i,j \in V} v_i \cdot v_j = 0$$

$$v_i \cdot v_i = 1 \qquad \forall i \in V$$

$$v_i \in \{1, -1\} \quad \forall i \in V. \tag{8}$$

We obtain a semidefinite programming relaxation for the max-$\frac{n}{2}$-uncut problem by relaxing constraint (8) to: $v_i \in \mathcal{R}^n$, $\forall i$. We denote the value of the relaxation of T on the graph G as $T_R(G)$.

Lemma 2. *Let $G = (V, A)$ and ϵ, δ be positive constants. Suppose the maximum acyclic subgraph of G is $(\frac{1}{2} + \delta)|A|$. If $P_R(G) \geq (1 - \epsilon)|A|$, then $T_R(G) \geq (1 - 2\epsilon - 2\delta)|A|$.*

Proof. For each edge $ij \in A$, we have:

$$\sum_{h<\ell} v_i^h \cdot v_j^\ell + v_i^{h-1} \cdot v_j^{\ell-1} - v_i^h \cdot v_j^{\ell-1} - v_i^{h-1} \cdot v_j^\ell = \tag{9}$$

$$\sum_{h<\ell} (v_i^h - v_i^{h-1}) \cdot (v_j^\ell - v_j^{\ell-1}) \leq$$

$$\sum_{h\leq\frac{n}{2},\ell\leq\frac{n}{2}} (v_i^h - v_i^{h-1}) \cdot (v_j^\ell - v_j^{\ell-1}) + \tag{10}$$

$$\sum_{h>\frac{n}{2},\ell>\frac{n}{2}} (v_i^h - v_i^{h-1}) \cdot (v_j^\ell - v_j^{\ell-1}) + \tag{11}$$

$$\sum_{h\leq\frac{n}{2},\ell>\frac{n}{2}} (v_i^h - v_i^{h-1}) \cdot (v_j^\ell - v_j^{\ell-1}). \tag{12}$$

For each edge, we refer to the quantity (9) as the forward value for that edge with respect to $P_R(G)$. The same term summed instead over $h > \ell$ is referred to as the *backward* value of the edge with respect to $P_R(G)$. We can simplify the terms above. Let $v_i = v_i^{\frac{n}{2}}$.

$$\sum_{h\leq\frac{n}{2},\ell\leq\frac{n}{2}} (v_i^h - v_i^{h-1}) \cdot (v_j^\ell - v_j^{\ell-1}) = \frac{1}{4}(v_i + v_0) \cdot (v_j + v_0),$$

$$\sum_{h>\frac{n}{2},\ell>\frac{n}{2}} (v_i^h - v_i^{h-1}) \cdot (v_j^\ell - v_j^{\ell-1}) = \frac{1}{4}(v_0 - v_i) \cdot (v_0 - v_j),$$

$$\sum_{h\leq\frac{n}{2},\ell>\frac{n}{2}} (v_i^h - v_i^{h-1}) \cdot (v_j^\ell - v_j^{\ell-1}) = \frac{1}{4}(v_i + v_0) \cdot (v_0 - v_j).$$

Since $P_R(G) \geq (1 - \epsilon)|A|$, we have:

$$\sum_{ij\in A} \sum_{h>\frac{n}{2},\ell\leq\frac{n}{2}} (v_i^h - v_i^{h-1}) \cdot (v_j^\ell - v_j^{\ell-1}) = \sum_{ij\in A} \frac{1}{4}(v_0 - v_i) \cdot (v_0 + v_j) \leq \epsilon|A|.$$

The above inequality says that the *backward* value of the vectors $\{v_i\}$ (i.e. quantity (5)) is at most the *backward* value of $P_R(G)$. By Lemma 1, the difference of the edges crossing the cut in the forward direction and the edges crossing the cut in the backward direction is at most $2\delta|A|$.

$$\sum_{ij \in A} \frac{1}{4}(v_i + v_0) \cdot (v_0 - v_j) \; - \; \sum_{ij \in A} \frac{1}{4}(v_0 - v_i) \cdot (v_0 + v_j) =$$

$$\sum_{ij \in A} \frac{1}{2}(v_i \cdot v_0 - v_j \cdot v_0) \leq 2\delta |A|.$$

This implies that the forward value cannot exceed the backward value by more than $2\delta |A|$. Thus, we can bound the forward value as follows:

$$\sum_{ij \in A} \sum_{h \leq \frac{n}{2}, \ell > \frac{n}{2}} (v_i^h - v_i^{h-1}) \cdot (v_j^\ell - v_j^{\ell-1}) = \sum_{ij \in A} \frac{1}{4}(v_0 - v_i) \cdot (v_0 + v_j) \leq (\epsilon + 2\delta)|A|.$$

This implies that if we sum the quantities (10) and (11) over all edges in A, then the total value of this sum is at least $(1 - 2\epsilon - 2\delta)|A|$. The sum of (10) and (11) taken over all the edges is:

$$\sum_{ij \in A} \frac{1 + v_i \cdot v_j}{2}. \tag{13}$$

□

3   Balanced Bisections of Random Graphs

A *bisection* of a graph is a partition of the vertices into two equal (or with cardinality differing by one if n is odd) sets. We use a related definition in this section.

Definition 1. *A γ-bisection of a graph for $\gamma \leq \frac{1}{2}$ is the set of edges that cross a cut in which each set of vertices has size at least γn.*

Suppose we choose an undirected random graph on n vertices in which every edge is present with probability $p = \frac{2d}{n}$. The expected degree of each vertex is $2d$ and the expected number of edges is dn. We will call such a class of graphs G_p.

Lemma 3. *For any fixed positive constants ϵ, γ, if we choose a graph $G \in G_p$ on n vertices for a sufficiently large n with $p = \frac{2d}{n}$ and $d = \omega(1)$, then the minimum γ-bisection contains at least $(1 - \epsilon)\gamma(1 - \gamma)2nd$ edges with high probability.*

Proof. We will use the principle of deferred decisions. First, we will choose a $\gamma n, (1 - \gamma)n$ partition of the vertices. Thus $\gamma(1 - \gamma)n^2$ edges from the complete graph on n vertices cross this cut. Then we can choose the random graph G by picking each edge with probability $p = \frac{2d}{n}$. The expected number of edges from G crossing the cut is $\mu = (\gamma(1 - \gamma)n^2)(\frac{2d}{n}) = \gamma(1 - \gamma)2dn$. For each edge in the complete graph that crosses the cut, we have the indicator random variable X_i such that $X_i = 1$ if the edge crosses the cut and $X_i = 0$ if the edge does not cross the cut. Let $X = \sum X_i$, i.e. X is the random variable for the number of edges that cross the cut. By Chernoff Bound, we have:

$$\Pr[X < (1 - \epsilon)\gamma(1 - \gamma)2dn)] < e^{\frac{\epsilon^2 \gamma(1-\gamma)2dn}{2}}.$$

We can union bound over all the possible γ-bisections. There are less than 2^n ways to divide the vertices so that at least γn are in each set. Thus, the probability that the minimum γ-bisection of G is less than a $(1 - \epsilon)$ fraction of its expectation is:

$$\Pr[\min \ \gamma\text{-bisection}(G) < (1 - \epsilon)\gamma(1 - \gamma)2nd] < \frac{2^n}{e^{\frac{\epsilon^2 \gamma(1-\gamma)2dn}{2}}}.$$

Let $d = \omega(1)$. Then for any fixed positive constants γ, ϵ, this probability will be arbitrarily small for sufficiently large n. □

4   A Contradictory Cut

In this section, we will prove our main theorem. Suppose we choose a directed random graph on n vertices in which every edge in the complete directed on n vertices is included with probability p. Let $p = \frac{d}{n}$ and let $d = \omega(1)$. We will call this class of graphs $\boldsymbol{G_p}$. Note that if we randomly choose a graph from $\boldsymbol{G_p}$, the underlying undirected graph is randomly chosen from G_p.

Theorem 1. *For sufficiently large n, $d = \omega(1)$, and $p = \frac{d}{n}$, if we randomly choose a graph $G \in \boldsymbol{G_p}$, then with high probability, the ratio $P_R(\boldsymbol{G})/P(\boldsymbol{G}) \leq$ 1.64.*

Let E represent the edges in the complete undirected graph K_n for some fixed n. Let $A \subseteq E$ represent the edges in an undirected graph G chosen at random from G_p. Let ϵ_1 be a small positive constant whose value can be arbitrarily small for sufficiently large n. We weight the edges in E as follows:

$$w_{ij} = -\frac{n}{(1 - \epsilon_1)2d} \quad \text{if} \quad ij \in A,$$

$$w_{ij} = 1 \quad \text{if} \quad ij \in E - A.$$

We will refer to this weighted graph as G'.

Lemma 4. *The minimum γ-bisection of G' has negative value with high probability.*

Proof. By Lemma 3 with high probability the minimum γ-bisection of G has at least $(1 - \epsilon_1)\gamma(1 - \gamma)2nd$ edges. Thus, with high probability the total weight of the edges in the minimum γ-bisection of G' is at most:

$$\gamma(1 - \gamma)n^2 - (1 - \epsilon_1)\gamma(1 - \gamma)2nd + (1 - \epsilon_1)\gamma(1 - \gamma)2nd(-\frac{n}{(1 - \epsilon_1)2d}) =$$

$$\gamma(1 - \gamma)\left(n^2 - (1 - \epsilon_1)2nd + (1 - \epsilon_1)2nd(-\frac{n}{(1 - \epsilon_1)2d})\right) =$$

$$\gamma(1 - \gamma)\left(-(1 - \epsilon_1)2nd\right) < 0.$$

□

Lemma 5. *Let* $\{v_i\}$, $i \in V$ *be a set of unit vectors that satisfy the following constraints:*

$$\sum_{i,j \in V} v_i \cdot v_j = 0 \tag{14}$$

$$\sum_{ij \in A} \frac{1 + v_i \cdot v_j}{2} \geq (1 - \epsilon_2)|A|. \tag{15}$$

If $\epsilon_2 < .36$, *then we can find a* γ-*bisection of* G' *with a strictly positive value.*

To prove Lemma 5, we will use the following theorem from [GW95].

Theorem 2.7 [GW] Let $W_- = \sum_{i<j} w_{ij}^-$, where $x^- = \min(0, x)$. Then

$$\{E[W] - W_-\} \geq \alpha \left\{ \frac{1}{2} \sum_{i<j} w_{ij}(1 - v_i \cdot v_j) - W_- \right\}.$$

Proof of Lemma 5: We will use Goemans-Williamson's random hyperplane algorithm to show that we can find a cut that is roughly balanced and has a strictly positive value. Let W represent the total weight of the edges that cross the cut obtained from a random hyperplane. Let W_- denote the sum of the negative edges weights, i.e. $W_- = A$. Applying Theorem 2.7 from [GW], we have:

$$E[W] \geq \alpha \left\{ \frac{1}{2} \sum_{i<j} w_{ij}(1 - v_i \cdot v_j) - W_- \right\} + W_-$$

$$\geq \alpha \left\{ \sum_{i<j:w_{ij}>0} w_{ij} \frac{1 - v_i \cdot v_j}{2} + \sum_{i<j:w_{ij}<0} |w_{ij}| \frac{1 + v_i \cdot v_j}{2} \right\} + W_-.$$

We want to calculate the value of $\sum_{i<j:w_{ij}>0} \frac{1 - v_i \cdot v_j}{2}$. By condition (14), we have that $\sum_{i,j \in V} v_i \cdot v_j = 0$ and therefore $\sum_{i<j} \frac{1 - v_i \cdot v_j}{2} = \frac{n^2 - 2n}{4}$.

$$\sum_{i<j:w_{ij}>0} \frac{1 - v_i \cdot v_j}{2} = \sum_{i<j} \frac{1 - v_i \cdot v_j}{2} - \sum_{i<j:w_{ij}<0} \frac{1 - v_i \cdot v_j}{2}$$

$$= \sum_{i<j} \frac{1 - v_i \cdot v_j}{2} - \frac{nd}{2} + \sum_{i<j:w_{ij}<0} \frac{v_i \cdot v_j}{2}$$

$$\geq \sum_{i<j} \frac{1 - v_i \cdot v_j}{2} - \frac{nd}{2} + \frac{(1 - 2\epsilon_2)nd}{2}$$

$$= \frac{n^2 - 2n}{4} - \epsilon_2 nd.$$

Now we have:

$$E[W] \geq \alpha \left\{ \left(\frac{n^2 - 2n}{4} - \epsilon_2 nd \right) + \frac{n}{(1-\epsilon_1)2d}(1-\epsilon_2)nd \right\} - \frac{n}{(1-\epsilon_1)2d}nd.$$

For large enough n, we can choose ϵ_1 to be arbitrarily small. So $E[W]$ can be bounded from below by a value arbitrarily close to the following:

$$\left(\frac{\alpha}{4} + \frac{\alpha}{2} - \frac{1}{2} - \frac{\alpha\epsilon_2}{2} \right)n^2 - o(n^2) \geq (.1585 - \frac{\alpha\epsilon_2}{2})n^2 - o(n^2). \qquad (16)$$

If the value of ϵ_2 is such that the quantity on line (16) is strictly greater than βn^2 for some positive constant β, then we will have a contradiction for sufficiently large n. Note that if this value is at least βn^2, then each side of the cut contains at least $\sqrt{\beta}n$ vertices, so it is a $\sqrt{\beta}$-bisection. This value will be strictly positive as long as $\epsilon_2 < .36$. Thus, it must be the case that $\epsilon_2 > .36$. $\qquad \square$

Proof of Theorem 1: We fix positive constants γ, ϵ_1. Suppose we choose a random directed graph G as prescribed and let the graph $G = (V, A)$ be the undirected graph corresponding to the underlying undirected graph of G. We then weight the edges in the graph K_n as discussed previously and obtain G'. By Lemma 4, the minimum γ-bisection of G' is negative with high probability. Thus, with high probability equation (15) hold only when $\epsilon_2 > .36$.

Suppose the maximum acyclic subgraph of G, i.e. $P(G)$ is $(\frac{1}{2} + \delta)|A|$ for some positive constant δ. Then the value of $P_R(G)$ is upper bounded by the maximum value for some set of unit vectors $\{v_i\}$ of (10), (11), and (12) summed over all edges in A. Note that this is equivalent to the quantity in (13) (which is no more than $.64|A|$) plus the quantity in (4). By Lemma 1, the difference between (4) and (5) must be no more than $2\delta|A|$. Thus, we can upper bound the value of $P_R(G)$ by $.64|A| + (2\delta + \frac{1}{2}(.36 - 2\delta))|A| = (.82 + \delta)|A|$. Thus, with high probability, we have:

$$\frac{P_R(G)}{P(G)} \leq \frac{.82 + \delta}{.5 + \delta} \leq \frac{.82}{.5} = 1.64.$$

$\qquad \square$

5   Discussion

In this paper, we make a connection between cuts and vertex ordering of graphs in order to obtain a new semidefinite programming relaxation for the linear ordering problem. We show that the relaxation is "good" on random graphs chosen with uniform edge probability, i.e. the integrality gap is strictly less than 2 for most of these graphs. We note that we can extend this theorem to show that this relaxation is "good" on graphs that have no small γ-bisections for some constant $\gamma > 0$.

In [HZ01], Halperin and Zwick give a .8118-approximation for a related problem that they call the max $\frac{n}{2}$-directed-uncut problem. Given a directed graph, the goal of this problem is to find a bisection of the vertices that maximizes the weight of the edges that cross the cut in the forward direction plus the weight of the edges that do not cross the cut. We note that a weaker version of Theorem 1 follows from their .8118-approximation algorithm. This is because their semidefinite program for the max $\frac{n}{2}$-directed uncut problem is the sum over all edges of terms (10), (11), and (12). If for some directed graph $G = (V, A)$, $P_R(G)$ has value at least $(1 - \epsilon)|A|$, then the value of their semidefinite programmming relaxation also has at least this value. Thus, if ϵ is arbitrarily small, we can obtain a directed uncut of value close to .8118 of the edges, which is a contradiction for a random graph with uniform edge probability. In this paper, our goal was to give a self-contained proof of this theorem.

We would like to comment on the similarity of this work to the work of Poljak and Delorme [DP93] and Poljak and Rendl [PR95] on the maxcut problem. Poljak showed that the class of random graphs with uniform edge probability could be used to demonstrate an integrality gap of 2 for several well-studied polyhedral relaxations for the maxcut problem [Pol92]. These same graphs can be used to demonstrate an integrality gap of 2 for several widely-studied polyhedral relaxations for the linear ordering problem [NV01]. The similarity of these results stems from the fact that the polyhedral relaxations for the maxcut problem are based on odd-cycle inequalities and the polyhedral relaxations for the linear ordering problem are based on cycle inequalities. Poljak and Delorme subsequently studied an eigenvalue bound for the maxcut problem that is equivalent to the bound provided by the semidefinite programming relaxation used in the Goemans-Williamson algorithm [GW95]. Despite the fact that random graphs with uniform edge probability exhibit worst-case behaviour for several polyhedral relaxations for the maxcut problem, Delorme and Poljak [DP93] and Poljak and Rendl [PR95] experimentally showed that the eigenvalue bound provides a "good" bound on the value of the maxcut for these graphs. This experimental evidence was the basis for their conjecture that the 5-cycle exhibited a worst-case integrality gap of 0.88445 ... for the maxcut semidefinite relaxation [DP93,Pol92]. The gap demonstrated for the 5-cycle turned out to be very close to the true integrality gap of .87856 ... [FS].

Acknowledgments

I would like to thank Prahladh Harsha, Santosh Vempala, and Michel Goemans for many helpful discussions.

References

[CGW03] Moses Charikar, Venkatesan Guruswami, and Anthony Wirth. Clustering with qualitative information. In *Proceedings of the 44th Annual IEEE Symposium on Foundations of Computer Science (FOCS)*, pages 524–533, Boston, 2003.

[DP93]   Charles Delorme and Svatopluk Poljak. The performance of an eigenvalue bound in some classes of graphs. *Discrete Mathematics*, 111:145–156, 1993. Also appeared in *Proceedings of the Conference on Combinatorics*, Marseille, 1990.

[FG95]   Uriel Feige and Michel X. Goemans. Approximating the value of two prover proof systems with applications to MAX-2-SAT and MAX DICUT. In *Proceedings of the Third Israel Symposium on Theory of Computing and Systems*, pages 182–189, 1995.

[FJ97]   Alan Frieze and Mark R. Jerrum. Improved approximation algorithms for MAX-k-Cut and MAX BISECTION. *Algorithmica*, 18:61–77, 1997.

[FS]   Uriel Feige and Gideon Schechtman. On the optimality of the random hyperplane rounding technique for MAX-CUT. *Random Structures and Algorithms*. To appear.

[GW95]   Michel X. Goemans and David P. Williamson. Improved approximation algorithms for maximum cut and satisfiability problems using semidefinite programming. *Journal of the ACM*, 42:1115–1145, 1995.

[GW04]   Michel X. Goemans and David P. Williamson. Approximation algorithms for MAX-3-CUT and other problems via complex semidefinite programming. *STOC 2001 Special Issue of Journal of Computer and System Sciences*, 68:442–470, 2004.

[HZ01]   Eran Halperin and Uri Zwick. A unified framework for obtaining improved approximation algorithms for maximum graph bisection problems. In *Proceedings of Eighth Conference on Integer Programming and Combinatorial Optimization (IPCO)*, pages 210–225, Utrecht, 2001.

[Kar72]   Richard M. Karp. Reducibility among combinatorial problems. In *Complexity of Computer Computations*, pages 85–104. Plenum Press, New York, 1972.

[KMS98]   David R. Karger, Rajeev Motwani, and Madhu Sudan. Improved graph coloring via semidefinite programming. *Journal of the ACM*, 45(2):246–265, 1998.

[New00]   Alantha Newman. Approximating the maximum acyclic subgraph. Master's thesis, Massachusetts Institute of Technology, Cambridge, MA, June 2000.

[NV01]   Alantha Newman and Santosh Vempala. Fences are futile: On relaxations for the linear ordering problem. In *Proceedings of Eighth Conference on Integer Programming and Combinatorial Optimization (IPCO)*, pages 333–347, 2001.

[Pol92]   Svatopluk Poljak. Polyhedral and eigenvalue approximations of the max-cut problem. *Sets, Graphs and Numbers, Colloqiua Mathematica Societatis Janos Bolyai*, 60:569–581, 1992.

[PR95]   Svatopluk Poljak and Franz Rendl. Computing the max-cut by eigenvalues. *Discrete Applied Mathematics*, 62(1–3):249–278, September 1995.

Min-Max Multiway Cut

Zoya Svitkina* and Éva Tardos**

Cornell University, Department of Computer Science, Upson Hall, Ithaca, NY 14853
{zoya,eva}@cs.cornell.edu

Abstract. We propose the MIN-MAX MULTIWAY CUT problem, a variant
of the traditional MULTIWAY CUT problem, but with the goal of minimizing the maximum capacity (rather than the sum or average capacity)
leaving a part of the partition. The problem is motivated by data partitioning in Peer-to-Peer networks. The min-max objective function forces
the solution not to overload any given terminal, and hence may lead to
better solution quality.
We prove that the MIN-MAX MULTIWAY CUT is NP-hard even on trees,
or with only a constant number of terminals. Our main result is an
$O(\log^3 n)$-approximation algorithm for general graphs, and an $O(\log^2 n)$-
approximation for graphs excluding any fixed graph as a minor (e.g.,
planar graphs). We also give a $(2 + \epsilon)$-approximation algorithm for the
special case of graphs with bounded treewidth.

1 Introduction

The MIN-MAX MULTIWAY CUT problem is defined by an undirected graph $G = (V, E)$ with edge capacities $c(e) \geq 0$, and a set $X = \{x_1, ..., x_k\} \subseteq V$ of distinguished nodes called terminals. A *multiway cut* is a partition of V into disjoint
sets $S_1, ..., S_k$ ($\bigcup_i S_i = V$), so that for all $i \in \{1, ..., k\}$, $x_i \in S_i$. For a partition
we will use $\delta(S_i)$ to denote the capacity of the cut separating S_i from the other
sets $\bigcup_{j \neq i} S_j$, and the goal of the min-max multiway cut problem is to minimize
the maximum capacity $\max_i \delta(S_i)$.

The min-max multiway cut problem models the data placement problem
in a distributed database system or a Peer-to-Peer system. In a Peer-to-Peer
database, the information is stored on many servers. When a user query is issued,
it is directed to the appropriate server. A request for some data item v can
lead to further requests for other data. One important issue in such Peer-to-
Peer databases is to find a good distribution of data that minimizes requests to
any single server. We model this by a graph in which the non-terminal nodes
represent the data items and the terminals represent the servers. Nodes in the
partition S_i correspond to the data that will be stored on server i. Edges in the
graph correspond to the expected communication patterns, i.e., the edge (x_i, v)

---

* Supported in part by NSF grant CCR-032553.
** Supported in part by NSF grant CCR-032553, ITR grant 0311333, and ONR grant
N00014-98-1-0589.

K. Jansen et al. (Eds.): APPROX and RANDOM 2004, LNCS 3122, pp. 207–218, 2004.

represents the number of queries that users at server i issue for the data v, and the edge (v, w) represents the expected number of times that a request for data v will result in an induced request for data w. Communication costs are incurred when a query from one server is sent to another. The goal then is to distribute the data among the servers so as to minimize the communication cost incurred by any one of them.

The min-max multiway cut problem is closely related to the traditional multiway cut problem of [2]. The difference is in the objective function. Unlike the min-max multiway cut, in which we seek to minimize the maximum capacity $\max_i \delta(S_i)$, the multiway cut problem evaluates a partition by the sum of the capacities of all edges that connect the parts, thus minimizing the average capacity $\delta(S_i)$. Multiway cut has been used to model similar applications of storing files on a network, as well as other problems such as partitioning circuit elements among different chips [2]. In many situations, however, the min-max objective function may be a better representation of the solution quality. Although the multiway cut minimizes the average communication cost of the terminals, this cost may not be distributed uniformly among them, resulting in a very heavy load on some terminals and almost no load on others. The objective of minimizing the maximum load tries to alleviate this problem by ensuring that no terminal is overloaded.

Multiway cut problem is NP-hard, but there are very good approximation algorithms for it [2], [1], [6]. However, they do not translate directly into good approximations for min-max multiway cut, because even the optimal solution to one problem can be up to a factor of $k/2$ worse than the optimum for the other.

Our Results. For two terminals, min-max multiway cut reduces to the well-studied minimum s-t cut problem, and hence it can be solved in polynomial time. However, as we show, it is already NP-hard for the case of 4 terminals. As a result, we focus on designing approximation algorithms. In Section 2, we present an $O(\alpha \cdot \log n)$-approximation algorithm for min-max multiway cut, where $\alpha = \log^2 n$ for general graphs, and $\alpha = \log n$ for graphs excluding any fixed graph as a minor. The algorithm uses an α-approximation algorithm for a new graph cut problem, called MAXIMUM SIZE BOUNDED CAPACITY CUT (MaxSBCC). We use it as a subroutine in a procedure that resembles the greedy set cover algorithm, incurring an additional factor of $O(\log n)$ in the approximation guarantee for the min-max multiway cut problem. One of the features of our algorithm is that it is able to exhibit flexibility when assigning graph nodes to terminals: if the cut that is found for one terminal is later discovered to be bad for another terminal, then the nodes are reassigned in a way that is good for both.

We extend our algorithm to a generalization of the problem, in which there is a separate bound B_i for each terminal x_i, and the goal is to find a partition in which $\delta(S_i)$ does not exceed B_i. This generalization is useful when the different peers corresponding to the terminals have different communication capabilities, and can withstand different loads.

Turning to special cases of min-max multiway cut, we show that it is NP-complete even on trees, and develop a $(2 + \epsilon)$-approximation algorithm for trees and graphs with bounded treewidth. What makes the problem hard on trees is that an optimal solution does not necessarily assign connected components of the tree to each terminal (see Figure 3, in which the black nodes are the terminals, and the optimal solution must assign the white node in the middle to one of the leaves). As a result, even if we know which edges should be cut, it may be hard to determine how to divide the resulting components among the terminals. The key idea of our $(2 + \epsilon)$ approximation algorithm is to separate the stage of finding connected pieces of the graph from the stage of partitioning them among the terminals. Then, in the first stage, the problem of finding "good" pieces is solved optimally, and in the second stage these pieces are combined to form a 2-approximate solution. To make the dynamic programming algorithm of the first stage run in polynomial time, the edge capacities are rounded, leading to an overall $(2 + \epsilon)$-approximation.

2   Min-Max Multiway Cut in General Graphs

2.1   Approximation Algorithm

Our main goal in this section is to provide an approximation algorithm for the min-max multiway cut problem and its extension with nonuniform bounds on the capacities.

First we briefly recall the 2-approximation algorithm of Dahlhaus et al. [2], as it is useful to understand why it does not work for the min-max version of the problem. Assume that there is a multiway cut with maximum capacity at most B. The algorithm finds a minimum capacity cut (S_i, T_i) separating each terminal x_i from all other terminals. It is not hard to see that the minimum cuts with smallest source sides S_i are disjoint. Let S_0 be the nodes not in any S_i, and let $\delta_i = \delta(S_i)$ be the capacity of the cut (S_i, T_i). The cut (S_i, T_i) is of minimum capacity, so we must have $\delta_i \leq B$. The algorithm of [2] assigns each set S_i to terminal x_i, and assigns the remaining nodes S_0 to one of the terminals (the one with maximum δ_i). This yields a 2-approximation (or, more precisely, a $2(1 - 1/k)$ approximation) algorithm for the multiway cut problem, but it is only a $(k-1)$-approximation for the min-max multiway cut problem, as the part $S_i \cup S_0$ can have capacity as high as $\sum_{j \neq i} \delta(S_j) \leq (k - 1)B$.

The idea of our algorithm is to take cuts around each terminal that are larger in size than the minimum cut. Assume that we are given a bound B, and assume that there is a multiway cut where each side has capacity at most B. We will use a binary search scheme to optimize B. For a given value of B, we will need a subroutine for the following *maximum size bounded capacity cut* problem.

Definition 1. *Given a graph $G = (V, E)$ with two distinguished vertices s and t, weights on vertices $w(v)$, capacities on edges $c(e)$, and an integer B, the* MAX-IMUM SIZE BOUNDED CAPACITY CUT *(MaxSBCC) problem is to find an s-t cut (S, T) such that $\delta(S) \leq B$ and $w(S) = \sum_{v \in S} w(v)$ is maximized.*

The MaxSBCC problem can be shown to be NP-hard using a reduction from KNAPSACK. For $\alpha \geq 1$ and a constant $0 < \beta \leq 1$, let us define an (α, β)-

approximation algorithm for MaxSBCC as an algorithm that, given an instance of MaxSBCC with a bound B and an (unknown) optimal solution (S^*, T^*), produces in polynomial time an s-t cut (S', T'), such that $\delta(S') \leq \alpha B$ and $w(S') \geq \beta w(S^*)$.

First we show how to use any such approximation algorithm as a subroutine for solving the min-max multiway cut problem, and later we give a specific $(log^2 n, 1)$ algorithm. The idea is analogous to the greedy $\log n$-approximation for the set-cover problem. Starting from the set V of unassigned nodes of the graph, our algorithm iteratively finds (approximate) maximum size bounded capacity cuts around each terminal, and temporarily assigns nodes to terminals, until no unassigned nodes remain. One important difference is that our algorithm is not greedy, in the sense that assignment made to terminals in one iteration can be revised in later iterations if that becomes useful. The full algorithm is shown in Figure 1.

---

1. Initialize $S_i = \{x_i\}$ for $i = 1, ..., k$, and initialize the weights $w(v)$ for all $v \in V$ by setting $w(x_i) = 0$ for all i, and $w(v) = 1$ for all other nodes.
2. While $\bigcup_i S_i \neq V$
   - For each terminal $x_i \in X$,
     (a) Construct a graph G' labeling x_i as source s and contracting all other terminals into a single sink t.
     (b) Find an (α, β)-approximate MaxSBCC (S, T) in graph G' with bound B and weights $w(v)$. Note that the set S does not have to contain S_i and does not have to be disjoint from the other sets S_j for $j \neq i$.
     (c) Consider the intersection $I_j = S \cap S_j$ for each $j \neq i$. We need to delete this intersection either from S_j or from S. If $c(I_j, S_j \setminus I_j) < c(I_j, S \setminus I_j)$, then let $S_j = S_j \setminus I_j$; otherwise let $S = S \setminus I_j$.
     (d) Let $S_i = S_i \cup S$, and set the weights of all $v \in S$ to $w(v) = 0$.
3. Return $S_1, ..., S_k$.

Fig. 1. Min-max multiway cut algorithm

Theorem 1. *If there is an (α, β)-approximation algorithm for MaxSBCC, then the above algorithm is an $O(\alpha \log_{1+\beta} n)$-approximation for the* MIN-MAX MULTIWAY CUT *problem.*

The key to the analysis is to see that each iteration assigns a constant fraction of the remaining nodes. By assumption there is a multiway cut $(S_1^*, ..., S_k^*)$ with capacity B. For each terminal x_i, we use the approximation bound to claim that the application of the MaxSBCC assigned at least as many new nodes to x_i as a β fraction of the remaining nodes in S_i^*.

Lemma 1. *If there is a multiway cut with maximum capacity at most B, then in any iteration of the while loop, if U is the set of unassigned nodes in the* *beginning of the iteration, and U' is the set of unassigned nodes at the end of this iteration, then $|U'| \leq \frac{1}{1+\beta}|U|$.*

Proof. Let N_i be the set of previously unassigned nodes added to the set S_i in this iteration. Notice that step (2c) of the algorithm only reassigns nodes with zero weight, so N_i has the same weight as the solution to MaxSBCC S obtained in step (2b).

Consider some optimal solution $(S_1^*, ..., S_k^*)$ to the min-max multiway cut instance. Now partition U' into sets $U_1', ..., U_k'$, such that $U_i' = S_i^* \cap U'$. We claim that $w(N_i) \geq \beta \cdot |U_i'|$. To see this, notice that the nodes in U' have weight 1 throughout this iteration of the **while** loop, and since S_i^* is a piece of the optimal partition, $\delta(S_i^*) \leq B$. Therefore, in the ith iteration of the **for** loop, $(S_i^*, V \setminus S_i^*)$ is a feasible solution to the MaxSBCC problem, and $w(S_i^*) \geq w(U_i') = |U_i'|$. By the (α, β)-approximation guarantee, the algorithm for MaxSBCC must find a set with $w(N_i) \geq \beta \cdot |U_i'|$. Summing over all i, we obtain that $|U| - |U'| = \sum_{i=1}^{k} w(N_i) \geq \beta \cdot |U'|$, which proves the claim. □

Proof (of Theorem 1). By using binary search, we can assume that a bound B is given, and our algorithm will either prove that no multiway cut of maximum capacity at most B exists, or it will find a multiway cut with capacity at most $O(\alpha \log_{1+\beta} n)B$.

Throughout the algorithm, $x_i \in S_i$ for all i, and the sets S_i are always disjoint. So the algorithm finds a multiway cut, as required. By Lemma 1 the algorithm terminates in at most $\log_{1+\beta} n$ iterations of the **while** loop, if a min-max multiway cut of capacity at most B exists. If given an infeasible bound $B < B^*$, it may not stop after $\log_{1+\beta} n$ iterations, which proves that $B < B^*$. This shows that the algorithm runs in polynomial time. We will also use this bound to give an approximation guarantee for the algorithm.

We claim that for each S_i returned by the algorithm, $\delta(S_i) \leq \alpha \log_{1+\beta} n \cdot B$. To see this, notice that for each application of the MaxSBCC subroutine in (2b), the capacity of the set S returned is at most $\delta(S) \leq \alpha B$. By the choice made in step (2c), the transfer operation does not increase either $\delta(S_j)$ or $\delta(S)$. So in each iteration of the **while** loop, the capacity of each S_i increases by at most αB. Combined with the bound on the number of iterations, this observation concludes the proof. □

Feige and Krauthgamer [3] give an $O(\log^2 n)$ approximation algorithm for the problem of finding cuts with specified number of nodes, and an improved $O(\log n)$ approximation for the case when the input graph G is assumed not to contain a fixed graph as a minor (e.g., for planar graphs). We will use this algorithm to give an $(O(\log^2 n), 1)$ approximation algorithm for the MaxSBCC problem in general graphs and an improved $(O(\log n), 1)$ approximation algorithm in the special case. This will yield the following theorem.

Theorem 2. *There is an $O(\log^3 n)$-approximation algorithm for MIN-MAX MULTIWAY CUT problem on general graphs, and an $O(\log^2 n)$-approximation for graphs excluding any fixed graph as a minor (e.g., planar graphs).*

Proof. Feige and Krauthgamer [3] give an algorithm for finding cuts with specified sizes. For a graph G with n nodes and each number $d < n$, their algorithm finds a cut (S_d, T_d) with $|S_d| = d$ and capacity $\delta(S_d)$ within $\alpha = O(\log^2 n)$ of the minimum capacity for such a cut. For graphs excluding any fixed graph as a minor, this guarantee is improved to $\alpha' = O(\log n)$. The algorithm also works for finding s-t cuts on graphs with node weights and edge capacities.

We claim that the cut that corresponds to the largest value d^* such that $\delta(S_{d^*}) \leq \alpha B$ is an $(\alpha, 1)$-approximate MaxSBCC. By definition, its capacity is at most αB. And, if the optimal MaxSBCC had size $d' > d^*$, then, by the guarantee of the algorithm, $\delta(S_{d'})$ would be at most αB, contradicting our choice of d^*. The result then follows from Theorem 1. $\square$

It is interesting to note that the algorithm can also be used for a version of the multiway cut problem in which there is a separate bound B_i for each $\delta(S_i)$. To obtain the extension, we use the MaxSBCC algorithm in each iteration i of the for loop with bound B_i rather than B.

Theorem 3. *Assume we are given a graph G with k terminals, edge capacities, and k bounds $(B_1, \ldots, B_k)$. If there is a multiway cut $(S_1, \ldots, S_k)$ such that $\delta(S_i) \leq B_i$ for each i, then in polynomial time we can find a multiway cut $(S'_1, \ldots, S'_k)$ such that $\delta(S'_i) \leq O(\log^3 n) B_i$, and the bound improves by a factor of $\log n$ for graphs excluding any fixed graph as a minor.*

Remark. Calinescu, Karloff and Rabani [1] and subsequently Karger et al. [6] gave improved approximation algorithms for the multiway cut problem based on linear programming and rounding. It appears that this technique does not yield a good approximation for our problem. To see this, consider the graph which is a star with k terminals and a single additional node at the center, and assume the capacity of each edge is 1. There is no multiway cut where each part has capacity at most $B = 2$, or even approximately 2. By assigning the center of the star to terminal x_i, we can get a multiway cut where the capacity of each part S_j for $j \neq i$ is 1, while the capacity of S_i is $k - 1$. A linear programming relaxation would allow us to take a "linear combination" of these cuts, and thereby have each side have capacity at most 2.

2.2   NP-Completeness

We prove using a reduction from BISECTION that the MIN-MAX MULTIWAY CUT problem is NP-hard already on graphs with 4 terminals.

Theorem 4. MIN-MAX MULTIWAY CUT *is NP-hard for any fixed $k \geq 4$ even with unit-capacity edges.*

Proof. We will show that it is NP-hard for $k = 4$ using a reduction from BISECTION [4]. Our construction uses capacities, but we can replace each edge with multiple parallel paths. An instance of BISECTION consists of a graph $G = (V, E)$

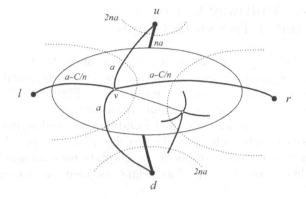

Fig. 2. Reduction from BISECTION to MIN-MAX MULTIWAY CUT

with an even number of vertices n, and an integer C. The question is whether or not there exists a partition of V into two sets X and Y, each of size $n/2$, such that the capacity of the cut $c(X, Y) \leq C$. Given G and C, we construct, in polynomial time, a graph F with 4 terminals and a bound B, so that F has a multiway cut with maximum capacity at most B if and only if G has a bisection with capacity at most C.

We obtain the graph $F = (V', E')$ by adding 4 new terminal nodes $X - \{u, d, l, r\}$ to G, and adding edges that connect nodes of G to the terminals (see Figure 2). E' includes E and the following additional edges, where a is chosen such that $2a > C$:

- Edge (u, d) of capacity na
- Edges (v, u) and (v, d), each of capacity a, for each $v \in V$.
- Edges (v, l) and (v, r), each of capacity $b = a - \frac{C}{n} > 0$, for each $v \in V$.

The bound is set to $B = 2na$.

Suppose F has a min-max multiway cut $(U \cup \{u\}, D \cup \{d\}, L \cup \{l\}, R \cup \{r\})$ where each part has capacity at most B. Then U and D must be empty, as $B = 2na \geq \delta(U \cup \{u\}) \geq 2na + 2b|U|$, just counting the edges to the terminals. So (L, R) is a cut of G, and let $C' = c(L, R)$ denote its capacity. The next observation is that $|L| = |R| = n/2$. To see this, suppose, for contradiction, that $|L| = k \geq \frac{n}{2} + 1$ (or similarly for $|R|$). Then

$$\delta(L \cup \{l\}) = 2ka + nb + C' \geq 2(\frac{n}{2} + 1)a + n(a - \frac{C}{n}) = 2na + (2a - C) > B,$$

where the last inequality follows from the choice of a. We conclude that the capacity C' of the bisection (L, R) must be at most C. This follows as the capacity of the cut $L \cup \{l\}$ is $na + nb + C' \leq B = 2na$, and by the choice of b this inequality implies $C' \leq C$. To show the opposite direction, given a bisection (X, Y) in G of capacity $C' \leq C$, we produce a min-max multiway cut $(\{u\}, \{d\}, X \cup \{l\}, Y \cup \{r\})$ of F, with each component's capacity at most B. $\square$

3   Min-Max Multiway Cut on Trees and Bounded Treewidth Graphs

Recall from the Introduction that in an optimal solution to the min-max multiway cut problem on trees the sets of nodes assigned to the terminals do not have to be connected. This can be seen in the example of Figure 3. All nodes except for the middle one are terminals, and all edges have capacity 1. The optimal solution cuts all the edges incident on the middle node and assigns it to one of the leaf (degree-one) terminals, achieving a value of 4. On the other hand, any solution that assigns connected parts of the graph to each terminal would leave the middle node connected to one of its neighbors, incurring a cost of 5.

Fig. 3. Example showing that in an optimal min-max multiway cut on a tree, the sets assigned to the terminals need not form connected components. The only non-terminal in this graph is the middle node

In Section 3.1 we use this observation to prove that the MIN-MAX MULTIWAY CUT problem in NP-hard on trees. Then we provide a $(2 + \epsilon)$-approximation on trees, and, finally, in Section 3.4 we extend it to graphs with bounded treewidth.

3.1   NP-Completeness

Theorem 5. MIN-MAX MULTIWAY CUT *is strongly NP-hard when the graph is a tree with weighted edges.*

Proof. We use a reduction from 3-PARTITION, which is known to be strongly NP-complete [5]. In 3-PARTITION, given a set $A = \{a_1, ..., a_{3m}\}$, a weight w_i for each $a_i \in A$, and a bound B, such that $\forall i \ B/4 < w_i < B/2$ and $\sum_{i=1}^{3m} w_i = mB$, we want to know if A can be partitioned into disjoint sets $S_1, ..., S_m$, such that for each j,

$$\sum_{a_i \in S_j} w_i = B.$$

Given an instance (A, B) of 3-partition, we construct an instance of min-max multiway cut as follows. The tree T consists of separate subtrees connected with zero-capacity edges. There will be $3m$ subtrees T_i, one for each element a_i, and m isolated terminals $x_1, ..., x_m$, one for each of the desired sets. Each T_i consists of six terminals and one non-terminal v_i, with edge capacities as in Figure 4.

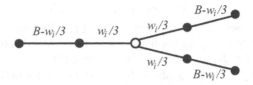

Fig. 4. Component T_i used in the NP-completeness reduction for MIN-MAX MULTIWAY CUT on trees

We claim that a min-max multiway cut of maximum capacity at most B exists if and only if the 3-partition instance is solvable. Notice that any min-max multiway cut with capacity at most B must cut all edges of T_i and assign all v_i's to the terminals $x_1, ..., x_m$, creating a partition of A. If a set of nodes S_j' is assigned to terminal x_j, then the capacity of the resulting part is $\sum_{v_i \in S_j'} w_i$. This implies that such a cut exists if and only if the 3-partition does.   □

3.2   Algorithm for Min-Max Multiway Cut on Trees

In this section we give a $(2+\epsilon)$-approximation algorithm for the min-max multiway cut on trees that have edge capacities.

The algorithm consists of two stages. In the first stage we consider a variant of the problem where we allow the algorithm to create extra parts in the partition that do not contain terminals. More precisely, we consider the following problem.

Definition 2. *The* TREE CUTTING *problem* (T, X, B) *for a tree* $T = (V, E)$, *terminals* $X = \{x_1, ..., x_k\} \subseteq V$, *and a bound* B *is to find a partition of* V *into connected subtrees* $T_1, ..., T_h$, *subject to the following constraints: (1) no two terminals are in the same connected component; and (2) for each connected component* T_i, $\delta(T_i) \leq B$. *The objective is to minimize* $\sum_i \delta(T_i)$.

In the next subsection we give a pseudo-polynomial time algorithm for this problem. Here we show how to use such an algorithm to get a $(2+\epsilon)$-approximation for the min-max multiway cut on trees.

Theorem 6. *Using a pseudo-polynomial time algorithm for the* TREE CUTTING *problem as a subroutine, we can give a polynomial time* $(2+\epsilon)$-*approximation for the* MIN-MAX MULTIWAY CUT *on trees.*

Proof. First we give a 2-approximation for min-max multiway cut that uses a pseudo-polynomial exact algorithm for tree cutting. Given a tree T with terminals X, we will binary search for a lower bound B^* to the min-max multiway cut optimum. Observe that connected components in a feasible solution to min-max multiway cut instance (T, X) of value B give a feasible solution to the tree cutting instance (T, X, B) of value at most kB. Therefore, if our optimal tree cutting solution $T_1, ..., T_h$ does not satisfy $\sum_{i=1}^h \delta(T_i) \leq kB$, then $B < B^*$. The algorithm groups the components T_i into k sets $S_1, ..., S_k$ of nodes greedily, assigning the terminals to different sets. Observe that if several components, say $T_1, ..., T_j$, are combined into a set S, then $\delta(S) \leq \sum_{i=1}^{j} \delta(T_i)$. Because $\sum_{i=1}^{h} \delta(T_i) \leq kB$, and each $\delta(S_i)$ is at least B, all components will be packed into the k sets. Also, since for all j, $\delta(T_j) \leq B$, for no i will $\delta(S_i)$ exceed $2B$.

Recall that our tree cutting algorithm runs in pseudopolynomial time. We obtain a polynomial-time $(2 + \epsilon)$-approximation algorithm via rounding. For a given ϵ, capacity bound B, and $m = |E|$, let

$$\alpha = \frac{\lceil m/\epsilon \rceil}{B}.$$

For each edge $e \in E$, scale the capacity so that $c'(e) = \lfloor \alpha c(e) \rfloor$. Also set $B' = \alpha B = \lceil m/\epsilon \rceil$. If there is a multiway cut with maximum capacity B in the original problem, then there is one of maximum capacity at most B' after rounding. Now we obtain a 2-approximate multiway cut $S_1, ..., S_k$ for the graph with capacities $c'(e)$ and bound B'. The running time is polynomial in $B' = \lceil m/\epsilon \rceil$ and n. The capacity of a part S_i of this partition is at most $\delta(S_i) \leq (2 + \epsilon) \cdot B$ using the original capacities. $\qquad \square$

3.3 Algorithm for the Tree Cutting Problem

We now describe an algorithm that solves optimally, in time polynomial in B and the size of the tree n, the tree cutting problem (which we used as a subroutine for the min-max multiway cut on trees). To simplify the presentation of the algorithm, assume, without loss of generality, that (1) T is rooted at a node r and all edges are directed away from the root; (2) T is binary. (To make the tree binary without affecting the solution, replace each node u that has $d > 2$ children with a $\lceil \log_2 d - 1 \rceil$-height complete binary subtree U with edge capacities $B + 1$, and attach u's children to the leaves of U, at most 2 per leaf.)

The tree cutting problem will be solved using dynamic programming. First consider the simpler problem with no terminals. To solve this problem, we construct a dynamic programming table $p(v, A)$ for all nodes $v \in V$ and integers $0 \leq A \leq B$, where the entry $p(v, A)$ is the minimum total capacity of edges in the subtree of T rooted at v that can be cut such that the total capacity of edges coming *out* (i.e., toward descendants) of v's component is at most A. We have the separate bound A because the remaining $B - A$ capacity will be used to cut the edges that are incident on v's component, but lie outside of its subtree. The values $p(v, A)$ can be computed in a single pass up the tree. If a node v has one child v_1, then cutting (v, v_1) implies that the component containing v_1 can have at most $B - c(v, v_1)$ capacity below v_1, so the total capacity obtained this way is $c(v, v_1) + p(v_1, B - c(v, v_1))$. If we do not cut the edge (v, v_1), then we get $p(v_1, A)$. This leads to the following recurrence for the case that v has a single child v_1.

$$p(v, A) = \begin{cases} p(v_1, A) & \text{if } c(v, v_1) > A \\ \min\{p(v_1, A), c(v, v_1) + p(v_1, B - c(v, v_1))\} & \text{otherwise.} \end{cases}$$

Now suppose that the internal node v has two children, v_1 and v_2. Then the capacity A available for cutting edges below v has to be partitioned between the edges that belong to the left subtree (including, possibly, (v, v_1)), and the ones that belong to the right subtree (possibly including (v, v_2)). The algorithm tries all possibilities for such a partition $A_1 + A_2 = A$. Then, given A_i, it decides independently for each child node v_i whether or not to cut (v, v_i), using an expression similar to the one above.

Next we extend the algorithm to make sure that all terminals in the tree are separated. For this, a binary variable t is added to the parameters of the table. The value of t limits the options available to the above simpler algorithm in each step. It will either require that a given component *not* contain a terminal $(t=0)$, or it will not impose such a restriction $(t = *)$. The idea is that if a node is connected to some ancestor which is a terminal, then it may not be connected to any descendants which are terminals, so in this case we will use a table entry with $t = 0$. Also, care has to be taken that a node is not connected to two terminals which are both descendants.

Theorem 7. *The optimal solution to the* TREE CUTTING *problem can be computed in time polynomial in the size of the graph and the bound B.*

3.4   Bounded Treewidth Graphs

Finally, we extend the $(2 + \epsilon)$-approximation algorithm for min-max multiway cut to work on graphs with bounded treewidth (see [7] for an introduction to tree decomposition). The only change is in solving the tree cutting problem.

First we note that the algorithm from Section 3.3 can work on trees with degree greater than two. The only potential difficulty is to generate the optimal guesses of $A_1, ..., A_d$, $\sum A_i = A$, in polynomial time. Given the values of the subproblems at the leaves, the problem of finding the best partition to obtain the value $p(v, A)$ is essentially a knapsack problem, and hence is solved by optimally solving the problem for each suffix $j, ..., d$ of the set of node's children, with j going from d to 1, and using the optimal result for $j, ..., d$ in order to solve the problem for $j - 1, ..., d$.

Now we sketch how to extend this algorithm to handle graphs with bounded treewidth. Suppose that we are given a graph $G = (V, E)$ and a decomposition tree T for it, such that for each node u in T, there is a set $V_u \subseteq V$ associated with it. The size of each V_u is bounded by a constant b. Let us root T at some node r and assign a height $h(u)$ to each node $u \in T$ so that r is the highest. Now we can associate with each vertex $v \in V$ a *label*, which is defined as $\max\{h(u)|v \in V_u\}$. The algorithm will again build a dynamic programming table. In this case a table entry will be $p(u, \{H_1, ..., H_h\}, \{A_1, ..., A_h\}, \{t_1, ..., t_h\})$, where u is a node in T; H_i's form a partition of V_u that has h components, for some $h \in [1, ..., b]$, with the meaning that different H_i's will be subsets of different components in the solution; A_i is a bound on the total capacity of edges (v, w) such that v is in the same component as H_i, w is not, and $label(v) > label(w)$; and t_i is a variable for H_i that specifies, as before, whether this component is allowed to contain terminals. The computation proceeds in a bottom-up fashion on the tree T. When a node u of T and its child node u_1 are considered, all allowed combinations of vertices in V_u and V_{u_1} are evaluated, subject to the constraints imposed by H_i's and t_i's, and the best one is chosen. If u has multiple children, then each A_i is divided among the children in the same way as before. The running time of the algorithm is exponential in b, but, given that b is a fixed constant, it remains polynomial (or pseudopolynomial, as before).

Theorem 8. *The above algorithm computes the optimal solution to the* TREE CUTTING *problem in graphs with bounded treewidth in time polynomial in the size of the graph and the bound B. As a consequence, we get a $(2 + \epsilon)$-approximation for the* MIN-MAX MULTIWAY CUT *in graphs with bounded treewidth.*

Acknowledgements

We would like to thank Johannes Gehrke and Ashwin Machanavajjhala for suggesting the problem to us and for many insightful conversations. The maximum size bounded capacity cut problem (or rather its minimization version) was formulated as joint work with Ara Hayrapetyan, David Kempe, and Martin Pál. The reduction from KNAPSACK is due to David Kempe.

References

1. G. Calinescu, H. Karloff, and Y. Rabani. *An improved approximation algorithm for multiway cut.* STOC 1998.
2. E. Dahlhaus, D.S. Johnson, C.H. Papadimitriou, P.D.Seymour, and M. Yannakakis. *The complexity of multiterminal cuts.* SIAM Journal on Computing, 23(4):864–894, 1994.
3. U. Feige and R. Krauthgamer. *A polylogarithmic approximation of the minimum bisection.* SIAM Journal on Computing, 31(4):1090–1118, 2002.
4. M. R. Garey, D. S. Johnson, and L. Stockmeyer. *Some Simplified NP-Complete Graph Problems.* Theoretical Computer Science, (1):237–267, 1976.
5. M. R. Garey and D. S. Johnson 1979. *Computers and Intractability.* W. H. Freeman and Company, New York.
6. D. R. Karger, P. Klein, C. Stein, M. Thorup, N. E. Young. *Rounding algorithms for a geometric embedding of minimum multiway cut.* STOC 1999.
7. J. Kleinberg and E. Tardos. *Algorithms Design.* Addison-Wesley, to appear.

The Chromatic Number
of Random Regular Graphs

Dimitris Achlioptas[1] and Cristopher Moore[2]

[1] Microsoft Research, Redmond, WA 98052, USA
optas@microsoft.com
[2] University of New Mexico, NM 87131, USA
moore@cs.unm.edu

Abstract. Given any integer $d \geq 3$, let k be the smallest integer such that $d < 2k \log k$. We prove that with high probability the chromatic number of a random d-regular graph is k, $k + 1$, or $k + 2$.

1 Introduction

In [10], Łuczak proved that for every real $d > 0$ there exists an integer $k = k(d)$ such that w.h.p.[1] $\chi(\mathcal{G}(n, d/n))$ is either k or $k + 1$. Recently, these two possible values were determined by the first author and Naor [4].

Significantly less is known for random d-regular graphs $\mathcal{G}_{n,d}$. In [6], Frieze and Łuczak extended the results of [9] for $\chi(\mathcal{G}(n, p))$ to random d-regular graphs, proving that for all integers $d > d_0$, w.h.p.

$$\left| \chi(\mathcal{G}_{n,d}) - \frac{d}{2 \log d} \right| = \Theta \left(\frac{d \log \log d}{(\log d)^2} \right) .$$

Here we determine $\chi(\mathcal{G}_{n,d})$ up to three possible values for all integers. Moreover, for roughly half of all integers we determine $\chi(\mathcal{G}_{n,d})$ up to two possible values. We first replicate the argument in [10] to prove

Theorem 1. *For every integer d, there exists an integer $k = k(d)$ such that w.h.p. the chromatic number of $\mathcal{G}_{n,d}$ is either k or $k + 1$.*

We then use the second moment method to prove the following.

Theorem 2. *For every integer d, w.h.p. $\chi(\mathcal{G}_{n,d})$ is either k, $k+1$, or $k+2$, where k is the smallest integer such that $d < 2k \log k$. If, furthermore, $d > (2k-1) \log k$, then w.h.p. $\chi(\mathcal{G}_{n,d})$ is either $k + 1$ or $k + 2$.*

The table below gives the possible values of $\chi(\mathcal{G}_{n,d})$ for some values of d.

d	4	5	6	7, 8, 9	10	100	1,000,000
$\chi(\mathcal{G}_{n,d})$	3, 4	3, 4, 5	4, 5	4, 5, 6	5, 6	18, 19, 20	46523, 46524

---

[1] Given a sequence of events $\mathcal{E}_n$, we say that $\mathcal{E}$ holds *with positive probability* (w.p.p.) if $\liminf_{n \to /\infty} \Pr[\mathcal{E}_n] > 0$, and *with high probability* (w.h.p.) if $\liminf_{n \to \infty} \Pr[\mathcal{E}_n] = 1$.

K. Jansen et al. (Eds.): APPROX and RANDOM 2004, LNCS 3122, pp. 219–228, 2004.
© Springer-Verlag Berlin Heidelberg 2004

1.1   Preliminaries and Outline of the Proof

Rather than proving our results for $\mathcal{G}_{n,d}$ directly, it will be convenient to work with random d-regular multigraphs, in the sense of the configuration model [5]; that is, multigraphs $\mathcal{C}_{n,d}$ generated by selecting a uniformly random configuration (matching) on dn "vertex copies." It is well-known that for any fixed integer d, a random such multigraph is simple w.p.p. As a result, to prove Theorem 1 we simply establish its assertion for $\mathcal{C}_{n,d}$.

To prove Theorem 2 we use the second moment method to show

Theorem 3. *If $d < 2k \log k$, then w.p.p. $\chi(\mathcal{C}_{n,d}) \leq k + 1$.*

Proof of Theorem 2. For integer k let $u_k = (2k - 1) \log k$ and $c_k = 2k \log k$. Observe that $c_{k-1} < u_k < c_k$. Thus, if k is the smallest integer such that $d < c_k$, then either i) $u_k < d < c_k$ or ii) $u_{k-1} < c_{k-1} < d \leq u_k < c_k$.

A simple first moment argument (see e.g. [11]) implies that if $d > u_k$ then w.h.p. $\chi(\mathcal{C}_{n,d}) > k$. Thus, if $u_k < d < c_k$, then w.h.p. $\mathcal{C}_{n,d}$ is non-k-colorable while w.p.p. it is $(k + 1)$-colorable. Therefore, by Theorem 1, w.h.p. the chromatic number of $\mathcal{C}_{n,d}$ (and therefore $\mathcal{G}_{n,d}$) is either $k+1$ or $k+2$. In the second case, we cannot eliminate the possibility that $\mathcal{G}_{n,d}$ is w.p.p. k-colorable, but we do know that it is w.h.p. non-$(k - 1)$-colorable. Thus, similarly, it follows that $\chi(\mathcal{G}_{n,d})$ is w.h.p. k, $k + 1$ or $k + 2$.   □

Throughout the rest of the paper, unless we explicitly say otherwise, we are referring to random multigraphs $\mathcal{C}_{n,d}$. We will say that a multigraph is k-colorable iff the underlying simple graph is k-colorable. Also, we will refer to multigraphs and configurations interchangeably using whichever form is most convenient.

2   2-Point Concentration

In [10], Łuczak in fact established two-point concentration for $\chi(\mathcal{G}(n, d/n))$ for all $\epsilon > 0$ and $d = O(n^{1/6-\epsilon})$. Here, mimicking his proof, we establish two-point concentration for $\chi(\mathcal{G}_{n,d})$ for all $\epsilon > 0$ and $d = O(n^{1/7-\epsilon})$.

Our main technical tool is the following martingale-based concentration inequality for random variables defined on $\mathcal{C}_{n,d}$ [12, Thm 2.19]. Given a configuration C, we define a *switching* in C to be the replacement of two pairs $\{e_1, e_2\}$, $\{e_3, e_4\}$ by $\{e_1, e_3\}$, $\{e_2, e_4\}$ or $\{e_1, e_4\}$, $\{e_3, e_2\}$.

Theorem 4. *Let X_n be a random variable defined on $\mathcal{C}_{n,d}$ such that for any configurations C, C' that differ by a switching*

$$|X_n(C) - X_n(C')| \leq b \ ,$$

for some constant $b > 0$. Then for every $t > 0$,

$$\Pr\big[X_n \leq \mathbf{E}[X_n] - t\big] < e^{-\frac{t^2}{dnb^2}} \quad and \quad \Pr\big[X_n \geq \mathbf{E}[X_n] + t\big] < e^{-\frac{t^2}{dnb^2}} \ .$$

Theorem 1 will follow from the following two lemmata. The proof of Lemma 1 is a straightforward union bound argument and is relegated to the full paper.

Lemma 1. *For any $0 < \epsilon < 1/6$ and $d < n^{1/6-\epsilon}$, w.h.p. every subgraph induced by $s \leq nd^{-3(1+2\epsilon)}$ vertices contains at most $(3/2 - \epsilon)s$ edges.*

Lemma 2. *For a given function $\omega(n)$, let $k = k(\omega, n, p)$ be the smallest k such that*

$$\Pr[\chi(\mathcal{C}_{n,d}) \leq k] \geq 1/\omega(n) .$$

With probability greater than $1 - 1/\omega(n)$, all but $8\sqrt{nd \log \omega(n)}$ vertices of $\mathcal{C}_{n,d}$ can be properly colored using k colors.

Proof. For a multigraph G, let $Y_k(G)$ be the minimal size of a set of vertices S for which $G - S$ is k-colorable. Clearly, for any k and G, switching two edges of G can affect $Y_k(G)$ by at most 4, as a vertex cannot contribute more than itself to $Y_k(G)$. Thus, if $\mu_k = \mathbf{E}[Y_k(\mathcal{C}_{n,d})]$, Theorem 4 implies

$$\Pr[Y_k \leq \mu_k - \lambda\sqrt{n}] < e^{-\frac{\lambda^2}{16d}} \quad \text{and} \quad \Pr[Y_k \geq \mu_k + \lambda\sqrt{n}] < e^{-\frac{\lambda^2}{16d}} . \tag{1}$$

Define now $u = u(n, p, \omega(n))$ to be the least integer for which $\Pr[\chi(G) \leq u] \geq 1/\omega(n)$. Choosing $\lambda = \lambda(n)$ so as to satisfy $e^{-\lambda^2/(16d)} - 1/\omega(n)$, the first inequality in (1) yields

$$\Pr[Y_u \leq \mu_u - \lambda\sqrt{n}] < 1/\omega(n) \leq \Pr[\chi(G) \leq u] = \Pr[Y_u = 0] .$$

Clearly, if $\Pr[Y_u \leq \mu_u - \lambda\sqrt{n}] < \Pr[Y_u = 0]$ then $\mu_u < \lambda\sqrt{n}$. Thus, the second inequality in (1) implies $\Pr[Y \geq 2\lambda\sqrt{n}] < 1/\omega(n)$ and, by our choice, $\lambda = 4\sqrt{d \log \omega(n)}$. $\qquad\square$

Proof of Theorem 1. The result is trivial for $d = 1, 2$. Given $d \geq 3$, let $k = k(d, n) \geq 3$ be the smallest integer for which the probability that $\mathcal{C}_{n,d}$ is k-colorable is at least $1/\log \log n$. By Lemma 2, w.h.p. there exists a set of vertices S such that all vertices outside S can be colored using k colors and $|S| < 8\sqrt{nd \log \log \log n} < \sqrt{nd} \log n \equiv s_0$. From S, we will construct an increasing sequence of sets of vertices $\{U_i\}$ as follows. $U_0 = S$; for $i \geq 0$, $U_{i+1} = U_i \cup \{w_1, w_2\}$, where $w_1, w_2 \notin U_i$ are adjacent and each of them has some neighbor in U_i. The construction ends, with U_t, when no such pair exists.

Observe that the neighborhood of U_t in the rest of the graph, $N(U_t)$, is always an independent set, since otherwise the construction would have gone on. We further claim that w.h.p. the graph induced by the vertices in U_t is k-colorable. Thus, using an additional color for $N(U_t)$ yields a $(k + 1)$-coloring of the entire multigraph, concluding the proof.

We will prove that U_t is, in fact, 3-colorable by proving that $|U_t| \leq s_0/\epsilon$. This suffices since by Lemma 1 w.h.p. every subgraph H of b or fewer vertices has average degree less than 3 and hence contains a vertex v with $\deg(v) \leq 2$. Repeatedly invoking Lemma 1 yields an ordering of the vertices in H such that each vertex is adjacent to no more than 2 of its successors. Thus, we can start with the last vertex in the ordering and proceed backwards; there will always be at least one available color for the current vertex. To prove $|U_t| \leq 2s_0 \log n$ we observe that each pair of vertices entering U "brings in" with it at least 3 new edges. Therefore, for every $j \geq 0$, U_j has at most $s_0 + 2j$ vertices and at least $3j$ edges. Thus, by Lemma 1, w.h.p. $t < 3s_0/(4\epsilon)$. $\square$

3   Establishing Colorability in Two Moments

Let us say that a coloring σ is *nearly–balanced* if its color classes differ in size by at most 1, and let X be the number of nearly–balanced k-colorings of $\mathcal{C}_{n,d}$. Recall that $c_k = 2k \log k$. We will prove that for all $k \geq 3$ and $d < c_{k-1}$ there exist constants $C_1, C_2 > 0$ such that for all sufficiently large n (when dn is even),

$$\mathbf{E}[X] > C_1 \, n^{-(k-1)/2} \, k^n \left(1 - \frac{1}{k} \right)^{dn/2} , \tag{2}$$

$$\mathbf{E}[X^2] < C_2 \, n^{-(k-1)} \, k^{2n} \left(1 - \frac{1}{k} \right)^{dn} . \tag{3}$$

By the Cauchy-Schwartz inequality (see e.g. [7, Remark 3.1]), we have $\Pr[X > 0] > \mathbf{E}[X]^2/\mathbf{E}[X^2] > C_1^2/C_2 > 0$, and thus Theorem 3.

To prove (2), (3) we will need to bound certain combinatorial sums up to constant factors. To achieve this we will use the following Laplace-type lemma, which generalizes a series of lemmas in [2–4]. Its proof is standard but somewhat tedious, and is relegated to the full paper.

Lemma 3. *Let ℓ, m be positive integers. Let $\mathbf{y} \in \mathbb{Q}^m$, and let M be a $m \times \ell$ matrix of rank r with integer entries whose top row consists entirely of 1's. Let s, t be nonnegative integers, and let $\mathbf{v}_i, \mathbf{w}_j \in \mathbb{N}^\ell$ for $1 \leq i \leq s$ and $1 \leq j \leq t$, where each $\mathbf{v}_i$ and $\mathbf{w}_j$ has at least one nonzero component, and where moreover $\sum_{i=1}^s \mathbf{v}_i = \sum_{j=1}^t \mathbf{w}_j$. Let $f : \mathbb{R}^\ell \to \mathbb{R}$ be a positive twice-differentiable function. For $n \in \mathbb{N}$, define*

$$S_n = \sum_{\{\mathbf{z} \in \mathbb{N}^\ell : M \cdot \mathbf{z} = \mathbf{y}n\}} \frac{\prod_{i=1}^s (\mathbf{v}_i \cdot \mathbf{z})!}{\prod_{j=1}^t (\mathbf{w}_j \cdot \mathbf{z})!} f(\mathbf{z}/n)^n$$

and define $g : \mathbb{R}^\ell \to \mathbb{R}$ as

$$g(\zeta) = \frac{\prod_{i=1}^s (\mathbf{v}_i \cdot \zeta)^{(\mathbf{v}_i \cdot \zeta)}}{\prod_{j=1}^t (\mathbf{w}_j \cdot \zeta)^{(\mathbf{w}_j \cdot \zeta)}} f(\zeta)$$

where $0^0 \equiv 1$. Now suppose that, conditioned on $M \cdot \zeta = \mathbf{y}$, g is maximized at some ζ^* with $\zeta_i^* > 0$ for all i, and write $g_{\max} = g(\zeta^*)$. Furthermore, suppose that the matrix of second derivatives $g'' = \partial^2 g / \partial \zeta_i \, \partial \zeta_j$ is nonsingular at ζ^*.

Then there exist constants $A, B > 0$, such that for any sufficiently large n for which there exist integer solutions $\mathbf{z}$ to $M \cdot \mathbf{z} = \mathbf{y}n$, we have

$$A \le \frac{S_n}{n^{-(\ell+s-t-r)/2} g_{\max}^n} \le B \ .$$

For simplicity, in the proofs of (2) and (3) below we will assume that n is a multiple of k, so that nearly–balanced colorings are in fact exactly balanced, with n/k vertices in each color class. The calculations for other values of n differ by at most a multiplicative constant.

4   The First Moment

Clearly, all (exactly) balanced k-partitions of the n vertices are equally likely to be proper k-colorings. Therefore, $\mathbf{E}[X]$ is the number of balanced k-partitions, $n!/(n/k)!^k$, times the probability that a random d-regular configuration is properly colored by a fixed balanced k-partition.

To estimate this probability we will label the d copies of each vertex, thus giving us $(dn-1)!!$ distinct configurations, and count the number of such configurations that are properly colored by a fixed balanced k-partition. To generate such a configuration we first determine the number of edges between each pair of color classes. Suppose there are b_{ij} edges between vertices of colors i and j for each $i \ne j$. Then a properly colored configuration can be generated by i) choosing which b_{ij} of the dn/k copies in each color class i are matched with copies in each color class $j \ne i$, and then ii) choosing one of the $b_{ij}!$ matchings for each unordered pair $i < j$. Therefore, the total number of properly colored configurations is

$$\prod_{i=1}^{k} \frac{(dn/k)!}{\prod_{j \ne i} b_{ij}!} \cdot \prod_{i<j} b_{ij}! = \frac{(dn/k)!^k}{\prod_{i<j} b_{ij}!} \ .$$

Summing over all choices of the $\{b_{ij}\}$ that satisfy the constraints

$$\forall i : \sum_{j} b_{ij} = dn/k \ , \tag{4}$$

we get

$$\mathbf{E}[X] = \frac{n!}{(n/k)!^k} \frac{1}{(dn-1)!!} \sum_{\{b_{ij}\}} \frac{(dn/k)!^k}{\prod_{i<j} b_{ij}!}$$

$$= 2^{dn/2} \frac{n!}{(n/k)!^k} \frac{(dn/k)!^k}{(dn)!} \sum_{\{b_{ij}\}} \frac{(dn/2)!}{\prod_{i<j} b_{ij}!} \ .$$

By Stirling's approximation $\sqrt{2\pi n}\,(n/e)^n < n! < \sqrt{4\pi n}\,(n/e)^n$ we get

$$\mathbf{E}[X] > D_1 \frac{2^{dn/2}}{k^{(d-1)n}} \sum_{\{b_{ij}\}} \frac{(dn/2)!}{\prod_{i<j} b_{ij}!} \ , \tag{5}$$

where $D_1 = 2^{-(k+1)/2} d^{(k-1)/2}$.

To bound the sum in (5) from below we use Lemma 3. Specifically, $\mathbf{z}$ consists of the variables b_{ij} with $i < j$, so $\ell = k(k-1)/2$. For $k \geq 3$, the k constraints (4) are linearly independent, so representing them as $M \cdot \mathbf{z} = \mathbf{y}n$ gives a matrix M of rank k. Moreover, they imply $\sum_{i<j} b_{ij} = dn/2$, so adding a row of 1's to the top of M and setting $y_1 = d/2$ does not increase its rank. Integer solutions $\mathbf{z}$ exist whenever n is a multiple of k and dn is even. We set $s = 1$ and $t = \ell$; the vector $\mathbf{v}_1$ consists of 1's and the $\mathbf{w}_j$ are the ℓ basis vectors. Finally, $f(\zeta) = 1$. Thus, $\ell + s - t - r = -(k-1)$ and

$$g(\zeta) = \frac{(d/2)^{d/2}}{\prod_{j=1}^{\ell} \zeta_k^{\zeta_k}} = \frac{1}{\prod_{j=1}^{\ell} (2\zeta_j/d)^{\zeta_j}} = e^{(d/2)H(2\zeta/d)} ,$$

where H is the entropy function $H(\mathbf{x}) = -\sum_{j=1}^{\ell} x_j \log x_j$.

Since g is convex it is maximized when $\zeta_j^* = d/(2\ell)$ for all $1 \leq j \leq \ell$, and g'' is nonsingular. Thus, $g_{max} = (k(k-1)/2)^{d/2}$ implying that for some $A > 0$ and all sufficiently large n

$$\mathbf{E}[X] > D_1 \frac{2^{dn/2}}{k^{(d-1)n}} \times A n^{-(k-1)/2} \left(\frac{k(k-1)}{2} \right)^{dn/2}$$

$$= D_1 A n^{-(k-1)/2} k^n \left(1 - \frac{1}{k} \right)^{dn/2} .$$

Setting $C_1 = D_1 A$ completes the the proof.

5   The Second Moment

Recall that X is the sum over all balanced k-partitions of the indicators that each partition is a proper coloring. Therefore, $\mathbf{E}[X^2]$ is the sum over all pairs of balanced k-partitions of the probability that both partitions properly color a random d-regular configuration. Given a pair of partitions σ, τ, let us say that a vertex v is in class (i, j) if $\sigma(v) = i$ and $\tau(v) = j$. Also, let a_{ij} denote the number of vertices in each class (i, j). We call $A = (a_{ij})$ the *overlap matrix* of the pair σ, τ. Note that since both σ and τ are balanced

$$\forall i : \sum_j a_{ij} = \sum_j a_{ji} = n/k . \tag{6}$$

We will show that for any fixed pair of k-partitions, the probability that they both properly color a random d-regular configuration depends only on their overlap matrix A. Denoting this probability by $q(A)$, since there are $n!/\prod_{ij} a_{ij}!$ pairs of partitions giving rise to A, we have

$$\mathbf{E}[X^2] = \sum_A \frac{n!}{\prod_{ij} a_{ij}!} q(A) \tag{7}$$

where the sum is over matrices A satisfying (6).

Fixing a pair of partitions σ and τ with overlap matrix A, similarly to the first moment, we label the d copies of each vertex thus getting $(dn-1)!!$ distinct configurations. To generate configurations properly colored by both σ and τ we first determine the number of edges between each pair of vertex classes. Let us say that there are $b_{ijk\ell}$ edges connecting vertices in class (i,j) to vertices in class (k,ℓ). By definition, $b_{ijk\ell} = b_{k\ell ij}$, and if both colorings are proper, $b_{ijk\ell} = 0$ unless $i \neq k$ and $j \neq \ell$. Since the configuration is d-regular, we also have

$$\forall i,j : \sum_{k \neq i, \ell \neq j} b_{ijk\ell} = da_{ij} . \tag{8}$$

To generate a configuration consistent with A and $\{b_{ijk\ell}\}$ we now i) choose for each class (i,j), which $b_{ijk\ell}$ of its da_{ij} copies are to be matched with copies in each class (k,ℓ) with $k \neq i$ and $\ell \neq j$, and then ii) choose one of the $b_{ijk\ell}!$ matchings for each unordered pair of classes $i < k$, $j \neq \ell$. Thus,

$$q(A) = \frac{1}{(dn-1)!!} \sum_{\{b_{ijk\ell}\}} \left(\prod_{ij} \frac{(da_{ij})!}{\prod_{k \neq i, \ell \neq j} b_{ijk\ell}!} \cdot \prod_{i<k, j \neq \ell} b_{ijk\ell}! \right)$$
$$= 2^{dn/2} \frac{\prod_{ij}(da_{ij})!}{(dn)!} \sum_{\{b_{ijk\ell}\}} \frac{(dn/2)!}{\prod_{i<k, j \neq \ell} b_{ijk\ell}!} , \tag{9}$$

where the sum is over the $\{b_{ijk\ell}\}$ satisfying (8). Combining (9) with (7) gives

$$\mathbf{E}[X^2] = 2^{dn/2} \sum_{\{a_{ij}\}} \sum_{\{b_{ijk\ell}\}} \frac{n!}{\prod_{ij} a_{ij}!} \frac{\prod_{ij}(da_{ij})!}{(dn)!} \frac{(dn/2)!}{\prod_{i<k, j \neq \ell} b_{ijk\ell}!} . \tag{10}$$

To bound the sum in (10) from above we use Lemma 3. We let $\mathbf{z}$ consist of the combined set of variables $\{a_{ij}\} \cup \{b_{ijk\ell} : i < k, j \neq \ell\}$, in which case its dimensionality ℓ (not to be confused with the color ℓ) is $k^2 + (k(k-1))^2/2$. We represent the combined system of constraints (6), (8) as $M \cdot \mathbf{z} = \mathbf{y}n$. The k^2 constraints (8) are, clearly, linearly independent while the $2k$ constraints (6) have rank $2k-1$. Together these imply $\sum_{ij} a_{ij} = 1$ and $\sum_{i<k, j \neq \ell} b_{ijk\ell} = d/2$, so adding a row of 1's to the top of M does not change its rank from $r = k^2 + 2k - 1$. Integer solutions $\mathbf{z}$ exist whenever n is a multiple of k and dn is even. Finally, $f(\zeta) = 2^{d/2}$, $s = k^2 + 2$ and $t = k^2 + 1 + (k(k-1))^2/2$, so $\ell + s - t - r = -2(k-1)$.

Writing α_{ij} and $\beta_{ijk\ell}$ for the components of ζ corresponding to a_{ij}/n and $b_{ijk\ell}/n$, respectively, we thus have

$$g(\zeta) = 2^{d/2} \frac{1}{\prod_{ij} \alpha_{ij}^{\alpha_{ij}}} \frac{\prod_{ij}(d\alpha_{ij})^{d\alpha_{ij}}}{d^d} \frac{(d/2)^{d/2}}{\prod_{i<k, j \neq \ell} \beta_{ijk\ell}^{\beta_{ijk\ell}}}$$
$$= \frac{1}{\prod_{ij} \alpha_{ij}^{\alpha_{ij}}} \frac{d^{d/2} \prod_{ij} \alpha_{ij}^{d\alpha_{ij}}}{\prod_{i<k, j \neq \ell} \beta_{ijk\ell}^{\beta_{ijk\ell}}} . \tag{11}$$

In the next section we maximize $g(\zeta)$ over $\zeta \in \mathbb{R}^\ell$ satisfying $M \cdot \zeta = \mathbf{y}$. We note that g'' is nonsingular at the maximizer we find below, but we relegate the proof of this fact to the full paper.

6   A Tight Relaxation

Maximizing $g(\zeta)$ over $\zeta \in \mathbb{R}^\ell$ satisfying $M \cdot \zeta = \mathbf{y}$ is greatly complicated by the constraints

$$\forall i, j : \sum_{k \neq i, \ell \neq j} \beta_{ijk\ell} = d\alpha_{ij} \ . \tag{12}$$

To overcome this issue we i) reformulate $g(\zeta)$ and ii) relax the constraints, in a manner such that the maximum value remains unchanged while the optimization becomes much easier.

The relaxation amounts to replacing the k^2 constraints (12) with their sum divided by 2, i.e., with the single constraint

$$\sum_{i < k, j \neq \ell} \beta_{ijk\ell} = d/2 \ . \tag{13}$$

But attempting to maximize (11) under this single constraint is, in fact, a bad idea since the new maximum is much greater. Instead, we maximize the following equivalent form of $g(\zeta)$

$$g(\zeta) = \frac{1}{\prod_{ij} \alpha_{ij}^{\alpha_{ij}}} \frac{d^{d/2} \prod_{ij} \alpha_{ij}^{\sum_{k \neq i, \ell \neq j} \beta_{ijk\ell}}}{\prod_{i < k, j \neq \ell} \beta_{ijk\ell}^{\beta_{ijk\ell}}} \ , \tag{14}$$

derived by using (12) to substitute for the exponents $d\alpha_{ij}$ in the numerator of (11). This turns out to be enough to drive the maximizer back to the subspace $M \cdot \zeta = \mathbf{y}$.

Specifically, let us hold $\{\alpha_{ij}\}$ fixed and maximize $g(\zeta)$ with respect to $\{\beta_{ijk\ell}\}$ using the method of Lagrange multipliers. Since $\log g$ is monotonically increasing in g, it is convenient to maximize $\log g$ instead. If λ is the Lagrange multiplier corresponding to the constraint (13), we have for all $i < k, j \neq \ell$:

$$\lambda = \frac{\partial}{\partial \beta_{ijk\ell}} \log g(\zeta) = \frac{\partial}{\partial \beta_{ijk\ell}} (\beta_{ijk\ell} \log(\alpha_{ij}\alpha_{k\ell}) - \beta_{ijk\ell} \log \beta_{ijk\ell})$$

$$= \log \alpha_{ij} + \log \alpha_{k\ell} - \log \beta_{ijk\ell} - 1$$

and so

$$\forall i < k, j \neq l : \beta_{ijk\ell} = C\alpha_{ij}\alpha_{k\ell}, \text{ where } C = e^{-\lambda - 1} \ . \tag{15}$$

Clearly, such $\beta_{ijk\ell}$ also satisfy the original constraints (12), and therefore the upper bound we obtain from this relaxation is in fact tight.

To solve for C we sum (15) and use (13), getting

$$\frac{2}{C} \sum_{i < k, j \neq \ell} \beta_{ijk\ell} = \frac{d}{C} = \sum_{i \neq k, j \neq \ell} \alpha_{ij}\alpha_{k\ell} = 1 - \frac{2}{k} + \sum_{ij} \alpha_{ij}^2 \equiv p \ .$$

Thus $C = d/p$ and (15) becomes

$$\forall i < k, j \neq l : \beta_{ijk\ell} = \frac{d\alpha_{ij}\alpha_{k\ell}}{p} \tag{16}$$

Observe that $p = p(\{a_{ij}\})$ is the probability that a single edge whose endpoints are chosen uniformly at random is properly colored by both σ and τ, if the overlap matrix is $a_{ij} = \alpha_{ij}n$. Moreover, the values for the $b_{ijk\ell}$ are exactly what we would obtain, in expectation, if we chose from among the $\binom{n}{2}$ edges with replacement, rejecting those improperly colored by σ or τ, until we had $dn/2$ edges – in other words, if our graph model was $G(n,m)$ with replacement, rather than $\mathcal{G}_{n,d}$.

Substituting the values (16) in (14) and applying (13) yields the following upper bound on $g(\zeta)$:

$$g(\zeta) \leq \frac{1}{\prod_{ij}\alpha_{ij}^{\alpha_{ij}}} \frac{d^{d/2} \prod_{ij}\alpha_{ij}^{(d/p)\alpha_{ij}\sum_{i\neq k,j\neq\ell}\alpha_{k\ell}}}{(d/p)^{\sum_{i<k,j\neq\ell}\beta_{ijk\ell}} \prod_{i<k,j\neq\ell}(\alpha_{ij}\alpha_{k\ell})^{(d/p)\alpha_{ij}\alpha_{k\ell}}}$$

$$= \frac{1}{\prod_{ij}\alpha_{ij}^{\alpha_{ij}}} \frac{d^{d/2}}{(d/p)^{d/2}} \left(\frac{\prod_{ij}a_{ij}^{\alpha_{ij}\sum_{i\neq k,j\neq\ell}\alpha_{k\ell}}}{\prod_{i\neq k,j\neq\ell}\alpha_{ij}^{\alpha_{ij}\alpha_{k\ell}}} \right)^{d/p}$$

$$= \frac{p^{d/2}}{\prod_{ij}\alpha_{ij}^{\alpha_{ij}}}$$

$$\equiv g_{G(n,m)}(\{\alpha_{ij}\}) \ .$$

In [4, Thm 5], Achlioptas and Naor showed that for $d < c_{k-1}$ the function $g_{G(n,m)}$ is maximized when $\alpha_{ij} = 1/k^2$ for all i,j. In this case $p = (1 - 1/k)^2$, implying

$$g_{\max} \leq k^2 p^{d/2} = k^2 \left(1 - \frac{1}{k}\right)^d$$

and, therefore, that for some constant C_2 and sufficiently large n

$$\mathbf{E}[X^2] \leq C_2\, n^{-(k-1)}\, k^{2n} \left(1 - \frac{1}{k}\right)^{dn} \ .$$

7   Directions for Further Work

A Sharp Threshold for Regular Graphs. It has long been conjectured that for every $k > 2$, there exists a critical constant c_k such that a random graph $G(n, m = cn)$ is w.h.p. k-colorable if $c < c_k$ and w.h.p. non-k-colorable if $c > c_k$. It is reasonable to conjecture that the same is true for random regular graphs, i.e. that for all $k > 2$, there exists a critical integer d_k such that a random graph $\mathcal{G}_{n,d}$ is w.h.p. k-colorable if $d \leq d_k$ and w.h.p. non-k-colorable if $d > d_k$. If this is true, our results imply that for d in "good" intervals (u_k, c_k) w.h.p. the chromatic number of $\mathcal{G}_{n,d}$ is precisely $k + 1$, while for d in "bad" intervals (c_{k-1}, u_k) the chromatic number is w.h.p. either k or $k + 1$.

Improving the Second Moment Bound. Our proof establishes that if X, Y are the numbers of balanced k-colorings of $\mathcal{G}_{n,d}$ and $G(n, m = dn/2)$, respectively, then $\mathbf{E}[X]^2/\mathbf{E}[X^2] = \Theta(\mathbf{E}[Y]^2/\mathbf{E}[Y^2])$. Therefore, any improvement on the upper bound for $\mathbf{E}[Y^2]$ given in [4] would immediately give an improved positive-probability k-colorability result for $\mathcal{G}_{n,d}$.

In particular, Moore has conjectured that the function $g_{G(n,m)}$ is maximized by matrices with a certain form. If true, this immediately gives an improved lower bound, c_k^*, for k-colorability satisfying $c_{k-1}^* \to u_k - 1$. This would shrink the union of the "bad" intervals to a set of measure 0, with each such interval containing precisely one integer d for each $k \geq k_0$.

3-Colorability of Random Regular Graphs. It is easy to show that a random 6-regular graph is w.h.p. non-3-colorable. On the other hand, in [1] the authors showed that 4-regular graphs are w.p.p. 3-colorable. Based on considerations from statistical physics, Krząkała, Pagnani and Weigt [8] have conjectured that a random 5-regular graph is w.h.p. 3-colorable. The authors (unpublished) have shown that applying the second moment method to the number of balanced 3-colorings cannot establish this fact (even with positive probability).

Acknowledgments

C. Moore is grateful to Tracy Conrad, Alex Russell, and Martin Weigt for helpful conversations, and is supported by NSF grant PHY-0200909.

References

1. D. Achlioptas and C. Moore, Almost all graphs of degree 4 are 3-colorable. *Journal of Computer and System Sciences*, 67(4):441–471, 2003.
2. D. Achlioptas and C. Moore, The asymptotic order of the k-SAT threshold. *Proc. 43rd Foundations of Computer Science*, 779–788, 2002.
3. D. Achlioptas and C. Moore, On the two-colorability of random hypergraphs. *Proc. 6th RANDOM*, 78–90, 2002.
4. D. Achlioptas and A. Naor, The two possible values of the chromatic number of a random graph. *Proc. 36th Symp. on the Theory of Computing*, 2004.
5. B. Bollobás, *Random graphs*. Academic Press, London-New York, 1985.
6. A. Frieze and T. Łuczak, On the independence and chromatic numbers of random regular graphs. *J. Combin. Theory Ser. B*, 54:123-132, 1992.
7. S. Janson, T. Łuczak and A. Ruciński, *Random Graphs*. Wiley & Sons, 2000.
8. F. Krząkała, A. Pagnani and M. Weigt, Threshold values, stability analysis and high-q asymptotics for the coloring problem on random graphs. Preprint, cond-mat/0403725. *Physical Review E*, to appear.
9. T. Łuczak, The chromatic number of random graphs. *Combinatorica*, 11(1):45–54, 1991.
10. T. Łuczak, A note on the sharp concentration of the chromatic number of random graphs. *Combinatorica*, 11(3):295–297, 1991.
11. M. Molloy, *The Chromatic Number of Sparse Random Graphs*. Master's thesis, Faculty of Mathematics, University of Waterloo, 1992.
12. N.C. Wormald, Models of random regular graphs. Surveys in Combinatorics, J.D. Lamb and D.A. Preece, eds. London Mathematical Society Lecture Note Series, 276:239–298. Cambridge University Press, 1999.

Estimating the Distance
to a Monotone Function⋆

Nir Ailon, Bernard Chazelle, Seshadhri Comandur, and Ding Liu

Department of Computer Science, Princeton University, Princeton NJ 08544, USA
{nailon,chazelle,csesha,dingliu}@cs.princeton.edu

Abstract. In standard property testing, the task is to distinguish between objects that have a property $\mathcal{P}$ and those that are ε-far from $\mathcal{P}$, for some $\varepsilon > 0$. In this setting, it is perfectly acceptable for the tester to provide a negative answer for every input object that does not satisfy $\mathcal{P}$. This implies that property testing in and of itself cannot be expected to yield any information whatsoever about the distance from the object to the property. We address this problem in this paper, restricting our attention to monotonicity testing. A function $f : \{1, \ldots, n\} \mapsto \mathbf{R}$ is at distance ε_f from being monotone if it can (and must) be modified at $\varepsilon_f n$ places to become monotone. For any fixed $\delta > 0$, we compute, with probability at least $2/3$, an interval $[(1/2 - \delta)\varepsilon, \varepsilon]$ that encloses ε_f. The running time of our algorithm is $O(\varepsilon_f^{-1} \log \log \varepsilon_f^{-1} \log n)$, which is optimal within a factor of $\log \log \varepsilon_f^{-1}$ and represents a substantial improvement over previous work. We give a second algorithm with an expected running time of $O(\varepsilon_f^{-1} \log n \log \log \log n)$.

1 Introduction

Since the emergence of property testing in the nineties [12,8], great progress has been made on a long list of combinatorial, algebraic, and geometric testing problems; see [11,6,4] for surveys. Property testing is a relaxation of the standard decision problem: Given a property $\mathcal{P}$, instead of determining exactly whether a given input object satisfies $\mathcal{P}$ or not, we require an exact answer only if the object satisfies the property or if it is far from doing so. This subsumes a notion of distance: Typically the object is said to be ε-far from $\mathcal{P}$ if at least a fraction ε of its description must be modified in order to enforce the property. The largest such ε is called the distance of the object to $\mathcal{P}$. In this setting, the tester can say "no" for every input object that does not satisfy $\mathcal{P}$, which precludes the leaking of any information regarding the distance of the object to the property.

This weakness has led Parnas, Ron, and Rubinfeld [10] to introduce the concept of *tolerant* property testing. Given $0 \leq \varepsilon_1 < \varepsilon_2 \leq 1$, a tolerant tester must accept all inputs that are not ε_1-far from $\mathcal{P}$ and reject all of those that are ε_2-far (and output anything it pleases otherwise). A related problem studied in [10]

⋆ This work was supported in part by NSF grants CCR-998817, CCR-0306283, ARO Grant DAAH04-96-1-0181.

K. Jansen et al. (Eds.): APPROX and RANDOM 2004, LNCS 3122, pp. 229–236, 2004.

is that of estimating the actual distance of the object to the property within prescribed error bounds. In the model considered, all algorithms are randomized and err with probability at most $1/3$. (or equivalently any arbitrarily small constant).

Testing the monotonicity of functions has been extensively studied [1–3, 5, 7, 9]. In the one-dimensional case, given a function $f : \{1, \ldots, n\} \mapsto \mathbf{R}$, after querying $O(\log n)/\varepsilon$ function values, we can, with probability at least $2/3$, accept f if it is monotone and reject it if it is ε-far from being monotone [3]. These methods do not provide for tolerant property testing, however. Very recently, Parnas, Ron and Rubinfeld [10] designed sublinear algorithms for tolerant property testing and distance approximation for two problems: function monotonicity and clustering. If ε_f denotes the distance of f to monotonicity, their algorithm computes an estimate $\hat{\varepsilon}$ for ε_f that satisfies $(1/2)\varepsilon_f - \delta \le \hat{\varepsilon} \le \varepsilon_f + \delta$ with high probability. The query complexity and running time of their algorithm are both $\tilde{O}((\log n)^7/\delta^4)$ (the $\tilde{O}$ notation hides a factor of $(\log \log n)^{O(1)}$). The algorithm maintains and queries a data structure called an "index-value tree." Since the running time is sublinear, the tree is stored implicitly and only relevant portions are constructed whenever necessary, using random sampling to make approximate queries on the tree. Their construction is sophisticated and highly ingenious, but all in all quite involved.

We propose a simpler, faster, algorithm that is nearly optimal. Given any fixed $\delta > 0$, it outputs an interval $[(1/2-\delta)\varepsilon, \varepsilon]$ that encloses ε_f with probability at least $2/3$. The running time is $O(\varepsilon_f^{-1} \log \log \varepsilon_f^{-1} \log n)$, which is optimal within a factor of $\log \log \varepsilon_f^{-1}$. (The optimality proof is quite simple and omitted from this version.) One thing to note is the different use of δ: in our algorithm it is part of the multiplicative factor, whereas in [10] it is an additive term. To achieve the same multiplicative factor as in our algorithm, the additive term needs to be $\Theta(\delta\varepsilon_f)$. This makes the running time of Parnas et al.'s algorithm $\tilde{O}((\log n)^7/\varepsilon_f^4)$, for any fixed δ.

The starting point of our algorithm is the property tester of Ergun et al. [3], which relies on a key fact: There exist at least $\varepsilon_f n$ "critical" integers $i \in \{1, \ldots, n\}$; for i to be critical means that it is the (left or right) endpoint of an interval at least half of whose elements are in violation with i. Here i is said to violate j if either $i < j$ and $f(i) > f(j)$ or $i > j$ and $f(i) < f(j)$. By proving an upper bound on the number of critical integers, we are able to define a "signature" distribution for f which reflects its distance ε_f fairly accurately. Specifically, two functions with distances to monotonicity off by a factor of 2 (roughly) will have signatures that are distinguishable in time $O(\varepsilon_f^{-1} \log n)$. This provides us with a tolerant property tester for monotonicity. We can turn it into a distance approximator by using a one-way searching strategy, which we discuss below. Just as in [10], our algorithm extends to higher dimension.

We also present an improvement of our one-dimensional algorithm for small enough values of ε. We show how to estimate ε_f in time $O(\varepsilon_f^{-1} \log n \log \log \log n)$. Unlike in our previous algorithm, the number of steps in this one is itself a random variable; therefore, the running time is to be understood in the expected sense over the random bits used by the algorithm.

2   Estimating Distance to Monotonicity

Given two functions $f, g : \{1, \ldots, n\} \mapsto \mathbf{R}$, let $d(f, g) = \text{Prob}[f(x) \neq g(x)]$ denote the distance between f and g, where $x \in \{1, \ldots, n\}$ is chosen uniformly at random. We define $\varepsilon_f = \min_{g \in \mathcal{M}} d(f, g)$, where $\mathcal{M}$ is the set of monotone functions from $\{1, \ldots, n\}$ to $\mathbf{R}$.

Theorem 1. *For any fixed $\delta > 0$, we can compute an interval $[(1/2 - \delta)\varepsilon, \varepsilon]$ that encloses ε_f with probability at least $2/3$. The running time is $O(\varepsilon_f^{-1} \log \log \varepsilon_f^{-1} \log n)$.*

It is not entirely clear from the theorem that amplifying the probability of success can be achieved by simply repeating the algorithm enough times and taking a majority vote. What if we get different candidate intervals every time? We do not. As will soon become obvious, majority voting does, indeed, boost the probability of success arbitrarily close to 1.

It is easy to reduce the search for such an interval to a "distance separation" decision problem. Suppose that, given any $\varepsilon > 0$, one can tell in $O(\varepsilon^{-1} \log n)$ time and with probability at least $2/3$ whether $\varepsilon_f > \varepsilon$ or $\varepsilon_f < (1/2 - \delta)\varepsilon$. If $(1/2 - \delta)\varepsilon \leq \varepsilon_f \leq \varepsilon$, the algorithm can report anything. For each $k = 1, 2, \ldots$, we run the algorithm $c \log(k + 1)$ times with ε set to $\varepsilon_k = (1/2 - \delta)^k$, where c is a large enough constant, and we take a majority vote. We continue until we hear the report that $\varepsilon_f > \varepsilon_\ell$. By Chernoff's bound, the probability that $\varepsilon_{\ell+1} \leq \varepsilon_f \leq \varepsilon_{\ell-1}$ is at least $1 - \sum_{k \geq 0} O(1/ck^2) > 2/3$. The running time of $\sum_{1 \leq k \leq \ell} O(\log(k+1)) \varepsilon_k^{-1} \log n$, which is $O(\varepsilon_f^{-1} \log \log \varepsilon_f^{-1} \log n)$ time, as claimed.

This does not quite do the job. Indeed, we are now left with the knowledge that ε_f falls in the interval $[\varepsilon_{\ell+1}, \varepsilon_{\ell-1}]$, which unfortunately is too big for our purposes. It is enclosed in the interval $[\varepsilon_0/5, \varepsilon_0]$, for some $0 < \varepsilon_0 < 1$, which we must now shrink to the right size. To do this we simply use the previous "distance separation" algorithm for the values $(1 - k\delta)\varepsilon_0$, for $0 \leq k \leq 1/\delta$. This allows us to pinpoint ε_f within an interval of the form $[(1/2 - O(\delta))\varepsilon, \varepsilon]$. Rescaling δ gives us the desired result. It thus suffices to prove the following lemma:

Lemma 1. *For any fixed $\varepsilon, \delta > 0$, we can decide, in time $O(\varepsilon^{-1} \log n)$ and with probability at least $2/3$, whether $\varepsilon_f > \varepsilon$ or $\varepsilon_f < (1/2 - \delta)\varepsilon$. If $(1/2 - \delta)\varepsilon \leq \varepsilon_f \leq \varepsilon$, the algorithm can report anything.*

2.1   A Separation Oracle

As mentioned in the introduction, the key to estimating the distance to monotonicity is to approximate the number of "critical" integers (to be defined in the next section). To identify a critical integer i, we need to find an interval starting or ending at i such that there are many violations with i in the interval. This is done through random sampling, to ensure a sublinear running time. The motivation for the following definitions on joint distributions of random variables will be made clear later in this section.

Let $\mathcal{D}$ be the joint distribution of m independent 0/1 random variables $x_1, \ldots, x_m$, which can be sampled independently. If $\mathbf{E}\, x_i \leq a$ for all i, then $\mathcal{D}$ is called a-*light*; else it is a-*heavy*. We describe an algorithm light-test which, given any $a < b$, determines whether a distribution is a-light or b-heavy.

Lemma 2. *If $\mathcal{D}$ is either a-light or b-heavy, for some fixed $a < b$, then with probability 2/3 we can tell which is the case in $O(bm/(b-a)^2)$ time.*

Proof. Call light-test$(\{x_1, \ldots, x_m\}, c_0)$, where c_0 is chosen so that $c_1 \stackrel{\text{def}}{=} c_0(b - a)^2/b$ is a large enough constant. The algorithm runs in time proportional to $\sum_{k \geq 0} c_0 k(m/2^k) = O(c_0 m)$. To see why it works, we begin with a simple observation. Suppose that $\mathbf{E}\, x_i > b$, then at the k-th recursive call we sample x_i (if at all) exactly $c_0 k$ times; therefore, by Chernoff's bounds,

$$\text{Prob}[\hat{x}_i \leq (a+b)/2] = 2^{-\Omega(c_1 k)}$$

The same upper bound holds for the probability that $\hat{x}_i > (a+b)/2$, assuming that $\mathbf{E}\, x_i \leq a$. Suppose now that :

- $\mathcal{D}$ IS b-HEAVY: Let x_i be such that $\mathbf{E}\, x_i > b$. At the k-th recursion call, the probability that S' is empty is $2^{-\Omega(c_1 k)}$. Summing up over all k bounds the likelihood of erring by 1/3.
- $\mathcal{D}$ IS a-LIGHT: The probability that any given $\hat{x}_i$ exceeds $(a+b)/2$ is at most 1/3 (conservatively) and so erring any time before the size of S is recursively reduced to below c_1 is $\sum_{c_1 \leq k < |S|} 2^{-\Omega(k)} = 2^{-\Omega(c_1)} < 1/6$. After that stage, the probability of reaching a b-heavy verdict is at most $O(c_1(\log c_1) 2^{-\Omega(c_1)}) < 1/6$. $\square$

---

light-test (S, k)

For each $x \in S$, sample it k times and compute the average $\hat{x}$;
Form $S' = \{ x \in S \mid \hat{x} > (a+b)/2 \}$.
If $|S'| = 0$, then output "a-light".
If $|S'| \geq |S|/2$, then output "b-heavy".
light-test$(S', k + c_0)$

2.2   Distance Separation: The Facts

Given $0 < \delta < 1/2$, the integer i is called δ-*big* if there exists $j > i$ such that

$$\left| \left\{ i \leq k \leq j \mid f(k) < f(i) \right\} \right| \geq (1/2 - \delta)(j - i + 1)$$

or, similarly, $j < i$ such that

$$\left| \left\{ j \leq k \leq i \mid f(k) > f(i) \right\} \right| \geq (1/2 - \delta)(i - j + 1).$$

Intuitively, integer i is big if $f(i)$ violates monotonicity with an abundance of witnesses. In the following we show that when δ is small, the number of δ-big integers approximates $\varepsilon_f n$ to within a factor of roughly 2.

Lemma 3. (i) At least $\varepsilon_f n$ integers are 0-big; (ii) no more than $(2 + 4\delta/(1 - 2\delta))\varepsilon_f n$ integers are δ-big.

Proof. Note that, for any $i < j$ such that $f(i) > f(j)$, either i or j (or both) is 0-big. Therefore, if we were to remove all the 0-big integers from the domain $\{1, \ldots, n\}$, the function f would become monotone; hence (i).

To prove (ii), let C be a set of $\varepsilon_f n$ integers in the domain of f over which the function can be modified to become monotone. An element i of C is called *high-critical* (resp. *low-critical*) if there is $j \notin C$ such that $j > i$ and $f(j) < f(i)$ (resp. $j < i$ and $f(j) > f(i)$). Note that the two definitions are exclusive. For each δ-big i, we choose a unique witness j_i to its bigness (which one does not matter). If $j_i > i$, then i is called *right-big*; else it is *left-big*. (Obviously, the classification depends on the choice of witnesses.)

To bound the number of right-bigs, we charge low-criticals with a credit scheme. (Then we apply a similar procedure to charge left-bigs.) Initially, each element of C is assigned 1 credit. For each right-big $i \notin C$ among $n, \ldots, 1$ in this order, *spread* one credit among all the low-criticals k such that $i \leq k \leq j_i$ and $f(k) < f(i)$. We use the word "spread" because we do not simply drop one credit into one account. Rather, viewing the accounts as buckets and credits as water, we pour one unit of water one infinitesimal drop at a time, always pouring the next drop into the least filled bucket. (There are other ways to describe this charging scheme, none of them quite as poetic.)

We now show that no low-critical ever receives an excess of $2 + 4\delta/(1 - 2\delta)$ credits. Suppose by contradiction that this were the case. Let i be the right-big that causes the low-critical k's account to reach over $2 + 4\delta/(1 - 2\delta)$. By construction i is not low-critical; therefore, the excess occurs while right-big i is charging the l low-criticals k such that $i < k \leq j_i$ and $f(k) < f(i)$. Note that, because $i \notin C$, any k satisfying these two conditions is a low-critical and thus gets charged. With the uniform charging scheme (remember the water?), this ensures that all of these l low-criticals have the same amount of credits by the time they reach the excess value, which gives a total greater than $l(2 + 4\delta/(1 - 2\delta))$. By definition of right-bigness, $l \geq (1/2 - \delta)(j_i - i + 1)$. But none of these accounts could be charged before step j_i; therefore,

$$(1/2 - \delta)(j_i - i + 1)(2 + 4\delta/(1 - 2\delta)) < j_i - i + 1,$$

which is a contradiction.

We handle left-bigs in a similar way by running now from left to right, ie, $i = 1, \ldots, n$. Since no integer can be both left-critical and right-critical, part (ii) of the lemma follows. $\qquad\square$

2.3   Distance Separation: The Algorithm

We need one piece of terminology before describing the distance separation algorithm. Given an interval in $[u, v]$, we define two 0/1 random variables $\alpha[u, v]$

and $\beta[u,v]$: given random $i \in [u,v] \cap \{1,\ldots,n\}$, $\alpha[u,v] = 1$ (resp. $\beta[u,v] = 1$) iff $f(u) > f(i)$ (resp. $f(i) > f(v)$). With probability at least $2/3$, distance-separation (f,ε,δ). reports that $\varepsilon_f > \varepsilon$ (resp. $\varepsilon_f < (1/2 - \delta)\varepsilon$) if it is, indeed the case, and anything it wants if $(1/2 - \delta)\varepsilon \le \varepsilon_f \le \varepsilon$.

---

distance-separation (f,ε,δ)

Pick $s = \lceil(1 + \delta/2)\varepsilon^{-1}\ln 2\rceil$ random $i \in \{1,\ldots,n\}$.
For each $1 \le k \le (5/\delta)\ln n$, define $x_{2k-1}^{(i)} = \alpha[i, i + (1 + \delta/4)^k]$
and $x_{2k}^{(i)} = \beta[i - (1 + \delta/4)^k, i]$.
Let $\mathcal{D}$ be the distribution of $(x_1^{(1)}, x_2^{(1)}, \ldots, x_1^{(2)}, x_2^{(2)}, \ldots, x_1^{(s)}, x_2^{(s)}, \ldots)$.
If $\mathcal{D}$ is $(1/2 - \delta/4)$-heavy, then output "$\varepsilon_f > \varepsilon$".
If $\mathcal{D}$ is $(1/2 - \delta/3)$-light, then output "$\varepsilon_f < (1/2 - \delta)\varepsilon$".

The algorithm assumes that both δ and ε/δ are suitably small. The requirement on δ is nonrestrictive. To make ε small, however, we use an artifice: set $f(i) = +\infty$ for $i = n + 1, \ldots, O(n/\delta)$. We also need to assume that the algorithm used for distinguishing between light and heavy succeeds with probability at least $1 - \delta^2$ (instead of $2/3$); to do that iterate it $\log \delta^{-1}$ times and take a majority vote. To prove the correctness of the algorithm, it suffices to show that:

− If $\varepsilon_f > \varepsilon$, then $\mathcal{D}$ is $(1/2 - \delta/4)$-heavy with probability $1/2 + \Omega(\delta)$:

By Lemma 3 (i), more than εn integers are 0-big, so the probability of hitting at least one of them in the first step (and hence, of ensuring that $\mathcal{D}$ is $(1/2)/(1 + \delta/4)$-heavy) is at least $1 - (1 - \varepsilon)^s > 1/2 + \Omega(\delta)$.

− If $\varepsilon_f < (1/2 - \delta)\varepsilon$, then $\mathcal{D}$ is $(1/2 - \delta/3)$-light with probability $1/2 + \Omega(\delta)$:

By Lemma 3 (ii), the number of $\delta/3$-big integers is less than $(1 - \delta)\varepsilon n$; therefore, the probability of missing all of them (and hence, of ensuring that $\mathcal{D}$ is $(1/2 - \delta/3)$-light) is at least $(1 - (1 - \delta)\varepsilon)^s > 1/2 + \Omega(\delta)$.

By running the whole algorithm $O(1/\delta^2)$ times and taking a majority vote, we can boost the probability of success to $2/3$. By Lemma 2, the running time is $O(\varepsilon^{-1}\log n)$, as claimed (for fixed δ). This completes the proof of Lemma 1 and hence of Theorem 1.

2.4   A Faster Algorithm for Small Distances

We show in this section how to slightly improve the query complexity of the algorithm to

$$O(\min\{\log\log\varepsilon_f^{-1}, \log\log\log n\}\,\varepsilon_f^{-1}\log n).$$

The running time is now expected (over the random bits used by the algorithm). To do this, we need the following theorem:

Theorem 2. *We can compute an interval* $[\Omega(\varepsilon/\log n), \varepsilon]$ *that encloses* ε_f *with probability at least* $2/3$. *The expected running time is* $O(\varepsilon_f^{-1}\log n)$.

Using this theorem, it is clear that the factor $\log\log\varepsilon_f^{-1}$ in the distance estimation algorithm can be replaced by $\min\{\log\log\varepsilon_f^{-1}, \log\log\log n\}$. Indeed, instead of taking $k = 1, 2, 3, \ldots$, and running the separation oracle for each value of ε_k a number of times (ie, $c\log(k+1)$ times), we redefine ε_k to be $(1/2-\delta)^k\varepsilon$, where ε is the estimate returned by Theorem 2. Because the maximum value of k is now $O(\log\log n)$, the running time drops to $O(\min\{\log\log\varepsilon_f^{-1}, \log\log\log n\}\varepsilon_f^{-1}\log n)$.

To prove Theorem 2, we turn to a construction introduced by Goldreich et al. [7]. Define a subset P of pairs of integers: $(i, j) \in P$ if $j > i$, and $j - i$ is at most t, where t is the largest power of 2 that divides either i or j. This set has the following two properties:

- $|P| = \Theta(n\log n)$.
- For any $i < j$, there exists k $(i < k < j)$ such that both $(i, k) \in P$ and $(k, j) \in P$. This means, in particular, that for any violation (i, j) of f, there exists a "witness" (i, k) or (k, j) of the violation in the subset P.

Now, for a function f, let M be a maximum matching in the violation graph (the undirected graph whose vertex set is $\{1, \ldots, n\}$ and where i is connected to j if $i < j$ and $f(i) > f(j)$). It is known [7] that $|M| = \Theta(\varepsilon_f n)$; to be precise, $\frac{1}{2}\varepsilon_f n \leq |M| \leq \varepsilon_f n$. Let $Q \subseteq P$ be the set of violations of f in P. Consider the bipartite graph G with M on the left and Q on the right. Connect an edge between $(i, j) \in M$ and $(a, b) \in Q$ if $\{i, j\} \cap \{a, b\} \neq \emptyset$. By the second property above, and from the definition of a maximum matching, every node on the right has degree either 1 or 2, and every node on the left has degree at least 1; therefore, the cardinality of the right side is $\Omega(|M|)$. We would like to show that it is $O(|M|\log n)$. If we could do that, then by sampling from P and checking for violations, we could then estimate the size of Q and get the desired approximation. Unfortunately, it is not quite the case that the cardinality of the right side is always $O(|M|\log n)$. To fix this problem, we need to introduce some more randomness.

We slightly change the definition of P: for an integer $r \in [1, n]$ let P_r denote the subset of pairs defined as follows: $(i, j) \in P_r$ if $j - i$ is at most t, where t is the largest power of 2 that divides either $i + r$ or $j + r$. The set P_r still has the two properties above. In addition, if r is chosen uniformly at random then, for any i, the expected number of j such that $(i, j) \in P_r$ and j' such that $(j', i) \in P_r$ is $O(\log n)$. The expected number of edges of the corresponding bipartite graph G_r, therefore, is $O(|M|\log n)$. So the expected cardinality of the right side is $\alpha|P_r|$, where $\alpha \in [\Omega(\varepsilon_f/\log n), O(\varepsilon_f)]$. We sample P_r to form an estimation $\hat{\alpha}$ for α and return $\varepsilon = C\hat{\alpha}\log n$, for some large enough constant C, to prove

Theorem 2. The estimation follows the predictable scheme: (1) pick a random $r \in \{1, \ldots, n\}$; (2) pick a pair (i,j) uniformly at random from P_r; (3) if (i,j) is a violation of f, output `success`, otherwise `failure`. The success probability is precisely α, so repeating the sampling enough times sharpens our estimation to the desired accuracy, as indicated by the following fact.

Lemma 4. *Given a 0/1 random variable with expectation $\alpha > 0$, with probability at least 2/3, the value of $1/\alpha$ can be approximated with a relative constant error by sampling it $O(1/\alpha)$ times on average. Therefore, α can be approximated within the same error and the same expected running time.*

Proof. Run Bernoulli trials on the random variable and define Y to be the number of trials until (and including) the first 1. It is a geometric random variable with $\mathbf{E}\,Y = 1/\alpha$, and $\mathbf{var}\,(Y) = (1-\alpha)/\alpha^2 \leq (\mathbf{E}\,Y)^2$. By taking several samples of Y and averaging we get an estimate $\hat{1/\alpha}$ of $1/\alpha$. Using Chebyshev's inequality, a constant number of samples suffices to get a constant factor approximation. $\qquad\square$

References

1. Batu, T., Rubinfeld, R., White, P. *Fast approximate PCPs for multidimensional bin-packing problems*, Proc. RANDOM (1999), 245–256.
2. Dodis, Y., Goldreich, O., Lehman, E., Raskhodnikova, S., Ron, D., Samorodnitsky, A. *Improved testing algorithms for monotonicity*, Proc. RANDOM (1999), 97–108.
3. Ergun, F., Kannan, S., Kumar, S. Ravi, Rubinfeld, R., Viswanathan, M. *Spot-checkers*, Proc. STOC (1998), 259–268.
4. Fischer, E. *The art of uninformed decisions: A primer to property testing*, Bulletin of EATCS, 75: 97-126, 2001.
5. Fischer, E., Lehman, E., Newman, I., Raskhodnikova, S., Rubinfeld, R., Samorodnitsky, A. *Monotonicity testing over general poset domains*, Proc. STOC (2002), 474–483.
6. Goldreich, O. *Combinatorial property testing - A survey*, in "Randomization Methods in Algorithm Design," 45-60, 1998.
7. Goldreich, O., Goldwasser, S., Lehman, E., Ron, D., Samordinsky, A. *Testing monotonicity*, Combinatorica, 20 (2000), 301–337.
8. Goldreich, O., Goldwasser, S., Ron, D. *Property testing and its connection to learning and approximation*, J. ACM 45 (1998), 653–750.
9. Halevy, S., Kushilevitz, E. *Distribution-free property testing*, Proc. RANDOM (2003), 302–317.
10. Parnas, M., Ron, D., Rubinfeld, R. *Tolerant property testing and distance approximation*, ECCC 2004.
11. Ron, D. *Property testing*, in "Handbook on Randomization," Volume II, 597-649, 2001.
12. Rubinfeld, R., Sudan, M. *Robust characterization of polynomials with applications to program testing*, SIAM J. Comput. 25 (1996), 647–668.

Edge Coloring with Delays

Noga Alon[1,*] and Vera Asodi[2]

[1] Department of Mathematics
Raymond and Beverly Sackler Faculty of Exact Sciences
Tel Aviv University, Tel Aviv, Israel
nogaa@post.tau.ac.il
[2] Department of Computer Science
Raymond and Beverly Sackler Faculty of Exact Sciences
Tel Aviv University, Tel Aviv, Israel
veraa@post.tau.ac.il

Abstract. Consider the following communication problem, that leads to a new notion of edge coloring. The communication network is represented by a bipartite multigraph, where the nodes on one side are the transmitters and the nodes on the other side are the receivers. The edges correspond to messages, and every edge e is associated with an integer $c(e)$, corresponding to the time it takes the message to reach its destination. A proper k-edge-coloring with delays is a function f from the edges to $\{0, 1, ..., k-1\}$, such that for every two edges e_1 and e_2 with the same transmitter, $f(e_1) \neq f(e_2)$, and for every two edges e_1 and e_2 with the same receiver, $f(e_1) + c(e_1) \not\equiv f(e_2) + c(e_2) \pmod{k}$. Haxell, Wilfong and Winkler [10] conjectured that there always exists a proper edge coloring with delays using $k = \Delta + 1$ colors, where Δ is the maximum degree of the graph. We prove that the conjecture asymptotically holds for simple bipartite graphs, using a probabilistic approach, and further show that it holds for some multigraphs, applying algebraic tools. The probabilistic proof provides an efficient algorithm for the corresponding algorithmic problem, whereas the algebraic method does not.

1 Introduction

Motivated by the study of optical networks, Haxell, Wilfong and Winkler considered in [10] a communication network in which there are two groups of nodes: transmitters and receivers. Each transmitter has to send a set of messages, each of which should reach one receiver (more than one message per receiver is allowed). Each message has an associated *delay*, which is the time from the moment it is sent until it reaches its destination. The network is timed by a clock, so all times are integers. We wish to find a periodic setup of message sending for all transmitters, such that in each cycle all messages of all transmitters are sent, where each transmitter sends at most one message and each receiver gets at

* Research supported in part by a USA-Israeli BSF grant, by the Israel Science Foundation and by the Hermann Minkowski Minerva Center for Geometry at Tel Aviv University.

K. Jansen et al. (Eds.): APPROX and RANDOM 2004, LNCS 3122, pp. 237–248, 2004.
© Springer-Verlag Berlin Heidelberg 2004

most one message, at any given time unit. The objective is to find such a cycle of minimal length.

We can formalize this problem as follows: we represent the network by a bipartite multigraph $G = (V, E)$ with sides A and B, where the vertices of A are the transmitters and the vertices of B are the receivers. The edges correspond to the messages that are to be sent, and a function $c : E \to N$ associates each edge with its delay. The aim is to find the smallest number k for which there exists an edge coloring $f : E \to \{0, 1, \ldots, k-1\}$, such that for every two edges e_1 and e_2 that have a common vertex in A, $f(e_1) \neq f(e_2)$, and for every two edges e_1 and e_2 that have a common vertex in B, $f(e_1) + c(e_1) \not\equiv f(e_2) + c(e_2) \pmod{k}$.

The minimum number of colors is, clearly, at least Δ, where Δ is the maximum degree in G. Furthermore, there are simple examples in which $\Delta + 1$ colors are required, for example, any Δ-regular graph in which all delays but one are 0. Haxell, Wilfong and Winkler [10] raised the following conjecture.

Conjecture 1. Let $G = (V, E)$ be a bipartite multigraph with sides A and B and with maximum degree Δ, and let $c : E \to N$ be a delay function. Then there is a coloring $f : E \to \{0, 1, \ldots, \Delta\}$, such that for every two edges e_1 and e_2 that have a common vertex in A, $f(e_1) \neq f(e_2)$, and for every two edges e_1 and e_2 that have a common vertex in B, $f(e_1) + c(e_1) \not\equiv f(e_2) + c(e_2) \pmod{(\Delta + 1)}$.

In this paper, we show that the conjecture asymptotically holds for simple graphs. More precisely, we prove that if G is a simple bipartite graph with maximum degree Δ, then $\Delta + o(\Delta)$ colors suffice. Our proof uses a random coloring procedure, following the result of Kahn [12] and its extensions and variants by Häggkvist and Janssen [11] and Molloy and Reed [14, 15] on list edge coloring.

Using algebraic techniques, based on [2, 3, 7], we also show that the conjecture holds for some families of bipartite multigraphs such as all even length regular multi-cycles where the degree of regularity plus one is a prime. We further describe a generalization of the problem to non-bipartite multigraphs, and prove the conjecture in some cases.

In section 2 we outline the proof of the asymptotic result for simple bipartite graphs, in section 3 we present the algebraic proofs for some families of multigraphs. The final section 4 contains some concluding remarks.

2   Simple Bipartite Graphs

In this section we show that if G is a simple bipartite graph with maximum degree Δ, then there is a coloring with the required properties using $(1 + o(1))\Delta$ colors.

Theorem 1. *For every $\epsilon > 0$ there is a $\Delta_0 = \Delta_0(\epsilon)$ such that the following holds for every $\Delta > \Delta_0$:*
Let $G = (V, E)$ be a bipartite simple graph with sides A and B and with maximum degree Δ, and let $c : E \to N$ be a delay function. Then for $k = \lceil (1 + \epsilon)\Delta \rceil$ there

is a coloring $f : E \rightarrow \{0, 1, \ldots, k-1\}$, *such that for every two edges* e_1 *and* e_2 *that have a common vertex in* A, $f(e_1) \neq f(e_2)$, *and for every two edges* e_1 *and* e_2 *that have a common vertex in* B, $f(e_1) + c(e_1) \not\equiv f(e_2) + c(e_2) \pmod{k}$.

Throughout the section we omit all floor and ceiling signs, whenever these are not crucial. We use, throughout the paper, the notation $\tilde{O}(x)$ to denote, as usual, $O(x(\log x)^{O(1)})$.

Let $G = (V, E)$ be a simple bipartite graph with sides A and B, and assume without loss of generality that G is Δ-regular. Let $c : E \rightarrow N$ be the delay function, and let $k = (1 + \epsilon)\Delta$, where $\epsilon > 0$ is an arbitrarily small constant. We show that if Δ is sufficiently large, then there is a coloring with the desired properties using k colors.

We present a coloring procedure with three stages:

1. Choose a small set of reserved colors for every edge.
2. Iteratively color the edges as follows. In each iteration, assign every uncolored edge a random color from the unreserved colors that are still available for it. An edge retains the color assigned to it only if no adjacent edge is assigned a conflicting color.
3. Complete the coloring from the lists of reserved colors.

In the rest of this section we describe some of the details of each stage, and prove that, with positive probability, the procedure finds a coloring with the required properties. Due to space limitations, most of the technical details are postponed to the full version of the paper.

2.1 Notation

To simplify the description of the procedure, we extend the function c by defining for all $e = (u, v) \in E$ with $u \in A$ and $v \in B$, $c(e, u) = 0$ and $c(e, v) = c(e)$. Now the coloring f should satisfy the condition that for every two edges $e_1, e_2 \in E$ with a common vertex u, $f(e_1) + c(e_1, u) \neq f(e_2) + c(e_2, u)$, where all the operations on the colors are done modulo k.

During the iterative procedure, we denote by $L_e \subseteq \{0, 1, \ldots, k-1\}$ the set of all unreserved colors still available for e, for all $e \in E$. For all $v \in V$ let $L_v = \{ c + c((u, v), v) \mid c \in L_{(u,v)}, (u, v) \subset E \}$. For a set of colors C and a color c, let $C - c = \{ c' - c \mid c' \in C \}$. Clearly, $L_{(u,v)} \subseteq (L_u - c((u, v), u)) \cap (L_v - c((u, v), v))$, for all $(u, v) \in E$. Thus, every time a color c is removed from a list L_v, then $c - c(e, v)$ is removed from the lists L_e, for all edges e incident with v.

For all $v \in V$ and $c \in L_v$, denote by $T_{v,c}$ the set of all uncolored edges e incident with v, for which the color $c - c(e, v)$ is still available, i.e.

$$T_{v,c} = \{ e = (u, v) \in E \mid c - c(e, v) \in L_e \text{ and } e \text{ is uncolored } \}.$$

In the first stage we choose a set of reserved colors for every edge. We do it by choosing a set of reserved colors $Reserve_v$ for all $v \in V$, and then by defining for every edge $e = (u, v) \in E$, $Reserve_e = (Reserve_u - c(e, u)) \cap (Reserve_v - c(e, v))$.

During the iterative procedure, for all $v \in V$ and color c, we denote by $R_{v,c}$ the set of all uncolored edges $e = (u, v)$ incident with v, such that the color $c - c(e, v) + c(e, u)$ is reserved for u, i.e.

$$R_{v,c} = \{\, e = (u, v) \in E \mid c - c(e, v) + c(e, u) \in Reserve_u \text{ and } (u, v) \text{ is uncolored}\}.$$

For all i, $e \in E$, $v \in V$ and color c, let l_e^i, l_v^i, $t_{v,c}^i$ and $r_{v,c}^i$ denote the sizes of the sets L_e, L_v, $T_{v,c}$ and $R_{v,c}$, respectively, after i iterations. If after i iterations $c \notin L_v$ we define $t_{v,c}^i = 0$. For all the above, $i = 0$ refers to the time before the first iteration, but after the first stage of the procedure, in which the reserved colors are chosen.

2.2   Coloring Scheme Overview

We now describe the coloring procedure. Start by choosing a small set of reserved colors $Reserve_v$ for all $v \in V$, and by defining for all $e = (u, v) \in E$, $Reserve_e = (Reserve_u - c(e, u)) \cap (Reserve_v - c(e, v))$. We would like these colors to be available for e at the final stage, and hence we remove from the initial set L_e of every edge $e = (u, v) \in E$, all the colors that conflict with the colors reserved for u and v. Therefore, for all $e = (u, v) \in E$, the initial set L_e is defined by $L_e = \{0, 1, \dots, k - 1\} \setminus ((Reserve_u - c(e, u)) \cup (Reserve_v - c(e, v)))$. We choose the reserved colors so that for all $e \in E$, $l_e^0 \geq (1 + \frac{\epsilon}{2})\Delta$, but we also make sure that for all $e \in E$, $|Reserve_e| \geq \frac{\epsilon^2}{200}\Delta$, and for all $v \in V$ and color c, $r_{v,c}^0 \leq \frac{\epsilon}{5}\Delta$, so the coloring can be completed at the final stage.

The second stage consists of a constant $t = t(\epsilon)$ number of iterations. In each iteration, we assign to each edge $e \in E$ a color c chosen randomly from L_e. If there is a conflict, i.e. if there is an adjacent edge e' to which we assigned a color c' such that $c + c(e, v) = c' + c(e', v)$, where v is the common vertex of e and e', we uncolor e. If an edge $e = (u, v)$ retains the color c it was assigned, then adjacent edges cannot be colored with conflicting colors. Thus we remove $c + c(e, u)$ from L_u, $c + c(e, v)$ from L_v, and the corresponding colors from the lists $L_{e'}$ of the edges e' incident with u and v.

We start with $l_e^0 \geq (1 + \frac{\epsilon}{2})\Delta$, for all $e \in E$, $t_{v,c}^0 \leq \Delta$ and $r_{v,c}^0 \leq \frac{\epsilon}{5}\Delta$ for all $v \in V$ and color c. We show that the sizes of the sets L_e and $T_{v,c}$ decrease at roughly the same rate, and hence, for every i, the sizes $t_{v,c}^i$ are slightly smaller than the sizes l_e^i.

The i-th iteration is carried out as follows. We first remove colors from the sets L_e so that they all have the same size, which we denote by l_{i-1}. Now, for all $e = (u, v) \in E$ assign e a color c chosen randomly from L_e. The color c is removed from e if an adjacent edge is assigned a conflicting color, i.e. if there is an edge e_1 incident with u that is assigned $c + c(e, u) - c(e_1, u)$, or an edge e_2 incident with v that is assigned $c + c(e, v) - c(e_2, v)$. Therefore, the probability that e retains c is

$$(1 - \frac{1}{l_{i-1}})^{t_{u,c+c(e,u)}^{i-1} + t_{v,c+c(e,v)}^{i-1} - 2} \approx e^{-2}.$$

For every color $c \in L_v$, the edges e incident with v that may retain $c - c(e, v)$ are the edges in $T_{v,c}$. For every $e \in T_{v,c}$, the probability that e is assigned the color c and retains it is roughly $\frac{1}{l_{i-1}e^2}$. There is at most one edge $e \in T_{v,c}$ that retains $c - c(e, v)$. Therefore, these events are disjoint, and if one of them occurs we remove c from L_v. Hence, the probability that we remove c from L_v is roughly

$$\frac{t_{v,c}^{i-1}}{l_{i-1}e^2} \approx e^{-2}.$$

For an edge $e = (u, v)$ and a color $c \in L_e$, we remove c from L_e if we remove $c + c(e, u)$ from L_u or $c + c(e, v)$ from L_v. If these events were independent, the probability that we do not remove c from L_e would have been roughly $(1 - e^{-2})^2$. We later show that although there is some dependence between those events, the probability that we do not remove c from L_e is still close to $(1 - e^{-2})^2$. Hence,

$$E[l_e^i] \approx (1 - e^{-2})^2 l_e^{i-1}.$$

For every $v \in V$ and $c \in L_v$, if $e = (u, v) \in T_{v,c}$, we remove e from $T_{v,c}$ if e retains the color it is assigned, or if e cannot be colored with $c - c(e, v)$ anymore, since $c - c(e, v) + c(e, u)$ was removed from L_u. If these events were independent, the probability that e remains in $T_{v,c}$ would have been roughly $(1 - e^{-2})^2$, since the probability of each of these events is about e^{-2}. We show later that the dependence between the above events does not change this probability much. Therefore,

$$E[t_{v,c}^i] \approx (1 - e^{-2})^2 t_{v,c}^{i-1}.$$

As for the sets $R_{v,c}$, we remove an edge e from $R_{v,c}$ only if it retains the color it was assigned. Therefore,

$$E[r_{v,c}^i] \approx (1 - e^{-2}) r_{v,c}^{i-1}.$$

We execute the iterative procedure for t iterations, where we choose t so that with positive probability $r_{v,c}^t \leq \frac{\epsilon^2}{3200} \Delta$ for all $v \in V$ and color c. We can then prove, using the local lemma (see, e.g., [4] chapter 5, or the results in [16, 17, 9]), that in this case the coloring can be completed with the reserved colors.

Due to space limitations, the detailed proofs of the first and last stages are omitted, and will appear in the full vesions. In the next subsection we describe some of the details of the main part of our coloring scheme, the iterative procedure.

2.3   The Iterative Procedure

In each iteration, we assign every edge e a color c chosen randomly from L_e. We let e retain c only if it causes no conflicts, i.e. no adjacent edge e' is assigned $c + c(e, u) - c(e', u)$, where u is the common vertex of e and e'. If an edge $e = (u, v)$ retains a color c, we remove $c + c(e, u)$ from L_u and $c + c(e, v)$ from L_v. To simplify the proof, we remove some colors from some of the lists L_e at the beginning of each iteration, so that all the lists have the same size, and we sometimes remove colors from edges and from the lists L_v even when there is no conflict, in order to equalize all the probabilities. The details are explained below.

We show that the size of each L_e and of each $T_{v,c}$ decreases by a factor of approximately $(1 - e^{-2})^2$ in each iteration, and the size of each $R_{v,c}$ by a factor of approximately $1 - e^{-2}$. Each $T_{v,c}$ is initially slightly smaller than each L_e, and since these sizes decrease at approximately the same rate, the above holds at the beginning of every iteration.

Consider the i-th iteration, and assume $l_e^{i-1} \geq l_{i-1}$ for all $e \in E$, and $t_{v,c}^{i-1} \leq l_{i-1}$ for all $v \in V$ and color c. We first remove colors from some of the lists L_e so that they all have size exactly l_{i-1}. For each edge $e = (u, v)$ and color $c \in L_e$, the probability that no adjacent edge is assigned a conflicting color, if e is assigned c, i.e. that no edge e_1 incident with u is assigned $c + c(e, u) - c(e_1, u)$, and that no edge e_2 incident with v is assigned $c + c(e, v) - c(e_2, v)$, is

$$P(e, c) = \left(1 - \frac{1}{l_{i-1}}\right)^{t_{u,c+c(e,u)}^{i-1} + t_{v,c+c(e,v)}^{i-1} - 2} \geq \left(1 - \frac{1}{l_{i-1}}\right)^{2l_{i-1}-2} > e^{-2}.$$

We now use an equalizing coin flip, so that if e is assigned c and no adjacent edge is assigned a conflicting color, we still remove c from e with probability

$$Eq(e, c) = 1 - \frac{1}{e^2 P(e, c)} > 0.$$

This ensures that the probability that e retains c, conditional of e receiving c, is precisely

$$P(e, c)(1 - Eq(e, c)) = e^{-2}.$$

For every $v \in V$ and $c \in L_v$, there is at most one edge $e \in T_{v,c}$ that retains $c - c(e, v)$. Therefore, the events that e retains $c - c(e, v)$ for all $e \in T_{v,c}$ are pairwise disjoint, and hence the probability that we do not have to remove c from L_v, i.e. the probability that no edge $e \in T_{v,c}$ retains $c - c(e, v)$ is

$$Q(v, c) = 1 - \frac{t_{v,c}^{i-1}}{l_{i-1}e^2} \geq 1 - e^{-2}.$$

If we do not have to remove c from L_v, we still remove it with probability

$$Vq(v, c) = 1 - \frac{1 - e^{-2}}{Q(v, c)}.$$

This ensures that the probability that c remains in L_v is precisely

$$Q(v, c)(1 - Vq(v, c)) = 1 - e^{-2}.$$

Lemma 1. *For every $v \in V$ and color c, $E[r_{v,c}^i] = (1 - e^{-2})r_{v,c}^{i-1}$.*

Proof. An edge is removed from $R_{v,c}$ if and only if it retains the color it is assigned. This happens with probability e^{-2}. Therefore, $E[r_{v,c}^i] = (1 - e^{-2})r_{v,c}^{i-1}$. $\qquad\square$

Lemma 2. *For each $e \in E$, $E[l_e^i] \geq (1 - e^{-2})^2 l_{i-1}$.*

Lemma 3. *For every $v \in V$ and color $c \in L_v$, $E[t_{v,c}^i] \leq (1 - e^{-2})^2 t_{v,c}^{i-1} + e^{-2}$.*

The proofs of Lemmas 2 and 3 require a more careful analysis, and will appear in the full version.

In Lemmas 1, 2 and 3, we have shown that the expected size of every L_e and $T_{v,c}$ decreases in one iteration by a factor of approximately $(1 - e^{-2})^2$, and of every $R_{v,c}$ by a factor of approximately $1 - e^{-2}$. We wish to prove that with positive probability, the sizes of all these sets indeed decrease by approximately these factors. For this purpose we show that these values are highly concentrated. More precisely, we prove that there exists a constant $\beta > 0$, such that the following hold for all $1 \leq i \leq t$.

Lemma 4. *For each $e \in E$,*

$$Pr\left[|l_e^i - E[l_e^i]| > \log \Delta \sqrt{l_{i-1}}\right] \leq e^{-\beta \log^2 \Delta},$$

where the probability is over the random choices of the i-th iteration.

Lemma 5. *For each $v \in V$ and $c \in L_v$, if $t_{v,c}^{i-1} > \frac{l_{i-1}}{10}$ then*

$$Pr\left[|t_{v,c}^i - E[t_{v,c}^i]| > \log \Delta \sqrt{l_{i-1}}\right] \leq e^{-\beta \log^2 \Delta},$$

where the probability is over the random choices of the i-th iteration.

Lemma 6. *For each $v \in V$ and color c, if $r_{v,c}^{i-1} > \frac{\epsilon^2}{3200} \Delta$ then*

$$Pr\left[|r_{v,c}^i - E[r_{v,c}^i]| > \log \Delta \sqrt{r_{v,c}^i}\right] \leq e^{-\beta \log^2 \Delta},$$

where the probability is over the random choices of the i-th iteration.

The proofs of Lemmas 4, 5 and 6 use Talagrand's inequality (see, e.g. [4, 15]), and will appear in the full version.

To complete the proof we apply the local lemma to each iteration, to show that with positive probability, the size of every set is within a small error of the expected size.

Lemma 7. *With positive probability, the following hold after i iterations, for all $0 \leq i \leq t$:*

1. *For all $e \in E$*

$$l_e^i \geq (1 - e^{-2})^{2i} \left(1 + \frac{\epsilon}{2}\right) \Delta - \tilde{O}(\sqrt{\Delta}).$$

2. *For all $v \in V$ and color c*

$$t_{v,c}^i \leq (1 - e^{-2})^{2i} \Delta + \tilde{O}(\sqrt{\Delta}).$$

3. For all $v \in V$ and color c

$$r_{v,c}^i \leq (1 - e^{-2})^i \frac{\epsilon}{5} \Delta + \tilde{O}(\sqrt{\Delta}).$$

Proof. We prove the lemma by induction on i. Suppose the claim holds after $i - 1$ iterations. Note that by the induction hypothesis, $t_{v,c}^{i-1} \leq l_{i-1}$. Thus, we can apply all the lemmas that were proved in this section. For all $v \in V$ and color c, if $t_{v,c}^{i-1} \leq (1 - e^{-2})^{2i} \Delta$ or $r_{v,c}^{i-1} \leq (1 - e^{-2})^i \frac{\epsilon}{5} \Delta$ then the corresponding condition after i iterations trivially holds. Otherwise, we can apply Lemmas 5 and 6.

For all $e \in E$, let A_e be the event that l_e^i violates the first condition, and for all $v \in V$ and color c, let $B_{v,c}$ and $C_{v,c}$ be the events that $t_{v,c}^i$ and $r_{v,c}^i$ violate the second and third conditions, respectively. By Lemmas 4, 5 and 6, the probability of each one of these events is less than $e^{-\beta \log^2 \Delta}$ for some constant β. Every event A_e is mutually independent of all other events except for those that correspond to the vertices incident with e and their neighbours, and to edges incident with these vertices. Thus, it is mutually independent of all but $O(\Delta^2)$ of the other events. Every event $B_{v,c}$ or $C_{v,c}$ is mutually independent of all other events except for those that correspond to the edges incident with v, to the vertices incident with them and to the neighbours of these vertices. Thus, it is mutually independent of all but $O(\Delta^3)$ of the other events. Thus, by the local lemma, all the conditions hold with positive probability.

To complete the proof of the lemma we use the bounds on l_e^0, $t_{v,c}^0$ and $r_{v,c}^0$ given by the first stage of our procedure. $\square$

3   Algebraic Methods

In this section we prove Conjecture 1 for several families of graphs using algebraic methods. The relevance of results from additive number theory to this conjecture appears already in [10], where the authors apply the main result of Hall [8] to prove their conjecture for the multigraph with two vertices and d parallel edges between them. Here we consider several more complicated cases.

3.1   Even Multi-cycles

Consider the case where the graph G is a d-regular multigraph whose underlying simple graph is a simple cycle of even length. We show that if $d + 1$ is a prime then there exists a coloring with the required properties using $d + 1$ colors.

In our proof, we use the following theorem proved in [3]:

Theorem 2. (Combinatorial Nullstellensatz)
   Let F be an arbitrary field, and let $P = P(x_1, \ldots, x_n)$ be a polynomial in $F[x_1, \ldots, x_n]$. Suppose that the degree of P is $\sum_{i=1}^{n} t_i$, where each t_i is a nonnegative integer, and suppose the coefficient of $\prod_{i=1}^{n} x_i^{t_i}$ in P is nonzero. Then, if $S_1, \ldots, S_n$ are subsets of F with $|S_i| > t_i$, there are $s_1 \in S_1, s_2 \in S_2, \ldots, s_n \in S_n$ so that $P(s_1, \ldots, s_n) \neq 0$.

Let $G = (V, E)$ be a simple even cycle with, possibly, multiple edges. Let A and B be the sides of G, and denote the vertices of A by $a_1, a_2, \ldots, a_n$, and the vertices of B by $b_1, b_2, \ldots, b_n$, such that there are edges between a_i and b_i for all $1 \leq i \leq n$, between b_i and a_{i+1} for all $1 \leq i \leq n-1$, and between b_n and a_1. Since G is d regular, all the edges (a_i, b_i) have the same multiplicity s, and hence all the other edges have multiplicity $t = d - s$.

We now associate a polynomial P with the graph G. For every edge e, we have a variable x_e. For every two edges e_1 and e_2 which have a common vertex in A we have a term $(x_{e_1} - x_{e_2})$, and for every two edges e_1 and e_2 which have a common vertex in B we have a term $(x_{e_1} + c(e_1) - x_{e_2} - c(e_2))$. Thus, the polynomial P is defined by

$$P = \left[\prod_{e_1 \cap e_2 \cap A \neq \emptyset} (x_{e_1} - x_{e_2}) \right] \left[\prod_{e_1 \cap e_2 \cap B \neq \emptyset} (x_{e_1} + c(e_1) - x_{e_2} - c(e_2)) \right].$$

Since the graph is d-regular, every edge has $d-1$ terms for each one of its vertices. Hence, the total degree of P is $nd(d-1)$.

Proposition 1. *If $d+1$ is a prime then the coefficient of the monomial $\prod_{e \in E} x_e^{d-1}$ in P is nonzero modulo $d+1$.*

Proof. The monomial $\prod_{e \subset E} x_e^{d-1}$ is of maximum degree, and thus its coefficient in P is equal to its coefficient in the polynomial

$$Q = \left[\prod_{e_1 \cap e_2 \cap A \neq \emptyset} (x_{e_1} - x_{e_2}) \right] \left[\prod_{e_1 \cap e_2 \cap B \neq \emptyset} (x_{e_1} - x_{e_2}) \right].$$

In [7] (see also [2]) it is shown that, for any d-regular planar multigraph, the absolute value of this coefficient is equal to the number of proper d-edge-colorings. Every d-edge-coloring of G is obtained by partitioning the colors into a subset of size s and a subset of size t, and for every two connected vertices, choosing a permutation of the appropriate set. Thus the number of edge colorings with d colors is $\binom{d}{s}(s!)^n(t!)^n = d!(s!)^{n-1}(t!)^{n-1}$, which is nonzero modulo $d+1$ since $d+1$ is a prime. □

Corollary 1. *Let $G = (V, E)$ be an even length d-regular multi-cycle, where $d+1$ is a prime. Let (A,B) be a bipartition of G, and let $c : E \to N$ be a delay function. Then, there is a coloring $f : E \to \{0, 1, \ldots, d\}$ such that if e_1 and e_2 have a common vertex in A then $f(e_1) \neq f(e_2)$, and if e_1 and e_2 have a common vertex in B then $f(e_1) + c(e_1) \not\equiv f(e_2) + c(e_2) \pmod{(d+1)}$.*

Proof. By Theorem 2, with $S_e = \{0, 1, \ldots, d\}$ and $t_e = d - 1$ for every $e \in E$, there is a function $f : E \to \{0, 1, \ldots, d\}$ such that the value of P, when every x_e is assigned the value $f(e)$, is nonzero modulo $d+1$. Thus, for every two edges e_1 and e_2 that have a common vertex in A, $f(e_1) \neq f(e_2)$ since we have a term $(x_{e_1} - x_{e_2})$ in P, and for every two edges e_1 and e_2 that have a common vertex in B, $f(e_1) + c(e_1) \not\equiv f(e_2) + c(e_2) \pmod{(d+1)}$ since we have a term $(x_{e_1} + c(e_1) - x_{e_2} - c(e_2))$ in P. □

3.2   Multi-K_4

The problem can be generalized to non-bipartite multigraphs. In this case, we specify for every edge which endpoint is the transmitter and which is the receiver. Following the notations used in section 2, for every edge $e = (u, v)$, where u is the transmitter and v is the receiver, we define $c(e, u) = 0$, and $c(e, v)$ to be the delay associated with e.

We can further generalize the problem as follows. Given a multigraph $G = (V, E)$, where every edge $e = (u, v) \in E$ is associated with two integers $c(e, u)$ and $c(e, v)$, we aim to find the smallest number k such that there is a coloring $f : E \to \{0, 1, \ldots, k - 1\}$, satisfying the following. For every vertex v and every two edges $e_1 \neq e_2$ incident with v, $f(e_1) + c(e_1, v) \not\equiv f(e_2) + c(e_2, v) \pmod{k}$. In this case, $\Delta + 1$ colors do not always suffice, simply because $\chi'(G)$ may be as large as $\lfloor \frac{3}{2}\Delta \rfloor$, where $\chi'(G)$ is the edge chromatic number of G. The following conjecture seems plausible for this case.

Conjecture 2. Let $G = (V, E)$ be a multigraph, and suppose that every edge $e = (u, v) \in E$ is associated with two integers $c(e, u)$ and $c(e, v)$. Let $k = \chi'(G) + 1$. Then there exists a coloring $f : E \to \{0, 1, \ldots, k - 1\}$ such that for every vertex v and every two edges $e_1 \neq e_2$ incident with v, $f(e_1) + c(e_1, v) \not\equiv f(e_2) + c(e_2, v) \pmod{k}$.

Proposition 2. *Let $G = (V, E)$ be a d-regular multi-K_4, where every edge $e = (u, v) \in E$ is associated with two integers $c(e, u)$ and $c(e, v)$, and suppose $d+1$ is a prime. Then there is a coloring $f : E \to \{0, 1, \ldots, d\}$, satisfying the condition that for every vertex v and every two edges $e_1 \neq e_2$ incident with v, $f(e_1) + c(e_1, v) \not\equiv f(e_2) + c(e_2, v) \pmod{(d + 1)}$.*

The proof is similar to the proof of the result for even cycles.

Proof. First, note that if G is d-regular, then every pair of non-adjacent edges of K_4 have the same multiplicity in G. Denote these multiplicities by a, b and c. We associate the following polynomial with G:

$$P = \prod_{v \in V} \prod_{e_1 \neq e_2 \in E \ : \ v \in e_1 \cap e_2} (x_{e_1} + c(e_1, v) - x_{e_2} - c(e_2, v)).$$

P is a polynomial in $2d$ variables, with total degree $2d(d-1)$. Hence, by Theorem 2, if the coefficient of $\prod_{e \in E} x_e^{d-1}$ is nonzero modulo $d+1$, then there is a function $f : E \to \{0, 1, \ldots, d\}$, such that if every variable x_e is assigned $f(e)$ then the value of P is nonzero modulo $d + 1$, and therefore f satisfies the required properties. The coefficient of $\prod_{e \in E} x_e^{d-1}$ in P is equal to its coefficient in

$$Q = \prod_{v \in V} \prod_{e_1 \neq e_2 \in E \ : \ v \in e_1 \cap e_2} (x_{e_1} - x_{e_2}),$$

which is, by [7], the number of proper d-edge-colorings of G. Such a coloring is obtained as follows. First, choose a permutation of the colors and color the edges incident to a vertex u accordingly. Now for every pair of other vertices v and w, the set of colors that may be used to color the edges between v and w is the same set used for the edges between u and the forth vertex, and we only choose the permutation of this set. Thus, the number of proper d-edge-colorings is $d!a!b!c! \not\equiv 0 \pmod{(d+1)}$, since $d+1$ is a prime. $\square$

4   Concluding Remarks

In section 3 we proved Conjecture 1 for some graphs using algebraic techniques. This method can be used to prove the conjecture for several other graphs. However, it seems that in order to prove the conjecture for the general case, more ideas are needed. In the graphs for which we used Theorem 2, the theorem implies that there is a proper edge coloring with delays using $\Delta + 1$ colors, even if there is one forbidden color for every edge.

In section 2 we showed that Conjecture 1 asymptotically holds for simple bipartite graphs. It would be interesting to extend this proof to multigraphs as well.

In the probabilistic proof presented in section 2, we proved that $\Delta + o(\Delta)$ colors suffice, and did not make any effort to minimize the $o(\Delta)$ term. By modifying our proof slightly, we can show that $\Delta + \tilde{O}(\Delta^{2/3})$ colors suffice, and it seems plausible that a more careful analysis, following the discussion in [14], can even imply that $\Delta + \tilde{O}(\sqrt{\Delta})$ colors suffice.

The probabilistic proof can be extended to other variations of edge coloring. Instead of the delays, one can associate with every edge e and an endpoint v of e an injective function $g_{e,v}$ on the colors. Then, by our proof, there is an edge coloring f using $\Delta + o(\Delta)$ colors such that for every vertex v and any two edges $e_1 \neq e_2$ incident with v, $g_{e_1,v}(f(e_1)) \neq g_{e_2,v}(f(e_2))$.

The known results about the algorithmic version of the local lemma, initiated by Beck ([5], see also [1],[13], [6]), can be combined with our probabilistic proof in Section 2 to design a polynomial time algorithm that solves the corresponding algorithmic problem. In contrast, the algebraic proofs of Section 3 supply no efficient procedures for the corresponding problems, and it will be interesting to find such algorithms.

Acknowledgement

We would like to thank B. Sudakov and P. Winkler for fruitful discussions.

References

1. N. Alon, A Parallel algorithmic version of the Local Lemma, Random Structures and Algorithms, 2 (1991), 367-378.
2. N. Alon, Restricted colorings of graphs, in "Surveys in Combinatorics", Proc. 14th British Combinatorial Conference, London Mathematical Society Lecture Notes Series 187, edited by K. Walker, Cambridge University Press, 1993, 1-33.

3. N. Alon, Combinatorial Nullstellensatz, Combinatorics, Probability and Computing, 8 (1999), 7-29.
4. N. Alon and J. H. Spencer, The Probabilistic Method, Second Edition, Wiley, 2000.
5. J. Beck, An algorithmic approach to the Lovász Local Lemma, Random Structures and Algorithms, 2 (1991), 343-365.
6. A. Czumaj and C. Scheideler, Coloring Nonuniform Hypergraphs: A New Algorithmic Approach to the General Lovász Local Lemma, Random Structures and Algorithms, 17 (2000), 213-237.
7. M. N. Ellingham and L. Goddyn, List edge colourings of some 1-factorable multigraphs, Combinatorica, 16 (1996), 343-352.
8. M. Hall, A combinatorial problem on abelian groups, Proc. Amer. Math. Soc., 3 (1952), 584-587.
9. P. E. Haxell, A note on vertex list colouring, Combinatorics, Probability and Computing, 10 (2001), 345-348.
10. P. E. Haxell, G. T. Wilfong and P. Winkler, Delay coloring and optical networks, to appear.
11. R. Häggkvist and J. Janssen, New bounds on the list chromatic index of the complete graph and other simple graphs, Combinatorics, Probability and Computing, 6 (1997), 273-295.
12. J. Kahn, Asymptotically good list-colorings, Journal of Combinatorial Theory Series A, 73 (1996), 1-59.
13. M. Molloy and B. Reed, Further Algorithmic Aspects of the Local Lemma, Proceedings of the 30th Annual ACM Symposium on Theory of Computing, May 1998, 524-529.
14. M. Molloy and B. Reed, Near-optimal list colorings, Random Structures and Algorithms, 17 (2000), 376-402.
15. M. Molloy and B. Reed, Graph Colouring and the Probabilistic Method, Springer (2001).
16. B. Reed, The list colouring constants, Journal of Graph Theory, 31 (1999), 149-153.
17. B. Reed and B. Sudakov, Asymptotically the list colouring constants are 1, Journal of Combinatorial Theory Series B, 86 (2002), 27-37.

Small Pseudo-random Families of Matrices: Derandomizing Approximate Quantum Encryption

Andris Ambainis[1] and Adam Smith[2],[*]

[1] Institute for Advanced Study, Princeton, NJ, USA
ambainis@ias.edu
[2] MIT Computer Science and AI Lab, Cambridge, MA, USA
asmith@csail.mit.edu

Abstract. A *quantum encryption scheme* (also called *private quantum channel*, or *state randomization protocol*) is a one-time pad for quantum messages. If two parties share a classical random string, one of them can transmit a quantum state to the other so that an eavesdropper gets little or no information about the state being transmitted. *Perfect* encryption schemes leak no information at all about the message. *Approximate* encryption schemes leak a non-zero (though small) amount of information but require a shorter shared random key. Approximate schemes with short keys have been shown to have a number of applications in quantum cryptography and information theory [8].

This paper provides the first deterministic, polynomial-time constructions of quantum approximate encryption schemes with short keys. Previous constructions [8] are probabilistic – that is, they show that if the operators used for encryption are chosen at random, then with high probability the resulting protocol will be a secure encryption scheme. Moreover, the resulting protocol descriptions are exponentially long. Our protocols use keys of the same length as the probabilistic constructions; to encrypt n qubits approximately, one needs $n + o(n)$ bits of shared key [8], whereas $2n$ bits of key are necessary for perfect encryption [3]. An additional contribution of this paper is a connection between classical combinatorial derandomization and constructions of pseudo-random matrix families in a continuous space.

1 Introduction

A *quantum encryption scheme* (or *private quantum channel*, or *state randomization protocol*) allows Alice, holding a *classical* key[1], to scramble a quantum state and send it to Bob (via a quantum channel) so that (1) Bob, given the key, can recover Alice's state exactly and (2) an adversary Eve who intercepts the ciphertext learns nothing about the message, as long as she doesn't know the key. We do not assume any shared quantum states between Alice and Bob, nor any back channels from Bob to Alice[2].

---

[*] A.A. supported by NSF grant DMS-0111298. A.S. supported by a Microsoft Ph.D. Fellowship.
[1] Classical keys are inherently easier to store, distribute and manipulate, since they can be copied. More subtly, encryption with a shared quantum key is in many ways a dual problem to encryption with a classical key; see [8, 5] for more discussion.
[2] A back channel from Bob to Alice would allow using quantum key distribution to generate a long secret key. However, such interaction is often impossible, e.g. if Alice wants to encrypt stored data for her own later use.

K. Jansen et al. (Eds.): APPROX and RANDOM 2004, LNCS 3122, pp. 249–260, 2004.
© Springer-Verlag Berlin Heidelberg 2004

There are two variants of this definition. An encryption scheme is called *perfect* if Eve learns zero information from the ciphertext, and *approximate* if Eve can learn some non-zero amount of information. A perfect encryption ensures that the distributions (density matrices) of ciphertexts corresponding to different messages are exactly identical, while an approximate scheme only requires that they be very close; we give formal definitions further below. In the classical case, both perfect and approximate encryption require keys of roughly the same length – n bits of key for n bits of message. In the quantum case, the situation is different.

For perfect encryption, Ambainis et al. [3] showed that $2n$ bits of key are necessary and sufficient to encrypt n qubits. The construction consists of applying two classical one-time pads – one in the "standard" basis $\{|0\rangle, |1\rangle\}$ and another in the "diagonal" basis $\{\frac{1}{\sqrt{2}}(|0\rangle + |1\rangle), \frac{1}{\sqrt{2}}(|0\rangle - |1\rangle)\}$.

Approximate encryption was studied by Hayden, Leung, Shor and Winters [8]. They introduced an additional, useful relaxation: they showed that if the plaintext is not entangled with Eve's system to begin with, then one can get *approximate* quantum encryption using only $n + o(n)$ bits of key – roughly half as many as are necessary for perfect encryption[3]. The assumption that Eve's system is unentangled with the message is necessary for this result; otherwise roughly $2n$ bits are needed, even for approximate encryption. The assumption holds in the quantum counterpart of the one-time pad situation (one party prepares a quantum message and sends it to the second party, using the encryption scheme) as long as the message is not part of a larger cryptographic protocol.

Hayden et al. [8] showed that a *random* set of $2^{n+o(n)}$ unitary matrices leads to a good encryption scheme with high probability (to encrypt, Alice uses the key to choose one of the matrices from the set and applies the corresponding operator to her input). However, verifying that a particular set of matrices yields a good encryption scheme is not efficient; even writing down the list of matrices is prohibitive, since there are exponentially many of them.

This paper presents the first polynomial time constructions of approximate quantum encryption schemes (to relish the oxymoron: derandomized randomization protocols). The constructions run in time $O(n^2)$ when the message ρ consists of n qubits. That is, given the key and the input message, Alice can produce the output using $O(n^2)$ steps on a quantum computer. The key length we achieve is slightly better than that of the probabilistic construction of [8]. Our results apply to the trace norm on matrices; exact results are stated further below.

The main tools in our construction are small-bias sets [10] of strings in $\{0,1\}^{2n}$. Such sets have proved useful in derandomizing algorithms [10], constructing short PCPs [6] and the encryption of high-entropy messages [12]. Thus, one of the contributions of this paper is a connection between classical combinatorial derandomization and constructions of pseudo-random matrix families in a continuous space. Specifically, we connecti Fourier analysis over $\mathbb{C}^{\mathbb{Z}_2^{2n}}$ to Fourier analysis over the matrices $\mathbb{C}^{2^n \times 2^n}$. This parallels to some extent the connection between quantum error-correcting codes over n qubits and classical codes over $GF(4)^n$.

---

[3] The result of [8] highlights an error in the proof of a lower bound on key length of authentication schemes in [4]. The results of that paper remain essentially correct, but the definition of authentication requires some strengthening, and the proof of the lower bound is more involved.

Definitions. We assume that the reader is familiar with the basic notation of quantum computing (see [11] for an introduction). Syntactically, an approximate quantum encryption scheme is a set of 2^k invertible operators $\{E_\kappa | \kappa \in \{0,1\}^k\}$. The E_κ's may be unitary, but need not be: it is sufficient that one be able to recover the input ρ from the output $E_\kappa(\rho)$, which may live in a larger-dimensional space than ρ. Each E_κ takes n qubits as input and produces $n' \geq n$ qubits of output. If $n' = n$ then each operator E_κ corresponds to a unitary matrix U_κ, that is $E_\kappa(\rho) = U_\kappa \rho U_\kappa^\dagger$.

For an input density matrix[4] ρ, the density matrix of the ciphertext from the adversary's point of view is:

$$\mathcal{E}(\rho) = \mathbb{E}_\kappa[E_\kappa(\rho)] = \frac{1}{2^k} \sum_{\kappa \in \{0,1\}^k} E_\kappa(\rho)$$

When the scheme is length-preserving, this yields $\mathcal{E}(\rho) = \frac{1}{2^k} \sum_\kappa U_\kappa \rho U_\kappa^\dagger$.

Definition 1. *The set of operators $\{E_\kappa\}$ is an approximate quantum encryption scheme with leakage ϵ (also called "ϵ-randomizing scheme") for n qubits if*

for all density matrices ρ on n qubits: $D(\mathcal{E}(\rho), \frac{1}{2^{n'}}\mathbb{I}) = \left\| \mathcal{E}(\rho) - \frac{1}{2^{n'}}\mathbb{I} \right\|_{tr} \leq \epsilon.$

$$\text{(1)}$$

Here $\mathbb{I}$ refers to the identity matrix in dimension $2^{n'}$, and $D(\cdot, \cdot)$ refers to the trace distance between density matrices. The trace norm of a matrix σ is the trace of the absolute value of σ (equivalently, the sum of the absolute values of the eigenvalues). The *trace distance* between two matrices ρ, σ is the trace norm of their difference:

$$D(\rho, \sigma) \overset{\triangle}{=} \|\rho - \sigma\|_{tr} = \text{Tr}(|\rho - \sigma|)$$

This distance plays the same role for quantum states that statistical difference plays for probability distributions: the maximum probability of distinguishing between two quantum states ρ, σ via a single measurement is $\frac{1}{2} + \frac{1}{4}D(\rho, \sigma)$. One can also measure leakage with respect to other norms; see below.

Remark 1. This definition of quantum encryption implicitly assumes that the message state ρ is not entangled with the adversary's system. Without that assumption the definition above is not sufficient, and it is *not* possible to get secure quantum encryption using $n(1 + o(1))$ bits of key (roughly $2n$ bits are provably necessary[5]). Thus, this sort of construction is not universally applicable, and must be used with care.

Previous Work. Ambainis et al. [3] considered perfect encryption; this corresponds to the case where $\epsilon = 0$. The choice of matrix norm is irrelevant there, since $\mathcal{E}(\rho) = \frac{1}{2^{n'}}\mathbb{I}$. As mentioned above, they showed that $2n$ bits of key are necessary and sufficient. The

---

[4] Recall that for a pure state $|\phi\rangle$, the density matrix ρ is $|\phi\rangle\langle\phi|$.

[5] This folklore result appears more or less explicitly in both [4, 8]. Similar arguments show that n bits of key are necessary to encrypt n classical bits, even with access to quantum computers (but not interaction).

construction uses the key to choose one of 2^{2n} Pauli operators (defined below) and applies that to the input state.

Hayden et al. [8] showed that a set of $O(n2^n/\epsilon^2)$ unitary operators suffices. They showed this both for the trace norm, and for the "operator norm," discussed below. For the trace norm, they also showed that a random set of Pauli matrices of the same size would suffice. This means that for encrypting n qubits, they gave a non-polynomial-time, randomized scheme requiring $n + \log n + 2\log(1/\epsilon) + O(1)$ bits of key.

Our Results. We present three explicit, polynomial time constructions of approximate state randomization protocols for the trace norm. All are based on existing constructions of δ-biased sets [10, 2, 1], or on families of sets with small average bias. The three constructions are explained and proven secure in Sections 3.1, 3.2 and 3.3, respectively.

The first construction is length-preserving, and requires $n + 2\log n + 2\log(1/\epsilon) + O(1)$ bits of key, roughly matching the performance of the non-explicit construction. The second construction is length-increasing: it encodes n qubits into n qubits and $2n$ classical bits but uses a shorter key: only $n + 2\log(1/\epsilon)$ bits of key are required. Both of these constructions are quite simple, and are proven secure using the same Fourier-analytic technique.

The final construction has a more sophisticated proof, but allows for a length-preserving scheme with slightly better dependence on the number of qubits:

$$n + \min\{2\log n + 2\log(1/\epsilon), \log n + 3\log(1/\epsilon)\} + O(1)$$

bits of key. The right-hand term provides a better bound when $\epsilon > \frac{1}{n}$.

Randomization Schemes for Other Norms? Definition 1 measures leakage with respect to the trace norm on density matrices, $\|\cdot\|_{tr}$. This is good enough for encryption since the trace norm captures distinguishability of states. However, Hayden et al. [8] also considered randomization schemes which give guarantees with respect to a different norm, the operator norm.

A guarantee on the operator norm implies a guarantee for the trace norm, but schemes with the operator norm guarantee also have a host of less cryptographic applications, for example: constructing efficient quantum data hiding schemes in the LOCC (local operation and classical communication) model; exhibiting "locked" classical correlations in quantum states [8]; relaxed authentication of quantum states using few bits of key [9]; and transmitting quantum states over a classical channel using $n + o(n)$ bits of communication, rather than the usual $2n$ bits required for quantum teleportation [5].

More formally, for a $d \times d$ Hermitian matrix A with eigenvalues $\{\lambda_1, \lambda_2, ..., \lambda_d\}$, the *operator norm* (or ∞-norm) is the largest eigenvalue, $\|A\|_\infty = \max|\lambda_i|$, the Frobenius norm is the Euclidean length of the vector of eigenvalues, $\|A\|_2 = (\sum_i \lambda_i^2)^{1/2}$, and the trace norm is the sum of the absolute values of the eignvalues, $\|A\|_{tr} = \sum_i |\lambda_i|$. It is easy to see the chain of inequalities:

$$\|A\|_{tr} \leq \sqrt{d}\,\|A\|_2 \leq d\,\|A\|_\infty.$$

We can then state the condition for a map $\mathcal{E}$ to be ϵ-randomizing map for n qubits in three forms of increasing strength. For all input states ρ on n qubits:

$$\left\| \mathcal{E}(\rho) - \tfrac{1}{2^n}\mathbb{I} \right\|_{tr} \leq \epsilon; \qquad \left\| \mathcal{E}(\rho) - \tfrac{1}{2^n}\mathbb{I} \right\|_2 \leq \epsilon/\sqrt{2^n}; \qquad \left\| \mathcal{E}(\rho) - \tfrac{1}{2^n}\mathbb{I} \right\|_\infty \leq \epsilon/2^n.$$

Our constructions satisfy the definition with respect to the Frobenius norm, but they are not known to satisfy the stronger operator-norm definition. This suggests two interesting questions. First, is it possible to prove that the other applications of state randomization schemes require only a guarantee on the Frobenius norm? Second, is it possible to design explicit (i.e. polynomial-time, deterministic) randomization schemes that give good guarantees with respect to the operator norm?

The remainder of this paper describes our constructions and their proofs of security.

2 Preliminaries

Small-Bias Spaces. The bias of a random variable A in $\{0,1\}^n$ with respect to a string $\alpha \in \{0,1\}^n$ is the distance from uniform of the bit $\alpha \odot A$, where $\odot$ refers to the standard dot product on $\mathbb{Z}_2^n$:

$$\hat{A}(\alpha) = \mathbb{E}_A \left[(-1)^{\alpha \odot A} \right] = 2\Pr[\alpha \odot A = 0] - 1.$$

The function $\hat{A}$ is the Fourier transform of the probability mass function of the distribution, taken over the group $\mathbb{Z}_2^n$.

The bias of a set $S \in \{0,1\}^n$ with respect to α is simply the bias of the uniform distribution over that set. A set S is called δ-biased if the absolute value of its bias is at most δ for all $\alpha \neq 0^n$.

Small-bias sets of size polynomial in n and $1/\delta$ were first constructed by Naor and Naor [10]. Alon, Bruck et al. (ABNNR, [1]) gave explicit (i.e. deterministic, polynomial-time) constructions of δ-biased sets in $\{0,1\}^n$ with size $O(n/\delta^3)$. Constructions with size $O(n^2/\delta^2)$ were provided by Alon, Goldreich, et al. (AGHP, [2]). The AGHP construction is better when $\delta = o(1/n)$. In both cases, the ith string in a set can be constructed in roughly n^2 time (regardless of δ).

One can sample a random point from a δ-biased space over $\{0,1\}^n$ using either $\log n + 3\log(1/\delta) + O(1)$ bits of randomness (using ABNNR) or using $2\log n + 2\log(1/\delta)$ bits (using AGHP).

Small-Bias Set Families. One can generalize small bias to *families* of sets (or random variables) by requiring that on average, the bias of a random set from the family with respect to every α is low [7]. Specifically, the expectation of the *squared* bias must be at most δ^2. Many results on δ-biased sets also hold for δ-biased families, which are easier to construct.

Definition 2. *A family of random variables (or sets) $\{A_i\}_{i \in I}$ is δ-biased if*

$$\mathbb{E}_{i \leftarrow I} \left[\hat{A}_i(\alpha)^2 \right] \leq \delta^2 \text{ for all } \alpha \neq 0^n.$$

Note that this is *not* equivalent, in general, to requiring that the expected bias be less than δ. There are two important special cases:

1. If S is a δ-biased set, then $\{S\}$ is a δ-biased set family with a single member;
2. A family of linear spaces $\{C_i\}_{i \in I}$ is δ-*biased* if no particular word is contained in the dual $C_i^{\perp}$ of a random space C_i from the family with high probability. Specifically:

$$\hat{C}_i(\alpha) = \begin{cases} 0 \text{ if } \alpha \notin C_i^{\perp} \\ 1 \text{ if } \alpha \in C_i^{\perp} \end{cases}$$

Hence a family of codes is δ-biased if and only if $\Pr_{i \leftarrow I}[\alpha \in C_i^{\perp}] \leq \delta^2$, for every $\alpha \neq 0^n$. Note that to meet the definition, for linear codes the expected bias must be at most δ^2, while for a single set the bias need only be δ.

One can get a good δ-biased family simply by taking $\{C_i\}$ to be the set of all linear spaces of dimension k. The probability that any fixed non-zero vector α lies in the dual of a random space is exactly $\delta^2 = \frac{2^{n-k}-1}{2^n-1}$, which is at most 2^{-k}.

One can save some randomness in the choice of the space using a standard pairwise independence construction. View $\{0,1\}^n$ as $GF(2^n)$, and let $K \subseteq GF(2^n)$ be an additive subgroup of size 2^k. For every non-zero string a, let the space C_a be given by all multiples $a\kappa$, where $\kappa \in K$, and the product is taken in $GF(2^n)$. The family $\{C_a \mid a \in GF(2^n), a \neq 0\}$ has the same bias as the set of all linear spaces ($\delta < 2^{-k/2}$). To see this, let $\{\kappa_1, ..., \kappa_k\}$ be a basis of K (over $GF(2)$). A string α is in $C_a^{\perp}$ if and only if $\alpha \odot (a\kappa_1) = \cdots = \alpha \odot (a\kappa_k) = 0$. This is a system of k linearly independent constraints on a, and so it is satisfied with probability $\delta^2 = 2^{-k}$ when $a \leftarrow GF(2^n)$, and even lower probability when we restrict a to be non-zero. Choosing a set C_a from the family requires n bits of randomness.

Entropy of Quantum States. As with classical distributions, there are several ways to measure the entropy of a quantum density matrix. We'll use the analogue of collision entropy (a.k.a. Renyi entropy).

For a classical random variable A on $\{0,1\}^n$, the collision probability of two independent samples of X is $p_c = \sum_a \Pr[A = a]^2$. The Renyi entropy of A is $-\log p_c$.

For a quantum density matrix ρ, the analogous quantity is $-\log \text{Tr}(\rho^2)$. If the eigenvalues of ρ are $\{p_x\}$, then the eigenvalues of ρ^2 are $\{p_x^2\}$, and so $\text{Tr}(\rho^2)$ is exactly the collision probability of the distribution obtained by measuring ρ in a basis of eigenvectors. $\sqrt{\text{Tr}(\rho^2)}$ is called the Frobenius norm of ρ.

If ρ is the completely mixed state in d dimensions, $\rho = \frac{1}{d}\mathbb{I}$, then $\text{Tr}(\rho^2)$ is $1/d$. The following fact states that any other density matrix for which this quantity is small must be very close to $\mathbb{I}$. The fact follows by applying the (standard) inequality $\text{Tr}(|\Delta|)^2 \leq d\text{Tr}(\Delta^2)$ to the Hermitian matrix $\Delta = \rho - \mathbb{I}/d$.

Fact 1. *If ρ is d-dimensional quantum state and $\text{Tr}(\rho^2) \leq \frac{1}{d}(1+\epsilon^2)$, then $D(\rho, \frac{1}{d}\mathbb{I}) \leq \epsilon$.*

Pauli Matrices. The 2×2 Pauli matrices are generated by the matrices:

$$X = \begin{pmatrix} 0 & 1 \\ 1 & 0 \end{pmatrix} \qquad Z = \begin{pmatrix} 1 & 0 \\ 0 & -1 \end{pmatrix}$$

The Pauli matrices are the four matrices $\{\mathbb{I}, X, Z, XZ\}$. These form a basis for the space of all 2×2 complex matrices. Since $XZ = -ZX$, and $Z^2 = X^2 = 1$, the set generated by X and Z is given by the Pauli matrices and their opposites: $\{\pm\mathbb{I}, \pm X, \pm Z, \pm XZ\}$.

If u and v are n-bit strings, we denote the corresponding tensor product of Pauli matrices by $X^u Z^v$. That is, if we write $u = (u1, ..., u_n)$ and $v = (v_1, ..., v_n)$, then

$$X^u Z^v = X^{u_1} Z^{v_1} \otimes \cdots \otimes X^{u_n} Z^{v_n}.$$

(The strings x and z indicate in which positions of the tensor product X and Z appear, respectively.) The set $\{X_u Z_v \mid u, v \in \{0,1\}^n\}$ forms a basis for the $2^n \times 2^n$ complex matrices. The main facts we will need are given below:

1. Products of Pauli matrices obey the group structure of $\{0, 1\}^{2n}$ up to a minus sign. That is, $(X^u Z^v)(X^a Z^b) = (-1)^{a \odot v} X^{u \oplus a} Z^{v \oplus b}$.

2. Any pair of Pauli matrices either commutes or anti-commutes. Specifically, $(X^u Z^v)(X^a Z^b) = (-1)^{u \odot b + v \odot a}(X^a Z^b)(X^u Z^v)$.

3. The trace of $X^u Z^v$ is 0 if $(u, v) \neq 0^{2n}$ (and otherwise it is $\operatorname{Tr}(\mathbb{I}) = 2^n$).

4. $(X^u Z^v)^\dagger = Z^v X^u = (-1)^{u \odot v} X^u Z^v$

Pauli Matrices and Fourier Analysis. The Pauli matrices form a basis for the set of all $2^n \times 2^n$ matrices. Given a density matrix ρ, we can write $\rho = \sum_{u,v \in \{0,1\}^n} \alpha_{u,v} X^u Z^v$. This basis is orthonormal with respect to the inner product given by $\frac{1}{2^n} \operatorname{Tr}(A^\dagger B)$, where A, B are square matrices. That is, $\frac{1}{2^n} \operatorname{Tr}((X^u Z^v)^\dagger X^a Z^b) = \delta_{a,u} \delta_{b,v}$.

Thus, the usual arithmetic of orthogonal bases (and Fourier analysis) applies. One can immediately deduce certain properties of the coefficients $\alpha_{u,v}$ in the decomposition of a matrix ρ. First, we have the formula $\alpha_{u,v} = \frac{1}{2^n} \operatorname{Tr}(Z^v X^u \rho)$. Second, the squared norm of ρ is given by the squared norm of the coefficients, that is $\frac{1}{2^n} \operatorname{Tr}(\rho^\dagger \rho) = \sum_{u,v} |\alpha_{u,v}|^2$. Since ρ is a density matrix, it is Hermitian ($\rho^\dagger = \rho$). One can use this fact, and the formula for the coefficients $\alpha_{u,v}$, to get a compact formula for the Renyi entropy (or Frobenius norm) in terms of the decomposition in the Pauli basis:

$$\operatorname{Tr}(\rho^2) = \frac{1}{2^n} \sum_{u,v} |\operatorname{Tr}(X^u Z^v \rho)|^2.$$

3 State Randomization and Approximate Encryption

3.1 Encrypting with a Small-Bias Space

The ideal quantum one-time pad applies a random Pauli matrix to the input [3]. Consider instead a scheme which first chooses a $2n$-bit string from some set with small bias δ (we will set δ later to be $\epsilon 2^{-n/2}$). If the set of strings is B we have:

$$\mathcal{E}(\rho_0) = \frac{1}{|B|} \sum_{(a,b) \in B} X^a Z^b \rho_0 Z^b X^a = \mathbb{E}_{a,b}\left[X^a Z^b \rho_0 Z^b X^a\right]$$

That is, we choose the key from the set B, which consists of $2n$-bit strings. To encrypt, we view a $2n$-bit string as the concatenation (a, b) of two strings of n bits, and apply the corresponding Pauli matrix.

(The intuition comes from the proof that Cayley graphs based on ϵ-biased spaces are good expanders: applying a Pauli operator chosen from a δ-biased family of strings to ρ_0 will cause all the Fourier coefficients of ρ_0 to be reduced by a factor of δ, which implies that the "collision probability" (Frobenius norm) of ρ_0 also gets multiplied by δ. We expand on this intuition below.)

As a first step, we can try to see if a measurement given by a Pauli matrix $X^u Z^v$ can distinguish the resulting ciphertext from a totally mixed state. More explicitly, we perform a measurement which projects the ciphertext onto one of the two eigenspaces of the matrix $X^u Z^v$. We output the corresponding eigenvalue. (All Pauli matrices have two eigenvalues with eigenspaces of equal dimension. The eigenvalues are always either -1 and 1 or $-i$ and i.)

To see how well a particular Pauli matrix $X^u Z^v$ will do at distinguishing, it is sufficient to compute

$$|\mathrm{Tr}(X^u Z^v \mathcal{E}(\rho_0))|.$$

This is exactly the statistical difference between the Pauli measurement's outcome and a uniform random choice from the two eigenvalues. We can compute it explicitly:

$$
\begin{aligned}
\mathrm{Tr}(X^u Z^v \mathcal{E}(\rho_0)) &= \mathrm{Tr}\left(X^u Z^v \mathbb{E}_{(a,b) \in B}\left[X^a Z^b \rho_0 Z^b X^a\right]\right) \\
&= \mathbb{E}_{a,b}\left[\mathrm{Tr}(X^u Z^v X^a Z^b \rho_0 Z^b X^a)\right] \\
&= \mathbb{E}_{a,b}\left[\mathrm{Tr}(Z^b X^a X^u Z^v X^a Z^b \rho_0)\right] \\
&= \mathbb{E}_{a,b}\left[(-1)^{a \odot v + b \odot u}\right]\mathrm{Tr}(X^u Z^v \rho_0)
\end{aligned}
$$

Since $a \odot v + b \odot u$ is linear in the concatenated $2n$-bit vector (a, b), we can take advantage of the small bias of the set B to get a bound:

$$|\mathrm{Tr}(X^u Z^v \mathcal{E}(\rho_0))| \leq \delta |\mathrm{Tr}(X^u Z^v \rho_0)| \qquad \text{when } (u, v) \neq 0^{2n}$$

Equivalently: if we express ρ_0 in the basis of matrices $X^u Z^v$, then each coefficient shrinks by a factor of at least δ after encryption. We can now bound the distance from the identity by computing $\mathrm{Tr}(\mathcal{E}(\rho_0)^2)$:

$$
\begin{aligned}
\mathrm{Tr}(\mathcal{E}(\rho_0)^2) &= \frac{1}{2^n} \sum_{u,v} |\mathrm{Tr}(X^u Z^v \mathcal{E}(\rho_0))|^2 \\
&\leq \frac{1}{2^n} + \frac{\delta^2}{2^n} \sum_{(u,v) \neq 0^{2n}} |\mathrm{Tr}(X^u Z^v \rho_0)|^2 \leq \frac{1}{2^n}(1 + \delta^2 2^n \mathrm{Tr}(\rho_0^2))
\end{aligned}
$$

Setting $\delta = \epsilon 2^{-n/2}$, we get approximate encryption for all states (since $\mathrm{Tr}(\rho_0^2) \leq 1$). Using the constructions of AGHP [2] for small-bias spaces, we get a polynomial-time scheme that uses $n + 2 \log n + 2 \log(1/\epsilon)$ bits of key.

3.2 A Scheme with Shorter Key Length

We can improve the key length of the previous scheme using δ-biased *families* of sets. The tradeoff is that the resulting states are longer: the ciphertext consists of n qubits and $2n$ classical bits. In classical terms, the encryption algorithm uses additional randomness which is not part of the shared key; in the quantum computing model, however, that randomness is "free" if one is allowed to discard ancilla qubits.

Lemma 1. *If $\{A_i\}_{i\in\mathcal{I}}$ is a family of subsets of $\{0,1\}^{2n}$ with average square bias δ^2, then the operator*

$$\mathcal{E}(\rho_0) = \mathbb{E}_{i\in\mathcal{I}}\left[|i\rangle\langle i| \otimes \mathbb{E}_{ab\in A_i}\left[X^a Z^b \rho_0 Z^b X^a\right]\right]$$

is an approximate encryption scheme for n qubits with leakage ϵ whenever $\delta \leq \epsilon 2^{-n/2}$.

Before proving the lemma, we give an example using the small-bias set family from the preliminaries. View the key set $\{0,1\}^k$ as an additive subgroup K of the field $\mathbb{F} = GF(2^{2n})$. For every element $a \in \mathbb{F}$, define the set $C_a = \{a\kappa | \kappa \in K\}$. The family $\{C_a\}$ has bias $\delta < 2^{-k/2}$ (Section 2). The corresponding encryption scheme takes a key $\kappa \in \{0,1\}^k \subseteq GF(2^{2n})$:

$$\mathcal{E}(\rho_0; \kappa) = \begin{bmatrix} \text{Choose } \alpha \leftarrow_R GF(2^{2n}) \setminus \{0\} \\ \text{Compute the product } \alpha\kappa \in GF(2^{2n}) \\ \text{Write } \alpha\kappa \text{ as a concatenation } (a,b), \text{ where } a, b \in \{0,1\}^n \\ \text{Output the classical string } \alpha \text{ and the quantum state } X^a Z^b \rho_0 Z^b X^a \end{bmatrix}$$

With a quantum computer, random bits are not really necessary for choosing α; it is sufficient to prepare $2n$ EPR pairs and discard one qubit from each pair. For the scheme to be secure, the bias δ should be less than $\sqrt{\epsilon/2^n}$, and so the key only needs to be $n + 2\log(1/\epsilon)$ bits long. The main disadvantage is that the length of the ciphertext has increased by $2n$ classical bits.

Proof. As before, the proof will use elementary Fourier analysis over the hypercube $\mathbb{Z}_2^{2n}$, and intuition comes from the proof that Cayley graphs based on ϵ-biased set families are also expanders.

Think of the output of the encryption scheme as a single quantum state consisting of two systems: the first system is a classical string describing which member of the δ-biased family will be used. The second system is the encrypted quantum state. To complete the proof, it is enough to bound the collision entropy of the entire system by $\frac{1}{2^n|\mathcal{I}|}(1 + \epsilon^2)$.

For each $i \in \mathcal{I}$ (that is, for each member of the set family), let ρ_i denote the encryption of ρ_0 with a random operator from the set A_i. The first step of the proof is to show that the collision entropy of the entire system is equal to the average collision entropy of the states ρ_i.

Claim. $\text{Tr}(\mathcal{E}(\rho_0)^2) = \dfrac{1}{|\mathcal{I}|}\mathbb{E}_{i\leftarrow I}\left[\text{Tr}(\rho_i^2)\right]$

Proof. We can write $\mathcal{E}(\rho_0) = \frac{1}{|\mathcal{I}|} \sum_i |i\rangle\langle i| \otimes \rho_i$. Then we have

$$\mathrm{Tr}(\mathcal{E}(\rho_0)^2) = \frac{1}{|\mathcal{I}|^2} \sum_{i,j} \mathrm{Tr}((|i\rangle\langle i||j\rangle\langle j|) \otimes \rho_i\rho_j)$$

Since $\langle i||j\rangle = \delta_{i,j}$, we get $\mathrm{Tr}(\mathcal{E}(\rho_0)^2) = \frac{1}{|\mathcal{I}|^2} \sum_i \mathrm{Tr}(\rho_i^2)$, as desired. $\qquad\square$

Take any string $w = (u,v) \in \{0,1\}^{2n}$, where $u, v \in \{0,1\}^n$. Recall that $\hat{A}_i(u,v)$ is the ordinary Fourier coefficient (over $\mathbb{Z}_2^{2n}$) of the uniform distribution on A_i, that is $\hat{A}_i(u,v) = \mathbb{E}_{a\leftarrow A_i}[(-1)^{a\odot w}]$. From the previous proof, we know that

$$\mathrm{Tr}(X^u Z^v \rho_i) = \hat{A}_i(v,u) \cdot \mathrm{Tr}(X^u Z^v \rho_0).$$

We can now compute the average collision entropy of the states ρ_i. Using linearity of expectations:

$$\mathbb{E}_i\left[\mathrm{Tr}(\rho_i^2)\right] = \mathbb{E}_i\left[\frac{1}{2^n} + \frac{1}{2^n}\sum_{(u,v)\neq 0}|\mathrm{Tr}(X^u Z^v \rho_i)|^2\right]$$

$$= \frac{1}{2^n} + \frac{1}{2^n}\sum_{(u,v)\neq 0}\mathbb{E}_i\left[|\mathrm{Tr}(X^u Z^v \rho_i)|^2\right]$$

$$= \frac{1}{2^n} + \frac{1}{2^n}\sum_{(u,v)\neq 0}\mathbb{E}_i\left[\hat{A}_i(v,u)^2\right]|\mathrm{Tr}(X^u Z^v \rho_0)|^2$$

The expression $\mathbb{E}_i\left[\hat{A}_i(v,u)^2\right]$ is exactly the quantity bounded by the (squared) bias δ^2. As in the previous proof, the entropy $\mathrm{Tr}(\mathcal{E}(\rho_0)^2)$ is bounded by $\frac{1}{2^n|\mathcal{I}|}(1 + \delta^2 2^n \mathrm{Tr}(\rho_0^2))$. By our choice of δ, the entropy is at most $\frac{1}{2^n|\mathcal{I}|}(1 + \epsilon^2)$, and so $\mathcal{E}(\rho_0^2)$ is within trace distance ϵ of the completely mixed state. $\qquad\square$

3.3 Hybrid Construction

Let d be a prime between 2^n and 2^{n+1}. Then, it suffices to show how to randomize a state in a d-dimensional space $\mathcal{H}_d$ spanned by $|i\rangle$, $i \in \{0,1,\ldots,d-1\}$, since a state on n qubits can be embedded into $\mathcal{H}_d$. We define X and Z on this space by $X|j\rangle = |(j+1) \bmod d\rangle$ and $Z|j\rangle = e^{2\pi ij/d}|j\rangle$. Notice that $X^j Z^k = e^{2\pi i(jk)/d} Z^k X^j$ and $(X^j Z^k)^\dagger = Z^{-k} X^{-j}$. (The definitions of X and Z are different than in the previous sections, since we are operating on a space of prime dimension).

We start with a construction that uses $n+1$ bits of randomness and achieves approximate encryption for $\epsilon = 1$. (Notice that this is a non-trivial security guarantee. The trace distance between perfectly distinguishable states is 2. Distance 1 means that the state cannot be distinguished from $\frac{I}{d}$ with success probability more than 3/4.) We will then extend it to any $\epsilon > 0$, using more randomness.

Let

$$\mathcal{E}(\rho) = \frac{1}{d}\sum_{a=1}^{d-1} X^a Z^{a^2} \rho Z^{-a^2} X^{-a}.$$

Claim.

$$Tr(\mathcal{E}(\rho)^2) \le \frac{1}{d}(1 + Tr(\rho^2)).$$

Proof. Let $\rho' = \mathcal{E}(\rho)$.

$$Tr(\rho')^2 = \sum_{ij} \rho'_{ij}(\rho'_{ij})^* = \sum_{i} \rho'_{ii}(\rho'_{ii})^* + \sum_{i,j:i\ne j} \rho'_{ij}(\rho'_{ij})^*.$$

The first sum is equal to $d\frac{1}{d^2} = \frac{1}{d}$ because $\rho'_{ii} = \frac{1}{d}\sum_{k=1}^{d}\rho_{kk} = \frac{1}{d}$. To calculate the second sum, we split it into sums $S_t = \sum_i \rho'_{i,i+t}(\rho'_{i,i+t})^*$ for $t = 1, 2, \ldots, d-1$. (In the indices for ρ_{ij} and ρ'_{ij}, we use $i + t$ as a shortcut for $(i + t) \mod d$.) We have

$$\rho'_{i,i+t} = \frac{1}{d}\sum_{a=0}^{d-1} w^{a^2 t}\rho_{i-a,i-a+t},$$

where w is the dth root of unity.

$$\rho'_{i,i+t}(\rho'_{i,i+t})^* = \frac{1}{d^2}\left(\sum_{a=0}^{d-1}|\rho_{i+a,i+t+a}|^2 + \sum_{a,b,a\ne b} w^{(b^2-a^2)t}\rho_{i-a,i+t-a}(\rho_{i-b,i+t-b})^*\right)$$

Therefore,

$$S_t = \frac{1}{d}\sum_{i=1}^{d}|\rho_{i,i+t}|^2 + \frac{1}{d^2}\sum_{i\ne j}c_{i,j}\rho_{i,i+t}(\rho_{j,j+t})^*$$

where

$$c_{i,j} = \sum_a w^{((i+a)^2-(j+a)^2)t} = \sum_a w^{(i^2-j^2+2a(i-j))t} = w^{(i^2-j^2)t}\sum_a w^{a*2(i-j)t}.$$

Since d is a prime, $2(i - j)t$ is not divisible by d. Therefore, $\sum_a w^{a*2(i-j)t} = 0$, $c_{ij} = 0$, $S_t = \frac{1}{d}\sum_{i=1}^{d}|\rho_{i,i+t}|^2$ and

$$Tr((\rho')^2) = \frac{1}{d} + \frac{1}{d}\sum_{i\ne j}|\rho_{ij}|^2.$$

$\square$

By Fact 1, $D(\mathcal{E}(\rho), \frac{I}{d}) \le 1$.

We now improve this construction to any ϵ. Let B be an ϵ-biased set on $m = \lceil \log d \rceil$ bits. For $b \in \{0,1\}^m$, define a unitary transformation U_b as follows. Identify numbers $0, 1, \ldots, d-1$ with strings $x \in \{0,1\}^m$. Define $U_b|x\rangle = (-1)^{b\odot x}|x\rangle$, with $b \odot x$ being the usual (bitwise) inner product of b and x. (Note that U_b is just to the Z operator over a different group. It is the same Z operator used in the previous sections). Let

$$\mathcal{E}'(\rho) = \sum_{b\in B} U_b\rho U_b^\dagger \text{ and } \mathcal{E}''(\rho) = \mathcal{E}(\mathcal{E}'(\rho)).$$

We claim that $\mathcal{E}''$ is ϵ-approximate encryption scheme. W.l.o.g., assume that ρ is a pure state $|\psi\rangle = \sum_i c_i |i\rangle$. Then $\rho_{ij} = c_i c_j^*$. Let $\rho' = \frac{1}{|B|} \sum_{b \in B} U_b \rho U_b^\dagger$ be the result of encrypting ρ by $\mathcal{E}'$. Then,

$$\rho'_{xy} = \frac{1}{|B|} \sum_{b \in B} (-1)^{b \odot x + b \odot y} \rho_{xy} = \frac{1}{|B|} \sum_{b \in B} (-1)^{b \odot (x+y)} \rho_{xy}.$$

Since B is ϵ-biased, $|\rho'_{xy}| \leq \epsilon |\rho_{xy}|$ for any x, y, $x \neq y$. Therefore, $\sum_{x \neq y} |\rho'_{xy}| \leq \epsilon \sum_{x \neq y} |\rho_{xy}|$. Together with the Claim above and Fact 1, this implies that $\mathcal{E}''$ is ϵ-randomizing. The number of key bits used by $\mathcal{E}''$ is $n + \log |B| + O(1)$ which is $n + 2\log n + 2\log \frac{1}{\epsilon} + O(1)$ if the AGHP scheme is used and $n + \log n + 3\log \frac{1}{\epsilon} + O(1)$ if ABNNR is used. The first bound is the same as the one achieved by using small-bias spaces directly (Section 3.1). The second bound gives a better result when $\epsilon > \frac{1}{n}$.

Acknowledgements

We are grateful for helpful discussions with and comments from Claude Crépeau, Daniel Gottesman, Patrick Hayden, Debbie Leung, Sofya Raskhodnikova, Alex Samorodnitsky, and an anonymous referee.

References

1. Noga Alon, Jehoshua Bruck, Joseph Naor, Moni Naor, and Ronny Roth. Construction of asymptotically good low-rate error-correcting codes through pseudo-random graphs. IEEE Transactions on Information Theory, 38:509-516, 1992.
2. Noga Alon, Oded Goldreich, Johan Håstad, René Peralta. Simple Construction of Almost k-wise Independent Random Variables. Random Structures and Algorithms 3(3): 289-304.
3. Andris Ambainis, Michele Mosca, Alain Tapp, Ronald de Wolf. Private Quantum Channels. FOCS 2000: 547-553.
4. Howard Barnum, Claude Crépeau, Daniel Gottesman, Adam Smith, Alain Tapp. Authentication of Quantum Messages. FOCS 2002: 449-458.
5. Charles Bennett, Patrick Hayden, Debbie Leung, Peter Shor and Andreas Winter. Remote preparation of quantum states. ArXiv e-Print quant-ph/0307100.
6. Eli Ben-Sasson, Madhu Sudan, Salil P. Vadhan, Avi Wigderson. Randomness-efficient low degree tests and short PCPs via epsilon-biased sets. STOC 2003: 612-621.
7. Yevgeniy Dodis and Adam Smith. Encryption of High-Entropy Sources. Manuscript, 2003.
8. Patrick Hayden, Debbie Leung, Peter Shor and Andreas Winter. Randomizing quantum states: Constructions and applications. Comm. Math. Phys., to appear. Also ArXiv e-print quant-ph/0307104.
9. Debbie Leung, personal communication, 2004.
10. Joseph Naor, Moni Naor. Small-Bias Probability Spaces: Efficient Constructions and Applications. SIAM J. Comput. 22(4): 838-856 (1993).
11. Michael Nielsen, Isaac Chuang. *Quantum Computation and Quantum Information*. Cambridge University Press, 2000.
12. Alexander Russell, Hong Wang. How to Fool an Unbounded Adversary with a Short Key. EUROCRYPT 2002: 133-148.

The Sketching Complexity of Pattern Matching

Ziv Bar-Yossef, T.S. Jayram, Robert Krauthgamer, and Ravi Kumar

IBM Almaden Research Center, 650 Harry Road, San Jose, CA 95120, USA
{ziv,jayram,robi,ravi}@almaden.ibm.com

Abstract. We address the problems of pattern matching and approximate pattern matching in the sketching model. We show that it is impossible to compress the text into a small sketch and use only the sketch to decide whether a given pattern occurs in the text. We also prove a sketch size lower bound for approximate pattern matching, and show it is tight up to a logarithmic factor.

1   Introduction

Pattern matching is the problem of locating a given (smaller) pattern in a (larger) text. It is one of the most fundamental problems studied in computer science, having a wide range of uses in text processing, information retrieval, computational biology, compilers, and web search. These application areas typically deal with large amounts of data and therefore necessitate highly efficient algorithms in terms of time and space.

In order to save space, I/O, and bandwidth, large text files are frequently stored in compressed form. The naive method for locating patterns in compressed files is to first decompress the files, and then run one of the standard pattern matching algorithms on them. Amir and Benson [2] initiated the study of pattern matching in compressed files; their approach is to process the compressed text directly, without first decompressing it. Their algorithm, as well as all the subsequent work in this area [3, 21, 12, 24, 11, 22, 15], deal with *lossless* compression schemes, such as Huffman coding and the Lempel-Ziv algorithm. The main focus of these results is the speedup gained by processing the compressed text directly.

In this paper we investigate a closely related question: how succinctly can one compress a text file into a small "sketch", and yet allow locating patterns in the text using the sketch alone? In this context we consider not only lossless compression schemes but also *lossy* ones. In turn, we permit pattern matching algorithms that are randomized and can make errors with some small constant probability. Our main focus is not on the speed of the pattern matching algorithms but rather on the succinctness of the compression. Highly succinct compression schemes of this sort could be very appealing in domains where the text is a massive data set or when the text needs to be sent over a network.

A fundamental and well known model that addresses problems of this kind is the *sketching model* [8, 14], which is a powerful paradigm in the context of computations over massive data sets. Given a function, the idea is to produce a fingerprint (*sketch*) of the data that is succinct yet rich enough to let one

K. Jansen et al. (Eds.): APPROX and RANDOM 2004, LNCS 3122, pp. 261–272, 2004.

compute or approximate the function on the data. The parameters that play a key role in the applications are the size of the sketch, the time needed to produce the sketch and the time required to compute the function given the sketch.

Results. Our first main result is an impossibility theorem showing that in the worst-case, no sketching algorithm can compress the text by more than a constant factor and yet allow exact pattern matching. Specifically, any sketching algorithm that compresses any text of length n into a sketch of size s and enables determining from the sketch alone whether an input pattern of length $m = \Omega(\log n)$ matches the text or not with a constant probability of error requires $s \geq \Omega(n - m)$. We further show that the bound is tight, up to constant factors.

The proof of this lower bound turns out to be more intricate than one might expect. One of the peculiarities of the problem is that it exhibits completely different behaviors for $m \leq (1 - o(1)) \log n$ and $m \geq \log n$. In the former case, a simple compression of the text into a sketch of size 2^m is possible. We prove a matching lower bound for this range of m as well. These results are described in Section 3.

Our second main result is a lower bound on the size of sketches for *approximate pattern matching*, which is a relaxed version of pattern matching: (i) if the pattern occurs in the text, the output should be "a match"; (ii) if every substring of the text is at Hamming distance at least k from the pattern, the output should be "no match". An arbitrary answer is allowed if neither of the two holds. We prove that any sketching algorithm for approximate pattern matching, requires sketch size $\Omega(n/m)$, where n is the length of the text, m is the length of the pattern, and the Hamming distance at question is $k = \varepsilon m$, for a fixed $0 < \varepsilon < 1$. We further show that this bound is tight, up to a logarithmic factor. These results are described in Section 4.

Interestingly, Batu *et al.* [6] showed a *sampling* procedure that solves (a restricted version of) approximate pattern matching using $\tilde{O}(n/m)$ non-adaptive samples from the text. In particular, their algorithm yields a sketching algorithm with sketch size $\tilde{O}(n/m)$. This procedure was the main building block in their sub-linear time algorithm for weakly approximating the edit distance. The fact that our sketching lower bound nearly matches their sampling upper bound suggests that it might be hard to improve their edit distance algorithm, even in the sketching model.

Techniques. A sketching algorithm naturally corresponds to the communication complexity of a one-way protocol. Alice holds the text and Bob holds the pattern. Alice needs to send a single message to Bob (the "sketch"), and Bob needs to use this message as well as his input to determine whether there is a match or not[1].

---

[1] Usually, a sketching algorithm corresponds to the communication complexity of a simultaneous messages protocol, which is equivalent to summarizing each of the text and the pattern into a small sketch. However, in the context of pattern matching, it is reasonable to have a weaker requirement, namely, that only the text needs to be summarized.

The most classical problem which is hard for one-way communication complexity is the indexing function: Alice is given a string $x \in \{0,1\}^n$ and Bob is given an index $i \in \{1, \ldots, n\}$, and based on a single message from Alice, Bob has to output x_i. It is well known that in any protocol solving this problem, even a randomized one, Alice's message has to be of length $\Omega(n)$. Our lower bound for approximate pattern matching is proved by a reduction from the indexing function.

Our lower bound for exact pattern matching uses a reduction from a variant of the indexing function. In this variant, Alice gets a string $x \in \{0,1\}^n$; Bob gets an index $i \in [n]$ and also the $m-1$ bits preceding x_i in x; the goal is to output x_i. Using tools from information theory we prove an $\Omega(n-m)$ lower bound for this problem in the one-way communication complexity model.

Related Work. Pattern matching and approximate pattern matching have a rich history and extensive literature – see, for instance, the excellent resource page [20]. To the best of our knowledge, pattern matching, has not been considered in the sketching model. For approximate pattern matching, the only relevant result appears to be the above mentioned work of Batu *et al.* [6].

Sketching algorithms for various problems, such as estimation of similarity between documents [8, 7, 9], approximation of Hamming distance [19, 13] and edit distance [4] between strings, and computation of L_p distances between vectors [1, 14], have been proposed in the last few years. Sketching is also a useful tool for approximate nearest-neighbor schemes [19, 16], and it is related to low-dimensional embeddings and to locality-sensitive hash functions [16].

2   Preliminaries

2.1   Communication Complexity

A *sketching algorithm* is best viewed as a *public-coin one-way communication complexity protocol*. Two players, Alice and Bob, would like to jointly compute a two-argument function $f : \mathcal{X} \times \mathcal{Y} \to \mathcal{Z}$. Alice is given $x \in \mathcal{X}$ and Bob is given $y \in \mathcal{Y}$. Based on her input and based on randomness that is shared with Bob, Alice prepares a "sketch" $s_A(x)$ and sends it to Bob. Bob uses the sketch, his own input y, and the shared randomness to determine the value $f(x, y)$. For every input $(x, y) \in X \times Y$, the protocol is required to be correct with probability at least $1 - \delta$, where $0 < \delta < 1$ is some small constant. Typically, the error probability δ can be reduced by repeating the procedure several times independently (in parallel).

The main measure of cost of a sketching algorithm is the length of the sketch $s_A(x)$ on the worst-case choice of shared randomness and of the input x. Another important resource is the amount of randomness between Alice and Bob. Newman [23] shows that the amount of shared randomness can always be reduced to $O(\log \frac{n}{\delta'})$ at the cost of increasing the protocol's error probability by δ'. In one-way protocols, Alice can privately generate these $O(\log \frac{n}{\delta'})$ and send them to Bob along with the sketch $s_A(x)$.

Some of our lower bounds use a reduction from the standard *indexing problem*, which we denote by IND_t: Alice is given a string $x \in \{0,1\}^t$, Bob is given $j \in [t]$, and the goal is to output x_j. This problem has a lower bound of $t(1 - H_2(\delta))$ in the one-way communication complexity model [18, 5].

2.2   Pattern Matching and Approximate Pattern Matching

For a Boolean string $x \in \{0,1\}^n$ and integer $1 \leq j \leq n$, let x_i denote the jth bit in x. For integers $1 \leq i \leq j \leq n$, $[i,j]$ denotes the corresponding integer interval, $[n]$ the interval $[1,n] = \{1,\ldots,n\}$, and $x[i,j]$ denotes the substring of x that starts at position i and ends at position j. We define the pattern matching and approximate pattern matching problems in the communication model.

Let $0 \leq m \leq n$. In the (n,m) *pattern matching* problem, denoted $\text{PM}_{n,m}$, Alice gets a string $x \in \{0,1\}^n$ and Bob gets a string $y \in \{0,1\}^m$. The goal is to determine whether there exists an index $i \in [n-m+1]$ such that $x[i, i+m-1] = y$. For the purpose of lower bounds, we would consider the simple Boolean function defined above. However, some of the algorithms we present can additionally find the position of the match i, if it exists.

We denote the *Hamming distance* of two strings $x, y \in \{0,1\}^n$ by $\text{HD}(x,y) \overset{\text{def}}{=} |\{i \in [n] : x_i \neq y_i\}|$. A relaxed version of pattern matching is the (n,m,ε) *approximate pattern matching* problem, denoted $\text{APM}_{n,m,\varepsilon}$, in which Bob would like to determine whether there exists an index $i \in [n-m+1]$ such that $x[i, i+m-1] = y$, or whether for all $i \in [n]$, $\text{HD}(x[i, i+m-1], y) \geq \varepsilon m$, assuming that one of the two holds.

Notation. Throughout the paper we denote random variables in upper case. For a Boolean string $x \in \{0,1\}^n$, $|x|$ denotes the Hamming weight (i.e., the number of 1's) of x. log denotes a logarithm to the base 2; ln denotes the natural logarithm. $H_2(p) = -p \log p - (1-p) \log(1-p)$ is the binary entropy function.

3   Exact Pattern Matching

In this section we obtain a simple sketching algorithm for exact pattern matching and show almost matching lower bounds. Recall that we denote by $\text{PM}_{n,m}$ the problem in which Alice gets a string $x \in \{0,1\}^n$, Bob gets a string $y \in \{0,1\}^m$, and their goal is to find whether there exists an index $i \in [n - m + 1]$ such that $x[i, i + m - 1] = y$.

3.1   Upper Bounds

First, we show an efficient (randomized) sketching algorithm for the pattern matching problem, based on the Karp–Rabin hash function [17]. Next, we show a deterministic sketching algorithm for the Boolean version of the pattern matching problem.

Proposition 1. *For $m \leq n - \log n$, there is a one-sided error randomized sketching algorithm for the pattern matching problem* $\mathrm{PM}_{n,m}$ *using a sketch of size* $O(n - m)$.

Proof. The randomized algorithm is based on the Karp–Rabin method [17]. Let $t = n - m + 1$; we assume in the sequel that $t \leq n/3$, as otherwise the proof follows trivially by Alice sending x. Let $\mathrm{x}^1, \ldots, \mathrm{x}^t$ denote the sequence of t substrings of x of length m. Alice and Bob use the shared randomness to pick a (weak) 2-universal hash function $h : \{0,1\}^m \rightarrow [n^2]$. Alice sends to Bob $h(\mathrm{x}^1), \ldots, h(\mathrm{x}^t)$. Bob outputs "match found at i", if $h(\mathrm{x}^i) = h(\mathrm{y})$. If no such i exists, Bob outputs "no match found".

This is a one-sided error algorithm: if there is a match, it will surely be output. There is a possibility for a false match, though: when $\mathrm{x}^i \neq \mathrm{y}$, but $h(\mathrm{x}^i) = h(\mathrm{y})$. The probability for a false match is thus at most the probability h has a collision between y and any of $\{\mathrm{x}^1, \ldots, \mathrm{x}^t\}$. A union bound shows that since the range of h is large enough, the probability of a collision between y and any x^i is at most $O(1/n)$.

The scheme described above uses a sketch of size $O(t \log n) = O((n-m) \log n)$. A further improvement is possible using the Karp–Rabin hash function: $h(b) = (\sum_{i=1}^{m} b_i \cdot 2^{m-i}) \bmod p$, where p is a randomly chosen prime in the range $[n^3]$. The advantage of this hash function is that the value of $h(\mathrm{x}^{i+1})$ can be computed from the value of $h(\mathrm{x}^i)$ and from the two bits x_i and x_{i+m}: $h(\mathrm{x}^{i+1}) = ((h(\mathrm{x}^i) - \mathrm{x}_i \cdot 2^{m-1}) \cdot 2 + \mathrm{x}_{i+m}) \bmod p$. Thus, what Alice needs to send is only $h(\mathrm{x}^1)$, the first t bits of x, and the last t bits of x. Thus, the sketch size goes down to $2t + O(\log n) = O(n - m)$.

Proposition 2. *There is a deterministic sketching algorithm for the pattern matching problem* $\mathrm{PM}_{n,m}$ *using a sketch of size* 2^m.

Proof. In the deterministic algorithm Alice sends to Bob a characteristic vector of length 2^m specifying all the strings of length m that occur as substrings of x. Bob outputs "match found" if and only if y is one of the substrings indicated by the characteristic vector.

3.2 Lower Bounds

We show lower bounds on the sketch size for the pattern matching problem. The first one, Theorem 1, deals with the case $m \geq \Omega(\log n)$. The second one, Theorem 2, considers the case $m \leq O(\log n)$.

Theorem 1. *If $n \leq m + \delta 2^m$, then any δ-error randomized sketching algorithm for the pattern matching problem* $\mathrm{PM}_{n,m}$ *requires a sketch of size at least* $(n - m + 1) \cdot (1 - H_2(2\delta))$, *where $H_2(\cdot)$ is the binary entropy function.*

Proof. Using Yao's Lemma [25], it suffices to exhibit a distribution μ over instances of $\mathrm{PM}_{n,m}$, and prove that any deterministic sketching algorithm that computes $\mathrm{PM}_{n,m}$ correctly with probability at least $1 - \delta$ when running over inputs chosen according to μ requires a sketch of size at least $(n - m + 1) \cdot (1 - H_2(2\delta))$.

The distribution μ is defined as follows. Alice is given a uniformly chosen bitstring $X \in \{0, 1\}^n$. Bob is given a uniformly chosen substring of X of length $m - 1$ concatenated with the bit 1.

The distributional lower bound w.r.t. μ is proven via a reduction from the following version of the indexing function, which we denote by $\text{IND}_{n,k}$: Alice is given a string $x \in \{0, 1\}^n$, and Bob is given an index $k + 1 \leq j \leq n$ and a string $y \in \{0, 1\}^k$, which is promised to be equal to the substring $x[j - k, j - 1]$. The goal is to compute x_j.

Let ν be the following distribution over instances of $\text{IND}_{n,k}$. Alice gets a uniformly chosen bitstring X in $\{0, 1\}^n$; Bob gets an index J, which is chosen independently and uniformly in the interval $[k + 1, n]$, and also the bits $X_{J-k}, \ldots, X_{J-1}$.

The following lemma shows the reduction.

Lemma 1. *Any deterministic sketching algorithm Π that computes $\text{PM}_{n,m}$ with error probability at most δ on instances drawn according to μ yields a deterministic sketching algorithm Π' that computes $\text{IND}_{n,m-1}$ with error probability at most 2δ on instances drawn according to ν and using exactly the same sketch size.*

Proof. In the indexing algorithm Π', given an input $x \in \{0, 1\}^n$, Alice sends whatever message she would have sent on this input in the algorithm Π. Given his input (j, y), where $m \leq j \leq n$ and $y \in \{0, 1\}^{m-1}$, Bob simulates the role of Bob in the pattern matching algorithm Π on the input $y \circ 1$, where $\circ$ denotes the concatenation of strings. If the output in Π is "match found", Bob outputs "1" and otherwise he outputs "0".

It is easy to verify that when the input given to Π' is distributed according to ν, then the input given to Π in the reduction is distributed according to μ. We can thus assume that Π errs with probability at most δ.

Fix an input $(x, (j, y))$ for $\text{IND}_{n,m-1}$, for which the protocol Π is correct on $(x, y \circ 1)$. If $\text{IND}_{n,m-1}(x, (j, y)) = 1$, then the string $y \circ 1$ is a substring of x (at position $j - m + 1$), and thus Π' will output "1", as needed. Suppose then that $\text{IND}_{n,m-1}(x, (j, y)) = 0$. Clearly, $y \circ 1$ does not equal to the substring of x that starts at position $j - m + 1$. Therefore, Π' outputs "0", unless there is some other substring of x that happens to equal to $y \circ 1$. We next prove that the latter occurs with low probability.

Define E to be the set of instances $(x, (j, y))$, for which there exists some $i \neq j$, so that the substring $x[i, i + m - 1]$ equals $y \circ 1$. The proof of Lemma 1 would follow easily once we prove that $\Pr(E) \leq \delta$, since Π' errs on an input $(x, (i, y))$ only if either it belongs to E or if Π errs on it.

Proposition 3. $\Pr(E) \leq \delta$.

Proof. $\Pr(E)$ can be rewritten as $\Pr(\exists i \neq J : X[i, i + m - 1] = X[J - m + 1, J - 1] \circ 1)$. In order to bound $\Pr(E)$, it would suffice to show that for all choices of

$m \leq j \leq n$, $\Pr(\exists i \neq j : X[i, i+m-1] = X[j-m+1, j-1 \circ 1]) \leq \delta$. So for the rest of the argument, fix such a j.

Define $t \stackrel{\text{def}}{=} j - m + 1$. We will show that for all $i \neq t$, $\Pr(X[i, i+m-1] = X[t, j-1] \circ 1) \leq 1/2^m$. It would then follow from the union bound that $\Pr(\exists i \neq j : X[i, i+m-1] = X[j-m+1, j-1 \circ 1]) \leq (n-m)/2^m \leq \delta$.

Fix any $i \neq t$. Suppose, for example, $i < t$ (the case $i > t$ is dealt with similarly). $X[i, i+m-1] = X[t, j-1] \circ 1$ if and only if $X[i, i+m-2] = X[t, j-1]$ and $X_{i+m-1} = 1$. It is easy to verify that the two events are independent and that the probability of the second is $1/2$. Thus, it suffices to show $\Pr(X[i, i+m-2] = X[t, j-1]) = 1/2^{m-1}$.

We will denote $X^i = X[i, i+m-2]$ and $X^t = X[t, j-1]$. Let p be the length of the longest prefix of X^i that does not overlap X^t (p can be anywhere between 1 and $m-1$). Divide X^i and X^t into $s \stackrel{\text{def}}{=} \lceil (m-1)/p \rceil$ non-overlapping blocks of size p each (except, maybe, for the last one which is shorter). Call the corresponding substrings $X_1^i, \ldots, X_s^i$ and $X_1^t, \ldots, X_s^t$, respectively. Note that $X_q^i = X_{q-1}^t$ for $q = 2, \ldots, s$. $X^t = X^i$ if and only if $X_q^t = X_q^i$ for all $q = 1, \ldots, s$, or equivalently, if and only if all the s blocks of X^t equal to X_1^i (except, maybe, the last one which needs to be a prefix of X_1^i). The latter event occurs with probability $1/2^{m-1}$, since the X^t and X_1^i are disjoint substrings of X, and the bits of X are chosen independently at random.

Recall that we assumed the probability Π errs on $(x, (i, y))$ is at most δ. Using Proposition 3 and applying a union bound, we get that the error probability of Π' is at most 2δ, completing the proof of Lemma 1.

Next, we obtain a lower bound on the one-way communication complexity of the modified indexing function. The proof is based on information-theoretic arguments.

Lemma 2. *Any one-way deterministic protocol that computes* $\text{IND}_{n,k}$ *with error probability at most* ε *on inputs chosen according to* ν *requires at least* $(n - k)(1 - H_2(\varepsilon))$ *bits of communication.*

Proof. Fix any such deterministic protocol, and let $A(\cdot)$ denote the function Alice applies on her input in this protocol to determine the message she sends to Bob. Bob outputs an answer based on the message from Alice and based on his input. Thus, using the random variables $A(X)$, J, and $X_{J-k}, \ldots, X_{J-1}$, Bob is able to predict the random variable X_J with probability of error at most ε. This is exactly the scenario captured by a classical result from information theory, called *Fano's inequality* (cf. [10]), which implies $H_2(\varepsilon) \geq H(X_J \mid A(X), J, X_{J-k}, \ldots, X_{J-1})$. Here $H(Z \mid Y)$ denotes the conditional Shannon entropy of the random variable Z given the random variable Y (cf. [10]). By definition, the conditional entropy $H(Z \mid Y)$ equals to $\sum_y H(Z \mid Y = y) \cdot \Pr(Y = y)$, where $H(Z \mid Y = y)$ is the entropy of the conditional distribution of Z given the event $\{Y = y\}$. Expanding over the random variable J, we thus have:

$$H_2(\varepsilon) \geq \frac{1}{n-k} \sum_{j=k+1}^{n} H(X_J \mid A(X), X_{J-k}, \ldots, X_{J-1}, J = j)$$

$$= \frac{1}{n-k} \sum_{j=k+1}^{n} H(X_j \mid A(X), X_{j-k}, \ldots, X_{j-1}).$$

Conditioning one variable on another can only reduce its entropy. Therefore, we can lower bound the j-th term on the righthand side by the conditional entropy $H(X_j \mid A(X), X_1, \ldots, X_{j-1})$ (we added the random variables $X_1, \ldots, X_{j-k-1}$ to the conditioning). We thus have:

$$H_2(\varepsilon) \geq \frac{1}{n-k} \sum_{j=k+1}^{n} H(X_j \mid A(X), X_1, \ldots, X_{j-1})$$

$$\geq \frac{1}{n-k} H(X_{k+1}, \ldots, X_n \mid A(X), X_1, \ldots, X_k).$$

The last transition follows from the chain rule for entropy. Another application of this rule implies that $H(X_{k+1}, \ldots, X_n \mid A(X), X_1, \ldots, X_k) = H(X_1, \ldots, X_n, A(X)) - H(A(X)) - H(X_1, \ldots, X_k \mid A(X))$. Since $A(X)$ fully depends on $X = (X_1, \ldots, X_n)$, we have

$$H(X_1, \ldots, X_n, A(X)) = H(X).$$

But, $H(X) = n$, as X has a uniform distribution on $\{0,1\}^n$. Since conditioning reduces entropy, $H(X_1, \ldots, X_k \mid A(X)) \leq H(X_1, \ldots, X_k) = k$. To conclude:

$$H_2(\varepsilon) \geq \frac{1}{n-k} \cdot (n - H(A(X)) - k).$$

Therefore, $H(A(X)) \geq (n-k)(1 - H_2(\varepsilon))$. Since $H(A(X))$ is always a lower bound on the length of $A(X)$, the lemma follows.

The proof of Theorem 1 now follows from Lemma 2, Lemma 1, and Yao's lemma [25].

For the case when $m \leq O(\log n)$, we have the following lower bound.

Theorem 2. *If $n \geq 2^{m/3} \cdot m$, then any δ-error randomized sketching algorithm for the pattern matching problem $PM_{n,m}$ requires a sketch of size at least $2^{m/3-1}$. $(1 - H_2(\delta))$.*

Proof. The lower bound follows from a reduction from the indexing problem, IND_t. The reduction works as follows. Let m be such that $t = 2^{m/3-1}$) and let $n \geq 2^{m/3} \cdot m$. Given her input $x \in \{0,1\}^t$, Alice maps it first into a string $x' \in \{0,1,\$\}^{n/3}$; $\$$ is a special symbol, which we call a "marker". Let $i_1, \ldots, i_k$ be the positions in which x has a 1. Then $x' = i_1 \$ i_2 \$ \ldots i_k \$\$ \ldots \$$ where each integer i_j is written in binary using $\log t$ bits and the trailing $\$$'s are used to make sure the string is of length $n/3$. Bob maps his input $j \in [t]$ into the string

$y' = j\$$ in $\{0, 1, \$\}^{m/3}$. Note that since a marker can match only a marker, $x_j = 1$ if and only if the pattern y' matches x'.

In order to complete the reduction, we need to describe a mapping ϕ from strings over the ternary alphabet into (longer) strings over a binary alphabet, so that a pattern y' matches a substring of x' over the ternary alphabet if and only if $\phi(y')$ matches a substring of $\phi(x')$. ϕ could be, for example, the following mapping: 0 maps to 010, 1 maps to 101, and \$ maps to 111.

4   Approximate Pattern Matching

In this section we give a nearly-tight sketching lower bound for approximate pattern matching. Once again we use the lower bound for the indexing function to obtain our lower bound. Recall that we denote by $\text{APM}_{n,m,\varepsilon}$ the problem in which Alice gets a string $x \in \{0, 1\}^n$, Bob gets a string $y \in \{0, 1\}^m$, and their goal is to determine whether there exists an index $i \in [n]$ such that $x[i, i+m-1] = y$, or whether for all $i \in [n]$, $\text{HD}(x[i, i + m - 1], y) \geq \varepsilon m$, assuming that one of the two holds.

Theorem 3. *If* $n \leq 2^{O(m)}$, *then for any constant* $0 < \varepsilon < 1/8$, *any randomized sketching algorithm for* $\text{APM}_{n,m,\varepsilon}$ *requires a sketch of size* $\Omega(\frac{n}{m})$.

Proof. The proof works by a reduction from the indexing function IND_t; we assume, for simplicity, m divides n, and let $t = n/m$. We first need to fix a collection $z_1, \ldots, z_{2t}$ of $2t$ binary strings in $\{0, 1\}^m$ with the following property: for any $m/2 \leq s \leq m$, and for any i, j, the prefix of z_i of length s and the suffix of z_j of length s have a large Hamming distance, namely, $\text{HD}(z_i[1, s], z_j[m - s + 1, m]) \geq \frac{m}{8}$. A simple probabilistic argument can show that such a collection exists as long as $t \leq 2^{\gamma m}$, for some constant $0 < \gamma < 1$ [2]. We will call the last t strings in the collection also $o_1, \ldots o_t$.

Suppose Π is a sketching algorithm for $\text{APM}_{n,m,\varepsilon}$. We use it to construct a sketching algorithm Π' for IND_t. Given her indexing input $x \in \{0, 1\}^t$, Alice maps it into a string $u \in \{0, 1\}^n$, which is formed by concatenating t strings of length m each. The j-th string is z_j if $x_j = 0$ and it is o_j if $x_j = 1$. Bob maps his input $i \in [t]$ into the string $v = o_i$.

Alice and Bob now run the algorithm Π on the inputs u and v. Bob decides that $x_i = 1$ if and only if it is determined that v approximately matches u.

If $x_i = 1$, then $v = o_i$ is a substring of u, and therefore the algorithm will be correct with high probability. If $x_i = 0$, then v is not one of the strings constituting u. We need to use now the property of the collection $z_1, \ldots, z_{2t}$ to show that no substring of u of length m has small Hamming distance from v.

Let $u^1, \ldots, u^t$ be the strings (which are taken from $z_1, \ldots, z_{2t}$) that constitute u. Suppose, to the contradiction, u has a substring α of length m such that

---

[2] Note that the corresponding Hamming distance is the summation of indicators for the events $(z_i)_\ell = (z_j)_{\ell+m-s}$ taken over $\ell = 1, \ldots, s$, and even if $i = j$, at least $s/2 \geq m/2$ of them are *independent*.

$HD(\alpha, v) \leq \varepsilon m \leq m/8$. The prefix of α overlaps some u^j and its suffix overlaps u^{j+1}. Call the overlapping prefix α_1 and the overlapping suffix α_2. At least one of α_1, α_2 has to be of size at least $m/2$. Suppose, for example, it is α_1, and let $s = |\alpha_1|$. Let v_1 be the prefix of v of length s. Since the total Hamming distance between α and v is at most $m/8$, also the Hamming distance between α_1 and v_1 is at most $m/8$. But that implies that the Hamming distance between the last s bits of u^j and the first s bits of v is at most $m/8$ even though $u^j \neq v$. This is a contradiction to the property of the collection $z_1, \ldots, z_{2t}$. We conclude that Π' is correct with high probability also when $x_i = 0$.

The lower bound now follows from the $\Omega(t)$ lower bound for indexing.

Theorem 4. *For any constant $\varepsilon > 0$, there is a randomized sketching algorithm for $APM_{n,m,\varepsilon}$ with sketch size $O(\frac{n}{m}\varepsilon^{-1} \log n)$.*

Proof. We may assume that the text size is at most $n' = (1+\varepsilon)m$, because Alice can divide the text x into substrings of length $(1 + \varepsilon)m$ having an overlap of m bits between successive substrings, and then Alice and Bob apply an $O(\frac{n'}{m} \log n)$ sketch independently for of these substrings. If the pattern matches the text, then at least one of these substrings must contain that pattern. If the text does not contain the pattern, then it suffices to have the algorithm err with probability at most $1/n^2$ on each of the $\frac{n}{\varepsilon m}$ substrings.

If the text size is $n \leq (1 + \varepsilon)m$, Alice simply computes $O(\log n)$ random inner products $\sum_j x_j r_j (\bmod 2)$ à la Kushilevitz, Ostrovsky and Rabani [19], and sends them to Bob. These inner products are tuned to determine whether the Hamming distance is at most $\varepsilon m/2$ or whether it is at least εm, namely, each bit r_j is chosen independently to be 1 with probability $1/(2\varepsilon m)$ and 0 with probability $1 - 1/(2\varepsilon m)$. Since each inner product results in a single bit, the sketch size is clearly $O(\log n)$.

It remains to show how Bob uses the results of these $O(\log n)$ inner products to determine, with high probability, whether $x[i, i + m - 1] = y$. Notice that in each inner product,

$$\Pr[r_j = 0 \text{ for all } j < i \text{ and for all } j \geq i + m] = (1 - 1/(2\varepsilon m))^{\varepsilon m} = \Omega(1).$$

That is, with probability at least some constant, any single inner product with x is actually also a random inner product with $x[i, i + m - 1]$, and hence *can be used* by Bob to estimate whether the Hamming distance $HD(x[i, i + m - 1], y)$ is at most $\varepsilon m/2$ or at least εm. The details of this estimate are exactly as in [19]; in short, in the latter case the probability that the inner product turns out to be 1 is higher additively by a constant than in the former. By a standard Chernoff bound, with high probability (say at least $1 - 1/n^3$), there are some $\Omega(\log n)$ inner products that Bob can use to estimate $HD(x[i, i + m - 1], y)$. The proof now follows by a union bound over the $n' \leq n$ possible values of i.

Remark. The above sketching algorithm is actually stronger than claimed in the theorem, as it determines, with high probability, whether there exists an index $i \in [n]$ such that $HD(x[i, i + m - 1], y) \leq \varepsilon m/2$, or whether for all $i \in [n]$, $HD(x[i, i + m - 1], y) \geq \varepsilon m$, assuming that one of the two holds.

References

1. N. Alon, Y. Matias, and M. Szegedy. The space complexity of approximating the frequency moments. *Journal of Computer and System Sciences*, 58(1):137–147, 1999.
2. A. Amir and G. Benson. Efficient two-dimensional compressed matching. In *Proceedings of IEEE Data Compression Conference (DCC)*, pages 279–288, 1992.
3. A. Amir, G. Benson, and M. Farach. Let sleeping files lie: Pattern matching in Z-compressed files. *J. of Computer and System Sciences*, 52(2):299–307, 1996.
4. Z. Bar-Yossef, T. S. Jayram, R. Krauthgamer, and R. Kumar. Approximating edit distance efficiently. Manuscript, 2004.
5. Z. Bar-Yossef, T. S. Jayram, R. Kumar, and D. Sivakumar. Information theory methods in communication complexity. In *Proceedings of the 17th Annual IEEE Conference on Computational Complexity*, pages 93–102, 2002.
6. T. Batu, F. Ergün, J. Kilian, A. Magen, S. Raskhodnikova, R. Rubinfeld, and R. Sami. A sublinear algorithm for weakly approximating edit distance. In *Proceedings of the 35th Annual ACM Symposium on Theory of Computing*, pages 316–324, 2003.
7. A. Broder, M. Charikar, A. Frieze, and M. Mitzenmacher. Min-wise independent permutations. *Journal of Computer and System Sciences*, 60(3):630–659, 2000.
8. A. Broder, S. C. Glassman, M. S. Manasse, and G. Zweig. Syntactic clustering of the web. *WWW6/Computer Networks*, 29(8–13):1157–1166, 1997.
9. M. Charikar. Similarity estimation techniques from rounding algorithms. In *Proceedings of the 34th Annual ACM Symposium on Theory of Computing*, pages 380–388, 2002.
10. T. M. Cover and J. A. Thomas. *Elements of Information Theory*. John Wiley & Sons, Inc., 1991.
11. E. de Moura, G. Navarro, N. Ziviani, and R. Baeza-Yates. Fast and flexible word searching on compressed text. *ACM Transactions on Information Systems*, 18(2):113–139, 2000.
12. M. Farach and M. Thorup. String matching in Lempel-Ziv compressed strings. *Algorithmica*, 20(4):388–404, 1998.
13. J. Feigenbaum, Y. Ishai, T. Malkin, K. Nissim, M. J. Strauss, and R. N. Wright. Secure multiparty computation of approximations. In *28th International Colloquium on Automata, Languages and Programming*, volume 2076 of *Lecture Notes in Computer Science*, pages 927–938. Springer, 2001.
14. J. Feigenbaum, S. Kannan, M. J. Strauss, and M. Viswanathan. An approximate L^1-difference algorithm for massive data streams. *SIAM J. Comput.*, 32(1):131–151, 2002/03.
15. P. Ferragina and G. Manzini. Opportunistic data structures with applications. In *Proceedings of the 41st Annual Symposium on Foundations of Computer Science*, pages 390–398. IEEE Computer Society, 2000.
16. P. Indyk and R. Motwani. Approximate nearest neighbors: Towards removing the curse of dimensionality. In *Proceedings of the 30th Annual ACM Symposium on Theory of Computing (STOC)*, pages 604–613, 1998.
17. R. M. Karp and M. O. Rabin. Efficient randomized pattern-matching algorithms. *IBM Journal of Research and Development*, 31(2):249–260, 1987.
18. I. Kremer, N. Nisan, and D. Ron. On randomized one-round communication complexity. *Computational Complexity*, 8(1):21–49, 1999.

19. E. Kushilevitz, R. Ostrovsky, and Y. Rabani. Efficient search for approximate near-est neighbor in high dimensional spaces. *SIAM Journal on Computing*, 30(2):457–474, 2000.
20. S. Lonardi. Pattern matching pointers. *Available* http://www.cs.ucr.edu/~stelo/pattern.html, 2004.
21. U. Manber. A text compression scheme that allows fast searching directly in the compressed file. *ACM Transactions on Information Systems*, 15(2):124–136, 1997.
22. G. Navarro and J. Tarhio. Boyer-Moore string matching over Ziv-Lempel com-pressed text. In *Proceedings of 11th Annual Symposium on Combinatorial Pattern Matching (CPM)*, volume 1848 of *Lecture Notes in Computer Science*, pages 166–180. Springer, 2000.
23. I. Newman. Private vs. common random bits in communication complexity. *Inf. Process. Lett.*, 39(2):67–71, 1991.
24. Y. Shibata, T. Matsumoto, M. Takeda, A. Shinohara, and S. Arikawa. A Boyer-Moore type algorithm for compressed pattern matching. In *Proceedings of 11th Annual Symposium on Combinatorial Pattern Matching (CPM)*, volume 1848 of *Lecture Notes in Computer Science*, pages 181–194. Springer, 2000.
25. A. C.-C. Yao. Lower bounds by probabilistic arguments. In *Proceedings of the 24th Annual IEEE Symposium on Foundations of Computer Science*, pages 420–428, 1983.

Non-Abelian Homomorphism Testing, and Distributions Close to Their Self-convolutions

Michael Ben Or[1], Don Coppersmith[2], Mike Luby[3], and Ronitt Rubinfeld[4]

[1] School of Computer Science and Engineering, The Hebrew University
Jerusalem, 91904, Israel
benor@cs.huji.ac.il
[2] IBM TJ Watson Research Center, Yorktown Heights, NY 10598, USA
dcopper@us.ibm.com
http://www.research.ibm.com/people/c/copper/
[3] Digital Fountain, 39141 Civic Center Dr., Ste. 300, Fremont, CA 94538
luby@digitalfountain.com
[4] MIT Computer Science and Artificial Intelligence Laboratory, Cambridge, MA 02139, USA
ronitt@csail.mit.edu

Abstract. In this paper, we study two questions related to the problem of testing whether a function is close to a homomorphism. For two finite groups G, H (not necessarily Abelian), an arbitrary map $f : G \to H$, and a parameter $0 < \epsilon < 1$, say that f is ϵ-close to a homomorphism if there is some homomorphism g such that g and f differ on at most $\epsilon|G|$ elements of G, and say that f is ϵ-far otherwise. For a given f and ϵ, a homomorphism tester should distinguish whether f is a homomorphism, or if f is ϵ-far from a homomorphism. When G is Abelian, it was known that the test which picks $O(1/\epsilon)$ random pairs x, y and tests that $f(x) + f(y) = f(x + y)$ gives a homomorphism tester. Our first result shows that such a test works for all groups G.

Next, we consider functions that are close to their self-convolutions. Let $A = \{a_g | g \in G\}$ be a distribution on G. The self-convolution of A, $A' = \{a'_g | g \in G\}$, is defined by $a'_x = \sum_{y,z \in G; yz=x} a_y a_z$. It is known that $A = A'$ exactly when A is the uniform distribution over a subgroup of G. We show that there is a sense in which this characterization is robust – that is, if A is close in statistical distance to A', then A must be close to uniform over some subgroup of G.

1 Introduction

In this paper, we focus on two questions that are related to the problem of testing whether a function is close to a homomorphism.

For two finite groups G, H (not necessarily Abelian), an arbitrary map $f : G \to H$, and a parameter $0 < \epsilon < 1$, say that f is ϵ-close to a homomorphism if there is some homomorphism g such that g and f differ on at most $\epsilon|G|$ elements of G. Define δ, the probability of group law failure, by

$$1 - \delta = \Pr_{x,y} [f(x) \times f(y) = f(x \times y)].$$

Define τ such that τ is the minimum ϵ for which f is ϵ-close to a homomorphism. In [4], it was shown that over Abelian groups, there is a constant δ_0, such that if $\delta \leq \delta_0$,

K. Jansen et al. (Eds.): APPROX and RANDOM 2004, LNCS 3122, pp. 273–285, 2004.
© Springer-Verlag Berlin Heidelberg 2004

then the one can upper bound τ in terms of a function of δ that is independent of $|G|$. This yields a homomorphism tester with query complexity that depends (polynomially) on $1/\epsilon$, but is independent of $|G|$. In particular, the writeup in [4] contains an improved argument by Coppersmith [5], which shows that $\delta_0 < 2/9$ suffices, and that τ is upper bounded by the smaller root of $x(1 - x) = \delta$ (yielding a homomorphism tester with query complexity linear in $1/\epsilon$). Furthermore, the bound on δ_0 was shown to be tight for general groups [5].

Our first result is to give a relationship between the probability of group law failure and the closeness to being a homomorphism that applies to general (non-Abelian) groups. We show that for $\delta_0 < 2/9$, then f is τ-close to a homomorphism where $\tau = (3 - \sqrt{9 - 24\delta})/12 \leq \delta/2$ is the smaller root of $3x - 6x^2 = \delta$. The condition on δ, and the bound on τ as a function of δ, are shown to be tight, and the latter improves that of [4].

Next, consider the following question about distributions that are close to their self-convolutions: Let $A = \{a_g | g \in G\}$ be a distribution on group G. The convolution of distributions A, B is

$$C = A * B, \quad c_x = \sum_{y,z \in G; \, yz=x} a_y b_z.$$

Let A' be the *self-convolution* of A, $A * A$, i.e. $a'_x = \sum_{y,z \in G; yz=x} a_y a_z$. It is known that $A = A'$ exactly when A is the uniform distribution over a subgroup of G. The question considered here is: when is A close to A'? In particular, if $dist(A, A') = \frac{1}{2} \sum_{x \in G} |a_x - a'_x| \leq \epsilon$ for small enough ϵ, what can be said about A? We show that A must be close to the uniform distribution over a subgroup of G, that is, for a distribution A over a group G, if $dist(A, A * A) \leq \epsilon \leq 0.0182$, then there is a subgroup H of G such that $dist(A, U_H) \leq 5\epsilon$, where U_H is the uniform distribution over H. On the other hand, we give an example of a distribution A such that $dist(A, A * A) \approx .1504$, but A is not close to uniform on any subgroup of the domain.

A weaker version of this result, with a somewhat more complicated proof, was used in the original proof of the homomorphism testing result in [4]. The earlier result was never published since the simpler and more efficient proof from [5] was substituted. Instead, a separate writeup of weaker versions of both of the results in this paper, by the current set of authors, was promised in [4]. This paper is the belated fulfillment of that promise, though the earlier results have been strengthened in the intervening time.

To give a hint of why one might consider the question on convolutions of distributions when investigating homomorphism testing, consider the distribution A_f achieved by picking x uniformly from G and outputting $f(x)$. It is easy to see that the error probability δ in the homomorphism test is at least $dist(A_f, A_f * A_f)$. Unfortunately, this last relationship is not in the useful direction. In Remark 2 of Section 3, we present a relationship between homomorphism testing and distributions close to their self-convolution.

Related Work: The homomorphism testing results can be improved in some cases: We have mentioned that $\delta_0 < 2/9$ is optimal over general Abelian groups [5]. However, using Fourier techniques, Bellare et. al. [1] have shown that for groups of the form $(\mathbf{Z}/2)^n$, $\delta_0 \leq 45/128$ suffices.

Several works have shown methods of reducing the number of random bits required by the homomorphism tests. That is, in the natural implementation of the homomorphism test, $2 \log |G|$ random bits per trial are used to pick x, y. The results of [7, 6, 3, 8] have shown that fewer random bits are sufficient for implementing the homomorphism tests. The recent work of [8] gives a homomorphism test for general (non-Abelian) groups that uses only $(1 + o(1)) \log_2 |G|$ random bits. Given a Cayley graph that is an expander with normalized second eigenvalue γ, and for the analogous definitions of δ, τ, they show that for $\frac{12\delta}{1-\gamma} < 1$, τ is upper bounded by $4\delta/(1 - \gamma)$.

2   Non-Abelian Homomorphism Testing

In this section, we show that the homomorphism test of [4] works over non-Abelian groups as well. As in the Introduction, we define δ, the probability of group law failure, by

$$1 - \delta = \Pr_{x,y} \left[f(x) \bowtie f(y) = f(x \bowtie y) \right].$$

We prove the following:

Theorem 1. *If $\delta < 2/9$ then f is τ-close to a homomorphism, where $\tau = [3 - \sqrt{9 - 24\delta}]/12 < \delta/2$ is the smaller root of $3x - 6x^2 = \delta$.*

The rest of this section is devoted to proving the theorem, and showing that the parameters of the theorem are tight.

$\Pr_{x,y}(\star)$ is the probability of $\star$ when x, y are independently selected from G with the uniform random distribution.

Given two finite groups G, H (not necessarily Abelian), and given an arbitrary map $f : G \rightarrow H$ (not necessarily a homomorphism) we will construct a map $g : G \rightarrow H$, which (under certain conditions on f) will be a group homomorphism and will agree with f on a large fraction of its domain G.

Given $f : G \rightarrow H$, with associated $\delta < 2/9$, we define $g : G \rightarrow H$ by

$$g(a) = majority_{x \in G} \left[f(a \times x) \times f(x)^{-1} \right].$$

That is, we evaluate the bracketed expression for each $x \in G$, and let $g(a)$ be the value most often attained. Define ϵ_a, ϵ and τ:

$$1 - \epsilon_a = \Pr_x \left[f(a \times x) f(x)^{-1} = g(a) \right]$$

$$\epsilon = \max_a \epsilon_a$$

$$1 - \tau = \Pr_x \left[f(x) = g(x) \right]$$

Lemma 1. *If $\delta < 2/9$ then $\epsilon_a \leq \hat{\epsilon}$ where $\hat{\epsilon}$ is the smaller root of $x - x^2 = \delta$.*

Proof: For $a \in G$, define

$$p_a = \Pr_{x,y \in G} \left[f(a \times x) \times f(x)^{-1} = f(a \times y) \times f(y)^{-1} \right].$$

By rearranging, we have

$$p_a = \Pr_{x,y \in G}\left[f(a \times y)^{-1} \times f(a \times x) = f(y)^{-1} \times f(x)\right]$$
$$\geq \Pr_{x,y \in G}\left[f(a \times y)^{-1} \times f(a \times x) = f(y^{-1} \times x) \wedge f(y)^{-1} \times f(x)\right.$$
$$\left. = f(y^{-1} \times x)\right].$$

Each of the latter two equations is a random instance of the test equation $f(u) \times f(v) \stackrel{?}{=} f(u \times v)$, or equivalently, $f(u)^{-1} \times f(u \times v) \stackrel{?}{=} f(v)$, so each holds with probability $1 - \delta$, and, by the union bound, they both hold simultaneously with probability at least $1 - 2\delta$. So we have

$$p_a \geq 1 - 2\delta > 5/9.$$

If we partition G into blocks

$$B_{a,z} = \{x \in G | f(a \times x) \times f(x)^{-1} = z\}$$

with relative sizes $b_{a,z} = |B_{a,z}|/|G|$, then

$$\sum_z b_{a,z} = 1$$

$$p_a = \sum_z b_{a,z}^2 \leq \max_z b_{a,z}$$

so that $\max_z(b_{a,z}) > 5/9$, and $g(a) = argmax_z(b_{a,z})$ is well defined. By definition, $1 - \epsilon_a = \max_z(b_{a,z}) > 5/9$. Since $1 - \epsilon_a > 1/2$, we also have

$$p_a \leq (1 - \epsilon_a)^2 + \epsilon_a^2$$

$$1 - 2\delta \leq 1 - 2\epsilon_a + 2\epsilon_a^2$$

$$\delta \geq \epsilon_a - \epsilon_a^2,$$

and since $\epsilon_a < 1/2$, we conclude that $\epsilon_a \leq \hat{\epsilon}$, the smaller root of $x - x^2 = \delta$. □

Corollary 1. *If $\delta < 2/9$ then $\epsilon_a < 1/3$ and $\epsilon < 1/3$.*

Lemma 2. *If $\delta < 2/9$ then g is a homomorphism.*

Proof:
    IDENTITY: $g(1) = 1$. Immediate since each value x gives $f(1 \times x) \times f(x)^{-1} = 1$.
    INVERSE: $g(a^{-1}) = g(a)^{-1}$. There is a one-one correspondence between x satisfying $f(a \times x) \times f(x)^{-1} = g(a)$ and y satisfying $f(a^{-1} \times y) \times f(y)^{-1} = g(a)^{-1}$, namely $y = a \times x$.
    PRODUCT: $g(a) \times g(b) = g(a \times b)$. Each of the following three equations holds with probability at least $1 - \epsilon > 2/3$ on random choice of y:

$$g(a) = f(a \times y) \times f(y)^{-1}$$

$$g(b) = f(y) \times f(b^{-1} \times y)^{-1}$$

$$g(a \times b) = f(a \times y) \times f(b^{-1} \times y)^{-1}$$

(In the definition of g, we substitute $y = x$ in the first equation, and $y = b \times x$ in the second and third.) By the union bound, all three equations hold simultaneously with probability at least $1 - 3\epsilon > 0$; that is, there is at least one value of y satisfying all three equations. Substituting one such value of y and combining these three equations, we conclude $g(a) \times g(b) = g(a \times b)$, as desired. $\square$

Lemma 3. $\tau \leq \delta + \epsilon$.

Proof:

$$\begin{aligned}
\tau &= \mathrm{Pr}_a \left[f(a) \neq g(a) \right] \\
&\leq \mathrm{Pr}_{a,x} \left[f(a) \neq f(a \times x) \times f(x)^{-1} \right] + \mathrm{Pr}_{a,x} \left[g(a) \neq f(a \times x) \times f(x)^{-1} \right] \\
&\leq \delta + Average_a(\epsilon_a) \leq \delta + \epsilon.
\end{aligned}$$

$\square$

Lemma 4. $\epsilon \geq 2(\tau - \tau^2)$.

Proof:

$$\mathop{\mathrm{Pr}}_{x,y} \left[f(x) \times f(y)^{-1} \neq g(x \times y^{-1}) \right]$$

is the average value of ϵ_a (over random choices of a), and so is bounded by ϵ. This group law failure will hold at least if either of these two mutually exclusive events occurs, since g is a homomorphism:

- $f(x) = g(x) \wedge f(y) \neq g(y)$;
- $f(x) \neq g(x) \wedge f(y) = g(y)$.

By the independence of x, y, each of the two events has probability $\tau(1 - \tau)$. So

$$\epsilon \geq \tau(1 - \tau) + (1 - \tau)\tau = 2(\tau - \tau^2).$$

$\square$

Corollary 2. If $\delta < 2/9$ then $\tau < \frac{3-\sqrt{3}}{6} < 0.2114$.

Proof: If $\delta < 2/9$ then $\epsilon < 1/3$, and Lemma 4 implies either $\tau < \frac{3-\sqrt{3}}{6} < 0.2114$ or $\tau > \frac{3+\sqrt{3}}{6} > 0.7886$. The latter is inconsistent with $\tau \leq \delta + \epsilon < 5/9$ (Lemma 3). $\square$

Lemma 5. If $\delta < 2/9$ then $\delta \geq 3\tau - 6\tau^2$.

Proof: Since g is a homomorphism, the inequality $f(x) \times f(y) \neq f(x \times y)$ (which has probability δ) will hold in at least the following three mutually exclusive events:

- $f(x) = g(x) \wedge f(y) = g(y) \wedge f(x \times y) \neq g(x \times y)$;
- $f(x) = g(x) \wedge f(y) \neq g(y) \wedge f(x \times y) = g(x \times y)$;
- $f(x) \neq g(x) \wedge f(y) = g(y) \wedge f(x \times y) = g(x \times y)$.

$\square$

Lemma 6. If $\delta < 2/9$ then τ is bounded by the smaller root of $3x - 6x^2 = \delta$.

Proof: Combine Corollary 2 ($\tau < 0.2114$) with Lemma 5 ($\delta \geq 3\tau - 6\tau^2$). $\square$
This finishes the proof of Theorem 1.

Example 1. The bound $\delta < 2/9$ is tight. The following example has $\delta = 2/9$ and $\epsilon = 1/3$, but $\tau = 1 - 1/3^{k-1}$ is arbitrarily close to 1.

$$f : \mathbf{Z}/3^k \to \mathbf{Z}/3^{k-1}$$

$$f(3\ell + d) = \ell, 0 \leq \ell < 3^{k-1}, d \in \{-1, 0, 1\}$$

More details are given in the full version [2].

Example 2. The bound $3\tau - 6\tau^2 \leq \delta$ is tight. Choose τ' with $0 < \tau' \leq 1/3$, choose N an arbitrarily large odd positive integer, and define $f : \mathbf{Z}/N \to \mathbf{Z}/2$ by

$$f(x) = 1 \Leftrightarrow \tau' < \frac{x}{N} < 2\tau'.$$

More details are given in the full version [2].

3 Convolutions of Distributions

In this section, we show that for a distribution A over a finite group G, if $|A - A * A| \leq \epsilon$ then A is δ-close to the uniform distribution over a subgroup of G.

We let capital letters A, B, C denote distributions over group G and subscripted uncapitalized letters a_x, b_y denote the probability of a particular element. X, Y, Z, H will be subsets of G.

We let U_S denote the uniform distribution on $S \subseteq G$.

We let $dist(A, B) = \frac{1}{2}|A - B|$. Note that distances satisfy the triangle inequality, i.e., $dist(A, C) \leq dist(A, B) + dist(B, C)$. Also it is easy to see that $dist(A * B, A * C) \leq dist(B, C)$.

It will also be convenient to consider a second kind of convolution,

$$C = A \bullet B, \; c_x = \sum_{y, z \in G; \, xy = z} a_y b_z.$$

When we have uniform distributions on subsets of equal size, the two convolutions enjoy the following relation, the proof of which is given in the full version [2]:

Lemma 7. *Let X, Y, Z be subsets of a finite group G, with $|X| = |Y| = |Z| = n$. Then*

$$dist(U_X, U_Y * U_Z) = dist(U_Y, U_Z \bullet U_X).$$

Remark 1. The lemma does not hold for arbitrary distributions, nor for uniform distributions on subsets of different sizes.

Overview of Proof: We will embed G in a larger group $F = G \times \mathbf{Z}/N$ for suitably large N, and consider a distribution B induced from A, namely $b_{(x,j)} = a_x/N$. This will alleviate problems later when we have to round to integers. We show that if $B' = B * B$ is close to B, then there is a set $X \subseteq F$ such that B is close to U_X. We next show that X must be close to a subgroup $\hat{H}$ of F, and further that this subgroup is of the form $\hat{H} = H \times \mathbf{Z}/N$. Then B is close to the uniform distribution on $\hat{H}$, and A is close to the uniform distribution on H. A bootstrap lemma allows us to claim that once A is moderately close to U_H, then it is very close.

Expanding the Group: Pick N suitably large. Define $F = G \times \mathbf{Z}/N$, with elements $\{(x, j) : x \in G, j \in \mathbf{Z}/N\}$ and group law $(x, j)(y, k) = (xy, j + k)$. The distribution B on F is given by $A \times U_{\mathbf{Z}/n}$, that is, $b_{(x,j)} = a_x(1/N)$. Defining $B' = B * B$ and $A' = A * A$, it is immediate that $dist(B, B') = dist(A, A')$.

B is Close to Uniform on a Subset: Our first theorem shows that if $B' = B * B$ is close to B, then there is a set $X \subseteq F$ such that B is close to U_X.

Theorem 2. *Let F be a finite group. Let B be a distribution on F for which no element has probability more than $1/N$. Let $1/8 > \epsilon > 0$ be a constant. If $dist(B, B * B) \leq \epsilon$ then there is a set $X \subseteq G$ such that $dist(B, U_X) \leq \epsilon'$ where $\epsilon' = 3\epsilon + O(1/N)$. Further, $dist(U_X, U_X * U_X) \leq 6\epsilon + O(\epsilon^2) + O(1/N)$.*

Proof: Let $B' = B * B$. In the rest of the proof, relabel the elements such that $b_1 \geq b_2 \geq b_3 \geq \ldots$, i.e., 1 corresponds to the element of F with the highest probability mass. For given $x \in G$, the N elements $b_{(x,k)}, k \in \mathbf{Z}/N$, are equal, so we arrange that they are contiguous in this ordering.

For $n \geq 1$, let $\mathrm{sum}_n = \sum_{j=1}^{n} b_j$ be the sum of the n highest probabilities. Also, let $\mathrm{sum}'_n = \sum_{j=1}^{n} b'_j$ be the sum of the probabilities with respect to B' of the n most likely elements with respect to B.

Let $\alpha_n = \mathrm{sum}_n^2 + n \sum_{j>n} b_j^2$. It is not hard to see that $\mathrm{sum}_n \geq \alpha_n$.

Claim. $\alpha_n \geq \mathrm{sum}'_n$.

Proof: [of claim] Equivalently,

$$(\sum_1^n b_j)^2 + n \sum_{j>n} b_j^2 \geq \sum_j = 1^n \sum_{xy=z_j} b_x b_y.$$

The right-hand side can be converted to the left-hand side by a carefully selected sequence of "edge swaps", replacing $b_k b_l + b_{l'} b_{k'}$ by $b_k b_{k'} + b_{l'} b_l$ when $k < l'$ and $k' < l$. Each edge swap does not decrease the sum. Details in the full version [2]. □

We will define an X such that $dist(B, U_X)$ is small. Pick τ with $1/4 \leq \tau \leq 3/4$; later we will specify $\tau = 3/5$. Select m with $\mathrm{sum}_{m-1} < \tau \leq \mathrm{sum}_m$. Set $h = b_m$. Set $p = \lfloor 1/h \rfloor$. Let the distribution $\hat{U}$ assign weight h to the first p elements, and a weight $1 - ph < h$ to the $(p + 1)$st element. Let $n = p$ if $b_{p+1} > \hat{u}_{p+1}$, and $n = p + 1$ otherwise. Let X consist of the first n elements, so that $dist(\hat{U}, U_X) < h = O(1/N)$. Also define $g = b_n$.

The distribution B differs from $\hat{U}$ in three places. For $i \leq m$, $b_i \geq \hat{u}_i$; define $\beta = \sum_{i \leq m}(b_i - \hat{u}_i)$. For $i > n$, $b_i \geq \hat{u}_i$; define $\delta = \sum_{i>n}(b_i - \hat{u}_i)$. For $m < i \leq n$, $b_i \leq \hat{u}_i$; we have $\beta + \delta = \sum_{m<i\leq n}(\hat{u}_i - b_i)$. So $dist(B, \hat{U}) = \beta + \delta$.

For $m < i \leq n$ we have $g \leq b_i \leq h$, so that $b_i^2 \leq (g + h)b_i - gh$. Similarly for $i > n$ we have $0 \leq b_i \leq g$, so that $b_i^2 \leq gb_i$. Recall also that $mh + \beta = \tau + O(1/N)$ and $nh = 1 + O(1/N)$. This enables the following computation:

$$\epsilon \geq \text{sum}_m - \text{sum}'_m \geq \text{sum}_m - \alpha_m$$
$$= (mh + \beta) - (mh + \beta)^2 - m \sum_{m+1}^n b_i^2 - m \sum_{i>n} b_i^2$$
$$\geq (mh + \beta)(1 - mh - \beta) - m(g + h) \sum_{m+1}^n b_i + m(n - m)gh - mg \sum_{i>n} b_i$$
$$= (mh + \beta)(1 - mh - \beta) - m(g + h)(nh - mh - \beta - \delta)$$
$$\quad + mngh - m^2gh - mg\delta$$
$$= (1 - mh - \beta)(\beta) - (1 - mh - \beta)(mg) + mh\delta + mg - m^2gh + O(1/N)$$
$$= (1 - \tau)(\beta) + mh\delta + mg\beta + O(1/N)$$
$$\geq (1 - \tau)\beta + \tau\delta - \beta\delta + O(1/N)$$
$$= [(1 - \tau)\beta + \tau\delta] + \frac{1}{4\tau(1-\tau)}\{[(1-\tau)\beta - \tau\delta]^2 - [(1-\tau)\beta + \tau\delta]^2\} + O(1/N)$$
$$\geq [(1 - \tau)\beta + \tau\delta] - \frac{1}{4\tau(1-\tau)}[(1-\tau)\beta + \tau\delta]^2 + O(1/N)$$
$$= u - \frac{u^2}{4\tau(1-\tau)} + O(1/N)$$

where $u = (1 - \tau)\beta + \tau\delta$. Now $\beta \leq \tau + O(1/N)$ and $\delta \leq 1 - \tau$ so that $u = (1 - \tau)\beta + \tau\delta \leq 2\tau(1 - \tau)$ is less than the larger root of

$$\frac{x^2}{4\tau(1 - \tau)} - x + \epsilon = 0,$$

namely

$$[(1 - \tau)\beta + \tau\delta] \leq 2\tau(1 - \tau) + O(1/N) \leq 2\tau(1 - \tau)[1 + \sqrt{1 - \epsilon/\tau(1 - \tau)}]$$

(since $\epsilon/\tau(1 - \tau) < (1/8)/(1/4 \times 3/4) < 1$), so it must be less than the smaller root:

$$[(1 - \tau)\beta + \tau\delta] \leq 2\tau(1 - \tau)[1 - \sqrt{1 - \epsilon/\tau(1 - \tau)}] + O(1/N),$$

remembering the error term.

Substituting $\tau = 3/5$, we have

$$2\beta + 3\delta \leq \frac{12}{5}\left[1 - \sqrt{1 - \frac{25}{6}\epsilon}\right] + O(1/N) = 5\epsilon + O(\epsilon^2) + O(1/N).$$

By the triangle inequality,

$$dist(B, U_X) \leq dist(B, \hat{U}) + dist(\hat{U}, U_X) = \beta + \delta + O(1/N).$$

One can calculate that if $\epsilon < 1/8$ then $\beta + \delta < 3\epsilon$, establishing the theorem. For later use, we also note that if $\epsilon < 0.0182$ then $\beta + \delta < 2.6\epsilon$.

Repeatedly applying the triangle inequality, we can obtain:

$$dist(B, U_X * U_X) \leq dist(B, B * B) + dist(B, B * U_X) + dist(B, U_X * U_X)$$
$$\leq \epsilon + 2(\beta + \delta).$$

To obtain $dist(U_X, U_X * U_X)$ we could apply the triangle inequality again. Instead, we recall that B differs from $\hat{U}$, and hence U_X, in three pieces, namely (up to errors of order $O(1/N)$) β (before m), $-\beta - \delta$ (between m and n), and δ (after n). The convolution $U_X * U_X$ is everywhere bounded by $n(1/n)^2 = 1/n$. So when we change B to U_X and monitor the change in $dist(*, U_X * U_X)$, the first change of β is driving us closer to $U_X * U_X$, while the other two changes of $\beta + \delta$ and δ might drive us further away. The net result is

$$dist(U_X, U_X * U_X) \leq dist(B, U_X * U_X) + \tfrac{1}{2}[-\beta + (\beta + \delta) + \delta] + O(1/N)$$
$$\leq \epsilon + 2\beta + 3\delta + O(1/N)$$
$$\leq \epsilon + \tfrac{12}{5}\left[1 - \sqrt{1 - \tfrac{25}{6}\epsilon}\right] + O(1/N)$$
$$= 6\epsilon + O(\epsilon^2) + O(1/N).$$

This establishes the second part of the theorem.$\square$

B Is Close to Uniform on a Subgroup of F: Next we show that if the uniform distribution on X is close in distance to its convolution with itself, then X is close to some subgroup $\hat{H}$ of F.

Theorem 3. *Let F be a finite group and X a subset of F. Let $\tau = dist(U_X, U_X * U_X) = dist(U_X, U_X \bullet U_X)$. If $\tau < 1/9$ then there is a subgroup $\hat{H}$ of F with $|X \backslash \hat{H}| + |\hat{H} \backslash X| \leq 3\tau|X|$.*

Proof: Let $n = |X|$. Let $V = U_X \bullet U_X$ so that

$$v_x = \frac{1}{n^2}|\{(y, z) : y, z \in X, xy = z\}|.$$

If e is the identity element of F, we see $v_e = \frac{1}{n}$, and $v_x \leq \frac{1}{n}$ for all $x \in G$.

We need to establish a triangle inequality on quantities such as $(v_e - v_w)$.

Lemma 8. *For $x, y \in F$, the quantities $(v_e - v_x)$, $(v_e - v_y)$, $(v_e - v_{xy})$ are nonnegative and satisfy the triangle inequalities:*

$$(v_e - v_x) + (v_e - v_y) \geq (v_e - v_{xy}) \tag{1}$$
$$(v_e - v_x) + (v_e - v_{xy}) \geq (v_e - v_y) \tag{2}$$
$$(v_e - v_y) + (v_e - v_{xy}) \geq (v_e - v_x) \tag{3}$$

Proof: If $v_x = k_x/n^2$, with $0 \leq k_x \leq n$, then there are k_x elements $z \in F$ such that both z and xz are in X; call such z "good elements" for x. There are $n - k_x$ elements z such that $z \in X$ and $xz \notin X$; there are $n - k_x$ elements z such that $z \notin X$ and $xz \in X$; call the latter two kinds of z "bad elements" for x. The number of bad elements for x is $2(n - k_x) = 2n^2(v_e - v_x)$. If z is neither good nor bad for x it is "neutral" for x.

For each $z \in F$, consider the three elements z, yz, xyz. If all three are in X, then we have found good elements for each of x, y, xy. (Namely, z is a good element for y and for xy, and yz is a good element for x.) If exactly two are in X, then we have found bad elements for exactly two of x, y, xy, and a good element for the other. (For example, if $z, xyz \in X$ and $yz \notin X$, then z is good for xy, z is bad for y, and yz is bad for x.) If exactly one of z, yz, xyz is in X, then we have bad elements for exactly two of x, y, xy and a neutral one for the other. If $z, yz, xyz \notin X$ then we have found neutral elements for all three. The important point is that the "bad elements" come in pairs, each time contributing to two of $2(n - k_x), 2(n - k_y), 2(n - k_{xy})$. Setting

$$p = |\{z : z \text{ bad for } xy \text{ and } y\}|$$
$$q = |\{z : z \text{ bad for } xy; yz \text{ bad for } x\}|$$
$$r = |\{z : z \text{ bad for } y; yz \text{ bad for } x\}|,$$

we find

$$2(n - k_x) = q + r$$
$$2(n - k_y) = p + r$$
$$2(n - k_{xy}) = p + q.$$

This establishes the triangle inequality among $2(n - k_x), 2(n - k_y), 2(n - k_{xy})$, and hence the lemma. $\square$

Next we show an "excluded middle" result for V.

Lemma 9. *Let* $\tau < 1/9$. *For all* $x \in F$, *either* $v_x \leq 3\frac{\tau}{n} < \frac{1}{3n}$ *or* $v_x \geq \frac{1-3\tau}{n} > \frac{2}{3n}$.

Proof: Assume the contrary: for some $x \in F$, $\frac{3\tau}{n} < v_x < \frac{1-3\tau}{n}$. Choose any $y \in X$. If $xy \in X$, use triangle inequality (3) to deduce

$$(v_e - v_y) + (v_e - v_{xy}) \geq (v_e - v_x)$$
$$|(U_X)_y - v_y| + |(U_X)_{xy} - v_{xy}| \geq (v_e - v_x) > \frac{1}{n} - (\frac{1}{n} - \frac{3\tau}{n}) = \frac{3\tau}{n}.$$

If $xy \notin X$, use triangle inequality (1) and $(U_X)_{xy} = 0$ to deduce

$$(v_e - v_x) + (v_e - v_y) \geq (v_e - v_{xy})$$
$$(v_e - v_y) + v_{xy} \geq v_x$$
$$|(U_X)_y - v_y| + |(U_X)_{xy} - v_{xy}| \geq v_x > \frac{3\tau}{n}.$$

Summing over $y \in X$,

$$\sum_{y \in X} [|(U_X)_y - v_y| + |(U_X)_{xy} - v_{xy}|] > n \left(\frac{3\tau}{n} \right) = 3\tau.$$

Then use

$$dist(U_X, V) = \sum_{y \in X} |(U_X)_y - v_y|$$

and

$$2dist(U_X, V) = \sum_{z \in F} |(U_X)_z - v_z| \geq \sum_{y \in X} |(U_X)_{xy} - v_{xy}|$$

to deduce

$$(1 + 2)dist(U_X, V) > 3\tau$$
$$3\tau = 3dist(U_X, V) > 3\tau,$$

contradicting our hypothesis and establishing the lemma. $\square$

The excluded middle gives us a natural candidate for our subgroup $\hat{H}$. The following is proved in the full version [2].

Lemma 10. *Let* $\tau < 1/9$. *Define* $\hat{H} = \{x \in F : v_x > \frac{2}{3n}\}$. *Then* $\hat{H}$ *is a subgroup of* F.

Finally we show that $\hat{H}$ is close to X. The proof is given in the full version [2].

Lemma 11. *With* $\tau < 1/9$ *and* $\hat{H}$ *as above, the symmetric difference between* X *and* $\hat{H}$ *satisfies:*

$$|\hat{H} \backslash X| + |X \backslash \hat{H}| \leq \frac{2\tau}{1 - 3\tau} |X| \leq 3\tau |X|.$$

This establishes the theorem. $\square$

B Is Close to a Subgroup $\hat{H}$: We can push these results back to the distributions B:

Theorem 4. *Let F be a finite group. Let B be a distribution on F. Let $X \subseteq F$ be a subset. Given ϵ, β, δ such that:*

- *$dist(B, B * B) \leq \epsilon$;*
- *$dist(B, U_X) \leq \beta + \delta$ (as in Theorem 2);*
- *$\tau = \epsilon + 2\beta + 3\delta < 1/9$,*

then there is a subgroup $\hat{H}$ of F with $dist(B, U_{\hat{H}}) < \beta + \delta + 2\tau/(1 - 3\tau)$. The same is true if we replace the first condition with

$$dist(B, B \bullet B) \leq \epsilon.$$

The proof uses Theorems 2 and 3, and is given in the full version [2].

Reverting to Original Group: The group $\hat{H}$ respects the block structure of $F = G \times \mathbf{Z}/N$, in the sense that for all $x \neq e \in G$ and $j, k \in \mathbf{Z}/N$, $v_{(x,j)} = v_{(x,k)}$, with v as defined in Lemma 8, so that $(x, j) \in \hat{H} \Leftrightarrow (x, k) \in \hat{H}$. Further, one can verify that $v_{(e,k)} \geq v_{(x,k)}$, so that if any $(x, k) \in \hat{H}$ with $x \neq e$, then the entire block $\{(e, k) : k \in \mathbf{Z}/N\}$ is in $\hat{H}$. (Note that if it were the case that there were no $(x, k) \in \hat{H}$ with $x \neq e$, then $H = \{e\}$, which is a subgroup of G.) This implies that $\hat{H}$ is of the form

$$\hat{H} = H \times \mathbf{Z}/N.$$

It is obvious that H is a subgroup of G, and that

$$dist(A, U_H) = dist(B, U_{\hat{H}}).$$

We tie in with Theorem 2.

Theorem 5. *Let A be a distribution on the finite group G. Let $dist(A, A * A) \leq \epsilon \leq 0.0182$. Then there is a subgroup $H \subseteq G$ with $dist(A, U_H) \leq 21\epsilon$.*

Proof: Pass to B and F, with

$$dist(B, B * B) \leq \epsilon.$$

A computation shows that with $\epsilon < 0.0182$, we have

$$2\beta + 3\delta \leq \tfrac{12}{5}\left[1 - \sqrt{1 - \tfrac{25}{6}\epsilon}\right] + O(1/N) \leq 5.1\epsilon$$
$$\beta + \delta \leq 2.6\epsilon.$$

From Theorem 2 we have a subset X with

$$dist(B, U_X) \leq \beta + \delta \leq 2.6\epsilon.$$

Then apply Theorem 4 with $\tau = \epsilon + 2\beta + 3\delta \leq 6.1\epsilon < 1/9$ to find the subgroup $\hat{H}$, and use the triangle inequality:

$$dist(B, U_{\hat{H}}) \leq dist(B, U_X) + dist(U_X, U_{\hat{H}}) \leq \beta + \delta + \frac{2\tau}{1 - 3\tau} < 2.6\epsilon + 3\tau \leq 21\epsilon.$$

Reverting to the original distribution,

$$dist(A, U_H) = dist(B, U_H) \leq 21\epsilon.$$

$\square$

Once we have bounds on $dist(A, A * A)$ (or $dist(A, A \bullet A)$) and a subgroup H with small $dist(A, U_H)$, we can improve the numerical estimates of $dist(A, U_H)$. The following is proved in the full version of the paper [2]:

Theorem 6. *Given a distribution A on G and a subgroup $H \subseteq G$ with*

$$dist(A, A * A) = \epsilon \leq 0.06$$

$$dist(A, U_H) = \rho \leq 0.4$$

then we can conclude

$$dist(A, U_H) \leq 5\epsilon.$$

Combining the last two results, we have:

Theorem 7. *Let A be a distribution on the finite group G. Let $dist(A, A * A) \leq \epsilon \leq 0.0182$. Then there is a subgroup $H \subseteq G$ with $dist(A, U_H) \leq 5\epsilon$.*

Proof: From Theorem 5 we have such a subgroup H with $dist(A, U_H) \leq 21\epsilon \leq 0.3822 < 0.4$. Since $0.0182 < 0.06$, Theorem 6 applies, giving $dist(A, U_H) \leq 5\epsilon$. $\square$

Example 3. Let $G = \mathbf{Z}/N$ with N a large prime integer. Let $a_n = \nu(240 - |n|)$ when $-200 < n < -1$ or $1 \leq n \leq 200$ and $a_n = 0$ otherwise, where $\nu = 1/55800$ is chosen to normalize A. Then $dist(A, A * A) \approx 0.1539$, but $dist(A, U_H) = 1 - O(1/N)$ for any subgroup H of $\mathbf{Z}/N$.

A gap remains.

Remark 2. The two notions explored in this paper (homomorphism testing, and distributions close to their self-convolution) are related. Given a map (not necessarily a homomorphism) $f : G \to H$ between two finite groups, we can construct the product group

$$G \times H = \{(x, y) : x \in G, y \in H\}$$

and a distribution A:

$$a_{(x,y)} = \begin{cases} 1/|G| & \text{if } y = f(x) \\ 0 & \text{otherwise} \end{cases}$$

Then we have the identity

$$dist(A, A * A) = \Pr_{x,y}[f(x) \times f(y) \neq f(x \times y)].$$

If f is close to a homomorphism g, then A is close to the uniform distribution on the subgroup $\{(x, g(x)) : x \in G\}$. Conversely, if f is likely to pass the homomorphism test, or equivalently, if A is such that $dist(A, A * A)$ is small enough, then f must be close to a homomorphism: The theorem from this section implies that A must be close to a distribution that is uniform over a subgroup S of $G \times H$. One can show that S is such that for each $a \in G$, $|S \cap \{(a,b) : b \in H\}| = 1$. Then S can be used to define a map $g : G \to H$ such that g is a homomorphism (this can be seen from the underlying group structure of S) and such that g agrees with f on most inputs.

But the correspondence when $dist(A, A * A)$ is larger is not exact: the map f given in Example 1 is not at all close to any homomorphism g on all of G, but there is a subgroup $\tilde{H}$ of $G \times H$ with $dist(A, U_{\tilde{H}}) = 2/3$, namely $\tilde{H} = \{(3\ell, \ell) : 0 \le \ell < 3^{k-1}\}$. The difference comes because g is required to be a homomorphism on all of G. We could relax the requirement, and notice that there is a large subgroup G' of G (namely $G' = \{3\ell\}$, with $|G'| = |G|/3$) and a homomorphism $g' : G' \to H$ that agrees with f on this subgroup. The subgroup $\tilde{H}$ of $G \times H$ is associated with G' and g'.

References

1. M. Bellare, D. Coppersmith, J. Hastad, M. Kiwi and M. Sudan, "Linearity Testing in Characteristic Two." FOCS 1995. *IEEE Transactions on Information Theory*, **Volume 42**, Number 6, 1782-1795, November 1996.
2. M. Ben Or, D. Coppersmith, M. Luby, R. Rubinfeld, "Non-Abelian Homomorphism Testing, and Distributions Close to their Self-Convolutions", ECCC Report Number TR04-052. http://eccc.uni-trier.de/eccc-reports/2004/TR04-052/index.html.
3. E. Ben Sasson, M. Sudan, S. Vadhan, A. Wigderson, "Randomness-efficient Low degree tests and short PCP's via Epsilon-biased sets", *In proceedings of the 35th STOC*, pp. 612-621, 2003.
4. M. Blum, M. Luby and R. Rubinfeld, "Self-Testing/Correcting with Applications to Numerical Problems," *J. Comp. Sys. Sci.* **Vol. 47**, No. 3, December 1993 (special issue on STOC 1990). Preliminary abstract appears in Proc. 22th ACM Symposium on Theory of Computing, 1990.
5. D. Coppersmith, Personal communication to the authors of [4]. December 1989.
6. J. Hastad, A. Wigderson, "Simple Analysis of Graph Tests for Linearity and PCP", to appear in *Random Structures and Algorithms*.
7. A. Samorodnitsky, L. Trevisan, "A PCP characterization of NP with optimal amortized query complexity", *In proceedings of 32nd STOC*, pp. 191-199, 2000.
8. A. Shpilka, A. Wigderson, "Derandomizing Homomorphism Testing in General Groups". To appear in *STOC 2004*.

Robust Locally Testable Codes
and Products of Codes

Eli Ben-Sasson[1] and Madhu Sudan[2]

[1] Radcliffe Institute for Advanced Study
34 Concord Avenue, Cambridge, MA 02138, USA
eli@eecs.harvard.edu
[2] MIT and Radcliffe IAS, The Stata Center Rm. G640
32 Vassar Street, Cambridge, MA 02139, USA
madhu@mit.edu

Abstract. We continue the investigation of locally testable codes, i.e., error-correcting codes for whom membership of a given word in the code can be tested probabilistically by examining it in very few locations. We give two general results on local testability: First, motivated by the recently proposed notion of *robust* probabilistically checkable proofs, we introduce the notion of *robust* local testability of codes. We relate this notion to a product of codes introduced by Tanner, and show a very simple composition lemma for this notion. Next, we show that codes built by tensor products can be tested robustly and somewhat locally, by applying a variant of a test and proof technique introduced by Raz and Safra in the context of testing low-degree multivariate polynomials (which are a special case of tensor codes).

Combining these two results gives us a generic construction of codes of inverse polynomial rate, that are testable with poly-logarithmically many queries. We note these locally testable tensor codes can be obtained from *any* linear error correcting code with good distance. Previous results on local testability, albeit much stronger quantitatively, rely heavily on algebraic properties of the underlying codes.

1 Introduction

Locally testable codes (LTCs) are error-correcting codes that admit highly efficient probabilistic tests of membership. Specifically, an LTC has a tester that makes a small number of oracle accesses into an oracle representing a given word w, accepts if w is a codeword, and rejects with constant probability if w is far from every codeword. LTCs are combinatorial counterparts of probabilistically checkable proofs (PCPs), and were defined in [18, 24, 2], and their study was revived in [20].

Constructions of locally testable codes typically come in two stages. The first stage is algebraic and gives local tests for algebraic codes, usually based on multivariate polynomials. This is based on a rich collection of results on "linearity testing" or "low-degree testing" [1, 3–9, 13, 14, 16–18, 20, 22, 24]. This first stage

K. Jansen et al. (Eds.): APPROX and RANDOM 2004, LNCS 3122, pp. 286–297, 2004.

either yielded codes of poor rate (mapping k information symbols to codewords of length $\exp(k)$) as in [14], or yielded codes over large alphabets as in [24]. To reduce the alphabet size, a second stage of "composition" is then applied. In particular, this is done in [20, 13, 11] to get code mapping k information bits to codewords of length $k^{1+o(1)}$, over the *binary alphabet*. This composition follows the lines of PCP composition introduced in [4], but turns out to be fairly complicated, and in most cases, even more intricate than PCP composition. The one exception is in [20, Section 3], where the composition is simple, but based on very specific properties of the codes used. Thus while the resulting constructions are surprisingly strong, the proof techniques are somewhat complex.

In this paper, we search for simple and general results related to local testing. A generic (non-algebraic) analysis of low-degree tests appears in [19], and a similar approach to PCPs appears in [15]. Specifically, we search for generic (non-algebraic) ways of getting codes, possibly over large alphabets, that can be tested by relatively local tests, as a substitute for algebraic ways. And we look for simpler composition lemmas. We make some progress in both directions. We show that the "tensor product" operation, a classical operation that takes two codes and produces a new one, when applied to linear codes gives codes that are somewhat locally testable (See Theorem 1). To simplify the second stage, we strengthen the notion of local testability to a "robust" one. This step is motivated by an analogous step taken for PCPs in [11], but is naturally formulated in our case using the "Tanner Product" for codes [20]. Roughly speaking, a "big" Tanner Product code of block-length n is defined by a "small" code of block-length $n' = o(n)$ and a collection of subsets $S_1, \ldots, S_m \subset [n]$, each of size n'. A word is in the big code if and only if its projection to every subset S_i is a word of the small code. Tanner Product codes have a natural local test associated with them: to test if a word w is a codeword of the big code, pick a random subset S_j and verify that w restricted to S_j is a codeword of the small code. The normal soundness condition would expect that if w is far from every codeword, then for a constant fraction of such restrictions, w restricted to S_j is not a codeword of the small code. Now the notion of robust soundness strengthens this condition further by expecting that if w is far from every codeword, then many (or most) projections actually lead to words that are *far* from codewords of the small code. In other words, a code is robust if global distance (from the large code) translates into (average) local distance (from the small code). A simple, yet crucial observation is that robust codes compose naturally. Namely, if the small code is itself locally testable by a robust test (with respect to a tiny code, of block-length $o(n')$), then distance from the large code (of block-length n) translates to distance from the tiny code, thus reducing query complexity while maintaining soundness. By viewing a tensor product as a robust Tanner product code, we show that a $(\log N / \log \log N)$-wise tensor product of *any* linear code of length $n = \operatorname{poly} \log N$ and relative distance $1 - \frac{1}{\log N} = 1 - \frac{1}{n^c}$, which yields a code of length N and polynomial rate, is testable with $\operatorname{poly}(\log N)$ queries (Theorem 2). Once again, while stronger theorems than the above have been known since [6], the generic nature of the result above might shed further light on the notion of local testability.

Organization. We give formal definitions and mention our main theorems in Section 2. In Section 3 we analyze the basic tester for tensor product codes. Finally in Section 4 we describe our composition and analyze some tests based on our composition lemma.

2   Definitions and Main Results

Throughout this paper Σ will denote a finite alphabet, and in fact a finite field. For positive integer n, let $[n]$ denote the set $\{1, \ldots, n\}$. For a sequence $x \in \Sigma^n$ and $i \in [n]$, we will let x_i denote the ith element of the sequence. The Hamming distance between strings $x, y \in \Sigma^n$, denoted $\Delta(x, y)$, is the number of $i \in [n]$ such that $x_i \neq y_i$. The relative distance between $x, y \in \Sigma^n$, denoted $\delta(x, y)$, is the ratio $\Delta(x, y)/n$.

A code C of length n over Σ is a subset of Σ^n. Elements of C are referred to as codewords. When Σ is a field, one may think of Σ^n as a vector space. If C is a linear subspace of the vector space Σ^n, then C is called a linear code. The crucial parameters of a code, in addition to its length and the alphabet, are its dimension (or information length) and its distance, given by $\Delta(C) = \min_{x \neq y \in C} \{\Delta(x, y)\}$. A linear code of dimension k, length n, distance d over the alphabet Σ is denoted an $[n, k, d]_\Sigma$ code. For a word $r \in \Sigma^n$ and a code C, we let $\delta_C(r) = \min_{x \in C} \{\delta(r, x)\}$. We say r is δ'-proximate to C (δ'-far from C, respectively) if $\delta_C(r) \geq \delta'$ ($\delta_C(r) \geq \delta'$, respectively).

Throughout this paper, we will be working with infinite families of codes, where their performance will be measured as a function of their length.

Definition 1 (Tester). *A tester T with query complexity $q(\cdot)$ is a probabilistic oracle machine that when given oracle access to a string $r \in \Sigma^n$, makes $q(n)$ queries to the oracle for r and returns an accept/reject verdict. We say that T tests a code C if whenever $r \in C$, T accepts with probability one; and when $r \notin C$, the tester rejects with probability at least $\delta_C(r)/2$. A code C is said to be locally testable with $q(n)$ queries if there is a tester for C with query complexity $q(n)$.*

When referring to oracles representing vectors in Σ^n, we emphasize the queries by denoting the response of the ith query by $r[i]$, as opposed to r_i. Through this paper we consider only non-adaptive testers, i.e., testers that use their internal randomness R to generate q queries $i_1, \ldots, i_q \in [n]$ and a predicate $P : \Sigma^q \to \{0, 1\}$ and accept iff $P(r[i_1], \ldots, r[i_q]) = 1$.

Our next definition is based on the notion of Robust PCP verifiers introduced by [11]. We need some terminology first.

Note that a tester T has two inputs: an oracle for a received vector r, and a random string s. On input the string s the tester generates queries $i_1, \ldots, i_q \in [n]$ and fixes circuit $C = C_s$ and accepts if $C(r[i_1], \ldots, r[i_q]) = 1$. For oracle r and random string s, define the robustness of the tester T on r, s, denoted $\rho^T(r, s)$, to be the minimum, over strings x satisfying $C(x) = 1$, of relative distance of $\langle r[i_1], \ldots, r[i_q] \rangle$ from x. We refer to the quantity $\rho^T(r) \stackrel{\text{def}}{=} \mathbf{E}_s[\rho^T(r, s)]$ as the expected robustness of T on r. When T is clear from context, we skip the superscript.

Definition 2 (Robust Tester). *A tester T is said to be c-robust for a code C if for every $r \in C$, the tester accepts w.p. one, and for every $r \in \Sigma^n$, $\delta_C(r) \leq c \cdot \rho^T(r)$.*

Having a robust tester for a code C implies the existence of a tester for C, as illustrated by the following proposition. It's proof appears in the full version of the paper [12].

Proposition 1. *If a code C has a c-robust tester T for C making q queries, then it is locally testable with $O(c \cdot q)$ queries.*

The main results of this paper focus on robust local testability of certain codes. For the first result, we need to describe the tensor product of codes.

Tensor Products and Local Tests. Recall that an $[n, k, d]_\Sigma$ linear code C may be represented by a $k \times n$ matrix M over Σ (so that $C = \{xM | x \in \Sigma^k\}$). Such a matrix M is called a generator of C. Given an $[n_1, k_1, d_1]_\Sigma$ code C_1 with generator M_1 and an $[n_2, k_2, d_2]_\Sigma$ code C_2 with generator M_2, their tensor product (cf. [21], [25, Lecture 6, Section 2.4]), denoted $C_1 \otimes C_2 \subseteq \Sigma^{n_2 \times n_1}$, is the code whose codewords may be viewed as $n_2 \times n_1$ matrices given explicitly by the set $\{M_2^T X M_1 | X \in \Sigma^{k_2 \times k_1}\}$. It is well-known that $C_1 \otimes C_2$ is an $[n_1 n_2, k_1 k_2, d_1 d_2]_\Sigma$ code.

Tensor product codes are interesting to us in that they are a generic construction of codes with "non-trivially" local redundancy. To elaborate, every linear code of dimension k does have redundancies of size $O(k)$, i.e., there exist subsets of $t = O(k)$ coordinates where the code does not take all possible Σ^t possible values. But such redundancies are not useful for constructing local tests; and unfortunately generic codes of length n and dimension k may not have any redundancies of length $o(k)$. However, tensor product codes are different in that the tensor product of an $[n, k, d]_\Sigma$ code C with itself leads to a code of dimension k^2 which is much larger than the size of redundancies which are $O(k)$-long, as asserted by the following proposition.

Proposition 2. *A matrix $r \in \Sigma^{n_2 \times n_1}$ is a codeword of $C_1 \otimes C_2$ if and only if every row is a codeword of C_1 and every column is a codeword of C_2.*

In addition to being non-trivially local, the constraints enumerated above are also redundant, in that it suffices to insist that all columns are codewords of C_2 and only k_2 (prespecified) rows are codewords of C_1. Thus the insistence that other rows ought to be codewords of C_1 is redundant, and leads to the hope that the tests may be somewhat robust. Indeed we may hope that the following might be a robust test for $C_1 \otimes C_2$.

> **Product Tester:** Pick $b \in \{1, 2\}$ at random and $i \in [n_b]$ at random. Verify that r with bth coordinate restricted to i is a codeword of C_{3-b}.

While it is possible to show that the above is a reasonable tester for $C_1 \otimes C_2$, it remains open if the above is a robust tester for $C_1 \otimes C_2$. (Note that the query complexity of the test is $\max\{n_1, n_2\}$, which is quite high. However if the test were robust, there would be ways of reducing this query complexity in many cases, as we will see later.)

Instead, we consider higher products of codes, and give a tester based on an idea from the work of Raz and Safra [23]. Specifically, we let C^m denote the code $\underbrace{C \otimes \cdots \otimes C}_{m}$. We consider the following test for this code:

> m-**Product Tester:** Pick $b \in [m]$ and $i \in [n]$ independently and uniformly at random. Verify that r with bth coordinate restricted to i is a codeword of C^{m-1}.

Note that this tester makes $N^{1-\frac{1}{m}}$ queries to test a code of length $N = n^m$. So its query complexity gets worse as m increases. However, we are only interested in the performance of the test for small m (specifically $m = 3, 4$). We show that the test is a robust tester for C^m for every $m \geq 3$. Specifically, we show

Theorem 1. *For a positive integer m and $[n, k, d]_\Sigma$-code C, such that $\left(\frac{d-1}{n}\right)^m \geq \frac{7}{8}$, m-Product Tester is 2^{16}-robust for C^m.*

This theorem is proven in Section 3. Note that the robustness is a constant, and the theorem only needs the fractional distance of C to be sufficiently large as a function of m. In particular a fractional distance of $1 - \frac{1}{O(m)}$ suffices. Note that such a restriction is needed even to get the fractional distance of C^m to be constant.

The tester however makes a lot of queries, and this might seem to make this result uninteresting (and indeed one doesn't have to work so hard to get a non-robust tester with such query complexity). However, as we note next, the query complexity of robust testers can be reduced significantly under some circumstances. To describe this we need to revisit a construction of codes introduced by Tanner [26].

Tanner Products and Robust Testing. For integers (n, m, t) an (n, m, t)-ordered bipartite graph is given by n left vertices $[n]$, and m right vertices, where each right vertex has degree t and the neighborhood of a right vertex $j \in [m]$ is ordered and given by a sequence $\ell_j = \langle \ell_{j,1}, \ldots, \ell_{j,t} \rangle$ with $\ell_{j,i} \in [n]$.

A Tanner Product Code (TPC), is specified by an $[n, m, t]$ ordered bipartite graph G and a code $C_{\text{small}} \subseteq \Sigma^t$. The product code, denoted TPC$(G = \{\ell_1, \ldots, \ell_m\}, C_{\text{small}}) \subseteq \Sigma^n$, is the set

$$\{r \in \Sigma^n \mid r|_{\ell_j} \stackrel{\text{def}}{=} \langle r_{\ell_{j,1}}, \ldots, r_{\ell_{j,t}} \rangle \in C_{\text{small}}, \ \forall j \in [m]\}.$$

Notice that the Tanner Product naturally suggests a test for a code. "Pick a random right vertex $j \in [m]$ and verify that $r|_{\ell_j} \in C_{\text{small}}$." Associating this test with such a pair (G, C_{small}), we say that the pair is c-robust if the associated test is a c-robust tester for TPC(G, C_{small}).

The importance of this representation of tests comes from the composability of robust tests coming from Tanner Product Codes. Suppose (G, C_{small}) is

c-robust and C_{small} is itself a Tanner Product Code, $\text{TPC}(G', C_{\text{small}}')$ where G' is an (d, m', t')-ordered bipartite graph and (G', C_{small}') is c'-robust. Then $\text{TPC}(G, C_{\text{small}})$ has an $c \cdot c'$-robust tester that makes only t' queries. (This fact is completely straightforward and proven in Lemma 3.)

This composition is especially useful in the context of tensor product codes. For instance, the tester for C^4 is of the form (G, C^3), while C^3 has a robust tester of the form (G', C^2). Putting them together gives a tester for C^4, where the tests verify appropriate projections are codewords of C^2. The test itself is not surprising, however the ease with which the analysis follows is nice. (See Lemma 4.) Now the generality of the tensor product tester comes in handy as we let C itself be C'^2 to see that we are now testing C'^8 where tests verify some projections are codewords of C'^4. Again composition allows us to reduce this to a C'^2-test. Carrying on this way we see that we can test any code of the form C^{2^t} by verifying certain projections are codewords of C^2. This leads to a simple proof of the following theorem about the testability of tensor product codes.

Theorem 2. *Let $\{C_i\}_i$ be any infinite family of codes with C_i a $[n_i, k_i, d_i]_{\Sigma_i}$ code, with $n_i = p(k_i)$ for some polynomial $p(\cdot)$. Further, let t_i be a sequence of integers such that $m_i = 2^{t_i}$ satisfies $d_i/n_i \geq 1 - \frac{1}{7m_i}$. Then the sequence of codes $\{C_i' = C_i^{m_i}\}_i$ is a sequence of codes of inverse polynomial rate and constant relative distance that is locally testable with polylogarithmic number of queries.*

This theorem is proven in Section 4. We remark that it is possible to get code families C_i such as above using Reed-Solomon codes, as well as algebraic-geometric codes.

3   Testing Tensor Product Codes

Recall that in this section we wish to prove Theorem 1. We first reformulate this theorem in the language of Tanner products.

Let G_m^n denote the graph that corresponds to the tests of C^m by the m-Product Tester, where $C \subseteq \Sigma^n$. Namely G_m^n has n^m left vertices labelled by elements of $[n]^m$. It has $m \cdot n$ right vertices labelled (b, i) with $b \in [m]$ and $i \in [n]$. Vertex (b, i) is adjacent to all vertices $(i_1, \ldots, i_m)$ such that $i_b = i$. The statement of Theorem 1 is equivalent to the statement that (G_m^n, C^{m-1}) is 2^{16}-robust, provided $\left(\frac{d-1}{n}\right)^m \geq \frac{7}{8}$. The completeness of the theorem follows from Proposition 2, which implies $C^m = \text{TPC}(G_m^n, C^{m-1})$. For the soundness, we first introduce some notation.

Consider the code $C_1 \otimes \cdots \otimes C_m$, where $C_i = [n_i, k_i, d_i]_{\Sigma}$ code. Notice that codewords of this code lie in $\Sigma^{n_1 \times \cdots \times n_m}$. The coordinates of strings in $\Sigma^{n_1 \times \cdots \times n_m}$ are themselves m-dimensional vectors over the integers (from $[n_1] \times \cdots \times [n_m]$). For $r \in \Sigma^{n_1 \times \cdots \times n_m}$ and $i_1, \ldots, i_m$ with $i_j \in [n_j]$, let $r[i_1, \ldots, i_m]$ denote the $\langle i_1, \ldots, i_m \rangle$-th coordinate of r. For $b \in [m]$, and $i \in [n_b]$, let $r_{b,i} \in \Sigma^{n_1 \times \cdots \times n_{b-1} \times n_{b+1} \times \cdots \times n_m}$ be the vector obtained by projecting r to coordinates whose bth coordinate is i, i.e., $r_{b,i}[i_1, \ldots, i_{m-1}] = r[i_1, \ldots, i_{b-1}, i, i_b, \ldots, i_{m-1}]$.

The following simple property about tensor product codes will be needed in our proof. It's proof appears in the full version of the paper [12].

Proposition 3. *For $b \in \{1, \ldots, m\}$ let C_b be an $[n_b, k_b, d_b]_\Sigma$ code, and let I_b be a set of cardinality at least $n_b - d_b + 1$. Let C_b' be the code obtained by the projection of C_b to I_b. Then every codeword c' of $C_1' \otimes \cdots \otimes C_m'$ can be extended to a codeword c of $C_1 \otimes \cdots \otimes C_m$.*

Recall that the m-Product tester picks a random $b \in [m]$ and $i \in [n]$ and verifies that $r_{b,i} \in C^{m-1}$. The robustness of this tester for oracle r on random string $s = (b, i)$ is given by $\rho(r, (b, i)) = \delta_{C^{m-1}}(r_{b,i})$, and its expected robustness is given by $\rho(r) = \mathbf{E}_{b,i}[\delta_{C^{m-1}}(r_{b,i})]$. We wish to show for every r that $\delta_{C^m}(r) \leq 2^{16} \cdot \rho(r)$.

We start by first getting a crude upper bound on the proximity of r to C^m and then we use the crude bound to get a tighter relationship. To get the crude bound, we first partition the random strings into two classes: those for which the robustness $\rho(r, (b, i))$ is large, and those for which it is small. More precisely, for $r \in \Sigma^{n^m}$ and a threshold $\tau \in [0, 1]$, define the τ-soundness-error of r to be the probability that $\delta_{C^{m-1}}(r_{b,i}) > \tau$, when $b \in [m]$ and $i \in [n]$ are chosen uniformly and independently. Note that the $\sqrt{\rho}$-soundness error of r is at most $\sqrt{\rho}$ for $\rho = \rho(r)$. We start by showing that r is $O(\tau + \epsilon)$-close (and thus also $O(\sqrt{\rho})$-close) to some codeword of C^m.

Lemma 1. *If the τ-soundness-error of r is ϵ for $\tau + 2\epsilon \leq \frac{1}{12} \cdot \left(\frac{d-1}{n}\right)^m$, then $\delta_{C^m}(r) \leq 16 \cdot \left(\frac{n}{d}\right)^{m-1} \cdot (\tau + \epsilon)$.*

Proof. For every $i \in [n]$ and $b \in [m]$, fix $c_{b,i}$ to be a closest codeword from C^{m-1} to $r_{b,i}$. We follow the proof outline of Raz & Safra [23] which when adapted to our context goes as follows: (1) Given a vector r and an assignment of codewords $c_{b,i} \in C^{m-1}$, we define an "inconsistency" graph G. (Note that this graph is *not* the same as the graph G_m^n that defines the test being analysed. In particular G is related to the word r being tested.) (2) We show that the existence of a large independent set in this graph G implies the proximity of r to a codeword of C^m (i.e., $\delta_{C^m}(r)$ is small). (3) We show that this inconsistency graph is sparse if the τ-soundness-error is small. (4) We show that the distance of C forces the graph to be special in that every edge is incident at least one vertex whose degree is large.

Definition of G. The vertices of G are indexed by pairs (b, i) with $b \in [m]$ and $i \in [n]$. Vertex (b_1, i_1) is adjacent to (b_2, i_2) if *at least* one of the following conditions hold:

1. $\delta_{C^{m-1}}(r_{b_1,i_1}) > \tau$.
2. $\delta_{C^{m-1}}(r_{b_2,i_2}) > \tau$.
3. $b_1 \neq b_2$ and c_{b_1,i_1} and c_{b_2,i_2} are inconsistent, i.e., there exists some element $j = \langle j_1, \ldots, j_m \rangle \in [n]^m$, with $j_{b_1} = i_1$ and $j_{b_2} = i_2$ such that $c_{b_1,i_1}[j^{(1)}] \neq c_{b_2,i_2}[j^{(2)}]$, where $j^{(c)} \in [n]^{m-1}$ is the vector j with its b_cth coordinate deleted.

Independent Sets of G and Proximity of r. It is clear that G has mn vertices. We claim next that if G has an independent set I of size at least $m(n-d)+d+1$ then r has distance at most $1 - (|I|/(mn))(1-\tau)$ to C^m.

Consider an independent set $I = I_1 \cup \cdots \cup I_m$ in G with I_b of size n_b being the set of vertices of the form $(b,i), i \in [n]$. W.l.o.g. assume $n_1 \geq \cdots \geq n_m$. Then, we have $n_1, n_2 > n - d$ (or else even if $n_1 = n$ and $n_2 = n - d$ we'd only have $\sum_b n_b \leq n + (m-1)(n-d)$). We consider the partial vector $r' \in \Sigma^{I_1 \times n \times \cdots \times n}$ defined as $r'[i, j_2, \ldots, j_m] = c_{1,i}[j_2, \ldots, j_m]$ for $i \in I_1$, and $j_2, \ldots, j_m \in [n]$. We show that r' can be extended into a codeword of C^m and that the extended word is close to r and this will give the claim.

First, we show that any extension of r' is close to r: This is straightforward since on each coordinate $i \in I_1$, we have r agrees with r' on $1 - \tau$ fraction of the points. Furthermore I_i/n is at least $|I|/(mn)$ (since n_1 is the largest). So we have that r' is at most $1 - (|I|/(mn))(1-\tau)$ far from r.

Now we prove that r' can be extended into a codeword of C^m. Let $C_b = C|_{I_b}$ be the projection (puncturing) of C to the coordinates in I_b. Let r'' be the projection of r' to the coordinates in $I_1 \times I_2 \times [n] \times \cdots \times [n]$. We will argue below that r'' is a codeword of $C_1 \otimes C_2 \otimes C^{m-2}$, by considering its projection to axis-parallel lines and claiming all such projections yield codewords of the appropriate code. Note first that the restriction of r' to any line parallel to the b-th axis is a codeword of C, for every $b \in \{2, \ldots, m\}$, since $r'_{1,i}$ is a codeword of C^{m-1} for every $i \in I_1$. Thus this continues to hold for r'' (except that now the projection to a line parallel to the 2nd coordinate axis is a codeword of C_2). Finally, consider a line parallel to the first axis, given by restricting the other coordinates to $\langle i_2, \ldots, i_m \rangle$, with $i_2 \in I_2$. We claim that for every $i_1 \in I_1$, $r''[i_1, \ldots, i_m] = c_{2,i_2}[i_1, \ldots, i_m]$. This follows from the fact that the vertices $(1, i_1)$ and $(2, i_2)$ are not adjacent to each other and thus implying that c_{1,i_1} and c_{2,i_2} are consistent with each other. We conclude that the restriction of r'' to every axis parallel line is a codeword of the appropriate code, and thus (by Proposition 2), r'' is a codeword of $C_1 \otimes C_2 \otimes C^{m-2}$. Now applying Proposition 3 to the code $C_1 \otimes C^{m-1}$ and its projection $C_1 \otimes C_2 \otimes C^{m-2}$ we get that there exists a unique extension of r'' into a codeword c' of the former. We claim this extension is exactly r' since for every $i \in I_1$, $c'_{1,i}[j,k] = r'[i,j,k]$. Finally applying Proposition 3 one more time, this time to the code C^m and its projection $C_1 \otimes C^{m-1}$, we find that $r' = c'$ can be extended into a codeword of the former. This concludes the proof of this claim.

Density of G. We now see that the small τ-soundness-error of the test translates into a small density γ of edges in G. Below, we refer to pairs (b,i) with $b \in [m]$ and $i \in [n]$ as "planes" (since they refer to $(m-1)$-dimensional planes in $[n]^m$) and refer to elements of $[n]^m$ as "points". We say a point $p = \langle p_1, \ldots, p_m \rangle$ lies on a plane (b,i) if $p_b = i$. Now consider the following test: Pick two random planes (b_1, i_1) and (b_2, i_2) subject to the constraint $b_1 \neq b_2$ and pick a random point p in the intersection of the two planes and verify that c_{b_1,i_1} is consistent with $r[p]$. Let κ denote the rejection probability of this test. We bound κ from both sides.

On the one hand we have that the rejection probability is at least the probability that we pick two planes that are τ-robust and incident to each other in G (which is at least $\frac{m\gamma}{m-1} - 2\epsilon$) and the probability that we pick a point on the intersection at which the two plane codewords disagree (at least $(d/n)^{m-2}$), times the probability that the codeword that disagrees with the point function is the first one (which is at least $1/2$). Thus we get $\kappa \geq \frac{d^{m-2}}{2(n)^{m-2}} \left(\frac{m\gamma}{m-1} - 2\epsilon \right)$.

On the other hand we have that in order to reject it must be the case that either $\delta_{C^{m-1}}(r_{b_1,i_1}) > \tau$ (which happens with probability at most ϵ) or $\delta_{C^{m-1}}(r_{b_1,i_1}) \leq \tau$ and p is such that r_{b_1,i_1} and c_{b_1,i_1} disagree at p (which happens with probability at most τ). Thus we have $\kappa \leq \tau + \epsilon$. Putting the two together we have $\gamma \leq \frac{m-1}{m} \left(2\epsilon + \frac{2n^{m-2}}{d^{m-2}}(\tau + \epsilon) \right)$.

Structure of G. Next we note that every edge of G is incident to at least one high-degree vertex. Consider a pair of planes that are adjacent to each other in G. If either of the vertices is not τ-robust, then it is adjacent to every vertex of G. So assume both are τ-robust.

W.l.o.g., let these be the vertices $(1, i)$ and $(2, j)$. Thus the codewords $c_{1,i}$ and $c_{2,j}$ disagree on the $(m-2)$-dimensional surface with the first two coordinates restricted to i and j respectively. Now let $S = \{\langle k_3, \ldots, k_m \rangle \mid c_{1,i}[j, k_3, \ldots, k_m] \neq c_{2,j}[i, k_3, \ldots, k_m]\}$ be the set of disagreeing tuples on this line. By the distance of C^{m-2} we know $|S| \geq d^{m-2}$. But now if we consider the vertex (b, k_b) in G for $b \in \{3, \ldots, m\}$ and k_b such that there exists $k_1, \ldots, k_{m-2}$ satisfying $k = (k_1, \ldots, k_{m-2}) \in S$, it must be adjacent at least one of $(1, i)$ or $(2, j)$ (it can't agree with both at the point (i, j, k). Furthermore, there exists d such k_b's for every $b \in \{3, \ldots, m\}$. Thus the sum of the degrees of $(1, i)$ and $(2, j)$ is at least $(m - 2)d$, and so at least one has degree at least $(m - 2)d/2$.

Putting It Together. From the last paragraph above, we have that the set of vertices of degree less than $(m - 2)d/2$ form an independent set in the graph G. The fraction of vertices of degree at least $(m - 2)d/2$ is at most $2(\gamma mn)/((m - 2)d)$. Thus we get that if $mn \cdot (1 - 2(\gamma mn)/((m - 2)d)) \geq m(n - d) + d + 1$, then r is δ-proximate to C^m for $\delta \leq \tau + (1 - \tau) \cdot 2(\gamma mn)/((m - 2)d)$. The lemma now follows by simplifying the expressions above, using the upper bound on γ derived earlier. The calculations needed to derive the simple expressions can be found in the full version of this paper [12].

Next we improve the bound achieved on the proximity of r by looking at the structure of the graph G_m^n (the graph underlying the m-Product tester) and its "expansion". Such improvements are a part of the standard toolkit in the analysis of low-degree tests based on axis parallel lines (see e.g., [7, 6, 16, 17] etc.) We omit the proof of this part from this version – it may be found in the full version of this paper [12]. We state the resulting lemma and show how Theorem 1 follows from it.

Lemma 2. *Let m be a positive integer and C be an $[n, k, d]_\Sigma$ code with the property $d^{m-1}/n^{m-1} \geq \frac{7}{8}$. If $r \in \Sigma^{n^m}$ and $c \in C^m$ satisfy $\delta(r, c) \leq \frac{1}{4}$ then $\delta(r, c) \leq 8\rho(r)$.*

Proof (Theorem 1). Let $c = 2^{14} \cdot \left(\frac{n}{d-1}\right)^{2m}$. We will prove that the m-Product Tester is c-robust for C^m. Note that $c \leq 2^{16}$ as required for the theorem, and $\sqrt{\frac{1}{c}} \leq \min\{\frac{1}{36} \cdot \left(\frac{d-1}{n}\right)^m, \frac{1}{128} \cdot \left(\frac{d}{n}\right)^{m-1}\}$ (as will be required below).

The completeness (that codewords of C^m have expected robustness zero) follows from Proposition 2. For the soundness, consider any vector $r \in \Sigma^{n^m}$ and let $\rho = \rho(r)$. If $\rho \geq 1/c$, then there there is nothing to prove since $\delta_{C^m}(r) \leq 1 \leq c \cdot \rho$. So assume $\rho \leq 1/c$.

Note that r has $\sqrt{\rho}$-soundness-error at most $\sqrt{\rho}$. Furthermore, by the assumption on ρ, we have $3\sqrt{\rho} \leq 3\sqrt{\frac{1}{c}} \leq \frac{1}{12} \cdot \left(\frac{d-1}{n}\right)^m$ and so, by Lemma 1, we have $\delta_{C^m}(r) \leq 16 \cdot \left(\frac{n}{d}\right)^{m-1} \cdot 2 \cdot \sqrt{\rho}$. Now using $\sqrt{\rho} \leq \sqrt{\frac{1}{c}} \leq \frac{1}{128} \cdot \left(\frac{d}{n}\right)^{m-1}$, we get $\delta_{C^m}(r) \leq \frac{1}{4}$. Let v be a codeword of C^m closest to r. We now have $\delta(r, v) \leq \frac{1}{4}$ and $\left(\frac{d}{n}\right)^{m-1} \geq \frac{7}{8}$, and so, by Lemma 2, we get $\delta_{C^m}(r) = \delta(r, v) \leq 8\rho$. This concludes the proof.

4 Tanner Product Codes and Composition

In this section we define the composition of two Tanner Product Codes, and show how they preserve robustness. We then use this composition to show how to test C^m using projections to C^2. All proofs of Lemmas in this Section appear in the full version of the paper [12].

Recall that a Tanner Product Code is given by a pair (G, C_{small}). We start by defining a composition of graphs that corresponds to the composition of codes.

Given an (N, M, D)-ordered graph $G = \{\ell_1, \ldots, \ell_M\}$ and an additional (D, m, d)-ordered graph $G' = \{\ell'_1, \ldots, \ell'_m\}$, their Tanner Composition, denoted $G \copyright G'$, is an $(N, M \cdot m, d)$-ordered graph with adjacency lists $\{\ell''_{j,j'} | j \in [M], j' \in [m]\}$, where $\ell''_{(j,j'),i} = \ell_{j,\ell'_{j',i}}$.

Lemma 3 (Composition). *Let G_1 be an (N, M, D)-ordered graph, and $C_1 \subseteq \Sigma^D$ be a linear code with $C = \mathrm{TPC}(G_1, C_1)$. Further, let G_2 be an (D, m, d)-ordered graph and $C_2 \subseteq \Sigma^d$ be a linear code such that $C_1 = \mathrm{TPC}(G_2, C_2)$. Then $C = \mathrm{TPC}(G_1 \copyright G_2, C_2)$ (giving a d-query local test for C). Furthermore if (G_1, C_1) is c_1-robust and (G_2, C_2) is c_2-robust, then $(G_1 \copyright G_2, C_2)$ is $c_1 \cdot c_2$-robust.*

We continue by recasting the results of Section 3 in terms of robustness of associated Tanner Products. Recall that G_m^n denotes the graph that corresponds to the tests of C^m by the m-Product Tester, where $C \subseteq \Sigma^n$.

Note that G_m^n can be composed with G_{m-1}^n and so on. For $m' < m$, define $G_{m,m'}^n = G_m^n$ if $m' = m - 1$ and define $G_{m,m'}^n = G_m^n \copyright G_{m-1,m'}^n$ otherwise. Thus we have that $C^m = \mathrm{TPC}(G_{m,m'}^n, C^{m'})$. The following lemma (which follows easily from Theorem 1 and Lemma 3 gives the robustness of $(G_{4,2}^n, C^2)$.

Lemma 4. *Let C be an $[n, k, d]_\Sigma$ code with $(d - 1/n)^4 \leq \frac{7}{8}$. Then $(G_{4,2}^n, C^2)$ is 2^{32}-robust.*

Finally we define graphs H_t^n so that $C^{2^t} = \mathrm{TPC}(H_t^n, C^2)$. This is easily done recursively by letting $H_2^n = G_{4,2}^n$ and letting $H_t^n = G_{4,2}^{n^{2^{t-2}}} \copyright H_{t-1}^n$ for $t > 2$. We now analyze the robustness of (H_t^n, C^2).

Lemma 5. *There exists a constant c such that the following holds: Let t be an integer and C be an $[n, k, d]_\Sigma$ code such that $d - 1 \geq (1 - \frac{1}{10m}) \cdot n$. Then (H_t^n, C^2) is c^t-robust.*

We are ready to prove Theorem 2.

Proof (Theorem 2). Let c be the constant given by Lemma 5. Fix i and let $C = C_i$, $n = n_i$ etc. (i.e., we suppress the subscript i below). Then C^m is an $[N, K, D]_q$ code, for $N = n^m$, $K = k^m$ and $D = d^m$. Since $d/n \geq 1 - \frac{1}{2m}$, we have C^m has relative distance $d^m/n^m \geq \frac{1}{2}$. Furthermore, the rate of the code is inverse polynomial, i.e., $N = n^m = (p(k))^m \leq \mathrm{poly}(k^m) = \mathrm{poly}(K)$. Finally, we have $C^m = \mathrm{TPC}(H_{\log_2 m}^n, C^2)$, where $(H_{\log_2 m}^n, C^2)$ is a $c^{\log_2 m}$-robust tester for C^m and this tester has query complexity $O(n^2)$. From Proposition 1 we get that there is a tester for C that makes $O(n^2 c^{O(\log_2 m)}) = \mathrm{poly} \log N$ queries. $\square$

Acknowledgments

We wish to thank Irit Dinur, Oded Goldreich and Prahladh Harsha for valuable discussions.

References

1. Noga Alon, Tali Kaufman, Michael Krivelevich, Simon Litsyn, and Dana Ron. Testing Reed-Muller codes. In *Proc. RANDOM 2003*, pages 188–199, 2003.
2. Sanjeev Arora. *Probabilistic checking of proofs and the hardness of approximation problems*. PhD thesis, University of California at Berkeley, 1994.
3. Sanjeev Arora, Carsten Lund, Rajeev Motwani, Madhu Sudan, and Mario Szegedy. Proof verification and the hardness of approximation problems. *Journal of the ACM*, 45(3):501–555, May 1998.
4. Sanjeev Arora and Shmuel Safra. Probabilistic checking of proofs: A new characterization of NP. *Journal of the ACM*, 45(1):70–122, January 1998.
5. Sanjeev Arora and Madhu Sudan. Improved low-degree testing and its applications. In *Proc. STOC97*, pages 485–495, El Paso, Texas, 4-6 May 1997.
6. László Babai, Lance Fortnow, Leonid A. Levin, and Mario Szegedy. Checking computations in polylogarithmic time. In *Proc. STOC91*, pages 21–32. 1991.
7. László Babai, Lance Fortnow, and Carsten Lund. Non-deterministic exponential time has two-prover interactive protocols. *Computational Complexity*, 1(1):3–40, 1991.
8. Mihir Bellare, Don Coppersmith, Johan Håstad, Marcos Kiwi, and Madhu Sudan. Linearity testing over characteristic two. *IEEE Transactions on Information Theory*, 42(6):1781–1795, November 1996.

9. Mihir Bellare, Shafi Goldwasser, Carsten Lund, and Alex Russell. Efficient probabilistically checkable proofs and applications to approximation. In *Proc. STOC93*, pages 294–304. ACM, New York, 1993.

10. Mihir Bellare and Madhu Sudan. Improved non-approximability results. In *Proc. STOC94*, pages 184–193, Montreal, Quebec, Canada, 23-25 May 1994.

11. Eli Ben-Sasson, Oded Goldreich, Prahladh Harsha, Madhu Sudan, and Salil Vadhan. Robust PCPs of proximity, shorter PCPs and applications to coding. In *Proc. STOC04*, (to appear), 2004.

12. Eli Ben-Sasson and Madhu Sudan. Robust Locally Testable Codes and Products of Codes. Available at Electronic Colloquium on Computational Complexity http://eccc.uni-trier.de/eccc-reports/2004/TR04-046/index.html

13. Eli Ben-Sasson, Madhu Sudan, Salil Vadhan, and Avi Wigderson. Randomness efficient low-degree tests and short PCPs via ϵ-biased sets. In *Proc. STOC03*, pages 612–621, 2003.

14. Manuel Blum, Michael Luby, and Ronitt Rubinfeld. Self-testing/correcting with applications to numerical problems. *Journal of Computer and System Sciences*, 47(3):549–595, 1993.

15. Irit Dinur and Omer Reingold. Assignment-Testers: Towards a Combinatorial Proof of the PCP-Theorem. *Manuscript*, 2004.

16. Uriel Feige, Shafi Goldwasser, Laszlo Lovasz, Shmuel Safra, and Mario Szegedy. Interactive proofs and the hardness of approximating cliques. *Journal of the ACM*, 43(2):268–292, 1996.

17. Katalin Friedl, Zsolt Hatsagi, and Alexander Shen. Low-degree tests. In *Proc. SODA94*, pages 57–64, 1994.

18. Katalin Friedl and Madhu Sudan. Some improvements to total degree tests. In *Proc. Israel STCS*, pages 190–198, Tel Aviv, Israel, 4-6 January 1995.

19. Oded Goldreich and Muli Safra. A Combinatorial Consistency Lemma with application to the PCP Theorem. In *SIAM Jour. on Comp.*, Volume 29, Number 4, pages 1132-1154, 1999.

20. Oded Goldreich and Madhu Sudan. Locally testable codes and PCPs of almost-linear length. In *Proc. FOCS02*, Vancouver, Canada, 16-19 November 2002.

21. F. J. MacWilliams and Neil J. A. Sloane. *The Theory of Error-Correcting Codes*. Elsevier/North-Holland, Amsterdam, 1981.

22. Alexander Polishchuk and Daniel A. Spielman. Nearly linear-size holographic proofs. In *Proc. STOC94*, pages 194–203, Montreal, Canada, 1994.

23. Ran Raz and Shmuel Safra. A sub-constant error-probability low-degree test, and a sub-constant error-probability PCP characterization of NP. In *Proc. STOC97*, pages 475–484. ACM Press, 1997.

24. Ronitt Rubinfeld and Madhu Sudan. Robust characterizations of polynomials with applications to program testing. *SIAM J. Comp.*, 25(2):252–271, 1996.

25. Madhu Sudan. Algorithmic introduction to coding theory. Lecture notes, Available from http://theory.csail.mit.edu/~madhu/FT01/, 2001.

26. R. Michael Tanner. A recursive approach to low complexity codes. *IEEE Transactions of Information Theory*, 27(5):533–547, September 1981.

A Stateful Implementation
of a Random Function
Supporting Parity Queries over Hypercubes

Andrej Bogdanov and Hoeteck Wee

Computer Science Division
University of California, Berkeley
{adib,hoeteck}@cs.berkeley.edu

Abstract. Motivated by an open problem recently suggested by Goldreich et al., we study truthful implementations of a random binary function supporting compound XOR queries over sub-cubes of the hypercube $\{0,1\}^n$. We introduce a relaxed model of an implementation, which we call a *stateful* implementation, and show how to implement the desired specification in this model. The main technical construction is an algorithm for detecting linear dependencies between n dimensional hypercubes, viewed as characteristic vectors in $\mathbb{F}_2^{\{0,1\}^n}$. Using coding theoretic techniques, we first exhibit a randomized algorithm for detecting such dependencies. We then show how a recent approach by Raz and Shpilka for polynomial identity testing in non-commutative models of computation can be applied to obtain a deterministic algorithm.

1 Introduction

In a recent paper, Goldreich, Goldwasser and Nussboim [3] initiated the study of efficient pseudorandom implementations of *huge random objects* – objects so big that they cannot be represented using bounded resources, such as randomness or time (in particular, these objects have size that is exponential in the running time of the applications), but for which we can obtain approximations good enough for many algorithmic and cryptographic applications. A celebrated example from cryptography is the construction of a pseudo-random function from any one way function [2]: Even though a truly random boolean function on n input bits cannot be specified by fewer than 2^n random bits, in many cryptographic applications this infeasible object can be approximated by a *pseudo-random function* that can be specified using only poly(n) random bits and evaluated on arbitrary inputs in poly(n) time.

1.1 Stateful and Stateless Implementations

Since the work of Goldreich et al. is somewhat motivated by cryptography, they consider only "stateless" implementations of huge random objects. An implementation refers to a polynomial-time (oracle) machine that computes the huge

K. Jansen et al. (Eds.): APPROX and RANDOM 2004, LNCS 3122, pp. 298–309, 2004.
© Springer-Verlag Berlin Heidelberg 2004

object; in the case of a random boolean function, the machine takes as input a string of n bits and outputs the value of the function at that point. In a stateless implementation I, the answer to a query posed to I cannot depend on queries previously seen by I and their answers. This property of the implementation is often important in cryptographic settings, where multiple parties share the same copy of the huge object in question. For example, a common technique in cryptography is to design protocols in the random oracle model, where each party has access to the same infinite sequence of random bits. To obtain an implementation of the protocol, one replaces the random oracle with a pseudo-random function. As we want the parties in the protocol to share the same pseudo-random function, it is important that the implementation of the random oracle by a pseudo-random function be independent of the queries seen by a particular party in the protocol; namely, the implementation must be stateless.

In addition to cryptography, Goldreich et al. also consider algorithmic applications of huge random objects. For example, they imagine a scenario where one wants to run experiments on, say, random codes. They observe that global properties of these codes, such as having large minimum distance, may not be preserved when the randomness of the code is replaced by a pseudo-random generator. This leads to the problem of implementing huge random objects that are guaranteed to preserve a certain property, such as codes with good minimum distance.

Unlike in the cryptographic setting, it is not clear that a stateless implementation (the only type allowed by the model of Goldreich et al.) gives any advantage over a "stateful" one. In other words, it may be possible to do more by generating the desired huge random object "on the fly" rather than subscribing to an implementation predetermined by the random tape of our machine. In particular, it would be interesting to know whether there exists a natural specification that allows a stateful implementation but not a stateless one. We suspect that the specification considered in this paper, suggested for study by Goldreich et al. – a random boolean function supporting XOR queries over hypercubes – may provide a separation between stateful and stateless *perfect implementations in the random oracle model*.

1.2   Random Functions Supporting Complex Queries

Goldreich et al. observe that, assuming the existence of one-way functions, if a specification can be close-implemented[1] in the random oracle model, then it can also be implemented by an ordinary probabilistic polynomial-time machine. Moreover, this transformation preserves truthfulness[2]: Namely, to obtain a truth-

---

[1] In fact, it is sufficient that the specification be pseudo-implementable. Note that the terms close-implementable, pseudo-implementable and truthful are technical terms defined in [3].

[2] Intuitively, truthfulness requires that an implementation of Type T objects generates only objects of Type T. In particular, a random function is not a truthful implementation of a random permutation even though they are indistinguishable to a computationally bounded adversary.

ful pseudo-implementation of a huge random object, it is sufficient to construct such an implementation in the random oracle model. This is a common technique in cryptography, used among other things in the construction of pseudo-random permutations [4]. Though the transformation is only shown to hold for stateless implementations, we observe that it also works for stateful ones (this is because the definition of pseudo-randomness in the context of pseudo-random functions allows from stateful adversaries).

In particular, this observation implies that random functions have a trivial truthful pseudo-implementation. However, one may ask whether it is possible to truthfully implement random functions supporting queries beyond evaluation on arbitrary inputs. One variant proposed by Goldreich et al. asks for the implementation of a random function $f : \{0,1\}^n \rightarrow \{0,1\}$, augmented with queries regarding the XOR of the values of f on arbitrary *intervals* of $\{0,1\}^n$ (with respect to the lexicographic ordering of n bit strings.) Note that a trivial implementation of f in the random oracle model cannot hope to answer such queries efficiently, as they may involve XORing exponentially many bits. However, Goldreich et al. show how, using a suitable data structure, one can obtain a stateless perfect implementation of f that answers queries in $O(n^2)$ time. As perfect implementations are always truthful, this construction yields a truthful pseudo-implementation by an ordinary machine.

We observe that a simpler construction for implementing random functions supporting interval-XOR queries with $O(n)$ running time can be achieved as follows: let $f' : \{0,1\}^n \rightarrow \{0,1\}$ be the random oracle, and return $f'(\alpha-1) \oplus f'(\beta)$ as the answer to the query (α, β), corresponding to the value $\bigoplus_{\alpha \leq x \leq \beta} f(x)$. Here, $\alpha-1$ denotes the n-bit binary string that immediately precedes α in lexicographic order, and we specify $f'(0^n - 1) = 0$. The underlying idea is a simple change of basis: instead of specifying a random function f by its values at all $x \in \{0,1\}^n$, we specify the value of f using the values $\bigoplus_{y \leq x} f(y)$. Note that this implementation makes only 2 queries into the random oracle per interval-XOR query, which is optimal (in an amortized sense). Unfortunately, this construction, unlike that by Goldreich et al., does not yield a truthful close-implementation of random functions supporting any symmetric interval query.

As a follow-up to their work on interval queries, Goldreich et al. propose the following more general question:

Open Problem. [3] Provide a truthful close-implementation in the random oracle model of the following specification. The specification machine defines a random function $f : \{0,1\}^n \rightarrow \{0,1\}$, and answers queries that succinctly describe a set S, taken from a specific class of sets, with the value $\bigoplus_{x \in S} f(x)$. A natural case is the class of sub-cubes of $\{0,1\}^n$; that is, a set S is specified by a pattern σ in $\{0,1,*\}^n$ such that S is the set of points in $\{0,1\}^n$ that match the pattern σ.

We suspect that the technique of Goldreich et al. for implementing a random function supporting interval queries does not extend to the case of hypercube queries, though we have not been able to show a negative result confirming our intuition. Instead, we show how to obtain a *stateful* perfect-implementation of this specification in the random oracle model.

1.3   Main Contributions

Our main contributions are the following:

1. We propose a notion of stateful implementations of huge random objects and reduce the problem of constructing a stateful implementation of a random binary function supporting compound XOR queries over sub-cubes of the hypercube $\{0,1\}^n$ to an algorithmic problem of hypercube linearity testing.

2. We then present two algorithms for hypercube linearity testing: a randomized algorithm using coding theoretic techniques, and a deterministic based on non-commutative polynomial identity testing. It follows from the first algorithm that there is a stateful close implementation of the afore-mentioned specification in the random oracle model, and from the second, a stateful perfect implementation.

In fact, the second algorithm subsumes the first, but we still include the latter as we feel that the technique used is fairly insightful and the analysis can in fact be used to construct quantum states of provably superpolynomial tree size.

2   Preliminaries

Let $a \in \{0,1,*\}^n$. The *hypercube*[3] $H(a)$ is the set of all $x \in \{0,1\}^n$ that match the string a, namely such that

$$x[i] = \begin{cases} 0, & \text{if } a[i] = 0 \\ 1, & \text{if } a[i] = 1 \\ 0 \text{ or } 1, & \text{if } a[i] = *. \end{cases}$$

As in Goldreich et al., we specify the huge random object in question by a computationally unbounded probabilistic Turing machine that halts with probability one. The huge random object is determined by the input-output relation of this machine when the random tape is selected uniformly at random from $\{0,1\}^\infty$.

A Random Function Supporting XOR Queries on Hypercubes
INPUT A query $a \in \{0,1,*\}^n$
RANDOM TAPE A sequence of functions $f_1, f_2, \ldots$, where $f_k : \{0,1\}^k \to \{0,1\}$
OUTPUT The value $\bigoplus_{x \in H(a)} f_{|a|}(x)$

We are interested in efficient implementations of this specification, namely ones that can be obtained by Turing machines that run in time polynomial in the length of the input.

---

[3] In [3] and in the introduction, we use the term sub-cubes of the hypercube $\{0,1\}^n$.

2.1   Stateful Implementations

Following Goldreich et al., we say machine I is a *(stateful) implementation* specification S with respect to a machine M of that makes queries to a protocol party, if (1) On the qth query x, I runs in time polynomial in q and x, and (2) The distribution D_S of the transcript $M(1^n) \leftrightarrow S$ is indistinguishable from the distribution D_I of $M(1^n) \leftrightarrow I$. Specifically:

1. If D_S and D_I are identical for every M, we say that I *perfectly implements* S;
2. If D_S and D_I have negligible $(n^{\omega(1)})$ statistical difference for all M that make $\mathrm{poly}(n)$ queries, we say that I *closely implements* S;
3. If D_S and D_I are indistinguishable for all M that run in polynomial time, we say that I *pseudo-implements* S.

An implementation I is *truthful* with respect to S if for every sequence of queries $x_1, \ldots, x_q$, the support of the distribution $I(x_1), \ldots, I(x_q)$ is contained in the support of $S(x_1), \ldots, S(x_q)$; namely, if the implementation never provides answers that are inconsistent with the specification. Note that perfect implementations are always truthful. An oracle Turing machine $I^?$ is an implementation of S *in the random oracle model* if the distribution of $M(1^n) \leftrightarrow S$ is indistinguishable from the distribution of $M(1^n) \leftrightarrow I^R$ over a random oracle R. As for stateless implementations (see Theorem 2.9 of [3]), we have the following:

Proposition 1. *Suppose that one-way functions exist. Then any specification that has a pseudo-implementation in the random oracle model also has a pseudo-implementation by an ordinary machine. Moreover, if the former implementation is truthful then so is the latter.*

Since perfect implementations are always truthful, for our purposes it will be sufficient to provide a perfect implementation of a random function supporting queries on hypercubes in the random oracle model. The heart of this implementation consists of an algorithm for the problem of hypercube linearity testing, or HYPERCUBE-LIN, which we describe next.

2.2   Reduction to Hypercube Linearity Testing

We call collection of hypercubes $H_1, \ldots, H_m \subseteq \{0,1\}^n$ *linearly dependent* if there exists a non-empty set $S \subseteq \{1, 2, \ldots, m\}$ such that for all points $x \in \{0,1\}^n$, x is contained in an even number of hypercubes amongst the subset of hypercubes $\{H_i \mid i \in S\}$. For any such set S, we write $\sum_{i \in S} H_i = 0$. Equivalently, we may view each hypercube H_i as a vector in $\mathbb{F}_2^{\{0,1\}^n}$, where $\mathbb{F}_2 = \{0,1\}$ is the two element field and $H_i[x] = 1$ if $x \in H_i$, and 0 otherwise. In this notation, linear independence between hypercubes translates into linear independence of the corresponding vectors.

HYPERCUBE-LIN: Given q strings $a_1, \ldots, a_q \in \{0, 1, *\}^n$, accept iff the hypercubes $H(a_1), \ldots, H(a_q)$ are linearly dependent.

Note that standard techniques for testing linear independence, such as Gaussian elimination, do not apply directly for this problem because we are dealing with vectors whose length is exponential in the size of the input. However, we will still be able to show the following:

Theorem 1. *There is a deterministic polynomial-time algorithm for the problem* HYPERCUBE-LIN.

In fact, we will begin with a randomized algorithm for HYPERCUBE-LIN, which nicely illustrates the coding theoretic nature of the hypercube linearity testing problem. We then argue that the test performed by the algorithm can be viewed, in some sense, as an application of polynomial identity testing. Even though we don't know, in general, how to derandomize polynomial identity testing, in our case the derandomization can be performed using a recent identity testing algorithm for non-commutative formulas of Raz and Shpilka [5].

Theorem 2. *There exists a stateful perfect implementation of a random function supporting XOR queries on hypercubes in the random oracle model.*

Proof. Let A be the algorithm from Theorem 1, and R be the random oracle. First, we consider the following (promise) search problem:

INPUT Strings $a_1, \ldots, a_q \in \{0,1,*\}^n$ and $b \in \{0,1,*\}^n$, such that the hypercubes $H(a_1), \ldots, H(a_q)$ are linearly independent
PROBLEM If $H(b)$ is linearly independent from $H(a_1), \ldots, H(a_q)$, output $\varnothing$. Otherwise, output coefficients $c_1, \ldots, c_q \in \mathbb{F}_2$ such that $H(b) = \sum_{i=1}^{q} c_i H(a_i)$.

It is not difficult to see that, by a self reduction argument, we can obtain a polynomial time algorithm A' for this problem using black box access to A (in fact, $2q$ invocations of A suffices). With this in hand, we implement a random function supporting XOR queries on hypercubes as follows: After seeing queries $a_1, \ldots, a_{q-1}$, the implementation keeps track of a subset of queries $\{a_k : k \in B\}$, where $B \subseteq [q-1]$ is chosen such that the subset of hypercubes $\{H(a_k) : k \in B\}$ form a basis for the set of vectors $\{H(a_k) \in \mathbb{F}_2^{\{0,1\}^n} : k \in [q-1]\}$. On query a_q, we run the algorithm A' on inputs $\{a_k : k \in B\}$ and a_q. If the algorithm returns $\varnothing$, then we return $R(|B|+1)$, which is a fresh random bit, and add a_q to the set B. Otherwise, the algorithm A' outputs coefficients $c_1, \ldots, c_{q-1} \in \mathbb{F}_2$, and we return the value $\sum_{i=1}^{|B|} c_i R(i)$.

We show this is a perfect implementation, by induction on q. Let us assume that the specification transcript and implementation transcript are statistically indistinguishable after $q-1$ queries. At query q, there are two possibilities: If $H(a_q)$ is linearly dependent in $H(a_1), \ldots, H(a_{q-1})$, then both in the implementation and in the specification the answer to the qth query is determined by the previous answers, so by the inductive hypothesis the new transcripts are indistinguishable. If $H(a_q)$ is linearly independent in $H(a_1), \ldots, H(a_{q-1})$, then the answer to the qth query in the specification is statistically independent from all previous answers. By construction, this is also true in the implementation, so again, the new transcripts are indistinguishable.

3   Algorithms for Hypercube Linearity Testing

In this section, we present two algorithms for HYPERCUBE-LIN, thereby completing our implementation of a random function supporting XOR queries on hypercubes.

3.1   A Randomized Algorithm for HYPERCUBE-LIN

Let $\mathbb{F} = \mathbb{F}_{2^s}$ be a field of characteristic 2 of size 2^s, where $s = \lceil \log 3n \rceil$. For each $a \in \{0, 1, *\}^n$, we define a polynomial $p_a = z_1 z_2 \cdots z_n$ over $\mathbb{F}[x_1, y_1, \ldots, x_n, y_n]$, where:

$$z_i = \begin{cases} x_i & \text{if } a_i = 0 \\ y_i & \text{if } a_i = 1 \\ x_i + y_i & \text{if } a_i = * \end{cases}$$

Note that $H(a)$ is a homogeneous polynomial of total degree exactly n, and that $p_a = \sum_{x \in H(a)} p_x$.

Lemma 1. *Let $\mathbb{F}$ be any field of characteristic two. Then $H(a_1) + \cdots + H(a_m) = 0$ if and only if $p_{a_1} + \cdots + p_{a_m}$ is the zero polynomial in $\mathbb{F}[x_1, y_1, \ldots, x_n, y_n]$.*

Proof. Each point x in $\{0, 1\}^n$ is represented by a unique monomial p_x from $\mathbb{F}[x_1, y_1, \ldots, x_n, y_n]$. Therefore, each point x in $\{0, 1\}^n$ appears an even number of times in $H(a_1), \ldots, H(a_m)$ iff each of the corresponding monomials p_x has an even integer coefficient in $p_{a_1} + \cdots + p_{a_m}$. In addition, since $\mathbb{F}$ has characteristic 2, the latter condition is equivalent to $p_{a_1} + \cdots + p_{a_m}$ being identically zero.

We may now define a binary encoding $C_{H(a)}$ of hypercubes $H(a)$, which is obtained by concatenating the Reed-Muller code associated with the polynomial p_a with the Hadamard code. More precisely, given any $a \in \{0, 1, *\}$, we define $C_{H(a)} : \{0, 1\}^{(2n+1)s} \to \{0, 1\}$ as follows:

$$C_{H(a)}(x_1, y_1, \ldots, x_n, y_n, \tau) = \langle p_a(x_1, y_1, \ldots, x_m, y_m), \tau \rangle$$

using the first $2ns$ bits of the input to $C_{H(a)}$ to pick $x_1, y_1, \ldots, x_n, y_n \in \mathbb{F}$, and the remaining s bits to pick $\tau \in \{0, 1\}^s$.

Lemma 2. *$\{C_{H(a)} \mid a \in \{0, 1, *\}^n\}$ is a binary encoding of hypercubes in $\{0, 1\}^n$ with the following properties:*

1. *(large distance) It has relative distance $1/3$.*
2. *(locally encodable) There is a $O(ns \log s)$ algorithm that computes $C_{H(a)}(r)$ on input a and $r \in \{0, 1\}^{(2n+1)s}$.*
3. *(linearity) For all $a, a' \in \{0, 1, *\}^n$, $C_{H(a)+H(a')} = C_{H(a)} + C_{H(a')}$.*

Proof. By the Schwartz-Zippel Lemma [6, 7], for any $a \neq a' \in \{0, 1, *\}^n$, p_a and $p_{a'}$ evaluate to different values on at least a $2/3$ fraction of values in $\mathbb{F}^{2n}$. Upon concatenating with the Hadamard code, the minimum relative distance becomes $1/3$. Next, observe that we can write down p_a and evaluate p_a at any input in time $O(ns \log s)$. It follows that we also compute $C_{H(a)}(r)$ in time $O(ns \log s)$. Finally, linearity follows from the fact that both the encoding as polynomials and the Hadamard code are linear.

Randomized Algorithm for HYPERCUBE-LIN

1. Fix $\ell = O(q)$. Choose $r_1, \ldots, r_\ell \in \{0,1\}^{(2n+1)s}$ uniformly at random.
2. Construct the $q \times \ell$ matrix M over $\mathbb{F}_2$, where $M_{ij} = C_{H(a_i)}(r_j)$, that is:

$$M = \begin{pmatrix} C_{H(a_1)}(r_1) & C_{H(a_1)}(r_2) & \ldots & C_{H(a_1)}(r_\ell) \\ C_{H(a_2)}(r_1) & C_{H(a_2)}(r_2) & \ldots & C_{H(a_2)}(r_\ell) \\ \vdots & \vdots & \ddots & \vdots \\ C_{H(a_q)}(r_1) & C_{H(a_q)}(r_2) & \ldots & C_{H(a_q)}(r_\ell) \end{pmatrix}$$

3. Compute the rank of M over $\mathbb{F}_2$. Accept if the rank is less than q; reject otherwise.

Proposition 2. *There is a* coRP-*algorithm for* HYPERCUBE-LIN *running in time* $O(q^3 + q^2 ns \log s)$.

Proof. Fix any $S \subseteq [q]$. It follows from linearity and Lemma 1 that

$$\sum_{i \in S} H(a_i) = 0 \quad \text{if and only if} \quad \sum_{i \in S} C_{H(a_i)}(r) = 0 \quad \forall\, r \in \{0,1\}^{(2n+1)s} \quad (1)$$

Therefore, if $H(a_1), \ldots, H(a_q)$ are linearly dependent, there is some S for which (1) holds. Then, the rows of M identified by S add up to 0, and thus M has rank less than q. On the other hand, if $H(a_1), \ldots, H(a_q)$ are not linearly dependent, then for all $S \subseteq [q]$, $\sum_{i \in S} C_{H(a_i)}$ is not the zero codeword[4], and thus

$$\Pr_{r_1, \ldots, r_\ell} \left[\sum_{i \in S} C_{H(a_i)}(r_j) = 0 \quad \forall\, j = 1, 2 \ldots, \ell \right] \leq \left(\frac{2}{3} \right)^\ell$$

Now, the probability that M has rank less than q is equal to the probability that there is some $S \subseteq [q]$ for which (1) holds, which by a union bound is at most $2^q \cdot (2/3)^\ell < 1/2$. Finally, computing M takes time $O(q\ell ns \log s)$, and performing Gaussian elimination on M takes time $O(q^2 \ell)$, which yields a total running time of $O(q^3 + q^2 ns \log s)$.

---

[4] In fact, $\sum_{i \in S} C_{H(a_i)}$ is not necessarily a codeword in the code defined in Lemma 2, but it is still a codeword in the code obtained by concatenating a Reed Muller code with a Hadamard code. Hence, the analysis for the relative distance of the code also shows that $\sum_{i \in S} C_{H(a_i)}$ has relative distance at least $1/3$ from the zero codeword.

3.2   A Deterministic Algorithm for HYPERCUBE-LIN

The randomized algorithm for hypercube linearity testing is based on Lemma 1, which allows us to reduce testing a particular dependency between hypercubes to a polynomial identity test. In the actual algorithm, the power of randomness is used twice. First, randomness allows us to efficiently perform the polynomial identity test from Lemma 1. Second, using standard amplification of success probabilities we can take a union bound over the exponentially many possible linear dependencies between hypercubes in the correctness proof for the randomized algorithm.

Even though polynomial identity testing remains notoriously hard to derandomize, Raz and Shpilka [5] recently found a deterministic algorithm that works for certain alternate and restricted models of computation. In particular, their algorithm works for formulas over arbitrary fields in non-commuting variables (i.e., formulas in the noncommutative ring $\mathbb{F}\{x_1, \ldots, x_n\}$), and for $\Sigma\Pi\Sigma$ *circuits* – depth three multilinear arithmetic circuits with a plus gate at the root.

The formal polynomial $p_1 + \ldots + p_m$ in Lemma 1 can be trivially computed by an $\Sigma\Pi\Sigma$ circuit. This immediately gives a deterministic way of testing whether a particular linear relation $H_1 + \ldots + H_m = 0$ is satisfied over $\mathbb{F}_2$. However, we are interested in testing whether any one of the exponentially many relations $c_1 H_1 + \ldots + c_m H_m = 0$ holds, or equivalently, if there exists coefficients $c_1, \ldots, c_m \in \mathbb{F}_2$, not all zero, such that $c_1 p_1 + \ldots + c_m p_m = 0$ as a polynomial in $\mathbb{F}_2[x_i, y_i]$. We will show that a construction, along the lines of Raz and Shpilka's algorithm, works for this problem as well. Instead of trying to come up with a general theorem for testing linear dependencies between polynomials, we will focus on polynomials representing hypercubes, though it may be possible to extend the analysis to a somewhat more general scenario.

Let $\mathbb{F}$ now be an arbitrary finite field, and $p_1, \ldots, p_m$ be polynomials in $\mathbb{F}[x_{ij}]$, where $i \in [n], j \in [m]$ and each $p_k, k \in [m]$ has the following form:

$$p_k(x_{ij}) = \prod_{i=1}^{n} \sum_{j=1}^{m} \alpha_{ij}^k x_{ij}, \qquad (2)$$

where $\alpha_{ij}^k \in \mathbb{F}$ are constants. It is easy to check that the polynomials in Lemma 1 are written in this form. We are interested in determining, in time polynomial in m and n, whether there exist coefficients $c_1, \ldots, c_m \in \mathbb{F}$ such that $c_1 p_1 + \ldots + c_m p_m$ is the zero polynomial over $\mathbb{F}$. Since this is a multilinear polynomial, this is equivalent to asking whether $p_1, \ldots, p_m$ are linearly dependent as elements of the vector space V over $\mathbb{F}$ generated by the monomials $x_{1j_1} \ldots x_{nj_n}$, where $j_1, \ldots, j_n$ range over $[m]$.

As in [5], the analysis works by induction on n. When $n = 1$, we have $p_k(x_{1j}) = \sum_{j=1}^{m} \alpha_{1j}^k x_{1j}$, so the vectors $p_k \in V$ are independent if and only if the matrix $M \in \mathbb{F}^{m \times m}$ with $M[j, k] = \alpha_{1j}^k$ has full rank. Using Gaussian elimination this can be checked in, say, $O(m^3 \log |F|)$ time.

We now show how to reduce a problem of degree n to one of degree $n - 1$. The trick is to look at the following partial expansion of the polynomials p_k:

$$p_k(x_{ij}) = \sum_{j_1, j_2 = 1}^{m} \alpha_{1j_1}^{k} \alpha_{2j_2}^{k} x_{1j_1} x_{2j_2} \cdot q_k(x_{(3...n)j}),$$

where q_k is given by

$$q_k(x_{(3...n)j}) = q_k(x_{3j}, x_{4j}, \ldots, x_{dj}) = \prod_{i=3}^{n} \sum_{j=1}^{m} \alpha_{ij}^{k} x_{ij}.$$

Now consider a linear combination $P(x_{ij}) = \sum_{k=1}^{m} c_k p_k(x_{ij})$. We can expand this as

$$P(x_{ij}) = \sum_{j_1, j_2 = 1}^{m} x_{1j_1} x_{2j_2} \sum_{k=1}^{m} \alpha_{1j_1}^{k} \alpha_{2j_2}^{k} \cdot c_k q_k(x_{(3...n)j}).$$

From this expansion, we see that $P \equiv 0$ in $\mathbb{F}[x_{ij}]$ if and only if for all $j_1, j_2 \in [m]$:

$$\sum_{k=1}^{m} \alpha_{1j_1}^{k} \alpha_{2j_2}^{k} \cdot c_k q_k(x_{(3...n)j}) \equiv 0 \text{ in } \mathbb{F}[x_{(3...n)j}]. \tag{3}$$

Consider the matrix $M \in \mathbb{F}^{m^2 \times m}$, whose rows are indexed by pairs $(j_1, j_2) \in [m] \times [m]$ such that $M[(j_1, j_2), k] = \alpha_{1j_1}^{k} \alpha_{2j_2}^{k}$. Choose a subset S of m rows such that the rows indexed by S span the row space of M. Then the constraints (3) are all satisfied if and only if

$$\text{For all } s \in S, \sum_{k=1}^{m} M[s, k] \cdot c_k q_k(x_{(3...n)j}) \equiv 0 \text{ in } \mathbb{F}[x_{(3...n)j}]. \tag{4}$$

We now want to rewrite the set of constraints (4) as a single constraint. For this, we introduce m additional formal variables y_s with $s \in S$. Constraints (4) are satisfied if and only if

$$\sum_{s \in S} y_s \sum_{k=1}^{m} M[s, k] \cdot c_k q_k(x_{(3...n)j}) \equiv 0 \text{ in } \mathbb{F}[y_s, x_{(3...n)j}]. \tag{5}$$

Finally, let

$$r_k(y_s, x_{(3...n)j}) = \left(\sum_{s \in S} M[s, k] y_s \right) q_k(x_{(3...n)j}),$$

and note that constraint (5) is satisfied if and only if $\sum_{k=1}^{m} c_k r_k(y_s, x_{(3...n)j}) \equiv 0$ in $\mathbb{F}[y_s, x_{(3...n)j}]$. To summarize:

Lemma 3. *The polynomials $p_1, \ldots, p_m$ are linearly independent in $\mathbb{F}[x_{ij}]$ if and only if the polynomials $r_1, \ldots, r_m$ are linearly independent in $\mathbb{F}[y_s, x_{(3...n)j}]$* [5].

---

[5] Moreover, the p_i satisfy a particular dependency $\sum c_i p_i \equiv 0$ if and only if the r_i satisfy the same dependency. This can be used to find linear dependencies between hypercubes, which makes the decision to search reduction in the proof of Proposition 1 unnecessary.

Now we can apply the induction hypothesis to the polynomials r_k, which have the prescribed form (2). The bottleneck in the running time of the algorithm is computing the linearly independent set S, which can be done in $O(m^4 \log |\mathbb{F}|)$ time by Gaussian elimination. This yields a total running time of $O(m^4 n \log |\mathbb{F}|)$, and concludes the proof of Theorem 1.

4   Extensions

4.1   A Connection with Quantum Computation

The following problem was brought to our attention by Aaronson in his recent work addressing skepticism of quantum computing [1]. The construction below implies the existence of explicit (constructible in deterministic polynomial-time) quantum states of provably superpolynomial tree size. The argument is similar to the proof of Proposition 2.

Proposition 3. *For every $\delta < 1/4$, there is a polynomial-time constructible $k \times n$ binary matrix M with $k = n^\delta$, such that a random $k \times k$ submatrix of M has rank at least $k - 1$ with constant probability.*

Aaronson's original argument requires M to have full rank with constant probability, but it is not difficult to see that his argument applies even if M has only rank $k - 1$ with constant probability.

Proof. Fix $\epsilon > 0$ and choose k_0 such that $k = k_0 \log k_0^{2+\epsilon}$. Consider the concatenation of a Reed-Solomon code with codeword size $k_0^{2+\epsilon}$ (over smallest possible alphabet size) with a Hadamard code. The resulting binary code C has length at most $n = k^{4+2\epsilon}$ and relative distance $1/2 - \eta$, where $\eta = 1/k^{1+\epsilon-o(1)}$. Pick a basis $e_1, \ldots, e_k$ of $\{0,1\}^k$ and take M to be the $k \times n$ matrix whose ith row is the codeword $C(e_i)$.

Now consider choosing a random $k \times k$ submatrix of M. Since $k = O(n^{1/4})$, with probability $1 - o(1)$, this is equivalent to choosing an independent set of k columns $M_1, \ldots, M_k$ of M with repetition. For any nonzero vector $x \in \{0,1\}^k$, the probability that all dot products $x^T M_1, \ldots, x^T M_k$ vanish is at most $(1/2 + \eta)^k$. It follows that the expected number of nonzero x for which all these dot products vanish is at most $2^k \cdot (1/2 + \eta)^k < \exp(2k^{-\epsilon}) = 1 + o(1)$. By Markov's inequality, the number of such x is less than two with probability at least, say, $1/3$. If this is the case, then the columns of M satisfy at most one linear relation, so that the rank of M is at most $k - 1$.

4.2   Towards a Stateless Implementation?

We discuss our conjecture regarding a possible separation between stateful and stateless perfect implementations in the random oracle model. We say that a set S of vectors in $\mathbb{F}_2^{2^n}$ admits an *efficient basis* B where B is a (standard) basis for $\mathbb{F}_2^{2^n}$ if every vector $v \in S$ can be written as the sum of poly(n) vectors in B and the basis representation can be computed in poly(n) time.

Proposition 4. *If S admits an efficient basis, then there is a perfect (stateless) implementation of a random function $f : \{0,1\}^n \to \{0,1\}$, and answers queries that succinctly describe a set S, taken from a specific class of sets, with the value $\bigoplus_{x \in S} f(x)$.*

It is easy to see that intervals of $\{0,1\}^n$ admit an efficient basis, namely the collection of vectors $\{v_i \in \mathbb{F}_2^{2^n}, i = 1, 2, \ldots, 2^n\}$ where v_i is the vector whose first i positions are ones and whose remaining positions are zeros. This observation underlies our construction for implementing random functions support interval-XOR queries in Section 1.2. On the other hand, we do not know if the hypercubes of $\{0,1\}^n$ admit an efficient basis.

Acknowledgements

We thank Luca Trevisan and the anonymous referees for comments on an earlier version of this paper.

References

1. Scott Aaronson. Multilinear formulas and skepticism of quantum computing. In *Proceedings of the 36th ACM Symposium on Theory of Computing*, 2004.
2. Oded Goldreich, Shafi Goldwasser, and Silvio Micali. How to construct random functions. *Journal of the ACM*, 33(4):210–217, 1986.
3. Oded Goldreich, Shafi Goldwasser, and Asaf Nussboim. On the implementation of huge random objects. In *Proceedings of the 44th IEEE Symposium on Foundations of Computer Science*, pages 68–79, 2003. Preliminary full version at http://www.wisdom.weizmann.ac.il/~oded/p_toro.html.
4. Michael Luby and Charles Rackoff. How to construct pseudorandom permutations from pseudoranom functions. *SICOMP*, 17:373–386, 1988.
5. Ran Raz and Amir Shpilka. Deterministic polynomial identity testing in non commutative models. In *Proceedings of the 17th Conference on Computational Complexity*, 2004. To appear.
6. J. T. Schwartz. Fast probabilistic algorithms for verification of polynomial identities. *Journal of the ACM*, 27(4):701–717, 1980.
7. R. E. Zippel. Probabilistic algorithms for sparse polynomials. In *Proceedings of EUROSAM 79*, pages 216–226, 1979.

Strong Refutation Heuristics for Random k-SAT

Amin Coja-Oghlan[1], Andreas Goerdt[2], and André Lanka[2]

[1] Humboldt-Universität zu Berlin, Institut für Informatik
Unter den Linden 6, 10099 Berlin, Germany
coja@informatik.hu-berlin.de
[2] Technische Universität Chemnitz, Fakultät für Informatik
Straße der Nationen 62, 09107 Chemnitz, Germany
{goerdt,lanka}@informatik.tu-chemnitz.de

Abstract. A simple first moment argument shows that in a randomly chosen k-SAT formula with m clauses over n boolean variables, the fraction of satisfiable clauses is at most $1 - 2^{-k} + o(1)$ as $m/n \to \infty$ almost surely. In this paper, we deal with the corresponding algorithmic *strong refutation problem*: given a random k-SAT formula, can we find a *certificate* that the fraction of satisfiable clauses is at most $1 - 2^{-k} + o(1)$ in polynomial time? We present heuristics based on spectral techniques that in the case $k = 3$, $m \geq \ln(n)^6 n^{3/2}$ and in the case $k = 4$, $m \geq C n^2$ find such certificates almost surely, where C denotes a constant. Our methods also apply to some hypergraph problems.

1 Introduction and Results

The *k-SAT problem* – given a set of k-clauses, i.e. disjunctions of k literals over a set of boolean variables, decide whether there exists an assignment of the variables that satisfies all clauses – is *the* generic NP-complete problem. In addition to the decision version, the optimization version MAX k-SAT – given a set of k-clauses, find an assignment that satisfies the maximum number of clauses – is of fundamental interest as well. However, Håstad [18] has shown that there is no polynomial time algorithm that approximates MAX k-SAT within a factor better than $1-2^{-k}$, unless $\mathcal{P} = \mathcal{NP}$. Hence, it is NP-hard to distinguish between instances of k-SAT in which a $(1-\varepsilon)$-fraction of the clauses can be satisfied, and instances in which every truth assignment satisfies at most a $(1 - 2^{-k} + \varepsilon)$-share of the clauses for any $\varepsilon > 0$. Indeed, Håstad's NP-hardness result is best possible, as by picking a random assignment, we can satisfy a $(1 - 2^{-k})$-fraction of the clauses in polynomial time.

These hardness results motivate the study of *heuristics* for k-SAT or MAX k-SAT that are successful at least on a large class of instances. From this point of view, the satisfiability problem is interesting in two respects. First, one could ask for heuristics for *finding* a satisfying assignment (in the case of k-SAT) or a "good" assignment (in the case of MAX k-SAT). This problem has been studied e.g. by Flaxman [11], who has shown that in a rather general model of random satisfiable formulas a satisfying assignment can be found in polynomial time almost surely (cf. also [22] for an extension to semirandom formulas). Secondly,

K. Jansen et al. (Eds.): APPROX and RANDOM 2004, LNCS 3122, pp. 310–321, 2004.

one can ask for heuristics that can *refute* a k-SAT instance, i.e. find a certificate that no satisfying assignment exists; of course, in the worst-case this problem is coNP-complete. In this paper, we deal with the second problem. More precisely, we present *strong refutation heuristics*, i.e. heuristics that certify that no assignment satisfying considerably more than the trivial $(1 - 2^{-k})$-fraction of the clauses exists. One motivation for studying this problem is the relationship between the existence of strong refutation heuristics and approximation complexity pointed out by Feige [8].

In order to analyze a heuristic rigorously, we need to specify on which type of instances the heuristic is supposed to work properly. In this paper, we consider a standard model of *random* instances of MAX k-SAT. Let $V = \{x_1, \ldots, x_n\}$ be a set of n boolean variables. Then, there are $(2n)^k$ possible k-clauses over the variables V. If $0 < p < 1$, then we let $\text{Form}_{n,k,p}$ be a random set of k-clauses obtained by including each of the $(2n)^k$ possible clauses with probability p independently. Hence, the expected number of clauses in $\text{Form}_{n,k,p}$ is $m = (2n)^k p$.

The combinatorial structure of random k-SAT formulas has attracted considerable attention. Friedgut [12] has shown that $\text{Form}_{n,k,p}$ exhibits a *sharp threshold behavior*: there exist numbers $c_k = c_k(n)$ such that $\text{Form}_{n,k,p}$ is satisfiable almost surely if $m < (1 - \varepsilon)c_k n$, whereas $\text{Form}_{n,k,p}$ is unsatisfiable almost surely if $m > (1 + \varepsilon)c_k n$. The asymptotic behavior of c_k as $k \to \infty$ has been determined by Achlioptas and Peres [3] (building on the work of Achlioptas and Moore [1]). Moreover, a simple first moment argument shows that the maximum number of clauses of $\text{Form}_{n,k,p}$ that can be satisfied by any assignment is at most $(1 - 2^k + o(1))m$ as $m/n \to \infty$. More precise results have been obtained by Achlioptas, Naor, and Peres [2].

Various types of resolution proofs for the non-existence of satisfying assignments have been investigated on $\text{Form}_{n,k,p}$. Ben-Sasson [5] has shown that tree-like resolution proofs have size $\exp(\Omega(n/\Delta^{1/(k-2)+\varepsilon}))$ almost surely, where $\Delta = n^{k-1}p$ and $0 < \varepsilon < 1/2$ is an arbitrary constant. Hence, tree-like resolution proofs are of exponential length even if the expected number of clauses is $n^{k-1-\delta}$ (i.e. $p = 1/n^{1+\delta}$) for any constant $\delta > 0$. Furthermore, [5, Theorem 2.24] shows that general resolution proofs almost surely have exponential size if $p \leq n^{-k/2-\delta}$ ($\delta > 0$ constant).

Goerdt and Krivelevich [17] have suggested a heuristic that uses spectral techniques for refuting $\text{Form}_{n,4,p}$ with $p = \ln(n)^7 n^{-2}$ (i.e. the expected number of clauses is $m = \ln(n)^7 n^2$). No efficient resolution-based refutation heuristic is known for this range of p. Removing the polylogarithmic factor, Feige and Ofek [9] and (independently) Coja-Oghlan, Goerdt, Lanka, and Schädlich [7] have shown that spectral techniques can be used to refute $\text{Form}_{n,4,p}$ if $p \geq Cn^{-2}$ for a sufficiently large constant $C > 0$. Moreover, Feige and Ofek [10] have shown that a heuristic that combines spectral techniques with extracting and refuting a XOR formula from $\text{Form}_{n,3,p}$ can refute $\text{Form}_{n,3,p}$ for $p \geq Cn^{-3/2}$ (i.e. $m = Cn^{3/2}$). This result improves on previous work by Friedman and Goerdt [13], and Goerdt and Lanka [16]. We emphasize that in all of the above cases, the values of p to which the refutation heuristics apply exceed the threshold when $\text{Form}_{n,k,p}$ becomes unsatisfiable almost surely by at least a factor of $n^{(k-2)/2}$.

The new aspect in the present paper is that we deal with *strong* refutation heuristics. That is, our aim are heuristics that on input $\text{Form}_{n,k,p}$ almost surely certify that not more than a $(1 - 2^{-k} + \varepsilon)$-fraction of the clauses can be satisfied, for any $\varepsilon > 0$. This aspect has not (at least not explicitly) been studied in the aforementioned references. For instance, resolution proofs cannot provide strong refutation. Moreover, the spectral heuristics studied so far [7, 9, 10, 13, 16, 17] only certify that every assignment leaves a $o(1)$-fraction of the clauses unsatisfied. With respect to MAX 3-SAT, we have the following result.

Theorem 1. *Suppose that $p \geq \ln(n)^6 n^{-3/2}$. Let $\varepsilon > 0$ be an arbitrarily small constant. There is a polynomial time algorithm 3-Refute that satisfies the following.*

- Correctness: *For any MAX 3-SAT instance φ, the output of 3-Refute(φ) is an upper bound on the number of satisfiable clauses.*
- Completeness: *If $\varphi = \text{Form}_{n,3,p}$, then 3-Refute$(\varphi) \leq (7 + \varepsilon)n^3 p$ almost surely.*

Since the number of clauses of $\text{Form}_{n,3,p}$ is $\sim 8n^3 p$ almost surely, 3-Refute does indeed certify almost surely that not more than a $\frac{7}{8} + \varepsilon$ fraction of the clauses can be satisfied. Note that the value of p required for Theorem 1 is by a factor of $\ln(n)^6$ larger than that required by the heuristic of Feige and Ofek [10] (which does not provide strong refutation). We sketch the proof of Theorem 1 in Section 3. The following result addresses MAX 4-SAT.

Theorem 2. *Suppose that $p \geq c_0 n^{-2}$ for a sufficiently large constant $c_0 > 0$. There is a polynomial time algorithm 4-Refute that satisfies the following.*

- Correctness: *For any MAX 4-SAT instance φ, the output of 4-Refute(φ) is an upper bound on the number of satisfiable clauses.*
- Completeness: *If $\varphi = \text{Form}_{n,4,p}$, then almost surely 4-Refute$(\varphi) \leq 15n^4 p + c_1 n^3 \sqrt{p}$, where $c_1 > 0$ is a constant.*

4-Refute almost surely certifies that not more than a $\frac{15}{16} + O((n\sqrt{p})^{-1})$ fraction of the clauses can be satisfied. The second order term $O((n\sqrt{p})^{-1})$ gets arbitrarily small as $n^2 p$ grows. Theorem 2 applies to the same range of p as the best previously known refutation heuristics [7, 9] for 4-SAT, but provides strong refutation. We describe the heuristic for Theorem 2 in Section 2.

The algorithms for Theorems 1 and 2 build on and extend the techniques proposed in [7, 15]. For instance, 4-Refute constructs several graphs from the input formula $\varphi = \text{Form}_{n,4,p}$. To each of these graphs, 4-Refute applies a subroutine that tries to certify that the graph has "low discrepancy"; i.e. every set of vertices spans approximately the expected number of edges. This subroutine in turn relies on computing the eigenvalues of a certain auxiliary matrix. Finally, if all graphs have passed the discrepancy check, then we conclude that the input formula φ does not admit an assignment that satisfies more than $15n^4 p + c_1 n^3 \sqrt{p}$

clauses. The MAX 3-SAT algorithm for Theorem 1 proceeds similarly, but is a bit more involved. Though in contrast to [7, 15] we obtain strong refutation heuristics, the algorithms and the proofs in the present paper are simpler.

We point out that the approach in [15] can be used to obtain a heuristic that almost surely certifies that at most a $(\frac{7}{8} + \varepsilon)$-fraction of the clauses of $Form_{n,3,p}$ can be satisfied, though this issue is not addressed explicitly in that paper. However, the methods in [15] only seem to apply to somewhat bigger values of the clause probability p (namely, $p \geq n^{-3/2+\delta}$, $\delta > 0$ fixed) than those addressed in Theorem 1.

The techniques that the algorithms 3-Refute and 4-Refute rely on yield heuristics for a variety of further hard computational problems, e.g. for hypergraph problems. Recall that a k-uniform hypergraph H consists of a set $V(H)$ of vertices and a set $E(H)$ of edges. The edges are subsets of $V(H)$ of cardinality k. An independent set in H is a set $S \subset V(H)$ such that there is no edge $e \in E(H)$ with $e \subset S$. The independence number $\alpha(H)$ is the number of vertices in a maximum independent set. Moreover, H is called κ-colorable, if there exists κ independent sets $S_1, \ldots, S_\kappa$ in H such that $S_1 \cup \cdots \cup S_\kappa = V(H)$. The chromatic number $\chi(H)$ is the least integer $\kappa \geq 1$ such that H is κ-colorable.

In analogy with the $Form_{n,k,p}$ model of random k-SAT instances, there is the $H_{n,k,p}$-model of random k-uniform hypergraphs: the vertex set of $H_{n,k,p}$ is $V = \{1, \ldots, n\}$, and each of the $\binom{n}{k}$ possible edges is present with probability $0 < p < 1$ independently. Krivelevich and Sudakov [21] have solved the combinatorial problem of determining the probable value of the independence number and of the chromatic number of random hypergraphs. The following two theorems deal with the algorithmic problem of refuting that a 3-uniform hypergraph has a large independent set, or that a 4-uniform hypergraph is κ-colorable. We sketch the heuristics for Theorems 3 and 4 in Section 4.

Theorem 3. Let $\varepsilon > 0$ be arbitrarily small but fixed. Suppose that $p = f/n^{3/2}$, where $\ln^6 n \leq f = o(n^{1/2})$. There is a polynomial time algorithm 3-RefuteInd that satisfies the following.

- Correctness: If H is a 3-uniform hypergraph, then 3-RefuteInd(H) either outputs "α is small" or "fail". If the answer is "α is small", then $\alpha(H) < \varepsilon n$.
- Completeness: On input $H = H_{n,3,p}$, 3-RefuteInd(H) outputs "α is small" almost surely.

Theorem 4. Let $\kappa \geq 2$ be an integer. Suppose that $p \geq c_0 \kappa^4 n^{-2}$ for some sufficiently large constant $c_0 > 0$. There is a polynomial time algorithm 4-RefuteCol that satisfies the following.

- Correctness: If H is a 4-uniform hypergraph, 4-RefuteCol(H) either outputs "not κ-colorable" or "fail". If the answer is "not κ-colorable", then $\chi(H) > \kappa$.
- Completeness: On input $H = H_{n,4,p}$, the output of 4-RefuteCol(H) is "not κ-colorable" almost surely.

2   Random MAX 4-SAT

In Section 2.2 we present the heuristic for Theorems 2. The main tool is a procedure for certifying that a random bipartite graph is of low discrepancy. This procedure is the content of Section 2.1.

2.1   Discrepancy in Random Bipartite Graphs

Throughout, we let $V_1 = \{v_1, \ldots, v_n\}$ and $V_2 = \{w_1, \ldots, w_n\}$ be two disjoint sets consisting of n labeled vertices each. We consider bipartite graphs G with bipartition (V_1, V_2), i.e. the vertex set of G is $V_1 \cup V_2$, and all edges of G have one vertex in V_1, and one in V_2. If $S_1 \subset V_1$ and $S_2 \subset V_2$, then we let $E_G(S_1, S_2)$ denote the set of edges in G that connect a vertex in S_1 with a vertex in S_2. Furthermore, $B_{n,p}$ denotes a random bipartite graph obtained by including each possible edge $\{v_i, w_j\}$ with probability p independently. The aim in this section is to prove the following proposition.

Proposition 5. *Suppose that $np \geq c_0$ for some sufficiently large constant $c_0 > 0$. There is a polynomial time algorithm* BipDisc *and a constant $c_1 > 0$ such that the following two conditions hold.*

1. *Let G be a bipartite graph with bipartition (V_1, V_2). On input G,* BipDisc *either outputs "low discrepancy" or "fail". If* BipDisc(G) *outputs "low discrepancy", then for any two sets $S_i \subset V_i$, $i = 1, 2$, we have*

$$\left| |S_1||S_2|p - |E_B(S_1, S_2)| \right| \leq c_1 \sqrt{|S_1||S_2|np} + n \exp(-np/c_1). \qquad (1)$$

2. BipDisc$(B_{n,p})$ *outputs "low discrepancy" almost surely.*

If $|S_1|, |S_2| = \Omega(n)$, then Eq. (1) entails that the number $|E_G(S_1, S_2)|$ of edges from S_1 to S_2 in G deviates from its expectation $|S_1||S_2|p$ "not too much". The crucial point is that BipDisc certifies that Eq. (1) holds for *all* sets S_1, S_2.

BipDisc is based on computing the eigenvalues of a certain auxiliary matrix. Given a graph B with bipartition (V_1, V_2), we let $A = A(B) = (a_{ij})_{i,j=1,\ldots,n}$ be the matrix with entries $a_{ij} = 1$ if $\{v_i, w_j\} \in E(B)$, and $a_{ij} = 0$ if $\{v_i, w_j\} \notin E(B)$. Let J denote an $n \times n$ matrix with all entries equal to 1. Then, we let $M = M(B) = pJ - A(B)$. Furthermore, let $\|M\| = \sup\{\|M\xi\| : \xi \in \mathbf{R}^n, \|\xi\| = 1\}$ denote the norm of M. On input B, $\|M\|$ can be computed in polynomial time up to an arbitrarily small additive error. The next lemma shows what $\|M\|$ has to do with discrepancy certification.

Lemma 6. *Let B be a graph with bipartition (V_1, V_2). Then, for any two sets $S_i \subset V_i$, $i = 1, 2$, the inequality $\left| |E_B(S_1, S_2)| - |S_1||S_2|p \right| \leq \sqrt{|S_1||S_2|} \cdot \|M(B)\|$ holds.*

Sketch of proof. Let ξ_i be the characteristic vector of S_i, i.e. the j'th entry of ξ_1 (respectively ξ_2) is 1 if $v_j \in S_1$ (respectively $w_j \in S_2$), and 0 otherwise.

Then $\|\xi_i\| = \sqrt{|S_i|}$. Hence, $|\langle M\xi_2, \xi_1\rangle| \leq \sqrt{|S_1||S_2|} \cdot \|M\|$. Moreover, a direct computation shows that $\langle M\xi_2, \xi_1\rangle = |S_1||S_2|p - |E_B(S_1, S_2)|$.   □

In the case $np \geq \ln(n)^7/n$, one can show that $\|M\| \leq O(\sqrt{np})$ almost surely (via the "trace method" from [14]). Hence, in this case, by Lemma 6 we could certify that (1) holds almost surely just by computing $\|M(B_{n,p})\|$. In the case $np = O(1)$, however, we almost surely have that $\|M(B_{n,p})\| \gg np$, i.e. $\|M(B_{n,p})\|$ is much too large to give the bound (1) (cf. [20] for a detailed discussion). Following an idea of Alon and Kahale [4], we avoid this problem by removing all edges that are incident with vertices whose degree is too high (at least $10np$, say). This leads to the following algorithm.

Algorithm 7. BipDisc(G)
Input: A bipartite graph $G = (V_1, V_2, E)$.
Output: Either "low discrepancy" or "fail".

1. If the number of vertices in G that have degree $> 10np$ is $> n\exp(-c_2np)$, then output "fail" and halt. Here $c_2 > 0$ is a sufficiently small constant.
2. If the number of edges in G that are incident with vertices of degree $> 10np$ is larger than $c_3n^2p\exp(-c_2np)$, where $c_3 > 0$ is a sufficiently large constant, then halt with output "fail".
3. Let G' be the graph obtained from G by deleting all edges that are incident with vertices of degree $> 10np$. Let $M = M(G')$. If $\|M\| > c_4\sqrt{np}$ for a certain constant c_4, then output "fail" and halt.
4. Output "G has low discrepancy".

The analysis of BipDisc is based on two lemmas.

Lemma 8. *There are constants $c_2, c_3 > 0$ such that almost surely $B = B_{n,p}$ has the following properties.*

1. *The set S of all vertices of degree $> 10np$ has cardinality $\leq n\exp(-c_2np)$.*
2. *The number of edges in B that are incident with at least one vertex in S is $\leq c_3n^2p\exp(-c_2np)$.*

Lemma 9. *There is a constant $c_4 > 0$ such that almost surely the random bipartite graph $B = B_{n,p}$ enjoys the following property. Let B' be the graph obtained from B by deleting all edges that are incident with vertices of degree $> 10np$ in B. Then, $\|M(B')\| \leq c_4\sqrt{np}$.*

Lemma 8 follows from a standard computation. The proof of Lemma 9 is based on estimates on the eigenvalues of random matrices from [4].

Proof of Proposition 5. Let $G = B_{n,p}$, and let $S_i \subset V_i$ for $i = 1, 2$. Moreover, let S be the set of vertices of degree $> 10np$ in G. Suppose that BipDisc(G) answers "low discrepancy". Then, by Lemma 6,

$$|E_G(S_1 \setminus S, S_2 \setminus S)| - |S_1 \setminus S||S_2 \setminus S|p| \leq c_4\sqrt{|S_1||S_2|np}.$$

Moreover, because of Step 2 of BipDisc, we have $|E_G(S_1, S_2)| - |E_G(S_1 \setminus S, S_2 \setminus S)| \leq c_3 n^2 p \exp(-c_2 np)$. Further,

$$|S_1||S_2|p - |S_1 \setminus S||S_2 \setminus S|p \leq 3np|S| \leq n \exp(-c_2 np/2),$$

as otherwise Step 1 would have failed. Thus, (1) holds for S_1, S_2. Finally, by Lemmas 8 and 9, BipDisc($B_{n,p}$) outputs "low discrepancy" almost surely.   □

2.2　The Refutation Heuristic for 4-SAT

Throughout this section, we let $V = \{x_1, \ldots, x_n\}$ be a set of n propositional variables. Moreover, we assume that $n^2 p \geq c_0$ for a large constant c_0.

Table 1. Clause types and unsatisfied clauses in the case of 4-SAT.

i	type	$A_i \subset V_1$	$B_i \subset V_2$	i	type	$A_i \subset V_1$	$B_i \subset V_2$
1	$x_1 \vee x_2 \vee x_3 \vee x_4$	$F \times F$	$F \times F$	9	$\bar{x}_1 \vee x_2 \vee x_3 \vee x_4$	$T \times F$	$F \times F$
2	$x_1 \vee x_2 \vee x_3 \vee \bar{x}_4$	$F \times F$	$F \times T$	10	$\bar{x}_1 \vee x_2 \vee x_3 \vee \bar{x}_4$	$T \times F$	$F \times T$
3	$x_1 \vee x_2 \vee \bar{x}_3 \vee x_4$	$F \times F$	$T \times F$	11	$\bar{x}_1 \vee x_2 \vee \bar{x}_3 \vee x_4$	$T \times F$	$T \times F$
4	$x_1 \vee x_2 \vee \bar{x}_3 \vee \bar{x}_4$	$F \times F$	$T \times T$	12	$\bar{x}_1 \vee x_2 \vee \bar{x}_3 \vee \bar{x}_4$	$T \times F$	$T \times T$
5	$x_1 \vee \bar{x}_2 \vee x_3 \vee x_4$	$F \times T$	$F \times F$	13	$\bar{x}_1 \vee \bar{x}_2 \vee x_3 \vee x_4$	$T \times T$	$F \times F$
6	$x_1 \vee \bar{x}_2 \vee x_3 \vee \bar{x}_4$	$F \times T$	$F \times T$	14	$\bar{x}_1 \vee \bar{x}_2 \vee x_3 \vee \bar{x}_4$	$T \times T$	$F \times T$
7	$x_1 \vee \bar{x}_2 \vee \bar{x}_3 \vee x_4$	$F \times T$	$T \times F$	15	$\bar{x}_1 \vee \bar{x}_2 \vee \bar{x}_3 \vee x_4$	$T \times T$	$T \times F$
8	$x_1 \vee \bar{x}_2 \vee \bar{x}_3 \vee \bar{x}_4$	$F \times T$	$T \times T$	16	$\bar{x}_1 \vee \bar{x}_2 \vee \bar{x}_3 \vee \bar{x}_4$	$T \times T$	$T \times T$

Let φ be a set of 4-clauses over V. To employ the procedure BipDisc from Section 2.1, we construct 16 bipartite graphs $G^{(1)}, \ldots, G^{(16)}$ from φ. Each $G^{(i)}$ is a graph with bipartition (V_1, V_2), where $V_i = V \times V \times \{i\}$ (i.e. V_1, V_2 are disjoint copies of $V \times V$). Each graph $G^{(i)}$ corresponds to one of the 16 possible ways to place the negation signs in a 4-clause: in $G^{(i)}$, the edge $\{(x_{i_1}, x_{i_2}, 1), (x_{i_3}, x_{i_4}, 2)\}$ is present iff the clause $l_{i_1} \vee l_{i_2} \vee l_{i_3} \vee l_{i_4}$ is contained in φ, where l_{i_j} is either x_{i_j} or $\bar{x}_{i_j}$, according to the negation signs in Table 1. For instance, the edge $\{(x_{i_1}, x_{i_2}, 1), (x_{i_3}, x_{i_4}, 2)\}$ is in $G^{(7)}$ iff the clause $x_{i_1} \vee \bar{x}_{i_2} \vee \bar{x}_{i_3} \vee x_{i_4}$ occurs in φ. Thus, each clause of φ induces an edge in one of the graphs $G^{(i)}$, and each edge results from a unique clause. The algorithm for Theorem 2 is as follows.

Algorithm 10. 4-Refute(φ)
Input: A set φ of 4-clauses over V.
Output: An upper bound on the number of satisfiable clauses.

1. If the number of clauses in φ is larger than $16n^4 p + n^3 \sqrt{p}$, then return the total number of clauses in φ as an upper bound and halt.
2. Compute the graphs $G^{(i)}$ and run BipDisc($G^{(i)}$) for $i = 1, \ldots, 16$. If the answer of BipDisc($G^{(i)}$) is "fail" for at least one i, then return the total number of clauses in φ and halt.
3. Return $15n^4 p + c_1 n^3 \sqrt{p}$, where c_1 is a sufficiently large consant.

Let us first prove that 4-Refute outputs an upper bound on the number of clauses that can be satisfied.

Lemma 11. *There is a constant $c_2 > 0$ such that the following holds. Let φ be a set of 4-clauses such that BipDisc($G^{(i)}$) answers "low discrepancy" for all i. Then there is no assignment that satisfies more than $|\varphi| - n^4 p + c_2 n^3 \sqrt{p}$ clauses.*

Proof. Consider an assignment that sets the variables $T \subset V$ to true, and $F = V \setminus T$ to false. We shall bound the number of edges in the graphs $G^{(i)}$ that correspond to unsatisfied clauses. Let $A_i \subset V_1$ and $B_i \subset V_2$ be the sets defined in Table 1 for $i = 1, \ldots, 16$. Then, in the graph $G^{(i)}$, the edges corresponding to unsatisfied clauses are precisely the A_i-B_i-edges. Thus, invoking Proposition 5, we have the following bound on the number of unsatisfied clauses:

$$\sum_{i=1}^{16} |E_{G_i}(A_i, B_i)| \geq \sum_{i=0}^{4} \binom{4}{i} \left(|F|^i |T|^{4-i} p - c_3 n^3 \sqrt{p} \right) \geq n^4 p - c_2 n^3 \sqrt{p},$$

where c_2, c_3 are suitable constants.   □

Proof of Theorem 2. The correctness follows from Lemma 11. Since by Chernoff bounds (cf. [19, p. 26]) the number of clauses in Form$_{n,4,p}$ is at most $16 n^4 p + o(n^3 \sqrt{p})$ almost surely, the completeness follows from Proposition 5.   □

3   Random MAX 3-SAT

While our refutation heuristic for 4-SAT is based on certifying that certain (bipartite) graphs are of low discrepancy, the heuristic for 3-SAT needs to certify that a couple of triple systems are of low discrepancy. In Section 3.1, we describe the procedure for certifying low discrepancy in triple systems. Then, in Section 3.2, we show how to employ this procedure in order to refute MAX 3-SAT strongly.

3.1   Discrepancy in Triple Systems

Let $V = \{x_1, \ldots, x_n\}$ be a fixed set of cardinality n. In this section, we consider *triple systems* over V, i.e. subsets $S \subset V \times V \times V$. If $V_1, V_2, V_3 \subset V$, then we let $(V_1, V_2, V_3)_S$ signify the set of triples $(v_1, v_2, v_3) \in S$ with $v_i \in V_i$ for $i = 1, 2, 3$. Let $\varepsilon > 0$ be a constant. We say that S has *low discrepancy with respect to ε* if the following holds for all $X \subseteq V$ with $\varepsilon n \leq |X| \leq (1 - \varepsilon) n$: letting $Y = V \setminus X$ and $\alpha = |X|/n$, we have

$$|(X, X, X)_S| = (1 + o(1)) \cdot \alpha^3 \cdot |S|,$$
$$|(X, X, Y)_S|, |(X, Y, X)_S|, |(Y, X, X)_S| = (1 + o(1)) \cdot \alpha^2 (1 - \alpha) \cdot |S|,$$
$$|(X, Y, Y)_S|, |(Y, X, Y)|, |(Y, Y, X)_S| = (1 + o(1)) \cdot \alpha (1 - \alpha)^2 \cdot |S|,$$
$$|(Y, Y, Y)_S| = (1 + o(1)) \cdot (1 - \alpha)^3 \cdot |S|.$$

For $0 < p < 1$, we obtain the random triple system $S_{n,p}$ by including each triple in V^3 with probability p independently. The aim of this section is to prove the following propostion.

Proposition 12. *Let $\varepsilon > 0$. There is a polynomial time algorithm* TripleDisc$_\varepsilon$ *that satisfies the following conditions.*

- *For each triple system $S \subset V^3$ the output of* TripleDisc$_\varepsilon(S)$ *is either "low discrepancy" or "fail". If the output is "low discrepancy", then S has low discrepancy w.r.t. ε.*
- *If $p \geq \ln(n)^6 n^{-3/2}$, then the output of* TripleDisc$_\varepsilon(S_{n,p})$ *is "low discrepancy" almost surely.*

To certify that the triple system $S \subset V^3$ is of low discrepancy, the algorithm TripleDisc constructs three *projection graphs* G_{ij}, $1 \leq i < j \leq 3$. The vertex set of G_{ij} is V, and the edge $\{x, y\}$ is present in G_{ij} iff there is a triple $(z_1, z_2, z_3) \in S$ with $x = z_i$ and $y = z_j$, or $x = z_j$ and $y = z_i$. Thus, if $S = S_{n,p}$, then the edge $\{x, y\}$ is present in G_{ij} with probability $p' \sim 2np$ independently of all other edges, so that G_{ij} is distributed as a binomial random graph $G_{n,p'}$.

We say that a graph $G = (V, E)$ has *low discrepancy w.r.t.* ε if for all $X \subset V$ of cardinality $\varepsilon n \leq |X| \leq (1 - \varepsilon)n$ we have

$$||E_G(X)| - |X|^2 n^{-2}|E|| \leq \varepsilon|E| \wedge ||E_G(X, V \setminus X)| - 2|X|(n - |X|)n^{-2}|E|| \leq \varepsilon|E|,$$

where $E_G(X)$ is the set of edges in G with both vertices in X, and $E_G(X, Y)$ is the set of edges in G with one vertex in X and the other in Y. One ingredient to the algorithm TripleDisc for Proposition 12 is to certify that the graphs G_{ij} are of low discrepancy. The following lemma provides us with a polynomial time algorithm for this problem.

Lemma 13. *Let $\varepsilon > 0$. Suppose that $p' \geq 1/n^{1/2}$. There is a polynomial time algorithm $\mathcal{A}$ that satisfies the following conditions.*

- Correctness: *For any graph $G = (V, E)$, the output of $\mathcal{A}(G)$ is either "low discrepancy" or "fail". If the output is "low discrepancy", then G has low discrepancy w.r.t. ε.*
- Completeness: *If $G = G_{n,p'}$, then the output of $\mathcal{A}(G)$ is "low discrepancy" almost surely.*

The proof of Lemma 13 is based on the relationship between graph discrepancy and eigenvalues (cf. [6]) and results on the eigenvalues of random symmetric matrices [14].

In order to certify that the triple system S has low discrepancy, it is, however, *not* sufficient to check that the projection graphs G_{ij} are of low discrepancy. Therefore, in addition to the projection graphs, one could consider the *product graph* $G_\pi = (V \times V, E_\pi)$, which is defined as follows: an edge $\{(a_1, b_1), (a_2, b_2)\}$ is in E_π iff there exists a $z \in V$ such that there are two different triples $(a_1, a_2, z), (b_1, b_2, z) \in S$. Note that in contrast to the projection graphs G_{ij}, the product graph G_π is not distributed as a binomial random graph (the edges do not occur independently). If the projection graphs G_{ij} and the product graph G_π all have low discrepancy, then S is of low discrepancy as well.

However, for the values of p in Proposition 12, we do not know a direct way to derive bounds on the eigenvalues of the adjacency matrix of the product graph (e.g. it seems difficult to apply the methods in [4, 9, 15]). Therefore, instead of dealing with the product graph and its adjacency matrix, we consider the matrix $\mathbf{A} = \mathbf{A}(S, p)$ defined as follows. For $0 < p < 1$ and $b_1, b_2, z \in V$ we let $B_{b_1 b_2 z} = B_{b_1 b_2 z}(S, p) = -1$ if $(b_1, b_2, z) \in S$, and $B_{b_1 b_2 z} = B_{b_1 b_2 z}(S, p) = p/(1 - p)$, otherwise. Then, the $n^2 \times n^2$-matrix $\mathbf{A} = \mathbf{A}(S, p) = (\mathbf{a}_{b_1 c_1, b_2 c_2})_{(b_1, c_1), (b_2, c_2) \in V^2}$ is given by

$$\mathbf{a}_{b_1 c_1, b_2 c_2} = \sum_{z \in V} (B_{b_1 b_2 z} \cdot B_{c_1 c_2 z} + B_{b_2 b_1 z} \cdot B_{c_2 c_1 z}) \text{ if } (b_1, b_2) \neq (c_1, c_2),$$

and $\mathbf{a}_{b_1 c_1, b_2 c_2} = 0$ if $(b_1, b_2) = (c_1, c_2)$. Since $\mathbf{A}$ is symmetric and real-valued, the matrix has n^2 real eigenvalues $\lambda_1 \geq \cdots \geq \lambda_{n^2}$. We let $\|\mathbf{A}\| = \max\{\lambda_1, -\lambda_{n^2}\}$ signify the norm of $\mathbf{A}$.

If $S \subset V^3$, $x \in V$, and $i \in \{1, 2, 3\}$, then the *degree of x in slot i* is $d_{x,i} = |\{(z_1, z_2, z_3) \in S : z_i = x\}|$. We say that S is *asymptotically regular* if $d_{x,i} = (1 + o(1))n^{-1}|S|$ for all x, i. Equipped with these definitions, we can state the following sufficient condition for S being of low discrepancy.

Lemma 14. *Let $f = pn^{3/2}$, and suppose that $\ln^6 n \leq f = o(n^{1/2})$. If S is a triple system that satisfies the following four conditions, then S is of low discrepancy w.r.t. $\varepsilon > 0$.*

1. *$s = |S| = f \cdot n^{3/2} \cdot (1 + o(1))$.*
2. *S is asymptotically regular.*
3. *The three projection graphs of S are of low discrepancy with respect to $\varepsilon > 0$.*
4. *We have $\|\mathbf{A}(S, p)\| \leq \ln^5 n \cdot f$.*

As by Lemma 13 we can check in polynomial time whether the conditions in Lemma 14 hold, we obtain the following algorithm.

Algorithm 15. TripDisc$_\varepsilon(S)$
Input: A set $S \subset V^3$. *Output:* Either "low discrepancy" or "fail".

1. Check whether Conditions 1–4 in Lemma 14 hold.
2. If so, output "low discrepancy". If not, return "fail".

In order to prove Proposition 12, it remains to establish that the algorithm is complete. A standard application of Chernoff bounds (cf. [19, p. 26]) shows that the random triple system $S = S_{n,p}$ with p as in Proposition 12 satisfies Conditions 1–2 in Lemma 14 almost surely. Moreover, the Condition 3 holds almost surely by Lemma 13. Thus, it suffices to show that Condition 4 holds almost surely. The rather technical proof of the following lemma is based on the trace method from [14].

Lemma 16. *Let $f \geq \ln(n)^6$, and let $p = fn^{-3/2}$. If $S = S_{n,p}$, then $\|\mathbf{A}(S, p)\| \leq \ln^5 n \cdot f$ almost surely.*

3.2   The Refutation Heuristic for 3-SAT

Let φ be a set of 3-clauses over the variable set $V = \{x_1, \ldots, x_n\}$. To apply the procedure `TripDisc` from Section 3.1, we construct 8 triple systems $S^{(1)}, \ldots, S^{(8)} \subset V^3$ from φ, each corresponding to one of the 8 possible ways to set the negation signs in a 3-clause. In the triple system $S^{(i)}$, the triple $(x_{i_1}, x_{i_2}, x_{i_3}) \in V^3$ is present iff the clause $l_{i_1} \vee l_{i_2} \vee l_{i_3}$ occurs in φ, where either $l_{i_j} = x_{i_j}$ or $l_{i_j} = \bar{x}_{i_j}$, according to the negation signs for the clause types (where clause types are defined similarly as in Section 2.2).

Algorithm 17. `3-Refute`(φ, ε)
Input: A set φ of 3-clauses over V.
Output: An upper bound on the number of satisfiable clauses.

1. Compute the triple systems $S^{(i)}$ and run `TripDisc`$_{\varepsilon/8}(S^{(i)})$ for $i = 1, \ldots, 8$. If the output is "fail" for at least one i, then return the total number of clauses in φ as an upper bound and halt.
2. Return $(7 + \varepsilon)n^3 p$.

The proof of Theorem 1 relies on a similar argument as the proof of Lemma 11.

4   Hypergraph Problems

Let $H = (V, E) = H_{n,4,p}$ be a random 4-uniform hypergraph with vertex set $V = \{1, \ldots, n\}$. Let κ be an integer, and suppose that $p \geq c_0 \kappa^4 n^{-2}$ for a sufficiently large constant c_0. The algorithm `4-RefuteCol` for Theorem 4 is randomized. On input H, the algorithm obtains a set $S \subset V^4$ of *ordered* 4-tuples as follows (recall that the edges E are not ordered). If $e = \{x_1, x_2, x_3, x_4\} \in E$, then there are $4! = 24$ possibilities to order the vertices x_1, x_2, x_3, x_4. Let $T(e)$ be the set of the 24 possible ordered tuples. Letting $p_0 = 1 - (1-p)^{1/24}$, we choose the set $\emptyset \neq X_e \subset T(e)$ of tuples that we include into S to represent e according to the distribution $P(X_e) = p_0^{|X_e|}(1-p_0)^{24-|X_e|}p^{-1}$. Thus, each edge $e \in E$ gives rise to at least one tuple in S. The choice of the sets X_e is independent for all $e \in E$. Furthermore, we include each tuple $(x_1, x_2, x_3, x_4) \in V^4$ such that $|\{x_1, x_2, x_3, x_4\}| < 4$ into S with probability p_0 independently. A trite computation shows that if $H = H_{n,4,p}$, then the resulting set $S = S(H)$ of 4-tuples is distributed so that every possible 4-tuple in V^4 is present with probability p_0 independently.

Let $V_1 = V \times V \times \{1\}$, $V_2 = V \times V \times \{2\}$ be two disjoint copies of V. Having computed $S = S(H)$, `4-RefuteCol` constructs a graph G with bipartition (V_1, V_2) in which the edge $\{(x_1, x_2, 1), (x_3, x_4, 2)\}$ is present iff $(x_1, x_2, x_3, x_4) \in S$. If $H = H_{n,4,p}$, then G is a random bipartite graph B_{n^2,p_0}. To this graph G, `4-RefuteCol` applies the procedure `BipDisc`. If `BipDisc` answers "low discrepancy", then `4-RefuteCol` answers "H is not κ-colorable". Otherwise, the output is "fail".

The heuristic `3-RefuteInd`(H) for Theorem 3 transforms the hypergraph H into a triple system and applies `TripleDisc`.

Acknowledgment

We are grateful to Uri Feige for helpful discussions.

References

1. Achlioptas, D., Moore, C.: The asymptotic order of the k-SAT threshold. Proc. 43rd FOCS (2002) 779–788
2. Achlioptas, D., Naor, A., Peres, Y.: The fraction of satisfiable clauses in a typical formula. Proc. 44th FOCS (2003) 362–370
3. Achlioptas, D., Peres, Y.: The threshold for random k-SAT is $2^k \ln 2 - O(k)$. Proc. 35th STOC (2003) 223–231
4. Alon, N., Kahale, N.: A spectral technique for coloring random 3-colorable graphs. SIAM J. Comput. **26** (1997) 1733–1748
5. Ben-Sasson, E.: Expansion in Proof Complexity. PhD thesis.
   http://www.eecs.harvard.edu/~eli/papers/thesis.ps.gz
6. Chung, F.K.R.: Spectral Graph Theory. American Mathematical Society 1997
7. Coja-Oghlan, A., Goerdt, A., Lanka, A., and Schädlich, F.: Certifying unsatisfiability of random $2k$-Sat instances using approximation techniques. Proc. 14th FCT (2003) 15–26
8. Feige, U.: Relations between average case complexity and approximation complexity. Proc. 24th STOC (2002) 534–543
9. Feige, U., Ofek, E.: Spectral techniques applied to sparse random graphs. Report MCS03-01, Weizmann Institute of Science (2003)
   http://www.wisdom.weizmann.ac.il/~erano/
10. Feige, U., Ofek, E.,: Easily refutable subformulas of large random 3CNF formulas. Weizmann Institute of Science (2003)
    http://www.wisdom.weizmann.ac.il/~erano/
11. Flaxman, A.: A spectral technique for random satisfiable 3CNF formulas. Proc. 14th SODA (2003) 357–363
12. Friedgut, E.: Necessary and sufficient conditions for sharp thresholds of graph properties and the k-SAT problem. Journal of the American Mathematical Society **12** (1999) 1017–1054
13. Friedman, J., Goerdt, A.: Recognizing more unsatisfiable random 3-Sat instances efficiently. Proc. 28th ICALP (2001) 310–321
14. Füredi, Z., Komlós, J.: The eigenvalues of random symmetric matrices. Combinatorica **1** (1981) 233–241
15. Goerdt, A., Jurdzinski, T.: Some results on random unsatisfiable k-SAT instances and approximation algorithms applied to random structures. Combinatorics, Probability and Computing **12** (2003) 245 – 267
16. Goerdt, A., Lanka, A.: Recognizing more random unsatisfiable 3-SAT instances efficiently. Electronic Notes in Discrete Mathematics **16** (2003)
17. Goerdt, A., Krivelevich., M.: Efficient recognition of random unsatisfiable k-SAT instances by spectral methods. Proc. 18th STACS (2001) 294–304
18. Håstad, J.: Some optimal inapproximability results. Journal of the ACM **48** (2001) 798–859
19. Janson, S., Łuczak, T., Ruciński, A.: Random graphs. John Wiley and Sons 2000
20. Krivelevich, M., Sudakov, B.: The largest eigenvalue of sparse random graphs. Combinatorics, Probability and Computing **12** (2003) 61–72
21. Krivelevich, M., Sudakov, B.: The chromatic numbers of random hypergraphs. Random Structures & Algorithms **12** (1998) 381–403
22. Vilenchik, D.: Finding a satisfying assignment for random satisfiable 3CNF formulas. M.Sc. thesis, Weizmann Institute of Science (2004)

Counting Connected Graphs and Hypergraphs via the Probabilistic Method

Amin Coja-Oghlan[1,*], Cristopher Moore[2,**], and Vishal Sanwalani[2]

[1] Humboldt-Universität zu Berlin, Institut für Informatik
Unter den Linden 6, 10099 Berlin, Germany
coja@informatik.hu-berlin.de
[2] University of New Mexico, Albuquerque NM 87131
{vishal,moore}@cs.unm.edu

Abstract. It is exponentially unlikely that a sparse random graph or hypergraph is connected, but such graphs occur commonly as the giant components of larger random graphs. This simple observation allows us to estimate the number of connected graphs, and more generally the number of connected d-uniform hypergraphs, on n vertices with $m = O(n)$ edges. We also estimate the probability that a binomial random hypergraph $H_d(n, p)$ is connected, and determine the expected number of edges of $H_d(n, p)$ conditioned on its being connected. This generalizes prior work of Bender, Canfield, and McKay [2] on the number of connected graphs; however, our approach relies on elementary probabilistic methods, extending an approach of O'Connell, rather than using powerful tools from enumerative combinatorics. We also estimate the probability for each t that, given $k = O(n)$ balls in n bins, every bin is occupied by at least t balls.

1 Introduction and Results

A *d-uniform hypergraph* H consists of a set V of vertices and a set E of edges, which are subsets of V of cardinality d. Thus, a 2-uniform hypergraph is just a graph. A vertex w is *reachable in H* from a vertex v if either $v = w$ or there is a sequence $e_1, \ldots, e_k$ of edges such that $v \in e_1$, $w \in e_k$, and $e_i \cap e_{i+1} \neq \emptyset$ for $i = 1, \ldots, k-1$. Reachability in H is an equivalence relation, and the equivalence classes are the *connected components* of H. We say that H is *connected* if there is only one component.

Connectedness is perhaps the most basic property of graphs and hypergraphs, and estimating the number of connected graphs or hypergraphs with a given number of vertices and edges is a fundamental combinatorial problem. In most applications, one is interested in asymptotic results, where the number of vertices/edges tends to infinity. The main result of this paper is a formula for the asymptotic number of connected d-uniform hypergraphs; or, equivalently, an estimate of the probability that a random d-uniform hypergraph is connected.

* Research supported by the Deutsche Forschungsgemeinschaft (DFG FOR 413/1-1).
** Supported by NSF PHY-0200909 and Los Alamos National Laboratory.

K. Jansen et al. (Eds.): APPROX and RANDOM 2004, LNCS 3122, pp. 322–333, 2004.
© Springer-Verlag Berlin Heidelberg 2004

Our results hold up to a constant multiplicative factor. We study both a model $H_d(n, m)$ in which the number of vertices and edges is fixed, and a binomial model $H_d(n, p)$ in which each possible edge appears with probability p; in the latter case we also calculate the expected number of edges conditioned on the graph being connected. Furthermore, we obtain a simple algorithm for generating a connected hypergraph uniformly at random.

We obtain these results using a new probabilistic approach. Rather than using powerful techniques of enumerative combinatorics such as generating functions and complex analysis, our calculations are for the most part elementary, and rely on the fact that a connected (hyper)graph of a given order and size is likely to occur as the giant component of a larger one. We believe that this approach is of interest in its own right, and can be applied to a number of problems in combinatorics and random (hyper)graph theory; to illustrate this, we also calculate the probability that every bin in a balls-and-bins experiment is occupied by at least t balls.

For the special case of graphs, i.e., for $d = 2$, the results we present in this paper (and, in some cases, stronger results) are already known. Compared to graphs, little is known about d-uniform hypergraphs where $d \geq 3$, although several papers deal with their component structure (cf. Section 2 for details).

Given a hypergraph $H = (V, E)$, its *order* is its number of vertices $|V|$, and its *size* is its number of edges $|E|$. The *degree* $d_H(v)$ of a vertex $v \in V$ is the number of edges $e \in E$ such that $v \in e$. Note that if H is d-uniform, its average degree is $d|E|/n$. Throughout, $V = \{1, \ldots, n\}$ is a fixed set of n labeled vertices and $d \geq 2$ is a fixed integer. We let $C_d(n, m)$ signify the number of connected d-uniform hypergraphs of order n and size m. Observe that if a hypergraph is connected, then its average degree is at least $d/(d-1)$.

Two natural random models present themselves. In $H_d(n, m)$, we select one of the $\binom{\binom{n}{d}}{m}$ sets of m possible edges at random. In $H_d(n, p)$, each of the $\binom{n}{d}$ edges appears independently with probability p, in which case the expected number of edges is $p\binom{n}{d} \sim pn^d/d!$ and the expected degree is $p\binom{n-1}{d-1} \sim pn^{d-1}/(d-1)!$. Thus we will compare $H_d(n, m)$ where $m = cn/d$ with $H_d(n, p)$ where $p = c(d-1)!/n^{d-1}$. While these two models are interchangeable in many respects, with respect to large deviations they are not; in particular, as we will see, their probabilities of connectedness differ by an exponential factor.

The following theorem determines the asymptotic probability of connectedness of $H_d(n, m)$, or equivalently the number of connected d-uniform hypergraphs, up to a constant factor in the regime where the average degree is $c > d/(d-1)$.

Theorem 1. *Let $c > d/(d-1)$ be a constant independent of n, and let $m = cn/d$. Let $a = a(c)$ be the unique solution in $(0, 1)$ of the equation*

$$1 - a = \exp\left(-ca \cdot \frac{1 - (1-a)^{d-1}}{1 - (1-a)^d}\right) \tag{1}$$

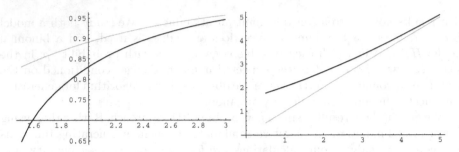

Fig. 1. Left figure: the functions $\Phi_3(c)$ (black) and $\Psi_3(c)$ (gray) in the range $1.5 < c < 3$. Right figure: the functions $\gamma_3(c)$ (black) and $c \mapsto c$ (gray) in the range $0.5 < c < 5$.

and set $\Phi_d(c) = a^{1-c}(1-a)^{(1/a)-1}(1-(1-a)^d)^{c/d}$. Then the probability that $H_d(n,m)$ is connected is $\Theta(\Phi_d(c)^n)$, and so $C_d(n,m) = \Theta\left(\Phi_d(c)^n \binom{\binom{n}{d}}{m}\right)$ (as $n \to \infty$).

Theorem 1 addresses the combinatorial problem of estimating the number of connected hypergraphs. But there is also a corresponding algorithmic problem: can we sample a connected hypergraph of a given order and size uniformly at random? We answer this question in the affirmative.

Theorem 2. Let $c > d/(d-1)$, and let $m = cn/d$. There is a randomized algorithm that samples a connected hypergraph of order n and size m uniformly at random in expected polynomial time.

Our next theorem gives an result analogous to Theorem 1 for $H_d(n,p)$.

Theorem 3. Let $c > 0$ be a constant, and let $p = c(d-1)!/n^{d-1}$. Let a be the unique solution in $(0,1)$ of the equation

$$1 - a = \exp\left(-c \cdot \frac{1-(1-a)^{d-1}}{a^{d-1}}\right) \tag{2}$$

and set $\Psi_d(c) = a(1-a)^{(1/a)-1}\exp\left(\frac{c}{d} \cdot \frac{1-a^d-(1-a)^d}{a^d}\right)$. Then the probability that $H_d(n,p)$ is connected is $\Theta(\Psi_d(c)^n)$. For $d = 2$, $\Psi_2(c) = 1 - \exp(-c)$.

To illustrate Theorems 1 and 3 we plot the functions $\Phi_3(c)$ and $\Psi_3(c)$ in Figure 1. In fact Ψ_3 is larger than Φ_3, i.e., $H_d(n,p)$ is exponentially more likely to be connected than $H_d(n,m)$ even when the two have the same (expected) average degree. The reason for this is that in $H_d(n,p)$ the number of edges is a random variable; we can think of $H_d(n,p)$ as first choosing a number of edges m' according to the binomial distribution $\mathrm{Bin}(\binom{n}{d},p)$, and then choosing a $H_d(n,m')$. Thus $H_d(n,p)$ can boost its probability of being connected by including a larger number m' of edges. Indeed, if we condition on $H_d(n,p)$ being connected, the conditional expectation of the number of edges will be significantly larger than $p\binom{n}{d}$. Our next theorem quantifies this observation.

Theorem 4. *As in Theorem 3, let $c > 0$ be a constant independent of n, let $p = c(d-1)!/n^{d-1}$, and let a be the unique solution in $(0,1)$ of Equation (2). Set*

$$\gamma_d(c) = c\left(\frac{1 - (1-a)^d}{a^d}\right).$$

The expected number of edges of $H_d(n,p)$ conditioned on the event that $H_d(n,p)$ is connected is $n\gamma_d(c)/d \pm o(n)$, so the expected average degree is $\gamma_d(c) \pm o(1)$. For $d=2$, $\gamma_2(c) = c \cdot (e^c + 1)/(e^c - 1) = c \cdot \coth(c/2)$.

The proof of Theorem 4 consists of maximizing the product of $\Phi_d(c = dm/n)^n$ times the binomial distribution $\text{Bin}(\binom{n}{d}, p)$. Figure 1 shows $\gamma_3(c)$. As $c \to \infty$, $\gamma_d(c) \to c$, since connectedness becomes a less unusual condition as c increases. As $c \to 0$, $\gamma_c(d) \to d/(d-1)$, the minimum average degree required for connectivity.

One ingredient of the proofs of Theorems 1–4 is the component structure of random hypergraphs $H_d(n,m)$ and $H_d(n,p)$. We say that $H_d(n,m)$ or $H_d(n,p)$ has a certain property *with high probability* (w.h.p.) if the probability that the property holds tends to 1 as $n \to \infty$. Suppose that the average degree $c = dm/n$ (resp. $c = pn^{d-1}/(d-1)!$ is a constant greater than $1/(d-1)$. Then, just as for graphs [6], w.h.p. there is a unique *giant component* of order $\Omega(n)$ in $H_d(n,p)$, and all other connected components are of order $O(\ln n)$ [7,13]. To prove Theorems 1–4, it is crucial to have a rather tight estimate on the order and size of the giant component.

We say that a random variable X defined on $H_d(n,m)$ or $H_d(n,p)$ is *concentrated in width y about x* if for each $\varepsilon > 0$ there is a number $C(\varepsilon)$ independent of n such that $\Pr[|X - x| > C(\varepsilon)y] < \varepsilon$. Further, we write $f(n) = \tilde{O}(g(n))$ if there is a constant $c > 0$ such that $|f(n)| \leq |g(n)| \ln^c n$ for all sufficiently large n.

Theorem 5. *Suppose that $c > 1/(d-1)$. Let $a = a(c)$ be the unique solution in $(0,1)$ of the equation*

$$1 - a = \exp\left(-c\left(1 - (1-a)^{d-1}\right)\right)$$

and let $b = b(c) = 1 - (1-a)^d$. Let $m = cn/d$ and let $p = c(d-1)!/n^{d-1}$. Then, the expected order (resp. size) of the giant component of $H_d(n,m)$ and of $H_d(n,p)$ is an $+ \tilde{O}(1)$ (resp. $bm + \tilde{O}(1)$). Moreover, the order (resp. size) of the giant component of both $H_d(n,m)$ and $H_d(n,p)$ is concentrated in width $\sqrt{n}$ about an (resp. bm).

Note that setting $d = 2$ in Theorem 5 recovers the classic result $1 - a = e^{-ca}$ for random graphs [6]. To further illustrate the applicability of our approach, we solve a type of occupancy problem. Suppose k balls are placed randomly in n bins; the following theorem estimates the probability $q_t(k,n)$ that every bin is occupied by at least t balls.

Theorem 6. *Let $t \geq 1$ be an integer, let $r > t$ be a constant independent of n, and let $k = rn$. Let $\alpha = \alpha(r)$ and $\beta = \beta(r)$ be the unique solutions in $(0,1)$ to the pair of equations*

$$\alpha = 1 - \sum_{j=0}^{t-1} \frac{e^{-r\alpha/\beta}(r\alpha/\beta)^j}{j!} \ , \quad \beta = 1 - \sum_{j=0}^{t-2} \frac{e^{-r\alpha/\beta}(r\alpha/\beta)^j}{j!}$$

Set $r_1 = r \cdot \frac{1-\beta}{\beta}\frac{1-\alpha}{\alpha}$, and let $\lambda = \lambda(r)$ be the unique solution of the equation $\sum_{j=1}^{t-1} \frac{\lambda^j}{(j-1)!} = r_1 \sum_{j=0}^{t-1} \frac{\lambda^j}{j!}$ unless $t = 1$, in which case set $\lambda = 1$. Then set

$$\phi(r) = \alpha^{1-r}(1-\alpha)^{\frac{1}{\alpha}+r(1-\frac{1}{\beta})-1}\left(r_1^{r_1}e^{-r_1}\prod_{j=0}^{t-1}\frac{1}{b_j^{b_j}(j!)^{b_j}}\right)^{1-1/\alpha}$$

where $b_j = \frac{\lambda^j/j!}{\sum_{\ell=0}^{t-1}\lambda^j/j!}$. Then, $q_t(k,n) = \Theta(\phi(r)^n)$ as $n \to \infty$.

For $t = 1$ in particular, Theorem 6 gives $\alpha = 1 - e^{-r\alpha}$, $\beta = 1$, and $r_1 = 0$, $\lambda = 1$, and $b_0 = 1$. Then $q_1(k,n) = \Theta\left((\alpha^{1-r}e^{r(\alpha-1)})^n\right)$. The proof of Theorem 6 is similar to the proofs of Theorems 1 and 3 and is omitted.

2   Related Work

The asymptotic number of connected *graphs* with a given number of edges and vertices was determined by Bender, Canfield, and McKay [2]. Their proof is based on generating functions, and is stronger than Theorem 1 in that it determines the number of connected graphs up to a multiplicative factor $1 + o(1)$ instead of a constant. The same authors have determined the probability that a random graph $G_{n,p}$ is connected as well as the expected number of edges of $G_{n,p}$ conditioned on connectedness [3], corresponding to the case $d = 2$ of Theorems 3 and 4; however, they give the asymptotic distribution of the number of edges conditioned on connectedness, not just its expectation. An algorithm for generating random connected graphs (the case $d = 2$ of Theorem 2) immediately follows from the results in [2, 3]. Pittel and Wormald [12] presented an alternative approach based on enumerating graphs of minimum degree 2 to derive slightly improved versions of the results in [2]. Among other things, O'Connell [11] has estimated the probability that a random graph $G_{n,p}$ is connected using a probabilistic approach. However, the result in [11] is somewhat less precise than [3]. Using large-deviation methods from statistical physics, Engel, Monasson, and Hartmann [4] have investigated exponentially unlikely component structures of random graphs, including connectedness and, more generally, a specified number of components per vertex.

The asymptotic order and size of the giant component of random graphs (the case $d = 2$ of Theorem 5) has been known since the pioneering work of Erdős and Renyi (cf. [6] for a unified treatment). Furthermore, Barraez, Boucheron, and Fernandez de la Vega [1] give a description of the asymptotic distribution of the order and size of the giant component of random graphs.

For d-uniform hypergraphs, $d \geq 3$, on the other hand, little seems to be known. Karonski and Łuczak [8] determined the number of connected hypergraphs of order n and size $m = n/(d-1) + k$ where $k = o(\ln n/\ln\ln n)$ (i.e., just above the number of edges necessary for connectedness) up to a factor of $1+o(1)$ via purely combinatorial techniques. Since Theorem 1 addresses the case $m = cn/d$ for $c > d/(d-1)$, our results and those of [8] are incomparable.

Schmidt-Pruzan and Shamir [13] have shown that in a very general model of random hypergraphs a *phase transition* phenomenon occurs: there is a certain average degree c^* such that for all $c < (1-\varepsilon)c^*$ the largest component of $H_d(n,m)$ with $m = cn/d$ has order $O(\ln n)$, whereas for $c > (1+\varepsilon)c^*$ there is a unique giant component of order $\Omega(n)$. In the case of d-uniform hypergraphs, the critical average degree is $c^* = 1/(d-1)$. Karonski and Łuczak [7] have studied this phase transition in greater detail. To the best of our knowledge, the expected order and size of the giant component as given in Theorem 5 have not been stated before (although our proof relies on standard methods).

With regards to the occupancy problem, in the case $t = 1$, more precise results for $q_1(k,n)$, can be obtained from asymptotic formulas for Stirling numbers of the second kind due to Temme [14] and Moser and Wyman [10]. However, these results rely on generating functions and complex analysis, while our approach is elementary. For $t > 1$, to the best of our knowledge the results stated in Theorem 6 are new.

3   Techniques and Outline

While most of the previous work relies on rather heavy machinery from enumerative combinatorics, our approach is quite different and purely probabilistic, and yields shorter and simpler proofs. For instance, to prove Theorem 1 the basic idea is as follows. Suppose that $m = cn/d$ for some $c > 1/(d-1)$, and let a, b be as in Theorem 5. Then, by Theorem 5, w.h.p. there is a unique giant component in $H_d(n,m)$, which—conditioned on its order and size—is a uniformly random connected hypergraph. As we shall see in Section 5, we can express the number $C_d(an, bm)$ of connected hypergraphs of order an and size bm in terms of the probability $\chi(an, bm)$ that the giant component of $H_d(n,m)$ has *precisely* order an and size bm. Then, in Section 6, we reduce the problem of computing $\chi(an, bm)$ to a balls-and-bins problem. Section 4 contains the proof of Theorem 5. The algorithm for Theorem 2 drops out of our proof of Theorem 1 in Section 6 immediately.

Our approach is related to that of O'Connell [11], but goes beyond it considerably. O'Connell shows that $\Pr[G_{n,p} \text{ is connected}] = \Psi_2(np)^{n+o(n)}$ by observing that w.h.p. the giant component of $G_{n,p}$ has order $an + o(n)$ where $1 - a = e^{-ac}$ (the $d = 2$ case of Theorem 5). However, due to the $o(n)$ error term, in [11] it is necessary to prove that a certain rate function is continuous. This proof relies heavily on the independence of the edges in $G_{n,p}$, and therefore is not easy to adapt to the $G_{n,m}$ model. By contrast, we do not need to work with rate functions, because we estimate the probability that the giant component of $G_{n,p}$ has *exactly* order an. In effect, we determine $\Pr[G_{n,p} \text{ is connected}]$ up to a constant, rather than subexponential, factor.

4   The Giant Component of $H_d(n,m)$ and $H_d(n,p)$

To estimate the expected order and size of the giant component of a random hypergraph $H_d(n,p)$, we establish an analogy between a Galton-Watson branching process with a suitable successor distribution and exploring the components of a random hypergraph. For graphs, a similar approach is well known (cf. [6]). To estimate the expected order/size of the giant component of a hypergraph up to an additive error of $\tilde{O}(1)$, we need to refine this approach slightly.

Let us first describe the branching process. During the process there are two kinds of "organisms": living and dead ones. In the beginning, there is only one organism, which is alive. In the i'th step, a living organism is chosen, produces a number Z_i of children, and dies. The Z_i's are independent and is distributed as a multiple $(d-1) \cdot \text{Po}(c)$ of a Poisson random variable with mean c. We say that the process *dies out* if at some step i the last living organism dies without producing any offspring. General results [5, p. 297] imply that the probability ρ that the process dies out is the unique solution in $(0,1)$ of the equation $\rho = \exp\left(c(\rho^{d-1} - 1)\right)$.

Now, for random hypergraphs $H = H_d(n,p)$, consider the following process, which explores the connected component of a vertex v. During the process, the vertices of H are either dead, alive, or neutral. Initially, only v is alive, and all other vertices are neutral. In each step, a living vertex w is chosen. We investigate all edges e of H that contain w, at least one neutral vertex, but no dead vertex (since a dead vertex indicates that that edge has already been explored). All neutral vertices contained in such edges are made live, and w dies. Let Z_w be the number of vertices made live from w. When there are no living vertices left, the set of dead vertices is precisely the connected component of v.

Recall that a random variable X *dominates* Y if $\Pr[Y \le t] \le \Pr[X \le t]$ for all t. Let $\text{Bin}(N,p)$ denote a binomial random variable with parameters N and p. Then, on the one hand, Z_w is dominated by $(d-1)\text{Bin}(\binom{n-1}{d-1}, p)$, since the degree of w in $H_d(n,p)$ is distributed as $\text{Bin}(\binom{n-1}{d-1}, p)$. On the other hand, Z_w dominates the following random variable S_k: with probability $1 - \tilde{O}(1/n)$, S_k has distribution $(d-1)\text{Bin}(\binom{n-k}{d-1}, p)$, and with the remaining probability $S_k = 0$. Using these estimates, a similar argument as in [6, proof of Theorem 5.4] yields the following.

Lemma 7. *Let a, b be as in Theorem 5. The expected number of vertices outside the giant component is $(1-a)n + \tilde{O}(1)$, and the expected number of edges outside the giant component is $(1-b)m + \tilde{O}(1)$.*

Lemma 7 establishes the first part of Theorem 5. In order to show that the order and size of the giant component of $H_d(n,p)$ are concentrated about their means, we show that both have variance $O(n)$, and apply Chebyshev's inequality. A similar approach works for $H_d(n,m)$.

5   The Number of Connected Hypergraphs

Suppose that $m = cn/d$ for some $c > 1/(d-1)$. Following the outline in Section 3, in this section we reduce the problem of counting connected hypergraphs to estimating the probability that the giant component of $H_d(n, m)$ has a given order and size. Let a, b be as in Theorem 5, and set

$$\psi(x, y) = \binom{n}{x}\binom{\binom{x}{d}}{y}\binom{\binom{n-x}{d}}{m-y}\binom{\binom{n}{d}}{m}^{-1}. \tag{3}$$

Let $c_d(an, bm) = C_d(an, bm)\binom{\binom{an}{d}}{bm}^{-1}$ be the probability that $H_d(an, bm)$ is connected. Then, $\psi(an, bm) \cdot c_d(an, bm)$ is the expected number of components of order an and size bm in $H_d(n, m)$. Furthermore, if H is a hypergraph, then we let $\mathcal{N}(H)$ (resp. $\mathcal{M}(H)$) denote the order (resp. size) of the component with the most vertices (resp. edges). Then letting $\chi(x, y)$ be the probability that $H = H_d(n, m)$ satisfies $\mathcal{N}(H) = x$ and $\mathcal{M}(H) = y$, we obtain

$$\chi(an, bm) \leq \psi(an, bm) \cdot c_d(an, bm) \tag{4}$$

(for $d = 2$, a similar inequality has been derived in [9]). Furthermore, $\psi(an, bm) \cdot c_d(an, bm) \cdot \Pr[\mathcal{N}(H_d(n - an, m - bm)) < an]$ is the probability that in $H_d(n, m)$ there is precisely one component of order an and size bm. Hence,

$$\psi(a, b) \cdot c_d(an, bm) \cdot \Pr[\mathcal{N}(H_d(n - an, m - bm)) < an] \leq \chi(an, bm). \tag{5}$$

Let us first get rid of the term $\Pr[\mathcal{N}(H_d(n - an, m - bm)) < an]$.

Lemma 8. $\Pr[\mathcal{N}(H_d(n - an, m - bm)) < an] \sim 1.$

Lemma 8 follows from the analogy between random hypergraphs and branching processes discussed in Section 4. Lemma 8, (4) and (5) yield

$$\chi(an, bm) \sim \psi(an, bm) \cdot c_d(an, bm) . \tag{6}$$

Thus, we have reduced the problem of computing $c_d(an, bm)$ to the problem of computing $\chi(an, bm)$. The following lemma, which we prove in Section 6, does this within a multiplicative constant.

Lemma 9. $\chi(an, bm) = \Theta((nm)^{-1/2}).$

Combining Lemma 9 and (6) gives $c_d(an, bm) = \Theta\left((nm)^{-1/2}\psi(an, bm)^{-1}\right)$. Let $\nu = an$ and $\mu = bm$. Expanding $\psi(an, bm)$ using Stirling's formula yields

$$c_d(\nu, \mu) = \frac{(ab(1-a)(1-b)nm)^{1/2}}{\Theta((nm)^{1/2})} \cdot a^{an}(1-a)^{(1-a)n} \tag{7}$$

$$\times \left(\frac{\binom{n}{d}bm}{\binom{an}{d}m}\right)^{bm} \left(\frac{\binom{n}{d}(1-b)m}{\binom{(1-a)n}{d}m}\right)^{(1-b)m}$$

Let $\zeta = d\mu/\nu$, i.e., the average degree of the giant component if it has order ν and size μ. Plugging the relation $(1-a)^d = 1 - b$ into (7), we obtain

$$\Theta(c_d(\nu, \mu)) = \left(a^{1-\zeta}(1-a)^{\frac{1-a}{a}}\left(1 - (1-a)^d\right)^{\zeta/d}\right)^\nu.$$

Finally, using $1 - a = \exp\left(-c\left(1 - (1 - a)^{d-1}\right)\right)$ and $b = 1 - (1 - a)^d$ from Theorem 5 and writing $c = dm/n = \zeta a/b$, we get

$$1 - a = \exp\left(-\zeta a \cdot \frac{1 - (1-a)^{d-1}}{1 - (1-a)^d}\right) \; ,$$

thereby proving Theorem 1.

To prove Theorem 3, we write an equation analogous to (3). If $c_d(n, p)$ is the probability that $H_d(n, p)$ is connected, then the expected number of connected components of order x is $\binom{n}{x}(1 - p)^{\binom{n}{d} - \binom{x}{d} - \binom{n-x}{d}} \cdot c_d(x, p)$. We also have the following lemma, analogous to Lemma 9.

Lemma 10. *Let a be as in Theorem 5. The giant component of $H_d(n, p)$ has order an with probability $\Theta(n^{-1/2})$.*

Theorem 3 then follows from Lemma 10 and algebraic manipulations similar to those for Theorem 1.

To prove Theorem 2, we observe that Lemma 9 yields a very simple randomized algorithm for generating a uniformly random connected hypergraph of order ν and size $\mu = \zeta\nu/d$ for constant $\zeta > d/(d - 1)$ in expected polynomial time. First, fix n, m, c such that $a(c)n = \nu$ and $b(c)m = \mu$ where $a = a(c)$, $b = b(c)$ are as in Theorem 5. Then construct a random hypergraph $H_d(n, m)$ and find its giant component $\mathcal{C}$. Since $\mathcal{C}$ is uniformly random conditioned on its order and size, we output it if it has order ν and size μ, and if not we repeat. By Lemma 9, this simple *rejection sampling* approach succeeds in expected time polynomial.

6   The Probability of Getting a Giant Component of a Given Order and Size

Lemma 9 is a rather uncommon statement about random hypergraphs, since it concerns the probability that the order and size of the giant component attain *exactly one particular value*. A more common type of result (which is, of course, interesting in its own right) would be to characterize the asymptotic distribution of e.g. $\mathcal{N}(G_{n,m})$. For instance, in [1] it is shown that $\mathcal{N}(G_{n,m})/\sqrt{n}$ has an asymptotically normal distribution. However, as this statement allows for additive errors of size $o(\sqrt{n})$, it does *not* imply Lemma 9. In this section, we shall prove that the following lemma on fluctuations in the classical balls-and-bins-problem implies Lemma 9. (Lemma 10 can be derived analogously.)

Lemma 11. *Suppose that we throw r balls into n bins uniformly at random and independently, where $r = \rho n$ for some constant $\rho > 0$. Let Z denote the number of empty bins. Then for each $C > 0$ there is a $\delta > 0$ such that for all z satisfying $|z - \mathrm{E}(Z)| \leq C\sqrt{n}$ we have $\Pr[Z = z] \geq \delta n^{-1/2}$.*

There are (at least) two ways to prove Lemma 11. On the one hand, one can derive the lemma from rather involved combinatorial results such as [14]. On the other hand, there is also a simple probabilistic argument using sums of independent random variables (details omitted).

Here is the basic idea for how to employ Lemma 11 to prove Lemma 9. We expose the edges of the random hypergraph $H_d(n, m)$ in two stages. First, we throw in $m_1 = (1 - \varepsilon)m$ random edges, thereby obtaining $H_1 = H_d(n, m_1)$. Here $\varepsilon > 0$ is a sufficiently small constant. Let $\mathcal{C}$ denote the giant component of H_1. Then, we obtain $H_d(n, m)$ by adding a set F of $m_2 = \varepsilon m$ further random edges to H_1. Let F_{ball} be the set of all edges in F that have $d-1$ vertices in $\mathcal{C}$ and whose d'th vertex either is an isolated vertex of H_1 or belongs to an isolated edge of H_1. Then, we first add the edges in $F \setminus F_{\text{ball}}$ to H_1. Finally, we add the edges in F_{ball}, thereby attaching isolated vertices and edges to the giant component. We shall model the number of isolated edges/vertices that we attach as a balls-and-bins experiment, the edges in F_{ball} being the balls, and the isolated edges/vertices of H_1 being the bins. By Theorem 5, $\mathcal{N}(H_d(n, m) = H_1 + F)$ and $\mathcal{M}(H_1 + F)$ are concentrated in width $\sqrt{n}$ about their means an, bm. Thus, in order to achieve e.g. that the order of $H_1 + F$ is precisely an, we need that the number of vertices attached to the giant component via F_{ball} is precisely $an - \mathcal{N}(H_1 + F \setminus F_{\text{ball}})$, and indeed Lemma 11 estimates the probability that this is the case. We need to consider both isolated vertices and edges, because we wish to prove a statement on both the order and the size of the giant component.

However, there are some details that we have glossed over so far. For instance, there could be edges in F that fall into $V \setminus \mathcal{C}$ and "eat up" isolated vertices that the edges in F_{ball} would like to use as bins. Moreover, probably there will be edges in F that contain vertices in the giant component of H_1 and several isolated vertices, etc. Therefore, we need to consider a finer partition of F. First, we add those edges that either land inside the giant component of H_1 or inside its complement entirely: let

$$F_1 = \{e \in F \mid e \subset \mathcal{C}\}, \quad F_2 = \{e \in F \mid e \subset V \setminus \mathcal{C}\}, \quad H_2 = H_1 + F_1 + F_2 \ .$$

If ε is small enough, then w.h.p. all components of $H_2 - \mathcal{C}$ will have order $O(\ln n)$, so that $\mathcal{C}$ is still the giant component of H_2. As a next step, we add

- the set F_3 of edges in F that join a component of order $> d$ of $H_2 - \mathcal{C}$ to $\mathcal{C}$,
- the set F_4 of all edges $e \in F$ that are incident with less than $d - 1$ vertices in $\mathcal{C}$,
- and the set F_5 of edges $e \in F$ that share precisely $d - 1$ vertices with another edge $e \neq e' \in F$.

All edges in $F_3 \cup F_4 \cup F_5$ join vertices in the giant component $\mathcal{C}$ of H_2 with vertices in $V \setminus \mathcal{C}$. Let $H_3 = H_2 + F_3 + F_4 + F_5$, and let $\mathcal{C}'$ be the vertex set of the giant component of H_3. Furthermore, let F' be the set of all edges in $e \in F \setminus (F_1 \cup \cdots \cup F_5)$ that join a component of order 1 or of order d (i.e. an isolated edge) of H_3 to $\mathcal{C}'$. By the choice of F_4 and F_5, F' has the following two properties.

i. Each $e \in F'$ has $d - 1$ vertices in $\mathcal{C} \subset \mathcal{C}'$.
ii. For any two edges $e, e' \in F'$ we have $e \cap \mathcal{C} \neq e' \cap \mathcal{C}$.

The edges in F' will be our "balls". Finally, let $F_6 = F \setminus (F_1 \cup \cdots \cup F_5 \cup F')$, and $H_4 = H_3 + F_6$. Then, the edges in F_6 connect $\mathcal{C}$ with an isolated vertex/edge of H_2, but we can't use these edges as "balls", because their isolated vertex/edge is also connected with $\mathcal{C}$ through another edge in F. Observe that all vertices of edges in F_6 lie inside the giant component of H_3, so that the vertex set of the giant component of H_4 is still $\mathcal{C}'$.

Let $\mathcal{Y}_1$ (resp. $\mathcal{Y}_2$) be the set of isolated vertices (resp. edges) of H_4, and set $Y_i = \#\mathcal{Y}_i$. Moreover, let F_1' (resp. F_2') be the set of all $e \in F'$ that connect an isolated vertex (resp. edge) of H_4 with $\mathcal{C}'$, and set $Z_i = \#F_i'$. Finally, let X_1 (resp. X_2) be the number of isolated vertices (resp. edges) that the edges in F_1' (resp. F_2') connect with $\mathcal{C}'$. The following observation is crucial for our argument.

Proposition 12. *If we condition on any specific outcome of (Y_1, Y_2, Z_1, Z_2), then X_1, X_2 are independent. Moreover, the conditional distribution of X_i is precisely the distribution of the number of non-empty bins when Z_i balls are thrown into Y_i bins uniformly at random and independently.*

Corollary 13. *Let $\mu_i(Z_i) = \mathrm{E}(X_i \mid Z_i = z_i)$. Conditioned on $Z_1 = z_1$ and $Z_2 = z_2$, both X_1 and X_2 are concentrated in width $\sqrt{n}$ about $\mu_1(z_1), \mu_2(z_2)$.*

Sketch of proof. Condition on the event that $Z_1 = z_1$, $Z_2 = z_2$, and $Y_1 = y_1$, $Y_2 = y_2$. By the proposition, X_1, X_2 are independent and distributed as in a balls-and-bins experiment. Therefore, a standard application of Azuma's inequality (cf. [6, p. 37]) shows that both X_1 and X_2 are concentrated in width $\sqrt{n}$. □

As a next step we prove that the number of balls and bins is of the same order of magnitude as n.

Lemma 14. *With probability $\geq 1 - n^{-10}$ we have $Y_1, Y_2, Z_1, Z_2 = \Omega(n)$.*

Observe that the number of vertices that the edges in F' attach to the giant component $\mathcal{C}'$ of H_4 is $X_1 + dX_2$. Furthermore, the number of edges (including the ones in F') joined to $\mathcal{C}'$ via F' is $Z_1 + Z_2 + X_2$.

Lemma 15. *Let $C > 0$ be a sufficiently large constant. Let $\mathcal{Z}$ denote the set of all pairs (z_1, z_2) such that the following holds:*

$$\Pr\big[|\mathcal{N}(H_4) - an - (\mu_1(z_1) + d\mu_2(z_2))| < C\sqrt{n} \text{ and} \tag{8}$$
$$|\mathcal{M}(H_4) - bm - (z_1 + z_2 + \mu_2(z_2))| < C\sqrt{n} \mid Z_1 = z_1, Z_2 = z_2\big] \geq 1/2 .$$

Then $\Pr[(Z_1, Z_2) \in \mathcal{Z}] \geq 99/100$.

Sketch of proof. Let $\lambda > 0$ be a sufficiently large constant. On the one hand, by Corollary 13, the probability that the number of isolated vertices or the number of isolated edges attached to $\mathcal{C}$ deviates by more than $\lambda\sqrt{n}$ from the expectation $\mu_1(Z_1), \mu_2(Z_2)$ is less than δ. On the other hand, by Theorem 5, the order and the size of the giant component of $H_d(n, m)$ are concentrated in width $\sqrt{n}$. □

Proof of Lemma 9 (sketch). To prove the lower bound $\Omega((nm)^{-1/2})$, condition on the following event $\mathcal{E}_1$: we have $Y_1, Y_2, Z_1, Z_2 = \Omega(n)$ and the outcome (z_1, z_2) of

(Z_1, Z_2) is such that (8) holds. By Lemmas 14 and 15, $\Pr[\mathcal{E}_1] \geq 9/10$. Moreover, by Lemma 15, the conditional probability that H_4 satisfies

$$|\mathcal{N}(H_4) - an - (\mu_1(z_1) + d\mu_2(z_2))|, \quad |\mathcal{M}(H_4) - bm - (z_1 + z_2 + \mu_2(z_2))| < C\sqrt{n} \quad (9)$$

is at least $1/2$. We additionally condition on the event $\mathcal{E}_2$ that (9) holds. Then, Proposition 12 and Lemma 11 entail

$$\Pr[X_1 + dX_2 = an - \mathcal{N}(H_4), \; Z_1 + Z_2 + X_2 = bm - \mathcal{M}(H_4)|\mathcal{E}_1 \cap \mathcal{E}_2] = \Theta(nm)^{-\frac{1}{2}}.$$

Furthermore, $\Pr[\mathcal{E}_1 \cap \mathcal{E}_2] \geq 9/20$, thereby proving the lower bound claimed in Lemma 9. The (even simpler) proof of the upper bound is omitted.   □

Acknowledgments

We are grateful to Tomasz Łuczak and to Nick Wormald for sending us copies of [7, 8, 12], and to Gregory Sorkin for bringing [9] to our attention. C.M. also thanks Tracy Conrad for helpful discussions.

References

1. Barraez, D., Boucheron, S., Fernandez de la Vega, W.: On the fluctuations of the giant component. Combinatorics, Probability and Computing **9** (2000) 287–304
2. Bender, E.A., Canfield, E.R., McKay, B.D.: The asymptotic number of labeled connected graphs with a given number of vertices and edges. Random Structures & Algorithms **1** (1990) 127–169
3. Bender, E.A., Canfield, E.R., McKay, B.D.: Asymptotic properties of labeled connected graphs. Random Structures & Algorithms **3** (1992) 183–202
4. Engel, A., Monasson, R., Hartmann, A.: On large deviation properties of Erdős-Rényi random graphs. Preprint arXiv, cond-mat/0311535 (2003)
5. Feller, W.: Introduction to probability theory and its applications. Wiley 1968
6. Janson, S., Łuczak, T, Ruciński, A.: Random Graphs, Wiley 2000
7. Karonski, M., Łuczak, T.: The phase transition in a random hypergraph. J. Comput. Appl. Math. **142** (2002) 125–135
8. Karonski, M., Łuczak, T.: The number of connected sparsely edged uniform hypergraphs. Discrete Math. **171** (1997) 153–168
9. Luczak, T.: On the number of sparse connected graphs. Random Structures & Algorithms **1** (1990) 171–173
10. Moser, L. and Wyman, M.: Stirling numbers of the second kind. Duke Mathematical Journal **25** (1958) 29–43
11. O'Connell, N.: Some large deviation results for sparse random graphs. Prob. Th. Relat. Fields **110** (1998) 277–285
12. Pittel, B., Wormald, N.C.: Counting connected graphs inside out. J. Combinatorial Theory Series B, to appear
13. Schmidt-Pruzan, J., Shamir, E.: Component structure in the evolution of random hypergraphs. Combinatorica **5** (1985) 81–94
14. Temme, N.M.: Asymptotic estimates of Stirling numbers. Studies in Applied Mathematics **89** (1993) 233–243

Improved Randomness Extraction
from Two Independent Sources

Yevgeniy Dodis[1,*], Ariel Elbaz[2,**], Roberto Oliveira[3,***], and Ran Raz[4,†]

[1] Department of Computer Science, New York University
dodis@cs.nyu.edu
[2] Department of Computer Science, Columbia University
arielbaz@cs.columbia.edu
[3] Department of Mathematics, New York University
oliveira@cims.nyu.edu
[4] Department of Computer Science, Weizmann Institute
ran.raz@weizmann.ac.il

Abstract. Given two independent weak random sources X, Y, with the same length ℓ and min-entropies b_X, b_Y whose sum is greater than $\ell + \Omega(\text{polylog}(\ell/\varepsilon))$, we construct a deterministic two-source extractor (aka "blender") that extracts $\max(b_X, b_Y) + (b_X + b_Y - \ell - 4\log(1/\varepsilon))$ bits which are ε-close to uniform. In contrast, best previously published construction [4] extracted at most $\frac{1}{2}(b_X + b_Y - \ell - 2\log(1/\varepsilon))$ bits. Our main technical tool is a construction of a strong two-source extractor that extracts $(b_X + b_Y - \ell) - 2\log(1/\varepsilon)$ bits which are ε-close to being uniform and *independent* of one of the sources (aka "strong blender"), so that they can later be reused as a seed to a seeded extractor. Our strong two-source extractor construction improves the best previously published construction of such strong blenders [7] by a factor of 2, applies to more sources X and Y, and is considerably simpler than the latter. Our methodology also unifies several of the previous two-source extractor constructions from the literature.

1 Introduction

IMPERFECT RANDOMNESS. Randomization has proved to be extremely useful and fundamental in many areas of computer science. Unfortunately, in many situations one does not have ideal sources of randomness, and therefore has to base a given application on *imperfect sources of randomness*.

Among many imperfect sources considered so far, perhaps the most general and realistic source is the *weak source* [29, 4]. The only thing guaranteed about a weak source is that no string (of some given length ℓ) occurs with probability more than 2^{-b}, where b is the so-called *min-entropy* of the source. We will call this

* Partially supported by the NSF CAREER and Trusted Computing Awards.
** Work supported by ISF grant.
*** Funded by a doctoral fellowship from CNPq, Brazil.
† Work supported by ISF grant.

K. Jansen et al. (Eds.): APPROX and RANDOM 2004, LNCS 3122, pp. 334–344, 2004.
© Springer-Verlag Berlin Heidelberg 2004

source (ℓ, b)-weak. Unfortunately, handling such weak sources is often necessary in many applications, as it is typically hard to assume much structure on the source beside the fact that it contains some randomness. Thus, by now a universal goal in basing some application on imperfect sources is to make it work with weak sources. The most direct way of utilizing weak sources would be to extract nearly perfect randomness from such a source. Unfortunately, it is trivial to see [4] that no deterministic function can extract even one random bit from a weak source, as long as $b \leq \ell - 1$ (i.e., the source is not almost random to begin with). This observation leaves two possible options. First, one can try to use weak sources for a given application without an intermediate step of extracting randomness from it. Second, one can try designing probabilistic extractors, and later justify where and how one can obtain the additional randomness needed for extraction.

USING A SINGLE WEAK SOURCE. A big and successful line of research [26, 24, 4, 5, 29, 2] following the first approach showed that a single weak source is sufficient to simulate any probabilistic computation of decision or optimization problems (i.e., problems with a "correct" output which are potentially solved more efficiently using randomization; That is, all problems in BPP). Unfortunately, most of the methods in this area are not applicable for applications of randomness, where the randomness is needed by the application itself, and not mainly for the purposes of efficiency. One prime example of this is cryptography. For example, secret keys have to be random, and many cryptographic primitives (such as public-key encryption) *must* be probabilistic. The problem of basing a cryptographic protocol on a single weak random source has only been studied in the setting of information-theoretic symmetric-key cryptography. In this scenario, the shared secret key between the sender and the recipient is no longer random, but comes from a weak source. As a very negative result, McInnes and Pinkas [12] proved that one cannot securely encrypt even a single bit, even when using an "almost random" $(\ell, \ell - 1)$-weak source. Thus, one cannot base symmetric-key encryption on weak sources. Dodis and Spencer [6] also consider the question of message authentication and show that one cannot (non-interactively) authenticate even one bit using $(\ell, \ell/2)$-weak source (this bound is tight as Maurer and Wolf [11] showed how to authenticate up to $\ell/2$ bits when $b > \ell/2$).

USING SEVERAL WEAK SOURCES. Instead, we will assume that we have several weak sources, each one independent from all the other weak sources. Specifically, we will try to extract nearly ideal randomness from two weak sources.

The question of extracting randomness from two or more independent random sources originated in the works of Sántha and Vazirani [20, 25, 27]. Chor and Goldreich [4] were the first to consider general weak sources of equal block length; let us say that the sources X and Y are (ℓ_X, b_X)-weak and (ℓ_Y, b_Y)-weak, and for simplicity we also assume $\ell_X = \ell_Y = \ell$. It is known [7] that in this setting it is possible to extract nearly all the entropy (i.e., nearly $b_X + b_Y - 2\log(1/\varepsilon)$ bits) from the two sources, where ε is the required distance from the uniform distribution. However, best known explicit constructions achieve much weaker parameters. Specifically, for one bit extractors Chor and Goldreich showed that the inner product function works provided $b_X + b_Y > \ell + 2\log(1/\varepsilon)$. The best multi-bit two-source extractor ([4], thm 14) is capable of extracting almost $\frac{1}{2}(b_X + b_Y - \ell - 2\log(1/\varepsilon))$ random bits, but [4] only show how to efficiently implement this extractor when $b_X + b_Y > 1.75\ell$. It was also known how to extract a non-constant bit (or a few bits) from two sources when $b_X = \ell(\frac{1}{2}+o(1))$, $b_Y = O(\log(\ell))$ [9,1], and this method can also be used to extract unbiased bits from such sources. In a setting of more than two weak sources, Barak et al. [3] recently show how to extract randomness from $(1/\delta)^{O(1)}$ independent $(\ell, \delta\ell)$-weak sources (for any $\delta > 0$), but this improvement does not apply to the case of two sources considered in this work.

Motivated by cryptographic applications, Dodis and Oliveira [7] recently considered the problem of extracting random bits from two independent sources which are (essentially) *independent* from one of the two sources (called the "seed"). They termed this two-source extractor a *strong blender*, since it is a common generalization of a two-source extractor (which they simply termed "blender") and a *strong extractor* [16], where the seed source is assumed to be public and truly uniform (and the objective of the latter is to make the seed as short as possible). They showed the existence of such strong blenders, whenever $b_X, b_Y > \log\ell + 2\log(1/\varepsilon)$, which are capable of extracting $(b_X - 2\log(1/\varepsilon))$ bits from X which are independent from the (ℓ, b_Y)-weak seed Y. On a constructive side, they showed that the constructions from [4] for two-source extractors (i.e., "blenders") and the related work of Vazirani [27] for the so-called SV sources [20] do extend to give strong blenders. The best such construction in [7] (based on a similar construction from [4]) is capable of extracting $\frac{1}{2}(b_X + b_Y - \ell - 2\log(1/\varepsilon))$ bits independent of one of the sources, but this is again efficient only when $b_X + b_Y > 1.75\ell$.

To summarize, while randomness extraction from two independent sources is possible, the known constructions and parameters seem far from optimal. In contrast, there are many efficient constructions of one-source randomness extractors (with truly random seeds) with nearly optimal parameters (see [10, 17, 21] and the references therein). Our work gives an improvement in the number of extracted bits, for the case $b_X + b_Y > \ell$, allowing for the first time the extraction of more than half the total randomness that is present in the sources.

OUR RESULTS. We give a simple two-source extractor capable of extracting $\max(b_X, b_Y) + (b_X + b_Y - \ell - 4\log(1/\varepsilon))$ bits whenever the sum of min-entropies is slightly greater than ℓ: $b_X + b_Y \geq \ell + \Omega(\text{polylog}(\ell/\varepsilon))$. Our construction is based on two observations. First, a strong blender with good parameters can give a two-source extractor (i.e., a "regular blender") with even better parameters, when combined with a seeded extractor (with good parameters). Specifically, if a strong blender extracts $k \geq \Omega(\text{polylog}(\ell/\varepsilon))$ nearly random bits Z which are independent from the weak seed Y, we can now apply a seeded extractor to Y, using the just extracted Z as the seed (since Z is "long enough" to be used with state-of-the-art extractors such as [17]). This allows us to extract a total of $(k+b_Y-2\log(1/\varepsilon))$ bits, which could be very high. While this obvious observation would already improve the state-of-the-art in the problem of two-source extraction, even when combined with the known strong blender constructions from

[7], our second observation is a much simpler (and better) construction of strong blenders. Specifically, we show how to extract $(b_X + b_Y - \ell - 2\log(1/\varepsilon))$ bits which are independent from one of the sources. Thus our strong blender construction: (1) improves by a factor of two the best previously published construction of [7]; (2) is very simple and efficient without any limitations on b_X and b_Y (as long as $b_X + b_Y \geq \ell + 2\log(1/\varepsilon)$); and (3) uses only trivial properties from linear algebra in its analysis. By choosing the initial "seed" source to be the one containing more min-entropy, we get our final two-source extractor.

OUR STRONG BLENDER. In fact, we give the following general technique for constructing our strong blender. Assume we wish to extract k nearly random bits from X using Y as a seed. Let $A_1, \ldots, A_k$ be some $\ell \times \ell$ matrices over $GF[2]$, specified later. View X and Y as ℓ-bit vectors, and output bits $((A_1 X) \cdot Y, \ldots, (A_k X) \cdot Y)$, where $\cdot$ denotes the inner product modulo 2, and $A_i X$ is the matrix-vector multiplication over $GF[2]$. Of course, the main question is which condition on matrices $A_1 \ldots A_k$ would allow us to maximize the value of k. As we show, all we need is to have every non-empty sum $A_S = \sum_{i \in S} A_i$ (where $S \neq \emptyset$) of these matrices to have "high" rank. In particular, by using a known linear algebra construction [14, 13, 19]) of such matrices, we can ensure that all the needed matrices A_S will have full rank, which will in turn give us the desired value $k = (b_X + b_Y - \ell - 2\log(1/\varepsilon))$. However, we also notice that by using several other simple, but sub-optimal choices of matrices $A_1 \ldots A_k$, we will obtain two of the three constructions from [7] as special cases, as well as several other constructions which appeared in the literature for related problems (i.e., [27]). Thus, our methodology elegantly unifies several of the previous approaches into one elementary framework.

2   Preliminaries

2.1   Basic Notation

We mostly employ standard notation. The symbol log is reserved for the base 2 logarithm. For a positive integer t, U_t denotes a random variable that is uniform over $\{0,1\}^t$ and independent of all other random variables under consideration. We also write $[t] \equiv \{1, 2, \ldots t\}$. For two random variables A, B taking values in the finite set $\mathcal{A}$, their *statistical distance* is $\|A - B\|_s \equiv \frac{1}{2} \sum_{a \in \mathcal{A}} |\Pr[A = a] - \Pr[B = a]|$, and the min-entropy of A is $H_\infty(A) \equiv \min_{a \in \mathcal{A}}(-\log(\Pr[A = a]))$. For a random variable A taking values in $\{0,1\}^t$, the *bias* of A is its statistical distance from U_t, $\|A - U_t\|_s$. The L_2 norm of a vector $v \in \mathcal{R}^t$ is $\|v\|_2 \equiv \sqrt{\sum_{i=1}^t (v_i)^2}$. For two strings s, t, their concatenation is denoted (s, t). For a vector $a \in \{0,1\}^t$ and $1 \leq i \leq t$, we say that $i \in a$ if $a_i = 1$. Also recall that an (ℓ, b)-source denotes some random variable X over $\{0,1\}^\ell$ with min-entropy $H_\infty(X) \geq b$.

2.2   Strong Blenders and Extractors

We start by defining the notion of a randomness extractor [16].

Definition 1 ([16]). *Let $b \geq 0$, $\varepsilon > 0$. A (b, ε)-extractor $\mathrm{EXT} : \{0,1\}^\ell \times \{0,1\}^k \to \{0,1\}^m$ is an efficient function such that for all (ℓ, b)-sources Y, we have*

$$\| \mathrm{EXT}(Y, U_k) - U_m \|_s \leq \varepsilon$$

Clearly, one of the main objectives in the area of building explicit extractors is to minimize the seed length k while still extracting nearly all the randomness from the source. Many nearly optimal constructions exist by now (see [15, 23, 10, 17, 22, 21] and the references therein). While none of them clearly beats the rest in all parameters, it is known (e.g., [17]) how to extract $m = k + b - 2\log(1/\varepsilon)$ nearly random bits (which is optimal) using a seed of length $k = O(\mathrm{polylog}(\ell/\varepsilon))$.

In this work, however, we are interested in (strong) randomness extraction from two weak sources. Correspondingly, we define the notion of two-source extractors (aka "blenders") and strong blenders. For simplicity, we assume both sources have the same length ℓ throughout the paper.

Definition 2. *A (b_X, b_Y, ε)-strong blender (SB) is an efficient function $\mathrm{BLE} : \{0,1\}^\ell \times \{0,1\}^\ell \to \{0,1\}^k$ such that for all (ℓ, b_X)-weak sources X and all (ℓ, b_Y)-weak sources Y, we have*

$$\| (Y, \mathrm{BLE}(X,Y)) - (Y, U_k) \|_s \leq \varepsilon$$

A (b_X, b_Y, ε)-two-source extractor is an efficient function $\mathrm{TWO\text{-}EXT} : \{0,1\}^\ell \times \{0,1\}^\ell \to \{0,1\}^m$ such that for all (ℓ, b_X)-weak sources X and all (ℓ, b_Y)-weak sources Y, we have

$$\| \mathrm{TWO\text{-}EXT}(X,Y) - U_m \|_s \leq \varepsilon$$

We now observe the following relation among these three primitives which follows immediately from the triangle inequality: namely, given X, Y, the output of a strong blender $\mathrm{BLE}(X,Y)$ can be used as a random seed for an extractor EXT, to extract the randomness from Y.

Lemma 1. *Let $\mathrm{BLE} : \{0,1\}^\ell \times \{0,1\}^\ell \to \{0,1\}^k$ be a (b_X, b_Y, ε)-strong blender and $\mathrm{EXT} : \{0,1\}^\ell \times \{0,1\}^k \to \{0,1\}^m$ be a (b_Y, ε)-extractor. Then the following function $\mathrm{TWO\text{-}EXT} : \{0,1\}^\ell \times \{0,1\}^\ell \to \{0,1\}^m$ is a $(b_X, b_Y, 2\varepsilon)$-two-source extractor:*

$$\mathrm{TWO\text{-}EXT}(X,Y) = \mathrm{EXT}(Y, \mathrm{BLE}(X, Y))$$

Our final two-source extractor will follow by applying the Lemma above to the strong blender which we construct in the next section (using any good seeded extractor, such as the one in [17]). Thus, from now on we concentrate on improved constructions of strong blenders.

3   Efficient Strong Blender Constructions

In this section we show a construction of an efficient strong blender. In 3.1 we give a general technique for constructing a strong blender, that requires a set of $\ell \times \ell$ matrices over $GF[2]$ with the property that the sum of every subset of matrices has high rank. In section 3.2 we give an explicit set of matrices (that appeared in [14, 13, 19]), with the property that the sum of any subset of matrices has full rank. This gives the best possible result for our technique. In the same section we also review some previous constructions for two-source randomness extraction as special cases of our technique.

3.1   General Technique for Constructing Strong Blenders

We now show how to extract many random bits from two weak sources X, Y of the same length ℓ, such that the bits are random even if one sees Y.

Let $A_1, \ldots, A_k$ be $\ell \times \ell$ matrices over $GF[2]$, such that for every nonempty subset $S \subseteq [k]$, the rank of $A_S \overset{\text{def}}{=} \sum_{i \in S} A_i$ is at least $\ell - r$, for $0 \leq r < \ell$ (we will want to have high rank and keep r as small as possible).

The proposed strong blender is

$$\mathrm{BLE}_A : \{0, 1\}^\ell \times \{0, 1\}^\ell \rightarrow \{0, 1\}^k$$
$$(x, y) \longmapsto \Big((A_1 x) \cdot y, \ldots, (A_k x) \cdot y \Big) \tag{1}$$

where $\cdot$ is the inner product mod 2 and $A_i x$ is a matrix-vector multiplication over $GF[2]$.

Theorem 1. *The function* BLE_A *is a* (b_X, b_Y, ϵ)-*SB with* $\log \frac{1}{\epsilon} = \frac{b_X + b_Y + 2 - (\ell + r + k)}{2}$, *that is*

$$\left\| (Y, \mathrm{BLE}_A(X, Y)) - (Y, U_k) \right\|_s \leq 2^{-\frac{b_X + b_Y + 2 - (\ell + r + k)}{2}}$$

Our main tool for proving theorem 1 is the Parity lemma (also known as the XOR lemma, see [8, 27]), which relates the bias of a k-bit random variable T with the sum of squared biases of every subset of the bits in T.

Lemma 2 (Parity lemma [8]). *For any* k-*bit random variable* T, $\| T - U_k \|_s$ *is upper bounded by*

$$\sqrt{\sum_{0^k \neq a \in \{0,1\}^k} \| T \cdot a - U_1 \|_s^2}$$

where $T \cdot a$ *is the inner product (mod 2) of* T *and* a.

We also use a relation between the rank of a matrix L and the min-entropy of LX.

Lemma 3. *Let* L *be an* $\ell \times \ell$ *matrix over* $GF[2]$. *If the rank of* L *is* $\ell - r$, *then for any random variable* X *over* $\{0, 1\}^\ell$, $H_\infty(LX) \geq H_\infty(X) - r$.

Proof. Note that for any $z \in \{0, 1\}^\ell$, there are at most 2^r different values $x \in \{0, 1\}^\ell$ such that $Lx = z$. Thus, $\max_{z \in \{0,1\}^\ell} \Pr[LX = z] \leq 2^r \cdot \max_{x \in \{0,1\}^\ell} \Pr[X = x]$.

Proof. [of Theorem 1] Following [4, 7], it is sufficient to consider the case when Y is uniformly distributed on some set S_Y, $|S_Y| = 2^{b_Y}$, and X is uniformly distributed on some set S_X, $|S_X| = 2^{b_X}$.

$$\|(Y \,,\, \mathrm{BLE}_A(X, Y)) - (Y \,,\, U_k)\| = \frac{1}{|S_Y|} \sum_{y \in S_Y} \|\mathrm{BLE}_A(X, y) - U_k\|_s$$

(by parity lemma)
$$\leq \sum_{y \in S_Y} \frac{1}{|S_Y|} \sqrt{\sum_{0^k \neq a \in \{0,1\}^k} (\|\mathrm{BLE}_A(X, y) \cdot a - U_1\|_s)^2}$$

(by concavity)
$$\leq \sqrt{\sum_{0^k \neq a \in \{0,1\}^k} \left[\frac{1}{|S_Y|} \sum_{y \in S_Y} (\|\mathrm{BLE}_A(X, y) \cdot a - U_1\|_s)^2 \right]}$$

We now proceed to upper bound each of the (above) bracketed terms by $2^{-(b_X + b_Y + 2 - \ell - r)}$; the final result then follows from simple addition. Fixing a value of $a = a_1 a_2 \ldots a_k \in \{0,1\}^k \setminus \{0^k\}$, and identifying a with the set of all $i \in [k]$ for which $a_i = 1$, we have

$$\mathrm{BLE}_A(X, y) \cdot a = \sum_{i \in a} \mathrm{BLE}_A(X, y)_i = \sum_{i \in a} (A_i X) \cdot y = \left(\left(\sum_{i \in a} A_i \right) X \right) \cdot y$$

where all sums are taken modulo 2.

Now define $A_a \stackrel{\mathrm{def}}{=} \sum_{i \in a} A_i \pmod 2$, and recall that the rank of A_a is at least $\ell - r$. As a proof tool, we introduce the $2^\ell \times 2^\ell$ matrix M whose rows and columns are labeled by the elements of $\{0,1\}^\ell$, and whose (x, y)th entry is $M_{x,y} = (-1)^{(A_a x) \cdot y}$ (for $x, y \in \{0,1\}^\ell$). We also let $\vec{M}_x$ be the xth row of M. The following two properties are the key to what follows.

1. *Every row of M is equal to at most 2^r other rows of M.* Indeed, if $x, x' \in \{0,1\}^\ell$ satisfy $\vec{M}_x = \vec{M}_{x'}$, then it must be that $A_a x = A_a x' \pmod 2$, i.e. $x + x' \pmod 2$ is in the kernel of A_a. Since the rank of A_a is at least $\ell - r$, the kernel of A_a has at most 2^r elements.

2. *Rows of M that are distinct are orthogonal (in the Euclidean inner product sense).* For if $x \neq x' \in \{0,1\}^\ell$ are such that $\vec{M}_x \neq \vec{M}_{x'}$, then $z \stackrel{\mathrm{def}}{=} A_a(x + x') \neq 0^\ell \pmod 2$, and the $\mathcal{R}$-valued Euclidean inner product of $\vec{M}_x$ and $\vec{M}_{x'} = (\vec{M}_x)^T (\vec{M}_{x'}) = \sum_{y \in \{0,1\}^\ell} (-1)^{z \cdot y} \pmod 2$ is 0.

Now let $\vec{D} \in \mathcal{R}^{2^\ell}$ be the vector defined by $\vec{D} \stackrel{\mathrm{def}}{=} \frac{1}{2|S_X|} \sum_{x \in S_X} \vec{M}_x$. It is not hard to see that the yth coordinate D_y of $\vec{D}$ is equal to $\frac{1}{2}(\Pr_{x \in S_X}[M_{x,y} = 1] - \Pr_{x \in S_X}[M_{x,y} = -1])$, so that $|D_y| = \|\mathrm{BLE}_A(X, y) \cdot a - U_1\|_s$. The quantity we wish to bound is thus

$$\frac{1}{|S_Y|} \sum_{y \in S_Y} (\|\mathrm{BLE}_A(X, y) \cdot a - U_1\|_s)^2 = \frac{1}{|S_Y|} \sum_{y \in S_Y} |D_y|^2 \leq \frac{1}{|S_Y|} \left\| \vec{D} \right\|_2^2$$

and $\left\| \vec{D} \right\|_2^2$ equals the Euclidean inner product $\vec{D}^T \vec{D}$ over $\mathcal{R}$

$$\left\| \vec{D} \right\|_2^2 = \frac{1}{4|S_X|^2} \sum_{x,x' \in S_X} (\vec{M}_x)^T (\vec{M}_{x'})$$

But properties 1. and 2. above show that for any $x \in S_X$

$$\sum_{x' \in S_X} (\vec{M}_x)^T (\vec{M}_{x'}) = \sum_{x' : \vec{M}_{x'} = \vec{M}_x} (\vec{M}_x)^T (\vec{M}_{x'}) + \sum_{x' : \vec{M}_{x'} \neq \vec{M}_x} (\vec{M}_x)^T (\vec{M}_{x'})$$

$$\leq 2^r \left\| \vec{M}_x \right\|_2^2 \leq 2^{r+\ell}$$

Hence $\left\| \vec{D} \right\|_2^2 \leq 2^{r+\ell}/4|S_X|$, and plugging this bound into (3.1) yields

$$\frac{1}{|S_Y|} \sum_{y \in S_Y} (\|\text{BLE}_A(X,y) \cdot a - U_1\|_s)^2 \leq \frac{2^{r+\ell}}{4|S_X||S_Y|} = 2^{-(b_X + b_Y + 2 - (\ell + r))}$$

which, as noted above, implies the theorem.

We emphasize that the above technique is entirely general: *any* set of matrices $A_1, \ldots, A_k$ such that the sum of any non-empty subset of matrices has rank of at least $\ell - r$, can be used to extract $k < b_X + b_Y + 2 - \ell - r$ bits from ℓ-bit weak sources X, Y with respective min-entropies b_X, b_Y.

The following two cases are of special interest (of course, they apply only when r is "small enough"; that is, when $r + k < b_X + b_Y + 2 - \ell$):

1. If the min-entropies $b_X + b_Y$ sum up to $\ell + \text{polylog}(\ell)$, we can extract polylogarithmic (in ℓ) number of bits with bias $2^{-\Omega(\text{polylog}(\ell))}$.
2. If the min-entropies $b_X + b_Y$ sum up to $(1 + c)\ell$, for any constant $0 < c \leq 1$, we can extract linear (in ℓ) number of bits with exponentially small bias.

3.2 Explicit Strong Blender Instantiations

Theorem 1 subsumes constructions introduced (sometimes implicitly) in previous works. These constructions are presented below, followed by our more randomness-efficient construction.

IDENTITY MATRIX. This is the simplest case when $k = 1$ and A_1 is the identity matrix. This gives $r = 0$ and implies that the inner product function is a strong blender with bias $\varepsilon = 2^{-\frac{(b_X + b_Y + 1 - \ell)}{2}}$, reproving the result from [7] (adapting the result from [4]).

CYCLIC SHIFT MATRICES. Vazirani [27] used cyclic shift matrices, to extract randomness from SV sources. When ℓ is a prime with 2 as a primitive root, the sum of any subset of matrices has rank at least $\ell - 1$ (that is $r = 1$, in our notation above). Formally, let A_i be a linear transformation matrix corresponding to a cyclic shift by $i - 1$ bits; That is, the jth row of A_i has a 1 in the $j - i + 1$ (mod ℓ) column and zero elsewhere. The required property of these matrices is given below.

Lemma 4 ([27]). *Let ℓ be a prime with 2 a primitive root modulo ℓ (i.e. 2 is a generator of Z_ℓ^*). Let $\vec{u} \in \{0,1\}^\ell \setminus \{0^\ell, 1^\ell\}$ be a vector (which is not the all 0's or the all 1's vector). Let A be an $\ell \times \ell$ matrix over $GF(2)$, such that the rows of A are the ℓ right-cyclic-shifts of $\vec{u}$. Then $rank(A) \geq \ell - 1$.*

By Theorem 1, a corresponding strong blender can extract k bits with bias $2^{-\frac{(b_X + b_Y + 1 - \ell - k)}{2}}$, or, equivalently, extract $k = (b_X + b_Y + 1 - \ell - 2\log(1/\varepsilon))$ bits with bias ε.

(NON-CYCLIC) RIGHT SHIFT MATRICES. Let $A_1, \ldots, A_k$ be linear transformation matrices such that A_i corresponds to the right shift by $i - 1$ bits. For any non-empty subset $S \subseteq [k]$, it is easy to see that the rank of $A_S = \sum_{i \in S} A_i$ is at least $\ell - k + 1$. By theorem 1 one can build a strong blender from those matrices that extracts k bits with bias $2^{-\frac{(b_X + b_Y - \ell - 2k)}{2}}$, or, equivalently, extract $k = (b_X + b_Y - \ell - 2\log(1/\varepsilon))/2$ bits with bias ε.

THE MATRICES FROM A GENERAL ERROR-CORRECTING CODE. The construction of [7] (adapted from [25]) also pertains to the present framework. Let C be an $[\ell, k, d]$ linear error correcting code (i.e., having dimension k and minimal distance d in $\{0,1\}^\ell$). Let $A_1, \ldots, A_k$ be diagonal matrices, the diagonal of A_i containing the coordinates of a codeword c_i encoding the i-th unit vector in $\{0,1\}^k$. By linearity, the diagonal of $A_S = \sum_{i \in S} A_i$ is also a codeword in C, which is non-zero when $S \neq \emptyset$, and thus has at least d ones, so that $rank(A_S) \geq d$. Then Theorem 1 applies with $r = \ell - d$, and the corresponding strong blender extracts k bits with bias $2^{-\frac{(b_X + b_Y + d - 2\ell - k)}{2}}$, or, equivalently, extract $k = (b_X + b_Y + d - 2\ell - 2\log(1/\varepsilon))$ bits with bias ε (notice, sometimes this k may not be achievable due to the coding constraints).

OUR CONSTRUCTION. We now present our new construction of strong blenders via Theorem 1. We show that there is a set of matrices $A_1, \ldots A_\ell$ whose non-empty sums have *full rank*, thus achieving the best possible result for using the technique in section 3.1. This latter fact follows from [14, 13, 19], and we reproduce its simple proof for completeness. We use the isomorphism $\{0,1\}^\ell \approx GF[2^\ell]$. Take any basis $x_1, x_2, \ldots, x_\ell$ of $GF[2^\ell]$. (There are many such bases for $GF[2^\ell]$, since it is isomorphic to the ℓ-dimensional vector space over $GF[2]$. One such basis, which is easy to use, is $1, x, x^2, \ldots, x^{\ell-1}$, where x is any primitive element of $GF[2^\ell]$; that is, an element that is not the root of any polynomial of degree $< \ell$ over $GF[2^\ell]$.) The matrices $\{A_i\}_{i=1}^\ell$ are defined such that matrix A_i corresponds to left-multiplication by the basis-element x_i:

$$A_i : y \in GF[2^\ell] \mapsto x_i y$$

Now for each non-empty set $S \subseteq [\ell]$, $x_S = \sum_{i \in S} x_i$ is a non-zero element in $GF[2^\ell]$, and therefore has an inverse x_S^{-1} in $GF[2^\ell]$. Let U_S be the $\ell \times \ell$ matrix over $GF[2]$ that corresponds to left-multiplication by x_S^{-1}:

$$U_S : y \in GF[2^\ell] \mapsto x_S^{-1} y$$

Now, it is easy to see that U_S is the matrix-inverse of $A_S = \sum_{i \in S} A_i$, since left-multiplication by matrix $U_S \sum_{i \in S} A_i$ corresponds to multiplication by the field identity element $x_S^{-1} \sum_{i \in S} x_i = 1$.
This immediately implies the desired conclusion:

Lemma 5. *Let $A_1, \ldots, A_\ell$ be matrices presented above. Then for all $\emptyset \neq S \subseteq [\ell]$, the rank of $A_S \overset{\text{def}}{=} \sum_{i \in S} A_i$ is ℓ.*

Plugging the first k of the matrices $A_1, \ldots, A_\ell$ into theorem 1 we deduce the existence of an *explicit* strong blender that can extract k bits with bias $2^{-\frac{(b_X + b_Y + 2 - \ell - k)}{2}}$, or, equivalently, extract $k = (b_X + b_Y + 2 - \ell - 2\log(1/\varepsilon))$ bits.

References

1. N. Alon. Tools from higher algebra. In *Handbook of Combinatorics*, R.L. Graham, M. Grötschel and L. Lovàsz, eds, North Holland (1995), Chapter 32, pp. 1749–1783.
2. A. Andreev, A. Clementi, J. Rolim, L. Trevisan. Dispersers, deterministic amplification, and weak random sources. In *SIAM J. on Comput.*, 28(6):2103–2116, 1999.
3. B. Barak, R. Impagliazzo and A. Wigderson. Extracting Randomness from Few Independent Sources. *Manuscript*, 2004.
4. B. Chor, O. Goldreich. Unbiased bits from sources of weak randomness and probabilistic communication complexity. *SIAM J. Comput.*, 17(2):230–261, 1988.
5. A. Cohen, A. Wigderson. Dispersers, deterministic amplification, and weak random sources. In *Proc. of FOCS*, pp. 14–19, 1989.
6. Y. Dodis, J. Spencer. On the (Non-)Universality of the One-Time Pad. In *Proc. of FOCS*, 2002.
7. Y. Dodis, R. Oliveira. On Extracting Private Randomness over a Public Channel. In *RANDOM-APPROX 252-263*, 2003.
8. O. Goldreich. Three XOR-Lemmas - An Exposition. *Electronic Colloquium on Computational Complexity (ECCC)*, 1995.
9. R.L. Graham, J.H. Spencer. A constructive solution to a tournament problem. In *Canad. Math. Bull.* 14, 45-48.
10. C. Lu, O. Reingold, S. Vadhan, A. Wigderson. Extractors: Optimal Up to Constant Factors. In *Proc. of STOC*, 2003.
11. U. Maurer, S. Wolf. Privacy Amplification Secure Against Active Adversaries. In *Proc. of CRYPTO*, Lecture Notes in Computer Science, Springer-Verlag, vol. 1294, pp. 307–321, 1997.
12. J. McInnes, B. Pinkas. On the Impossibility of Private Key Cryptography with Weakly Random Keys. In *Proc. of CRYPTO*, pp. 421–435, 1990.
13. R. Meshulam. Spaces of Hankel matrices over finite fields. *Linear Algebra and its Applications*, 218, 1995.
14. E. Mossel, A. Shpilka, L. Trevisan. On ϵ-biased Generators in NC^0. In *Proc. of FOCS*, 2003.
15. N. Nisan, A. Ta-Shma. Extracting Randomness: a survey and new constructions. In *JCSS*, 58(1):148–173, 1999.

16. N. Nisan, D. Zuckerman. Randomness is Linear in Space. In *JCSS*, 52(1):43–52, 1996.
17. R. Raz, O. Reingold, S. Vadhan. Extracting all the randomness and reducing the error in Trevisan's extractors. *Journal of Computer and System Sciences*, 2002.
18. L. Rónyai, L. Babai, M. Ganapathy. On the number of zero-patterns in a sequence of polynomials *Journal of the AMS*, 2002.
19. R. Roth. Maximum rank array codes and their application to crisscross error correction. *IEEE transactions on Information Theory*, 37, 1991.
20. M. Sántha, U. Vazirani. Generating Quasi-Random Sequences from Semi-Random Sources. *Journal of Computer and System Sciences*, 33(1):75–87, 1986.
21. R. Shaltiel. Recent developments in Explicit Constructions of Extractors. *Bulletin of the EATCS*, 77:67–95, 2002.
22. R. Shaltiel, C. Umans. Simple extractors for all min-entropies and a new pseudorandom generator. In *Proceedings of FOCS 2001*, pp.648-657, IEEE Computer Society, 2001.
23. L. Trevisan. Construction of Extractors Using PseudoRandom Generators. In Proc. of STOC, pp. 141–148, 1999.
24. U. Vazirani. Randomness, Adversaries and Computation. *PhD Thesis*, University of California, Berkeley, 1986.
25. U. Vazirani. Strong Communication Complexity or Generating Quasi-Random Sequences from Two Communicating Semi-Random Sources. *Combinatorica*, 7(4):375–392, 1987.
26. U. Vazirani, V. Vazirani. Random polynomial time is equal to slightly-random polynomial time. In *Proc. of 26th FOCS*, pp. 417–428, 1985.
27. U. Vazirani. Efficiency Considerations in using semi-random sources. In *Proceedings of the nineteenth annual ACM conference on Theory of computing*, pp. 160–168, 1987.
28. A. Wigderson. Open problems. Notes from *DIMACS Workshop on Pseudorandomness and Explicit Combinatorial Constructions*, 1999
29. D. Zuckerman. Simulating BPP Using a General Weak Random Source. *Algorithmica*, 16(4/5):367-391, 1996.

The Diameter of Randomly Perturbed Digraphs and Some Applications

Abraham D. Flaxman* and Alan M. Frieze**

Department of Mathematical Sciences
Carnegie Mellon University
Pittsburgh, PA, 15213, USA
abie@cmu.edu, alan@random.math.cmu.edu

Abstract. The central observation of this paper is that if ϵn random edges are added to any n-node connected graph or digraph then the resulting graph has diameter $\mathcal{O}(\log n)$ with high probability. We apply this to smoothed analysis of algorithms and property testing.
Smoothed Analysis: Recognizing strongly connected digraphs is a basic computational task in graph theory. Even for graphs with bounded out-degree, it is **NL**-complete. By XORing an arbitrary bounded out-degree digraph with a sparse random digraph $R \sim \mathbb{D}_{n,\epsilon/n}$ we obtain a "smoothed" instance. We show that, with high probability, a log-space algorithm will correctly determine if a smoothed instance is strongly connected. We also show that if **NL** $\not\subseteq$ almost-**L** then no heuristic can recognize similarly perturbed instances of (s,t)-connectivity.
Property Testing: A digraph is called k-linked if for every choice of $2k$ distinct vertices $s_1, \ldots, s_k, t_1, \ldots, t_k$, the graph contains k vertex disjoint paths joining s_r to t_r for $r = 1, \ldots, k$. Recognizing k-linked digraphs is **NP**-complete for $k \geq 2$. We describe a polynomial time algorithm for bounded degree digraphs which accepts k-linked graphs with high probability, and rejects all graphs which are at least ϵn edges away from being k-linked.

1 Introduction

The diameter of a graph G is the length of the longest shortest path in G. In other words, if $d(u,v)$ is the length of the shortest path from u to v in G, then the diameter of G is $\max_{u,v} d(u,v)$. A graph is connected (and a directed graph is strongly connected) if it has finite diameter. The central observation of this paper is that if ϵn random edges are added to any n-node connected graph then the diameter becomes $\mathcal{O}(\log n)$ with high probability (meaning with probability tending to 1 as $n \to \infty$ and abbreviated **whp**). This is also true for strongly connected directed graphs (digraphs) and for several ways of generating the random edges. For ease of exposition, we state this as a theorem only for a strongly connected digraph perturbed by adding a random digraph $R \sim \mathbb{D}_{n,\epsilon/n}$.

* Supported in part by NSF Grant CCR-0122581.
** Supported in part by NSF Grant CCR-0200945.

K. Jansen et al. (Eds.): APPROX and RANDOM 2004, LNCS 3122, pp. 345–356, 2004.

Here $R \sim \mathbb{D}$ means R is distributed according to distribution $\mathbb{D}$, and $\mathbb{D}_{n,p}$ is the distribution over n-node digraphs where each arc is included independently with probability p.

Theorem 1. *Let ϵ be any positive constant. For any strongly connected n-node digraph $\bar{D}$, let $D = \bar{D} + R$ where $R \sim \mathbb{D}_{n,\epsilon/n}$. Then* **whp** *the diameter of D is at most $100\epsilon^{-1} \log n$.*

Similar results hold for perturbations formed by adding ϵn edges selected at random with or without replacement, or by adding a random matching or assignment with ϵn edges, or by picking ϵn vertices and adding a uniformly random out-arc to each of them.

Theorem 1 is related to a class of problems regarding the possible change of diameter in a graph where edges are added or deleted, for example, the results of Alon, Gryárfás, and Ruszinkó in [2] on the minimum number of edges that must be added to a graph to transform it into a graph of diameter at most d. The study of these extremal diameter alteration questions was initiated by Chung and Garey in [12]. It is also related to the theorem of Bollobás and Chung on the diameter of a cycle plus a random matching [10].

1.1 Application: Smoothed Analysis

Recognizing strongly connected digraphs is a basic computational task, and the set of strongly connected digraphs is **NL**-complete [21]. Thus, if **NL** $\nsubseteq$ **L** then there is no log-space algorithm which recognizes strongly connected digraphs.

Perhaps this conclusion of worst-case complexity theory is too pessimistic. We will consider the performance of a simple heuristic which runs in randomized log-space. We will show that the heuristic succeeds on random instances **whp**. However, the "meaning" of this result depends on the probability space from which we draw the random instances. It seems reasonable to assume that any real-world digraph will contain some amount of randomness, so it is tempting to believe this result shows that in the real-world strong connectivity only requires log-space. Unfortunately, this is not valid if we use the "wrong" model for randomness. For example, the distribution $\mathbb{D}_{n,p}$ is pleasant for analysis, but basic statistics like degree sequence seem to differ from real-world graphs [14].

We will use a model of randomness that is more flexible. We will start with an arbitrary digraph $\bar{D}$ and perturb it by XORing it with a very sparse random graph $R \sim \mathbb{D}_{n,\epsilon/n}$. This produces a random instance which is "less random" than $\mathbb{D}_{n,p}$. The study of worst case instances with small random perturbations is called Smoothed Analysis.

Smoothed Analysis was introduced by Spielman and Teng in [27] and they discuss a perturbation model for discrete problems in [28]. They consider perturbing graphs by XORing the adjacency matrix with a random adjacency matrix, where each edge is flipped with some constant probability. Since the probability of an edge flip is constant, the perturbed instances are all dense graphs (i.e. a constant fraction of all possible edges appear). Independently, Bohman, Frieze and Martin [7] studied the issue of Hamiltonicity in a dense graph when random edges are *added*, and other graph properties were analyzed in this model by Bohman, Frieze, Krivelevich and Martin [8] and Krivelevich and Tetali [23].

We will also use an XOR perturbation, but we will make the probability of corruption much lower than [28]. Since we will have a linear number of edges present, it is appropriate for the perturbation to change about ϵn edges, which is the expected number of edges in $\mathbb{D}_{n,\epsilon/n}$.

Randomness and Strong Connectivity. Recognizing strongly connected digraphs is closely related to recognizing (s,t)-connectivity in digraphs, which is the canonical **NL**-complete problem. It is possible to recognize connectivity in undirected graphs with a randomized log-space algorithm using random walks [1]. Since the cover time of an arbitrary connected graph is bounded by $\mathcal{O}(n^3)$ [17], a random walk will visit every vertex in polynomial time **whp**. This approach will not work for arbitrary digraphs, however, since there the cover time can be exponential.

The diameter and connectivity of random graphs has been well-studied, see for example the books of Bollobás [9] and Janson, Łuczak, and Ruciński [20]. Perhaps closest in spirit to our investigation is the paper of Bollobás and Chung on the diameter of a Hamilton cycle plus a random matching [10] and the paper of Chung and Garey on the diameter of altered graphs [12]. Also, the component structure of random digraphs was studied by Karp in [22] and more recently by Cooper and Frieze [13].

A Heuristic for Recognizing Strong Connectivity. For each ordered pair of vertices (s,t) repeat the following procedure N_1 times: Starting from s, take N_2 steps in a random walk on the digraph. Here a random walk is the sequence of vertices $X_0, X_1, \ldots, X_t, \ldots$ visited by a particle which moves as follows: If $X_t = v$ then X_{t+1} is chosen uniformly at random from the out-neighbors of X_t. If the random walk with restarts ever reaches t, then the digraph contains an (s,t)-path, and we continue to the next pair of vertices.

If N_1 and N_2 are large enough, this algorithm is correct **whp**. For example, if there is a path from s to t and $N_1 = n^{2n}$, $N_2 = n$ we will discover the path **whp**. The values of N_1 and N_2 can be significantly improved for smoothed random instances

The main theorem of this section is that when N_1 and N_2 are suitable polynomials in n, this heuristic, which we will call Algorithm $\mathcal{A}$, is successful on perturbations of bounded out-degree instances **whp**. To prove this, we first show it is successful when the initial instance is a strongly connected digraph and the perturbation only adds edges. Then we extend this to show success when the initial instance is not necessarily strongly connected and the perturbation only adds edges. After this, it is simple to translate our results to the original perturbation model where edges are added and removed, since we can generate the perturbation in 2 rounds, by first deleting each existing edge with some probability, and then adding random edges to the resulting digraph.

We write $G_1 \oplus G_2$ to mean the XOR of digraphs G_1 and G_2, (which is to say $e \in G_1 \oplus G_2$ if and only if $e \in G_1$ and $e \notin G_2$ or vice versa.)

Theorem 2. *Let ϵ be any positive constant. For any n-node digraph $\bar{D}$ with maximum out-degree Δ, let $D = \bar{D} \oplus R$ where $R \sim \mathbb{D}_{n,\epsilon/n}$. Then* **whp** *Algorithm $\mathcal{A}$ is correct on D when $N_1 = n^{A_1 \epsilon^{-1} \log(10\Delta)}$ and $N_2 = B_1 \epsilon^{-1} \log n$.*

Here A_1 and B_1 are absolute constants. We find that $A_1 = 400$ and $A_2 = 200$ suffice, but we do not attempt to optimize these values.

If a strongly connected digraph has bounded out-degree and has diameter $\mathcal{O}(\log n)$ then a random walk of length $\mathcal{O}(\log n)$ has a $1/\operatorname{poly}(n)$ chance of going from s to t, and Algorithm $\mathcal{A}$ will succeed **whp** using values of N_1 and N_2 that can be realized in log-space. Unfortunately, even though our initial instances have bounded out-degree and (as indicated by Theorem 1) our perturbed instances have logarithmic diameter, the perturbation increases the maximum degree to $\Omega(\log n/\log\log n)$, so we must work a little harder to show that the random walk has a non-negligible probability of witnessing the path from s to t. (As an additional reward for this work, we find that Algorithm $\mathcal{A}$ can be derandomized by checking all paths from s of length $\mathcal{O}(\epsilon^{-1}\log n)$ and still only use log-space.)

The analysis of Algorithm $\mathcal{A}$ is further complicated by the possibility a $\bar{D}$ which is not strongly connected combining with R to produce a smoothed instance which is strongly connected. We handle this situation by arguing that for the smoothed instance to become strongly connected there cannot be too many small strong components of $\bar{D}$, and then the large components become strongly connected with short paths that a random walk is not too unlikely to travel, and finally, the small components, if they are connected to the large component, are "close" to it.

Why Study Instances with Bounded Out-Degree? It would be nice to extend our results to hold for perturbed copies of any digraph, instead of only digraphs with bounded out-degree. However, such a result is not possible for our heuristic. We demonstrate the necessity of our assumption that $\bar{D}$ has bounded out-degree by constructing a family of instances with growing degree on which Algorithm $\mathcal{A}$ does not succeed **whp**.

Theorem 3. *Let $f(n) \to \infty$, let ϵ be a positive constant less than 1, and let $R \sim \mathbb{D}_{n,\epsilon/n}$. Then for every sufficiently large n, there exists an n-node digraph $\bar{D}$ with maximum out-degree $\Delta = f(n)$ such that for $\bar{D} \oplus R$ the probability Algorithm $\mathcal{A}$ fails exceeds $1 - e^{-\epsilon} - o(1)$.*

The proof of Theorem 3 is omitted for lack of space.

Strong Connectivity Versus (s,t)-Connectivity. Strong connectivity is an **NL**-complete problem, but (s,t)-connectivity is "the" **NL**-complete problem. In Sipser's undergraduate text [26], the completeness of (s,t)-connectivity is proved in detail, while the completeness of strong connectivity is left as an exercise.

In light of this, it is natural to investigate the efficacy of heuristics on smoothed instances of (s, t)-connectivity. Here we find that there are instances on which Algorithm $\mathcal{A}$ fails **whp**. What is more, no log-space heuristic exists, provided a conjecture of complexity theory holds.

Theorem 4. *If* **NL** $\not\subseteq$ *almost-L then no log-space heuristic succeeds* **whp** *on smoothed instances of bounded out-degree (s, t)-connectivity.*

The proof consists of building a machine which simulates any nondeterministic log-space machine using the log-space heuristic for (s, t)-connectivity, were such a heuristic to exist. The proof is omitted for lack of space.

Smoothed Model Versus Semi-random Model. The semi-random model was introduced by Santha and Vazirani in [25]. In this model an adversary adaptively chooses a sequence of bits and each is corrupted independently with probability δ. They present this as a model for real-world random bits, such as the output of a Geiger counter or noisy diode, and consider the possibility of using such random bits in computation on worst-case instances. Blum and Spencer considered the performance of a graph coloring heuristic on random and semi-random instances in [6]. Subsequent work has uncovered an interesting difference between the random and semi-random instances in graph coloring. The work of Alon and Kahale [3] developed a heuristic which succeeds **whp** on random instances with constant expected degree, while work by Feige and Kilian [16] showed no heuristic can succeed on semi-random instances with expected degree $(1 - \epsilon) \log n$ (they also developed a heuristic for semi-random instances with expected degree $(1 + \epsilon) \log n$).

In the original semi-random model of Santha and Vazirani, an instance is formed by an adaptive adversary, who looks at all the bits generated so far, asks for a particular value for the next bit, and gets the opposite of what was asked for with probability δ. Several modifications are proposed in Blum and Spencer [6] and also in Feige and Krauthgamer [15]. However, all these variations maintain the adaptive aspect of the adversary's strategy, which at low density allows too much power; if the error probability $p = (1 - \epsilon) \log n / n$ then there will be roughly n^ϵ isolated vertices in $\mathbb{D}_{n,p}$ and the adversary will be able to encode a polynomial sized instance containing no randomness. Since we wish to consider extremely sparse perturbations, where the error probability $p = \epsilon/n$, we cannot allow an adversary as powerful as in the semi-random model. The XOR perturbation considered in this paper is equivalent to a natural weakening of the semi-random model: making the adversary oblivious.

1.2   Application: Property Testing

Property testing provides an alternative weakening of worst-case decision problems. It was formalized by Goldreich, Goldwasser, and Ron in [18]. The goal in property testing is to design an algorithm which decides whether a instance has a property or differs significantly from all instances which have that property. For example, a property tester for strong connectivity in bounded degree digraphs should accept all strongly connected instances and reject all instances that are ϵn edges away from being strongly connected. Note that Algorithm $\mathcal{A}$ (which is designed to work on smoothed random instances) can be converted into a property tester: given an instance $\bar{D}$ and a gap parameter ϵ, we can randomly perturb $\bar{D}$ ourselves by adding $\frac{\epsilon}{2}n$ random edges and then run Algorithm $\mathcal{A}$ on the perturbed version. This does not yield anything impressive for testing strong connectivity, since the undirected connectivity testing results of Goldreich and Ron in [19] can be applied to the directed case to produce a constant time tester. However, our perturbation approach also yields a property tester for a more difficult connectivity problem, k-linked.

A digraph is called k-linked if for every choice of $2k$ distinct vertices $s_1, \ldots, s_k$, $t_1, \ldots, t_k$, the graph contains k vertex disjoint paths joining s_1 to t_1, $\ldots$, s_k to t_k. Recognizing k-linked digraphs is **NP**-complete for $k \geq 2$. In the bounded degree property testing version of k-linked, we are given a constant ϵ and a digraph $\bar{D}$ with maximum out-degree Δ and our goal is to accept if $\bar{D}$ is k-linked and reject if $\bar{D}$ is more than ϵn arcs away from being k-linked, and we can do whatever if $\bar{D}$ is not k-linked, but is close to being so.

A Heuristic for Testing k-Linkedness. Given $\bar{D}$ and ϵ, we perturb $\bar{D}$ by flipping a coin for each node, and with probability $\epsilon/2$ adding a out-arc leading to a node chosen uniformly at random. Call this perturbed instance D. Then, for each choice of $2k$ distinct vertices, repeat the following procedure N_1 times: For $i = 1, \ldots, k$ starting at s_i take N_2 steps in a random walk on the graph. If all k random walks ever reach the correct k terminals via vertex disjoint paths, we continue to the next choice of $2k$ vertices. Otherwise reject.

Here k is assumed to be fixed, independent of the input.

Theorem 5. *Let ϵ be any positive constant. For any k-linked n-node graph $\bar{D}$ with maximum degree Δ, the algorithm above accepts* **whp**.

A few comments regarding the difference between this theorem and Theorems 1 and 2. Since we perturbed $\bar{D}$ by adding a 1-out, the maximum out-degree of D is bounded by $\Delta + 1$, so we are not faced with one of the additional challenges in Theorem 2 (which is addressed by Lemma 1); if k vertex-disjoint logarithmic length paths appear then we can immediately conclude that **whp** they will be discovered after a polynomial number of samples. What is more, the fact that such paths exist **whp** follows from a calculation analogous to the proof of Theorem 1, that explores disjoint neighborhoods around all $2k$ terminals simultaneously. Also, the analog of the most difficult part of Theorem 2, showing that Algorithm $\mathcal{A}$ is correct in the case where a disconnected $\bar{D}$ leads to a strongly connected D, is no longer necessary. In the property testing setting, we are not required to correctly recognize instances that lead to this situation. It seems as though it might be possible to carry out this most difficult part and obtain a heuristic for k-linked that works on smoothed instances, but the details remain elusive.

2   Proof of Theorem 1

We now show that if $\bar{D}$ is strongly connected then the diameter of $D = \bar{D} + R$ is $\mathcal{O}(\log n)$ **whp**.

Instead of adding a random digraph R with probability $p = \epsilon/n$, we will add two random digraphs R' and R'' with arcs appearing independently with probability p' and p'' respectively. We will take $p' = \frac{\epsilon n^{1/6}-1}{n^{7/6}-1} = (1-o(1))\epsilon/n$ and $p'' = n^{-7/6}$. The probability of an edge appearing in the union of R' and R'' is $p' + p'' - p'p'' = \epsilon/n$. So adding R' and R'' to $\bar{D}$ yields a digraph identically distributed with D.

We will show that **whp** D contains short paths of a special form, alternating between an appropriate number of edges from $\bar{D}$ and a random edge of R'. This is similar to the approach of Bollobás and Chung [10].

Let $S_0 = T_0 = \emptyset$, and let S_0' (resp. T_0') be the set of vertices s_0' (resp. t_0') for which there is a path of length at most $32\epsilon^{-1}\log n + 1$ from s to s_0' (resp. from t_0' to t) in $\bar{D}$.

Now, for $i \geq 1$, we identify a set S_i of neighbors of S_{i-1}' and then a set S_i' of neighbors of S_i, and do the same for T_i and T_i'.

Let S_i (resp. T_i) be the set of $s'' \notin \bigcup_{j=0}^{i-1} S_j \cup S_j'$ (resp. $t'' \notin \bigcup_{j=0}^{i-1} T_j \cup T_j'$) for which the arc (s', s'') (resp. (t'', t')) is in R' for some $s' \in S_{i-1}'$.

Let S_i' (resp. T_i') be the set of vertices s_i' (resp. t_i') for which there is a path in $\bar{D}$ of length at most

$$d = 5\epsilon^{-1} + 1$$

from some $s_i \in S_i$ to s_i' (resp. from t_i' to some $t_i \in T_i$). Let

$$\ell = \lceil \log_2 n \rceil.$$

Let $Z_S = \left| \bigcup_{j=0}^{\ell} (S_j \cup S_j') \right|$ and $Z_T = \left| \bigcup_{j=0}^{\ell} (T_j \cup T_j') \right|$. We will now show that

$$\Pr[Z_S \leq n^{2/3} \text{ or } Z_T \leq n^{2/3}] = o(n^{-2}). \tag{1}$$

We will argue this assertion only for Z_S. The part regarding Z_T is proved analogously.

Let $U_i = \bigcup_{j=0}^{i} (S_i \cup S_i')$. For $i \leq \ell$, suppose that $|U_i| \leq n^{2/3}$. Let s'' be some vertex not in U_i. Then the probability that there is an arc in R' connecting S_{i-1}' to s'' is $1 - (1 - p')^{|S_{i-1}'|} = (1 - o(1))\epsilon\frac{|S_{i-1}'|}{n}$.

We use the following Chernoff bounds from [20, Theorem 2.1] on the Binomial random variable $B(n,p)$

$$\Pr(B(n,p) \geq np + t) \leq \exp\left\{ -\frac{t^2}{2(np + t/3)} \right\} \tag{2}$$

$$\Pr(B(n,p) \leq np - t) \leq \exp\left\{ -\frac{t^2}{2np} \right\}. \tag{3}$$

Since the events $\{s'' \in S_i\}$ are all independent, we let $Z = |S_i|$ and see that Z is distributed as $B(n - o(n), (1 - o(1))\epsilon|S'_{i-1}|/n)$. So from (3) we have

$$\Pr\left[|S_i| \le \frac{\epsilon}{2}|S'_{i-1}| \,\Big|\, |U_i| \le n^{2/3}, |S'_{i-1}|\right] \le e^{-\epsilon|S'_{i-1}|/9}.$$

Now, for $|S_i| \le n^{2/3}$, we will show that $|S'_i|$ is likely to be at least $4\epsilon^{-1}|S_i|$. Note that given its size s, the set S_i is a random s-subset of the vertices not in U_i. Since $|U_i| \le n^{2/3}$, there are $n - o(n)$ vertices to choose from. And since the out-degree of $\bar{D}$ is bounded by Δ, there are at most Δ^{2d} vertices within distance $2d$ of a given vertex. We will call a randomly selected vertex a *failure* if it is at distance less than $2d$ from a previously selected vertex. Then the probability of failure of a particular vertex, conditioned on the success or failure of any other vertices is at most $(1 + o(1))|S_i|\Delta^{2d}/n$. So the number of failures N_f is stochastically dominated by $B(|S_i|, q)$, where $q = (1+o(1))|S_i|\Delta^{2d}n^{-1}$. Since we are interested in a wide range of values of $|S_i|$, we bound the same probability for large and small values of $|S_i|$. If $|S_i| \ge n^{5/12}$ we use (2) to obtain the following:

$$\Pr\left[N_f \ge |S_i|q + n^{1/3} \,\Big|\, |U_i| \le n^{2/3}, |S_i|\right]$$

$$\le \exp\left\{-\frac{n^{2/3}}{2(|S_i|q + n^{1/3}/3)}\right\} \quad \le \exp\left\{-\frac{n^{1/3}}{3\Delta^{2d}}\right\},$$

which relies on the fact that $|S_i| \le n^{2/3}$ and hence $|S_i|q \le (1 + o(1))\Delta^{2d}n^{1/3}$. If $|S_i| \le n^{5/12}$, then we use the union bound to show

$$\Pr\left[N_f \ge 18 \,\Big|\, |S_i|\right] \le \binom{|S_i|}{18}\left(\frac{|S_i|\Delta^{2d}}{n}\right)^{18} \le \frac{|S_i|^{36}\Delta^{36d}}{n^{18}} = \mathcal{O}(n^{-3}).$$

In either case, we have that the failure set has size $o(|S_i|)$ with probability $1 - \mathcal{O}(n^{-3})$. We now obtain a lower bound on the size of S'_i by ignoring all the failure vertices. Note that the neighborhoods of size d of the non-failure vertices are disjoint, and since $\bar{D}$ is strongly connected, there are at least $d - 1$ vertices within distance d of every vertex. So

$$|S'_i| \ge (1 - o(1))5\epsilon^{-1}|S_i|$$

with probability $1 - \mathcal{O}(n^{-3})$. Combining this with our bound on the probability that $|S_i| \le \frac{1}{2}\epsilon|S'_{i-1}|$ gives

$$\Pr\left[|S'_i| \le 2|S'_{i-1}| \,\Big|\, |U_i| \le n^{2/3}, |S'_{i-1}|\right] \le e^{-\epsilon|S'_{i-1}|/9} + \mathcal{O}(n^{-3}).$$

Again using the fact that $\bar{D}$ is strongly connected, we have $|S'_0| \geq 32\epsilon^{-1} \log n$, so we have

$$\Pr\left[|U_\ell| \leq n^{2/3}\right] \leq \Pr\left[\exists i : |U_i| \leq n^{2/3} \text{ and } |S'_i| \leq 2|S'_{i-1}|\right]$$

$$\leq \sum_{i=1}^{\ell} \Pr\left[|S'_i| \leq 2|S'_{i-1}| \,\Big|\, |U_i| \leq n^{2/3}, |S'_{i-1}| \geq 32\epsilon^{-1} \log n\right] = \mathcal{O}(n^{-3} \log n).$$

From the above we have that $U = \bigcup_{i=0}^{\ell} S_i \cup S'_i$ and $W = \bigcup_{i=0}^{\ell} T_i \cup T'_i$ have size at least $n^{2/3}$ with probability $1 - \mathcal{O}(n^{-3})$. If U, W intersect, then we know there is a path of length at most $2(d+1)\ell$. Otherwise, we expose the edges of R'' (which appear independently with probability $p'' = n^{-7/6}$) and find that

$$\Pr[R'' \cap U \times W = \emptyset] \leq (1 - p'')^{n^{4/3}} \leq e^{-n^{1/6}} = o(n^{-2}).$$

Since there are only $n(n-1)$ choices for s, t, the theorem follows. $\square$

3 Sketch of Proof of Theorem 2

3.1 When $\bar{D}$ Is Strongly Connected

By Theorem 1 we know that the diameter of D is $\mathcal{O}(\log n)$ **whp**. Unfortunately we cannot yet conclude that Algorithm $\mathcal{A}$ is successful **whp**. We must still argue that the probability of a random walk traversing the short path is not too small. In the original graph, $\bar{D}$, having a diameter of $\mathcal{O}(\log n)$ would imply an efficient algorithm, because the maximum out-degree is bounded. Our random perturbation has created some vertices with out-degree $\Omega(\log n / \log \log n)$, so we will have to work a little more.

Lemma 1. *Let* $D = \bar{D} + R$, *where* $\bar{D}$ *is an arbitrary digraph with maximum out-degree* Δ *and* $R \sim \mathbb{D}_{n,\epsilon/n}$. *Then* **whp** D *contains no path* P *of length* $\ell \leq \ell_0 = 100\epsilon^{-1} \log n$ *with* $\prod_{x \in P} \deg_D^+(x) \geq n^{100\epsilon^{-1} \log(10\Delta)}$.

The proof of Lemma 1 is omitted due to lack of space.

The correctness of Algorithm $\mathcal{A}$ in the case when $\bar{D}$ is strongly connected now follows from that fact that the probability a random walk follows a path P from s to t is precisely $\left(\prod_{x \in P} \deg_D^+(x)\right)^{-1}$.

3.2 When $\bar{D}$ Not Strongly Connected

The previous section shows that Algorithm $\mathcal{A}$ is correct **whp** for strongly connected digraphs $\bar{D}$. To prove that Algorithm $\mathcal{A}$ is correct **whp** when $\bar{D}$ is not strongly connected, we must do some more work.

Outline of approach

Consider the strong components of $\bar{D}$. If there are many components of size less than $\frac{1}{4}\epsilon^{-1}\log n$, then we show that **whp** one of them will be incident to no arcs of R and so D will not be strongly connected and Algorithm $\mathcal{A}$ will be correct. In the case where $\bar{D}$ consists mostly of larger strong components, we expose the random arcs R in two rounds. We argue that **whp** the strong components of $\bar{D}$ merge into a unique giant strong component containing at least $n - n^{16/17}$ vertices after the first round. Then we invoke Lemma 1 from the previous section to show that the random arcs from the second round gives the giant a low diameter. Then we deal with the vertices that belong to small strong components after the first round of random arcs have been added. These vertices might be connected to the giant in both directions and they might not, and there is not necessarily a sharp threshold for strong connectivity. However, we show that for some constants A_1 **whp** no such vertex is connected in either direction to the giant only by paths of length more than $A_1\epsilon^{-1}\log n$ i.e. such a vertex is close to the giant in some direction, or cannot be reached at all in this direction. Finally, by Lemma 1 we know that all the paths of length at most $A_1\epsilon^{-1}\log n$ have a non-negligible probability of being traversed by Algorithm $\mathcal{A}$ (take the bound in Lemma 1 and raise it to the power $A_1/100$). So we conclude that **whp** the graph is not strongly connected, in which case Algorithm $\mathcal{A}$ is correct, or the graph is strongly connected in such a way that Algorithm $\mathcal{A}$ is still correct.

The calculations required for this plan are omitted due to lack of space.

4   Proof of Theorem 5

We will show that if $\bar{D}$ is k-linked then **whp** D contains disjoint paths of length at most $100k\epsilon^{-1}\log n$ which witness its k-linkedness.

For a particular choice of $s_1, \ldots, s_k, t_1, \ldots, t_k$, let $Q_1, Q_2, \ldots, Q_k$ be vertex disjoint paths in $\bar{D}$ such that Q_r goes from s_r to t_r.

We order the paths from longest to shortest and define ℓ so that for $r \leq \ell$ each Q_r has length at least $100k\epsilon^{-1}\log n$. If $\ell = 0$ then there is nothing to prove, so suppose $\ell \geq 1$.

We use the same type of argument as in the proof of Theorem 1 to show the existence of short paths between s_r and t_r, but we work will all $r \leq \ell$ simultaneously to ensure that the paths we find are vertex disjoint. To this end, we define a sequence of collections of sets $S_{i,r}$ and $T_{i,r}$ for $i \geq 0$ and $1 \leq r \leq \ell$, (and define $S_{i,r} = Q_i$ for $i \geq 0$ and $r > \ell$). Let $S_{0,r}$ be the first $32k\epsilon^{-1}\log n$ vertices of Q_r and $T_{0,r}$ be the last $32k\epsilon^{-1}\log n$ vertices of Q_r, for $1 \leq r \leq \ell$.

We will call a node *useful* if it is not within $d = 5\epsilon^{-1}$ of any node which we have previously placed in any S or T set.

To define $S_{i,r}$, we check, for each node s' in $S_{i-1,r}$, if s' has a random out-arc between it and a useful node s''. If it does, we add s'' and all nodes reachable from s'' in d steps to $S_{i,r}$. We continue this until $|S_{i,r}|$ exceeds $2|S_{i-1,r}|$ (and declare failure if it does not).

$T_{i,r}$ is defined analogously, but the paths lead towards t_r instead of away. For a node t' in $T_{i-1,r}$, we look for useful nodes t'' where the random arc is directed from t'' to t', and put $2|T_{i-1,r}|$ of them in $T_{i,r}$.

We wish to argue that we can form sets $S_{i,r}$ and $T_{i,r}$ of the appropriate sizes while $|S_{i,r}| = |T_{i,r}| \leq n^{2/3}$. Since every time we discover a useful neighbor we add $d = 5\epsilon^{-1}$ nodes, we only see failure if for some i and r, we find that $S_{i,r}$ has less than $\frac{2}{5}\epsilon|S_{i,r}|$ useful neighbors. But at any step, there are at most $2kn^{2/3}$ nodes which have been placed in any S or T, and (since the degree is bounded by Δ) this means there must be at least $n - \Delta^d 2kn^{2/3}$ useful nodes. So each node in $S_{i,r}$ has a random out-arc to useful neighbor independently with probability at least $(1 - o(1))\epsilon$, and using the Chernoff bound (3), we have

$$\Pr\left[\text{The number of useful neighbors of } S_{i,r} \leq \frac{\epsilon}{2}|S_{i,r}|\right] \leq e^{-\epsilon^2|S_{i,r}|/9}.$$

Also, each useful node has a random out-arc to a node in $T_{i,r}$ independently with probability $\epsilon\frac{|T_{i,r}|}{n}$, so by (3) we have

$$\Pr\left[\text{The number of useful neighbors of } T_{i,r} \leq \frac{\epsilon}{2}|T_{i,r}|\right] \leq e^{-\epsilon^2|T_{i,r}|/9}.$$

Since $|S_{0,r}| = |T_{0,r}| = 32k\epsilon^{-1}\log n$, a union bound shows that the probability of failure is less than $2kn^{-3k}\log n$.

To conclude, we note that when $|S_{i,r}| = |T_{i,r}| \geq \frac{1}{2}n^{2/3}$, the probability that no node of $S_{i,r}$ has a random out-arc to a node of $T_{i,r}$ is at most

$$\left(1 - \frac{\epsilon}{2}n^{-1/3}\right)^{\frac{1}{2}n^{2/3}} < \exp\left\{-\frac{\epsilon}{4}n^{1/3}\right\} - o(n^{-3k}).$$

So the probability that there are not vertex disjoint paths of length less than $100\epsilon^{-1}k\log n$ from s_i to t_i is at most $o(n^{-2k})$. Thus by the union bound, we see that all choices of $2k$ nodes have short edge disjoint paths linking them **whp**. Since the random 1-out perturbation preserves the boundedness of out-degree, random walks will discover these paths with non-negligible probability.   □

References

1. R. Aleliunas, R. M. Karp, R. J. Lipton, L. Lovász, and C. Rackoff, Random walks, traversal sequences, and the complexity of maze problems, *Proc. 20th IEEE Symp. on the Foundations of Computer Science* (1979), 218-223.
2. N. Alon, A. Gyárfás, and M. Ruszinkó, Decreasing the diameter of bounded degree graphs, *J. Graph Theory* 35 (2000), 161-172.
3. N. Alon and N. Kahale, A spectral technique for coloring random 3-colorable graphs, *DIMACS TR-94-35*, 1994.
4. C. Banderier, K. Mehlhorn, and R. Beier. Smoothed analysis of three combinatorial problems, *Proc. 28th Intl. Symp. on the Mathematical Foundations of Computer Science* (2003).
5. R. Beier and B. Vöcking, Typical properties of winners and losers in discrete optimization, to appear in *STOC04* (2004).
6. A. Blum and J. Spencer, Coloring Random and Semi-Random k-Colorable Graphs, *Journal of Algorithms* 19 (1995) 204-234.

7. T. Bohman, A.M. Frieze and R. Martin, How many random edges make a dense graph Hamiltonian? *Random Structures and Algorithms* 22 (2003) 33-42.
8. T. Bohman, A.M. Frieze, M. Krivelevich and R. Martin, Adding random edges to dense graphs, Random Structures and Algorithms 24 (2004) 105-117.
9. B. Bollobás, Random Graphs, *Second Edition, Cambridge University Press* (2001)
10. B. Bollobás and F. Chung, The Diameter of a Cycle Plus a Random Matching, *SIAM Journal on Discrete Mathematics* 1(3) (1988) 328-333.
11. L. Bechetti, S. Leonardi, A. Marchetti-Spaccamela, G. Schaefer, and T. Vredeveld, Smoothening Helps: A probabilistic analysis of the multi-level feedback algorithm, *FOCS03* (2003).
12. F. Chung and M. Garey, Diameter bounds for altered graphs, *Journal of Graph Theory* 8 (1984) 511-534.
13. C. Cooper and A. M. Frieze, The size of the largest strongly connected component of a random digraph with a given degree sequence *Combinatorics, Probability and Computing* 13 (2004) 319-338.
14. M. Faloutsos, P. Faloutsos and C. Faloutsos, On powerlaw relationships of the Internet topology, *SIGCOMM 99* (1999) 251-262.
15. U. Feige and R. Krauthgamer, Finding and certifying a large hidden clique in a semirandom graph, *Random Structures and Algorithms*, 16:2 (2000) 195-208.
16. U. Feige and J. Kilian, Heuristics for Semirandom Graph Problems, *Journal of Computer and System Sciences* 63 (2001) 639-671.
17. U. Feige, A Tight Upper Bound on the Cover Time for Random Walks on Graphs, *Random Structures and Algorithms* 6 (1995) 51-54.
18. O. Goldreich, S. Goldwasser, D. Ron, Property testing and its connection to learning and approximation, *J. ACM* 45 (1998) 653-750.
19. O. Goldreich, D. Ron, Property testing in bounded degree graphs, *Proc. 29th STOC* (1997) 406-415.
20. S. Janson, T. Łuczak and A. Ruciński, *Random graphs*, Wiley-Interscience (2000).
21. N. D. Jones, Space-bounded reducibility among combinatorial problems, *Journal of Computer and System Sciences* 11 (1975) 68-75.
22. R. Karp, The transitive closure of a random digraph, *Random Structures and Algorithms* 1 (1990) 73-94.
23. M. Krivelevich and P. Tetali, personal communication.
24. N. Nisan, On read-once vs. multiple access randomness in logspace, *Theoretical Computer Science* 107:1 (1993) 135-144.
25. M. Santha and U. Vazirani, Generating quasi-random sequences from semi-random sources, *Journal of Computer and System Sciences* 33 (1986) 75-87.
26. M. Sipser, *Introduction to the Theory of Computation*, PWS (1996).
27. D. Spielman and S.H. Teng, Smoothed Analysis of Algorithms: Why The Simplex Algorithm Usually Takes Polynomial Time, *Proc. of the The Thirty-Third Annual ACM Symposium on Theory of Computing*, (2001) 296-305.
28. D. Spielman and S.H. Teng, Smoothed Analysis: Motivation and Discrete Models, *Proc. of WADS 2003, Lecture Notes in Computer Science*, Springer-Verlag (2003).
29. H. Vollmer and K.W. Wagner, Measure one results in computational complexity theory, *Advances in Algorithms, Languages, and Complexity* (1997) 285-312.

Maximum Weight Independent Sets and Matchings in Sparse Random Graphs
Exact Results Using the Local Weak Convergence Method

David Gamarnik[1], Tomasz Nowicki[2], and Grzegorz Swirszcz[3]

[1] IBM T.J. Watson Research Center, Yorktown Heights NY 10598, USA
gamarnik@watson.ibm.com
http://www.research.ibm.com/people/g/gamarnik
[2] IBM T.J. Watson Research Center, Yorktown Heights NY 10598, USA
nowicki@watson.ibm.com
[3] Institute of Mathematics, University of Warsaw
02–097 Warsaw, Banacha 2, Poland
swirszcz@us.ibm.com

Abstract. Let $G(n, c/n)$ and $G_r(n)$ be an n-node sparse random and a sparse random r-regular graph, respectively, and let $\mathcal{I}(n, c)$ and $\mathcal{I}(n, r)$ be the sizes of the largest independent set in $G(n, c/n)$ and $G_r(n)$. The asymptotic value of $\mathcal{I}(n, c)/n$ as $n \to \infty$, can be computed using the Karp-Sipser algorithm when $c \le e$. For random cubic graphs, $r = 3$, it is only known that $.432 \le \liminf_n \mathcal{I}(n, 3)/n \le \limsup_n \mathcal{I}(n, 3)/n \le .4591$ with high probability (w.h.p.) as $n \to \infty$, as shown in [FS94] and [Bol81], respectively.

In this paper we assume in addition that the nodes of the graph are equipped with non-negative weights, independently generated according to some common distribution, and we consider instead the maximum weight of an independent set. Surprisingly, we discover that for certain weight distributions, the limit $\lim_n \mathcal{I}(n, c)/n$ can be computed exactly even when $c > e$, and $\lim_n \mathcal{I}(n, r)/n$ can be computed exactly for some $r \ge 2$. For example, when the weights are exponentially distributed with parameter 1, $\lim_n \mathcal{I}(n, 2e)/n \approx .5517$ in $G(n, c/n)$, and $\lim_n \mathcal{I}(n, 3)/n \approx .6077$ in $G_3(n)$. Our results are established using the recently developed *local weak convergence* method further reduced to a certain *local optimality* property exhibited by the models we consider. We extend our results to maximum weight matchings in $G(n, c/n)$ and $G_r(n)$.

1 Introduction

Two models of random graphs considered in this paper are a sparse random graph $G(n, c/n)$ and a sparse random regular graph $G_r(n)$. The first is a graph on n nodes $\{0, 1, \ldots, n-1\} \equiv [n]$, where each undirected edge $(i, j), 0 \le i < j \le n-1$ is present in the graph with probability c/n, independently for all $n(n-1)/2$ edges. Here $c > 0$ is a fixed constant, independent of n. A random r-regular

K. Jansen et al. (Eds.): APPROX and RANDOM 2004, LNCS 3122, pp. 357–368, 2004.
© Springer-Verlag Berlin Heidelberg 2004

graph $G_r(n)$ is obtained by fixing a constant integer $r \geq 2$ and considering a graph selected uniformly at random from the space of all r-regular graphs on n nodes (graphs in which every node has degree r). A set of nodes V in a graph G is defined to be an independent set if no two nodes of V are connected by an edge. Let $\mathcal{I}(n, c)$ and $\mathcal{I}(n, r)$ denote the maximum cardinality of an independent set in $G(n, c/n)$ and $G_r(n)$ respectively. Suppose the nodes of a graph are equipped with some non-negative weights $W_i, 0 \leq i \leq n - 1$ which are generated independently according to some common distribution $F_w(t) = \mathbb{P}(W_i \leq t), t \geq 0$. Let $\mathcal{I}_w(n, c), \mathcal{I}_w(n, r)$ denote the maximum weight of an independent set in $G(n, c/n)$ and $G_r(n)$ respectively.

Let $\mathcal{M}(n, c)$ and $\mathcal{M}(n, r)$ denote the maximum cardinality of a matching in $G(n, c/n)$ and $G_r(n)$, respectively. It is known that $G_r(n), r \geq 3$ has a full matching w.h.p., that is $\mathcal{M}(n, r) = n/2$ ($\lfloor n/2 \rfloor$ for odd n) w.h.p. [JLR00]. If the edges of the graph are equipped with some non-negative random weights, then we consider instead the maximum weight of a matching $\mathcal{M}_w(n, c)$ and $\mathcal{M}_w(n, r)$ in graphs $G(n, c/n), G_r(n)$, respectively. The computation of $\mathcal{I}_w(n, c), \mathcal{I}_w(n, r)$, $\mathcal{M}_w(n, c)$, $\mathcal{M}_w(n, r)$ in the limit as $n \to \infty$ is the main subject of the present paper.

The asymptotic values of $\mathcal{M}(n, c)$ for all c were obtained by Karp and Sipser using a simple greedy type algorithm in [KS81]. The result extends to $\mathcal{I}(n, c)$ but only for $c \leq e$. It is an open problem to compute the corresponding limit for independent sets for the case $c > e$ or even to show that such a limit exists [Ald]. Likewise it is an open problem to compute the corresponding limit in random regular graphs.

The developments in this paper show that, surprisingly, proving the existence and computation of the limits $\lim_n \mathcal{I}(n, \cdot)/n, \lim_n \mathcal{M}(\cdot)/n$ is easier in the weighted case than in the unweighted case, at least for certain weight distributions. In particular, we compute the limits for independent sets in $G_r(n), r = 2, 3, 4$ and $G(n, c/n), c \leq 2e$, when the node weights are exponentially distributed, and we compute the limits for matchings in $G_r(n)$ and $G(n, c/n)$ for all r, c, when the edge weights are exponentially distributed. It was shown by the first author [Gam04] that the limit $\lim_n \mathcal{M}(n, \cdot)/n$ exists for every weight distribution with bounded support, though the non-constructive methods employed prevented the computation of the limits.

Our method of proof is based on a powerful *local weak convergence method* developed by Aldous [Ald92], [Ald01], Aldous and Steele [AS03], Steele [Ste02], further empowered by a certain *local optimality* observation derived in this paper. Local weak convergence is a recursion technique based on fixed points of distributional equations, which allows one to compute limits of some random combinatorial structures, see Aldous and Bandyopadhyay [AB] and Aldous and Steele [AS03] for recent surveys on these topics. In particular, the method is used to compute maximum weight matching on a random tree, when the weights are exponentially distributed. The tree structure was essential in [AS03] for certain computations and the approach does not extend directly to graphs like $G(n, c/n)$ with $c > 1$, where the convenience of a tree structure is lost due to the presence of a giant component. It was conjectured in [AS03] that a some long-range independence property might be helpful to deal with this difficulty. The present paper partially answers this qualitative conjecture in a positive way. We introduce a certain operator T acting on the space of distribution functions. We prove a certain local optimality property stating that, for example, for independent sets, whether a given node i belongs to the maximum weight independent set is asymptotically independent from the portion of the graph outside a constant size neighborhood of i, iff T^2 has a unique fixed point distribution. Moreover, when T^2 does have the unique fixed point, the size of the extremal object (say maximum weight independent set) can be derived from a fixed point of an operator T. The computations of fixed points is tedious, but simple in principle and the groundwork for that was already done in [AS03]. We hope that the long-range independence holds in other random combinatorial structures as well. In fact the issues of long-range independence were already considered by Aldous and Bandyopadhyay [AB], Bandyopadhyay [Ban03] using the notion of endogeny. This specific version of long-range independence turned out to be critical in Aldous [Ald01] for proving the $\zeta(2)$ limit of the random assignment problem.

The issue of long-range independence (usually called correlation decay instead) of random combinatorial objects is addressed in a somewhat different, statistical physics context in Mossel [Mos03], Brightwell and Winkler [BW03], Rozikov and Suhov [RS03], Martin [Mar03], Martinelli, Sinclair and Weitz [MSW03], where independent sets (hard-core model) and other combinatorial models are considered on regular trees, weighted by the Gibbs measure. In a different setting Talagrand [Tal03] proves a certain long-range independence property for the random assignment problem, where the usual min-weight matching is replaced by a partition function on the space of feasible matchings. He uses a rigorous mathematical version of the cavity method, which originated in physics, to prove that the spins are asymptotically independent as the size of the problem increases. The particular form of the long-range independence is similar to the one we obtain, and in fact the cavity method has some similarity with the local weak convergence method. which is also based on considering extremal objects (say independent sets) with one or several nodes excluded. Finally, we refer the reader to Hartmann and Weigt [HW01] who derive the same result as Karp and Sipser for independent sets using non-rigorous arguments from statistical physics.

We finish this section with some notational conventions. $\text{Exp}(\mu), \text{Pois}(\lambda)$ denote respectively exponential and Poisson distributions with parameters $\mu, \lambda > 0, 0 \leq z \leq 1$. Most of the proofs are too lengthy to fit under the page limitation and we refer the reader to [GNS03] for proofs.

2   Prior Work and Open Questions

It is known and simple to prove that the largest independent sets are $\Theta(n)$ w.h.p. in both $G(n, c/n)$ and $G_r(n)$ when c and r are any constants independent from n. Moreover, it is known that in random cubic graphs, w.h.p., $6\log(3/2) - 2 = .432\ldots \leq \liminf_n \mathcal{I}(n,3)/n \leq \limsup_n \mathcal{I}(n,3)/n \leq .4591$. The lower bound is due to Frieze and Suen [FS94], and the upper bound is due to Bollobas [Bol81]. The upper bound is generalized for any $r \geq 1$ and uses a very ingenious construction of random regular graphs via matching and random grouping, [Bol80], [JLR00]. It is natural to expect that the following is true, which unfortunately remains only a conjecture, appearing in several places, most recently in [Ald] and [AS03].

Conjecture 1 *For every $c > 0$ and $r \geq 3$ the limits*

$$\lim_{n\to\infty} \frac{\mathbb{E}[\mathcal{I}(n,c)]}{n}, \qquad \lim_{n\to\infty} \frac{\mathbb{E}[\mathcal{I}(n,r)]}{n}$$

exist.

The existence of these limits also implies the convergence to the same limits w.h.p. by applying Azuma's inequality, see [JLR00] for the statement and the applicability of this inequality.

The limits $\lim_n \mathbb{E}\mathcal{I}(n,c)/n, \lim_n \mathbb{E}\mathcal{M}(n,c)/n$ are known to exist for for independent sets when $c \leq e$ and for matching for all c and were first derived by Karp-Sipser [KS81]. Their results were strengthened later by by Aronson, Frieze and Pittel [APF98].

The algorithm underlying Karp and Sipser analysis unfortunately is not applicable to random regular graphs. Moreover, if the edges or the nodes of the graph $G(n, c/n)$ are equipped with weights then the Karp-Sipser algorithm can produce a strictly suboptimal solution and cannot be used in our setting of weighted nodes and edges. Also, when the edges of $G_r(n)$ are equipped with weights, the problem of computing maximum weight matching becomes nontrivial, as opposed to the unweighted case when the full matching exists w.h.p.

In a somewhat different domain of extremal combinatorics the following result was established by Hopkins and Staton [HS82]. A girth of a graph is the size of the smallest cycle. It is shown in [HS82] that the size of a largest independent set in an n-node graph with largest degree 3 and large girth is asymptotically at least $(7/18)n - o(n)$. The techniques we employ in this paper allow us to improved this lower bound.

3   Main Results

We begin by introducing the key technique for our analysis – recursive distributional equations and fixed point solutions. This technique was introduced by Aldous [Ald92], [Ald01] and was further developed in Aldous and Steele [AS03], Steele [Ste02], Aldous and Bandyopadhyay [AB]. Let W be a non-negative random variable with a distribution function $F_w(t) = \mathbb{P}(W \leq t)$. We consider four operators $T = T_{\mathcal{I},r}, T_{\mathcal{I},c}, T_{\mathcal{M},r}, T_{\mathcal{M},c}$ acting on the space of distribution functions $F(t), t \geq 0$, where $c > 0$ is a fixed constant and $r \geq 2$ is a fixed integer.

1. Given W distributed according to F_w (we write simply $W \sim F_w$), and given a distribution function $F = F(t)$, let $B_1, B_2, \ldots, B_r \sim F$ be generated independently. Then $T_{\mathcal{I},r} : F \to F'$, where F' is the distribution function of B' defined by

$$B' = \max(0, W - \sum_{1 \leq i \leq r} B_i). \tag{1}$$

2. Under the same setting as above, let $B_1, \ldots, B_m \sim F$, where m is a random variable distributed according to a Poisson distribution with parameter c, independently from W, B_i. Then $T_{\mathcal{I},c} : F \to F'$, where F' is the distribution function of B' defined by

$$B' = \max(0, W - \sum_{1 \leq i \leq m} B_i), \tag{2}$$

when $m \geq 1$ and $B' = W$ when $m = 0$. For simplicity we identify the sum above with zero when $m = 0$.

3. Let $W_1, \ldots, W_r \sim F_w$, $B_1, \ldots, B_r \sim F$. Then $T_{\mathcal{M},r} : F \to F'$, where F' is the distribution function of B' defined by

$$eq : Recursion Mr B' = \max_{1 \leq i \leq r} (0, W_i - B_i). \tag{3}$$

from W_i, B_i. Then $T_{\mathcal{M},c} : F \to F'$, where F' is the distribution function of B' defined by

$$B' = \max_{1 \leq i \leq m} (0, W_i - B_i), \tag{4}$$

when $m \geq 1$ and $B' = 0$ when $m = 0$. Again, for simplicity, we assume that the max expression above is zero when $m = 0$.

A distribution function F is defined to be a fixed point distribution of an operator T if $T(F) = F$.

We now state the main result of this paper. Recall, that a distribution function $F(t)$ is defined to be continuous (atom free) if for every x in its support $\lim_{\epsilon \to 0}(F(x + \epsilon) - F(x - \epsilon)) = 0$. Equivalently, for $B \sim F$ and every x, $\mathbb{P}(B = x) = 0$. We use $1\{\cdot\}$ to denote the indicator function.

Theorem 1. *Let F_w be a continuous non-negative distribution function. For $r \geq 1$ if the operator $T_{\mathcal{I},r-1}^2$ has a unique fixed point distribution function F^*, then, w.h.p.*

$$\lim_n \frac{\mathcal{I}_w(n, r)}{n} = \mathbb{E}[W \, 1\{W - \sum_{1 \leq i \leq r} B_i > 0\}], \tag{5}$$

where $W \sim F_w$, $B_i \sim F^*$, and W, B_i are independent. When $G_r(n)$ is replaced by $G(n, c/n)$, the same result holds for $T = T_{\mathcal{I},c}$, except the sum in the right-hand side of (5) is $\sum_{1 \leq i \leq m} B_i$ and $m \sim \text{Pois}(c)$.

Finally, the similar results hold for $\mathcal{M}_w(n, r)$ *and* $\mathcal{M}_w(n, c)$ *in* $G_r(n)$ *and* $G(n, c)$, *for* $T = T_{\mathcal{M}, r-1}$ *and* $T = T_{\mathcal{M}, c}$, *respectively, whenever the corresponding operator* T *is such that* T^2 *has the unique fixed point distribution* F^*. *The corresponding limits are*

$$\lim_n \frac{\mathcal{M}_w(n, r)}{n} = \frac{1}{2}\mathbb{E}[\sum_{1 \le i \le r} W_i 1\{W_i - B_i = \max_{1 \le j \le r}(W_j - B_j) > 0\}], \quad (6)$$

where $W_i \sim F_w$, $B_i \sim F^*$, *and*

$$\lim_n \frac{\mathcal{M}_w(n, c)}{n} = \frac{1}{2}\mathbb{E}[\sum_{i \le m} W_i 1\{W_i - B_i = \max_{j \le m}(W_j - B_j) > 0\}], \quad (7)$$

where $W_i \sim F_w, B_i \sim F^*, m \sim \text{Pois}(c)$.

The Theorem 1 is the core result of this paper. It will allow us to obtain several interesting corollaries, which we state below. The rest of the paper gives a coarse outline of the proof of Theorem 1. A detailed proof is found in [GNS03].

Theorem 2. *Suppose the weights of the nodes and edges of the graphs* $G = G_r(n)$ *and* $G = G(n, c)$ *are distributed as* $\text{Exp}(1)$. *Then*

1. $T^2_{\mathcal{I}, r-1}$ *has a unique fixed point distribution* F^* *iff* $r \le 4$. *In this case, w.h.p.*

$$\lim_n \frac{\mathcal{I}_w(n, r)}{n} = \frac{(1 - b)(r - rb + 2b + 2)}{4}, \quad (8)$$

where b *is the unique solution of* $b = 1 - (\frac{1+b}{2})^{r-1}$. *In particular, w.h.p.*

$$\lim_n \frac{\mathcal{I}_w(n, 2)}{n} = \frac{2}{3}, \quad \lim_n \frac{\mathcal{I}_w(n, 3)}{n} \approx .6077, \quad \lim_n \frac{\mathcal{I}_w(n, 4)}{n} \approx .4974, \quad (9)$$

2. $T^2_{\mathcal{I}, c}$ *has a unique fixed point distribution* F^* *iff* $c \le 2e$. *In this case, w.h.p.*

$$\lim_n \frac{\mathcal{I}_w(n, c)}{n} = (1 - b)(1 + \frac{c(1 - b)}{4}), \quad (10)$$

where b *is the unique solution of* $1 - b = e^{-\frac{c}{2}(1 - b)}$. *In particular, when* $c = 2e$, *this limit is* $\approx .5517$.

3. $T^2_{\mathcal{M}, r-1}$ *has a unique fixed point* F^* *for every* $r \ge 2$. *Moreover, w.h.p.*

$$\lim_n \frac{\mathcal{M}_w(n, r)}{n} = r(b^{r-1} + 1) \int_0^\infty te^{-t}(1 - e^{-t}(1 - b)))^{r-1}dt \quad (11)$$

$$- r \int_0^\infty te^{-t}(1 - e^{-t}(1 - b)))^{2r-2}dt,$$

where b *is the unique solution of* $b = 1 - \frac{1-b^r}{r(1-b)}$.

4. $T^2_{\mathcal{M},c}$ has a unique fixed point F^* for all $c > 0$. Moreover, for every $c > 0$ w.h.p.

$$\lim_n \frac{\mathcal{M}_w(n,c)}{n} = \frac{c}{2}(e^{cb-c} + 1) \int_0^\infty te^{-t-c(1-b)e^{-t}} dt \qquad (12)$$
$$- \frac{c}{2} \int_0^\infty te^{-t-2c(1-b)e^{-t}+c(1-b)^2e^{-2t}} dt,$$

where b is the unique solution of $1 - e^{-cb} = c(1-b)^2$.

The proof of Theorem 2, specifically the corresponding computations of the limits is quite lengthy and can be found in [GNS03]. The expression in (11) involving integrals is similar to the one found in [AS03] for maximum weight matching on a tree. It is a pleasant surprise, though, that the answers for independent sets are derived in closed form.

Part 2 of Theorem 2 leads to an interesting phase transition behavior. For $F_w = \mathrm{Exp}(1)$ our result says that the value of $c = 2e$ is a phase transition point for the operator $T^2_{\mathcal{I},c}$ where for $c \leq 2e$ the operator has a unique fixed point distribution, but for $c > 2e$ the fixed point distribution is not unique. It turns out (see Theorems 4, 5 below) that this phase transition is directly related to some long range independence/dependence property of the maximum weight independent sets in the underlying graph $G(n, c/n)$. Interestingly, no such phase transition occurs for maximum weight matchings.

The methods developed in this paper allow us the improve the following result of Hopkins and Staton [HS82]. Let $\mathcal{G}(n, r, d)$ denote the class of all (non-random) graphs on n nodes, with maximum degree r and girth at least d. For any $G \in \mathcal{G}(n, r, d)$ let $\mathcal{I}(G)$ denote the size of the largest independent set in G. Hopkins and Staton proved that $\liminf_{n,d} \min_{G \in \mathcal{G}(n,3,d)} \mathcal{I}(G)/n \geq 7/18 \approx .3887$. Our techniques allow us to obtain the following improvement.

Theorem 3.

$$\liminf_{n,d} \min_{G \in \mathcal{G}(n,3,d)} \frac{\mathcal{I}(G)}{n} \geq .3923.$$

For the proof of this theorem see [GNS03].

4   Fixed Points of the Operator T^2 and the Long-Range Independence

4.1   Maximum Weight Independent Sets and Matchings in Trees. Fixed Points of T^2 and the Bonus Function

We start by analyzing operator T – any of the four operators introduced in the previous section. Given two distribution functions F_1, F_2 defined on $[0, \infty)$, we say that F_2 stochastically dominates F_1 and write $F_1 \prec F_2$ if $F_1(t) \geq F_2(t)$ for every $t \geq 0$. A sequence of distribution functions F_n is defined to converge weakly to a distribution function F (written $F_n \Rightarrow F$) if $\lim_n F_n(t) = F(t)$ for every t which is a point of continuity of F. All the convergence of distributions is understood here in a weak sense.

Let 0 denote (for simplicity) the distribution function of a random variable X which is zero w.p.1. Let $W_r^{\max} = \max_{1 \le i \le r} W_i$ where $W_i \sim F_w$ are independent. Let also $W_c^{\max} = \max_{1 \le i \le m} W_i$, where $W_i \sim F_w$ are independent and $m \sim$ Pois(c). Denote by $F_{w,r}$ and $F_{w,c}$ the distribution functions of $W_r^{\max}$ and $W_c^{\max}$, respectively.

Proposition 1. *Fix $r \ge 1, c > 0$, a distribution function $F_w(t), t \ge 0$ and $T = T_{\mathcal{I},r}$ or $T_{\mathcal{I},c}$. As $s \to \infty$, the two sequences of distributions $T^{2s}(0), T^{2s}(F_w)$ weakly converge to some distribution functions F_{**}, F^{**}, respectively, which are fixed points of the operator T^2. For any distribution function $F_0 = F_0(t), t \ge 0$, $T^{2s}(0) \prec T^{2s+2}(F_0) \prec T^{2s}(F_w)$ and $T^{2s+1}(F_w) \prec T^{2s+3}(F_0) \prec T^{2s+1}(0)$ for all $s = 1, 2, \ldots$.*

If the operator T is such that T^2 has a unique fixed point (and $F_{**} = F^{**} \equiv F^*$), then for any distribution function $F_0 = F_0(t), T^s(F_0), s = 1, 2, \ldots$ converges to F^* as $s \to \infty$. In particular, $T^s(0), T^s(F_w) \to F^*$. Moreover, F^* is also the unique fixed point of T. When $T = T_{\mathcal{M},r}$ or $T = T_{\mathcal{M},c}$ the same result holds with $F_{w,r}$ and $F_{w,c}$ respectively replacing F_w.

Proof. See [GNS03].

We now switch to analyzing the maximum weight independent set problem on a tree. The derivation here repeats the development in [AS03] for maximum weight matching in random trees. We highlight important differences where appropriate.

Suppose we have a (non-random) finite tree H with nodes $0, 1, \ldots, h = |H| - 1$, with a fixed root 0. The nodes of this tree are equipped with some (non-random) weights $W_0, W_1, \ldots, W_h \ge 0$. For any node $i \in H$, let $H(i)$ denote the subtree rooted at i consisting of all the descendants of i. In particular, $H(0) = H$. Let $\mathcal{I}_{H(i)}$ denote the maximum weight of an independent set in $H(i)$ and let $B_{H(i)} = \mathcal{I}_{H(i)} - \sum_j \mathcal{I}_{H(j)}$, where the sum runs over nodes j which are children of i. If i has no children then $B_{H(i)}$ is simply $\mathcal{I}_{H(i)} = W_i$. Observe, that $B_{H(i)}$ is also a difference between $\mathcal{I}_{H(i)}$ and the maximum weight of an independent set in $H(i)$, which is not allowed to use node i. Clearly, $0 \le B_{H(i)} \le W_i$. The value $B_{H(i)}$ was considered in [AS03] in the context of maximum weight matchings and was referred to as a *bonus* of a node i in tree $H(i)$. W.l.g. denote by $1, \ldots, m$ the children of the root node 0.

Lemma 1.
$$B_{H(0)} = \max(0, W_0 - \sum_{1 \le i \le m} B_{H(i)}). \tag{13}$$

Moreover, if $W_0 > \sum_{1 \le i \le m} B_{H(i)}$ (that is if $B_{H(0)} > 0$) then the maximum weight independent set must contain node 0. If $W_0 < \sum_{1 \le i \le m} B_{H(i)}$ then the maximum weight independent set does not contain the node 0.

Proof. See [GNS03].

A similar development is possible for maximum weight matching. Suppose the edges of the tree H are equipped with weights $W_{i,j}$. Let $\mathcal{M}_{H(i)}$ denote the maximum weight of a matching in $H(i)$, and let $B_{H(i)}$ denote the difference between $\mathcal{M}_{H(i)}$ and the maximum weight of a matching in $H(i)$ which is not allowed to include any edge in $H(i)$ incident to i. Again $1, 2, \ldots, m$ are assumed to be the children of the root 0.

Lemma 2.

$$B_H = \max(0, \max_{1 \leq i \leq m} (W_{0,i} - B_{H(i)})). \tag{14}$$

Moreover, if $W_{0,i} - B_{H(i)} > W_{0,i'} - B_{H(i')}$ for all $i' \neq i$ and $W_{0,i} - B_{H(i)} > 0$, then every maximum weight matching contains edge $(0, i)$. If $W_{0,i} - B_{H(i)} < 0$ for all $i = 1, \ldots, m$, then every maximum weight matching does not contain any edge incident to 0.

Proof. See [GNS03].

4.2   Long-Range Independence

We now consider trees H of specific types. Given integers $r \geq 2, d \geq 2$ let $H_r(d)$ denote an r-regular finite tree with depth d. The root node 0 has degree $r - 1$, all the nodes at distance $\geq 1, \leq d - 1$ from the root have outdegree $r - 1$, and all the nodes at distance d from 0 are leaves. (Usually, in the definition of an r-regular tree, the root node is assumed to have degree r, not $r - 1$. The slight distinction here is done for convenience.) Also, given a constant $c > 0$, a Poisson tree $H(c, d)$ with parameter c and depth d is constructed as follows. The root node has a degree which is a random variable distributed according to $\text{Pois}(c)$ distribution. All the children of 0 have outdegrees which are also random, distributed according to $\text{Pois}(c)$. In particular, the children of 0 have total degrees $1 + \text{Pois}(c)$. Similarly, children of children of 0 also have outdegree $\text{Pois}(c)$, etc. We continue this process until either the process stops at some depth $d' < d$, where no nodes in level d' have any children, or until we reach level d. In this case all the children of the nodes in level d are deleted and the nodes in level d become leaves. We obtain a tree with depth $\leq d$. We call this a depth-d Poisson tree.

Let $H = H_r(d)$ or $H(c, d)$. Suppose the nodes and the edges of H are equipped with weights $W_i, W_{i,j}$, which are generated at random independently using a distribution function F_w. Fix any infinite sequences $\bar{w} = (w_1, w_2, \ldots) \in [0, \infty)^\infty$ and $\bar{b} = (b_1, b_2, \ldots) \in \{0, 1\}^\infty$. For every $i = 1, 2, \ldots, d$ let $i1, i2, \ldots, ij_i$ denote the nodes of H in level i (if any exist for $H(c, d)$). When $H = H_r(d), j_i = (r-1)^i$, of course. Let $(\mathcal{I}|(\bar{b}, \bar{w}))$ denote the maximum weight of an independent set V in H such that the nodes dj with $b_j = 1$ are conditioned to be in V, nodes dj with $b_j = 0$ are conditioned not to be in V, and the weights of nodes dj are conditioned to be equal to w_j for $j = 1, \ldots, j_d$. That is we are looking for maximum weight of an independent set among those which contain depth d leaves with $b_j = 1$, do not contain depth d leaves with $b_j = 0$, and with the weights of the leaves deterministically set by $\bar{w}$. For brevity we call it the maximum weight of an independent set with boundary condition $(\bar{b}, \bar{w})$. For the case $H = H(c, d)$, the boundary condition is simply absent when the tree does not contain any nodes in the last level d. $(\mathcal{I}_{H(ij)}|(\bar{b}, \bar{w}))$ are defined similarly for the subtrees $H(ij)$ spanned by nodes ij in level i. Given again $\bar{b}, \bar{w}$, let $(\mathcal{M}|(\bar{b}, \bar{w}))$ and $(\mathcal{M}_{H(ij)}|(\bar{b}, \bar{w}))$ denote, respectively, the maximum weight of a matching E in H and H_{ij}, such that the edges incident to nodes dj are conditioned to be in E when $b_j = 1$, edges incident to nodes dj are conditioned not to be in D when with $b_j = 0$, and the weights of the edges incident to nodes dj are conditioned to be equal to $w_j, j = 1, \ldots, j_d$ (of course, we refer to edges between nodes in levels $d - 1$ and d as there is only one edge per each node in level d).

For the case of independent sets, let $(B|(\bar{b}, \bar{w}))$ denote the bonus of the root node 0 given the boundary condition $(\bar{b}, \bar{w})$. Namely,

$$(B|(\bar{b}, \bar{w})) = (\mathcal{I}|(\bar{b}, \bar{w})) - \sum_{1 \le j \le j_1} (\mathcal{I}_{H(1j)}|(\bar{b}, \bar{w})).$$

For the case of matchings, let $(B|(\bar{b}, \bar{w}))$ also denote the bonus of the root node 0 given the boundary condition $(\bar{b}, \bar{w})$. Namely,

$$(B|(\bar{b}, \bar{w})) = (\mathcal{M}|(\bar{b}, \bar{w})) - \sum_{1 \le j \le j_1} (\mathcal{M}_{H(1j)}|(\bar{b}, \bar{w})).$$

It should be always clear from the context whether B is taken with respect to independent sets or matchings.

The following theorem establishes the crucial *long-range independence* property for the maximum weight independent sets and matchings in trees $H = H_r(d), H(c, d)$ when the corresponding operator T^2 has a unique fixed point. It establishes that the distribution of the bonus $(B|(\bar{b}, \bar{w}))$ of the root is asymptotically independent of the boundary condition $(\bar{b}, \bar{w})$ as d becomes large. Recall our convention that 0 denotes the distribution of a random variable which is equal to zero with probability one.

Theorem 4. *Given a distribution function F_w and a regular tree $H = H_r(d)$. Suppose the operator $T^2 = T^2_{\mathcal{I}, r-1}$ has the unique fixed distribution F^*. Then*

$$\sup_{\bar{b}, \bar{w}} \left| \mathbb{P}((B|(\bar{b}, \bar{w})) \le t) - F^*(t) \right| \to 0, \tag{15}$$

as $d \to \infty$. Similar assertion holds for $T = T_{\mathcal{M}, r}$, $T = T_{\mathcal{I}, c}$ and $T = T_{\mathcal{M}, c}$. For the cases $T = T_{\mathcal{M}, r}$ and $T_{\mathcal{M}, c}$, F_w is replaced with $F_{w,r}$ and $F_{w,c}$ respectively.

Let us compare this result with the developments in ([AS03]). In that paper maximum weight matching is considered on an n-node tree, drawn independently and uniformly from the space of all n^{n-2} labelled trees. The notion of a bonus is introduced and the recursion (14) is derived. However, since a tree structure is assumed to begin with, there is no need to consider the boundary conditions $(\bar{b}, \bar{w})$. Here we avoid the difficulty of the non-tree structure by proving the long-range independence property via the uniqueness of fixed points of T^2.

Proof. See [GNS03].

While it is not used in this paper, it is interesting that the uniqueness of the solution $T^2(F) = F$ is the tight condition for (15), as the following theorem indicates.

Theorem 5. *Under the setting of Theorem 5 suppose the operator T^2 has more than one fixed point distributions F^*. Then for every such F^**

$$\liminf_d \sup_{\bar{b}, \bar{w}} \left| \mathbb{P}((B|(\bar{b}, \bar{w})) \le t) - F^*(t) \right| > 0. \tag{16}$$

Proof. See [GNS03].

4.3 Applications to Maximum Weight Independent Sets and Matchings in $G_r(n)$ and $G(n, c/n)$

The goal of the current section is to demonstrate that Theorem 4 allows us to reduce the computation of the maximum weight independent set and the maximum weight matching in random graphs to a much simpler problem of finding those in trees. We highlight this key message of the paper as the following *local optimality* property: if the operator T^2 corresponding to a maximum weight combinatorial object (independent set or matching) in a sparse random graph has a unique fixed point, then for a randomly selected node (edge) of the graph, the event "the node (edge) belongs to the optimal object" and the distribution of the node (edge) weight, conditioned that it does, asymptotically depends only on the constant size neighborhood of the node and is independent from the rest of the graph. In other words, when T^2 has a unique fixed point, the maximum weight independent sets and matchings exhibit a long-range independence property. Our hope is that similar local optimality can be established for other random combinatorial structures. A complete proof of Theorem 1 is given in [GNS03]. Here we just give a high level idea. We observe that by symmetry $\mathbb{E}[\mathcal{I}_w(n, r)] = n\mathbb{E}[W_0 1\{0 \in V_r\}]$, where V_r is the independent set which achieves the maximum weight. We fix a large constant d and consider a size-d neighborhood of the node 1. With probability approaching unity this neighborhood is an r-regular tree. We then invoke Theorem 5 to argue that the quantity $\mathbb{E}[W_0 1\{0 \in V_r\}]$ is determined almost exactly just by the d-neighborhood of the node 1 and is given asymptotically as $\mathbb{E}[W_0 1\{W_0 - \sum_{1 \le i \le r} B_i > 0\}]$ where $B_1, \ldots, B_r$ are i.i.d distributed as F^* and F^* is the unique fixed point of the operator $T^2_{\mathcal{I}, r-1}$.

References

[AB]   D. Aldous and A. Bandyopadhyay, *A survey of max-type recursive distributional equations*, Preprint.

[Ald]   D. Aldous, *Some open problems*, http://stat-www.berkeley.edu/users/aldous/ Research/problems.ps.

[Ald92]   _____, *Asymptotics in the random assignment problem*, Probab.Th. Rel.Fields (1992), no. 93, 507–534.

[Ald01]   _____, *The $\zeta(2)$ limit in the random assignment problem*, Random Structures and Algorithms (2001), no. 18, 381–418.

[APF98]   J. Aronson, B. Pittel, and A. Frieze, *Maximum matchings in sparse random graphs: Karp-Sipser revisited*, Random Structures and Algorithms **12** (1998), 11–178.

[AS03]   D. Aldous and J. M. Steele, *The objective method: Probabilistic combinatorial optimization and local weak convergence*, Discrete Combinatorial Probability, H. Kesten Ed., Springer-Verlag, 2003.

[Ban03]   A. Bandyopadhyay, *Max-type recursive distributional equations*, University of California, Berkeley, 2003.

[Bol80]   B. Bollobas, *A probabilistic proof of an asymptotic formula for the number of regular graphs*, European J. Combinatorics **1** (1980), 311–316.

[Bol81]   _____, *The independence ratio of regular graphs*, Proc. Amer. Math. Soc. **83** (1981), no. 2, 433–436.

[BW03]   G.R. Brightwell and P. Winkler, *Gibbs extremality for the hard-core model on a Bethe lattice*, Preprint (2003).

[FS94]   A. Frieze and S. Suen, *On the independence number of random cubic graphs*, Random Structures and Algorithms **5** (1994), 649–664.

[Gam04]   D. Gamarnik, *Linear phase transition in random linear constraint satisfaction problems*, Probability Theory and Related Fields. **129** (2004), no. 3, 410–440.

[GNS03]   D. Gamarnik, T. Nowicki, and Grzegorz Swirscsz, *Maximum weight independent sets and matchings in sparse random graphs. Exact results using the local weak convergence method*, arXiv:math.PR/0309441 (2003).

[HS82]   G. W. Hopkins and W. Staton, *Girth and independence ratio*, Canad. Math. Bull. **25** (1982), 179–186.

[HW01]   A. K. Hartmann and M. Weigt, *Statistical mechanics perspective on the phase transition of vertex covering of finite-connectivity random graphs*, Theoretical Computer Science **265** (2001), 199–225.

[JLR00]   S. Janson, T. Luczak, and A. Rucinski, *Random graphs*, John Wiley and Sons, Inc., 2000.

[KS81]   R. Karp and M. Sipser, *Maximum matchings in sparse random graphs*, 22nd Annual Symposium on Foundations of Computer Science, 1981, pp. 364–375.

[Mar03]   J. Martin, *Reconstruction thresholds on regular trees*, Preprint (2003).

[Mos03]   E. Mossel, *Survey: information flow on trees*, Preprint (2003).

[MSW03]   F. Martinelli, A. Sinclair, and D. Weitz, *The Ising model on trees: boundary conditions and mixing time*, Proc. 44th IEEE Symposium on Foundations of Computer Science (2003).

[RS03]   U. A. Rozikov and U. M. Suhov, *A hard-core model on a Cayley tree: an example of a loss network*, Preprint (2003).

[Ste02]   J. M. Steele, *Minimal spanning trees for graphs with random edge lenghts*, Mathematics and Computer Science II. Algorithms, Trees, Combinatorics and Probabilities. (2002), 223–246.

[Tal03]   M. Talagrand, *An assignment problem at high temperature*, Annals of Probability **31** (2003), no. 2, 818–848.

Estimating Frequency Moments of Data Streams Using Random Linear Combinations

Sumit Ganguly

Indian Institute of Technology, Kanpur
sganguly@iitk.ac.in

Abstract. The problem of estimating the kth frequency moment F_k for any non-negative k, over a data stream by looking at the items exactly once as they arrive, was considered in a seminal paper by Alon, Matias and Szegedy [1,2]. The space complexity of their algorithm is $\tilde{O}(n^{1-\frac{1}{k}})$. For $k > 2$, their technique does not apply to data streams with arbitrary insertions and deletions. In this paper, we present an algorithm for estimating F_k for $k > 2$, over general update streams whose space complexity is $\tilde{O}(n^{1-\frac{1}{k-1}})$ and time complexity of processing each stream update is $\tilde{O}(1)$.

Recently, an algorithm for estimating F_k over general update streams with similar space complexity has been published by Coppersmith and Kumar [7]. Our technique is, (a) basically different from the technique used by [7], (b) is simpler and symmetric, and, (c) is more efficient in terms of the time required to process a stream update ($\tilde{O}(1)$ compared with $\tilde{O}(n^{1-\frac{1}{k-1}})$).

1 Introduction

A data stream can be viewed as a sequence of updates, that is, insertions and deletions of items. Each update is of the form $(l, \pm v)$, where, l is the identity of the item and v is the change in frequency of l such that $|v| \geq 1$. The items are assumed to draw their identities from the domain $[N] = \{0, 1, \ldots, N-1\}$. If v is positive, then the operation is an insertion operation, otherwise, the operation is a deletion operation. The frequency of an item with identity l, denoted by f_l, is the sum of the changes in frequencies of l from the start of the stream. In this paper, we are interested in computing the kth frequency moment $F_k = \sum_l f_l^k$, for $k > 2$ and k integral, by looking at the items exactly once when they arrive.

The problem of estimating frequency moments over data streams using randomized algorithms was first studied in a seminal paper by Alon, Matias and Szegedy [1,2]. They present an algorithm, based on sampling, for estimating F_k, for $k \geq 2$, to within any specified approximation factor ϵ and with confidence that is a constant greater than 1/2. The space complexity of this algorithm is $s = \tilde{O}(n^{1-\frac{1}{k}})$ (suppressing the term $\frac{1}{\epsilon^2}$) and time complexity per update is $\tilde{O}(n^{1-\frac{1}{k}})$, where, n is the number of distinct elements in the stream. This algorithm assumes that frequency updates are restricted to the form $(l, +1)$.

One problem with the sampling algorithm of [1,2] is that it is not applicable to streams with arbitrary deletion operations. For some applications, the ability to handle

K. Jansen et al. (Eds.): APPROX and RANDOM 2004, LNCS 3122, pp. 369–380, 2004.
© Springer-Verlag Berlin Heidelberg 2004

deletions in a stream may be important. For example, a network monitoring application might be continuously maintaining aggregates over the number of currently open connections per source or destination.

In this paper, we present an algorithm for estimating F_k, for $k > 2$, to within an accuracy of $(1 \pm \epsilon)$ with confidence at least 2/3. (The method can be boosted using the median of averages technique to return high confidence estimates in the standard way [1, 2].) The algorithm handles arbitrary insertions and legal deletions (i.e., net frequency of every item is non-negative) from the stream and generalizes the random linear combinations technique of [1, 2] designed specifically for estimating F_2. The space complexity of our method is $\tilde{O}(n^{1-\frac{1}{k-1}})$ and the time complexity to process each update is $\tilde{O}(1)$, where, functions of k and ϵ that do not involve n are treated as constants.

In [7], Coppersmith and Kumar present an algorithm for estimating F_k over general update streams. Their algorithm has similar space complexity (i.e., $\tilde{O}(n^{1-\frac{1}{k-1}})$) as the one we design in this paper. The principal differences between our work and the work in [7] are as follows.

1. *Different Technique.* Our method constructs random linear combinations of the frequency vector using randomly chosen roots of unity, that is, we construct the sketch $Z = f_l x_l$, where, x_l is a randomly chosen k^{th} root of unity. Coppersmith and Kumar construct random linear combinations $C = f_l x_l$, where, for $l \in [N]$, $x_l = -1/n^{1-\frac{1}{k-1}}$ or $1 - 1/n^{1-\frac{1}{k-1}}$ with probability $1 - 1/n^{1-\frac{1}{k-1}}$ and $1/n^{1-\frac{1}{k-1}}$ respectively.

2. *Symmetric and Simpler Algorithm.* Our technique is a symmetric method for all $k \geq 2$, and is a direct generalization of the sketch technique of Alon, Matias and Szegedy [1, 2]. In particular, for every $k \geq 2$, $\mathbf{E}[\operatorname{Re} Z^k] = F_k$. The method of Coppersmith and Kumar gives complicated expressions for estimating F_k, for $k \geq 4$. For $k = 4$, their estimator is $C^4 - B_n F_2^2$ (where, $B_n \approx n^{-4/3}(1 - n^{-2/3})^2$), and requires, in addition, an estimation of F_2 to within an accuracy factor of $(1 \pm n^{-1/3})$. The estimator expression for higher values of k (particularly, for powers of 2) are not shown in [7]. These expressions require auxiliary moment estimation and are quite complicated.

3. *Time Efficient.* Our method is significantly more efficient in terms of the time taken to process an arrival over the stream. The time complexity to process a stream update in our method is $\tilde{O}(1)$, whereas, the time complexity of the Coppersmith Kumar technique is $\tilde{O}(n^{1-\frac{1}{k-1}})$.

The recent and unpublished work in [11] presents an algorithm for estimating F_k, for $k > 2$ and for the append only streaming model (used by [1, 2]), with space complexity $\tilde{O}(n^{1-\frac{2}{k+1}})$. Although, the algorithm in [11] improves on the asymptotic space complexity of the algorithm presented in this paper, it cannot handle deletion operations over the stream. Further, the method used by [11] is significantly different from the techniques used in this paper, or from the techniques used by Coppersmith and Kumar [7].

Lower Bounds. The work in [1, 2] shows space lower bounds for this problem to be $\Omega(n^{1-5/k})$, for any $k > 5$. Subsequently, the space lower bounds have been strengthened to $\Omega(\epsilon^2 n^{1-(2+\epsilon)/k})$, for $k > 2$, $\epsilon > 0$, by Bar-Yossef, Jayram, Kumar and Sivaku-mar [3], and further to $\Omega(n^{1-2/k})$ by Chakrabarti, Khot and Sun [5]. Saks and Sun [14] show that estimating the L_p distance d between two streaming vectors to within a factor of d^δ requires space $\Omega(n^{1-2/p-4\delta})$.

Other Related Work. For the special case of computing F_2, [1, 2] presents an $O(\log n + \log m)$ space and time complexity algorithm, where, m is the sum of the frequencies. Random linear combinations based on random variables drawn from stable distributions were considered by [13] to estimate F_p, for $0 < p \le 2$. The work presented in [9] presents a sketch technique to estimate the difference between two streams based on the L_1 metric norm. There has been substantial work on the problem of estimating F_0 and related metrics (set expression cardinalities over streams) for the various models of data streams [10, 1, 4, 12].

The rest of the paper is organized as follows. Section 2 describes the method and Section 3 presents formal lemmas and their proofs. Finally we conclude in Section 4.

2   An Overview of the Method

In this section, we present a simple description of the algorithm and some of its proper-ties. The lemmas and theorems stated in this section are proved formally in Section 3. Throughout the paper, we treat k as a fixed given value larger than 1.

2.1   Sketches Using Random Linear Combinations of kth Roots of Unity

Let x be a randomly chosen root of the equation $x^k = 1$, such that each of the k roots is chosen with equal probability of $1/k$. Given a complex number z, its conjugate is denoted by $\bar{z}$. For any j, $1 \le j \le k$, the following basic property holds, as shown below.

$$\mathbf{E}[x^j] = \mathbf{E}[\bar{x}^j] = \begin{cases} 0 & \text{if } 1 \le j < k \\ 1 & \text{if } j = k. \end{cases} \tag{1}$$

Proof. Let $j = k$. Then, $\mathbf{E}[x^j] = \mathbf{E}[x^k] = \mathbf{E}[1] = 1$, since, x is a root of unity.

Let $1 \le j < k$ and let u be the elementary k^{th} root of unity, that is, $u = e^{2\pi\sqrt{-1}/k}$.

$$\mathbf{E}[x^j] = \frac{1}{k}\sum_{l=1}^{k}(u^l)^j = \frac{1}{k}\sum_{l=1}^{k}(u^j)^l = \frac{u^j}{k}\frac{(1 - u^{jk})}{(1 - u^j)}$$

where, the last equality follows from the sum of a geometric progression in the complex field. Since $u^k = 1$, it follows that $u^{jk} = 1$. Further, since u is the elementary k^{th} root of unity, $u^j = e^{2\pi j\sqrt{-1}/k} \ne 1$, for $1 \le j < k$. Thus, the expression $(1 - u^{jk})/(1 - u^j) = 0$. Therefore, $\mathbf{E}[x^j] = 0$, for $1 \le j < k$.

The conjugation operator is a 1-1 and onto operator in the field of complex numbers. Further, if x is a root of $x^k = 1$, then, $\bar{x}^k = \overline{x^k} = \bar{1} = 1$, and therefore, $\bar{x}$ is also a kth root of unity. Thus, the conjugation operator, applied to the group of kth roots of unity, results in a permutation of the elements in the group (actually, it is an isomorphism). It therefore follows that the sum of the jth powers of the roots of unity is equal to the sum of the jth powers of the conjugates of the roots of unity. Thus, $\mathbf{E}\left[\bar{x}^j\right] = \mathbf{E}\left[x^j\right]$.   □

Let Z be the random variable defined as $Z = \sum_{l \in [N]} f_l x_l$. The variable x_l, for each $l \in [N]$, is one of a randomly chosen root of $x^k = 1$. The family of variables $\{x_l\}$ is assumed to be $2k$-wise independent. The following lemma shows that Re Z^k is an unbiased estimator of F_k. Following [1, 2], we call Z as a *sketch*. The random variable Z can be efficiently maintained with respect to stream updates as follows. First, we choose a random hash function $\theta : [N] \rightarrow [k]$ drawn from a family of hash functions that is $2k$-wise independent. Further, we pre-compute the kth roots of unity into an array $A[1..k]$ of size k (of complex numbers), that is, $A[r] = e^{2 \cdot \pi \cdot r \cdot \sqrt{-1}/k}$, for $r = 1, 2, \ldots, k$. For every stream update (l, v), we update the sketch as follows.

$$Z = Z + v \cdot A[\theta(l)]$$

The space required to maintain the hash function $\theta = \tilde{O}(k)$, and the time required for processing a stream update is also $\tilde{O}(k)$.

Lemma 1. $\mathbf{E}\left[Re\ Z^k\right] = F_k$.

As the following lemma shows, the variance of this estimator is quite high.

Lemma 2. $\mathbf{Var}\left[Re\ Z^k\right] = O(k^{2k} F_2^k)$.

This implies that $\mathbf{Var}\left[\text{Re } Z^k\right]/(\mathbf{E}\left[\text{Re } Z^k\right])^2 = O(F_2^k/F_k^2)$, which could be as large as n^{k-2}. To reduce the variance we organize the sketches in a hash table.

2.2 Organizing Sketches in a Hash Table

Let $\phi : \{0, 1, \ldots, N-1\} \rightarrow [B]$ be a hash function that maps the domain $\{0, 1, \ldots, N-1\}$ into a hash table consisting of B buckets. The hash function ϕ is drawn from a family of hash functions $\mathcal{H}$ that is $2k$-wise independent. The random bits used by the hash family is independent of the random bits used by the family $\{x_l\}_{l \in \{0,1,\ldots,N-1\}}$, or, equivalently, the random bits used to generate ϕ and θ are independent. The indicator variable $y_{l,b}$, for any domain element $l \in \{0, 1, \ldots, N-1\}]$ and bucket $b \in [B]$, is defined as $y_{l,b} = 1$ if $\phi(l) = b$ and $y_{l,b} = 0$ otherwise. Associated with each bucket b is a sketch Z_b of the elements that have hashed to that bucket. The random variables, Y_b and Z_b are defined as follows.

$$Z_b = \sum_l f_l \cdot x_l \cdot y_{l,b}, \qquad Y_b = \text{Re } Z_b^k, \quad \text{and} \quad Y = \sum_{b \in [B]} Y_b$$

Maintaining the hash table of sketches in the presence of stream updates is analogous to maintaining Z. As discussed previously, let $\theta : \{0, 1, \ldots, N-1\} \rightarrow [k]$ denote a random hash function that is chosen from a $2k$-wise independent family of hash functions (and independently of the bits used by ϕ), and let $A[1 \ldots k]$ be an array whose j^{th} entry is $e^{2 \cdot \pi \cdot j \cdot \sqrt{-1}/k}$, for $j = 1, \ldots, k$. For every stream update (l, v), we perform the following operation.

$$Z_{\phi(l)} = Z_{\phi(l)} + v \cdot A[\theta(l)]$$

The time complexity of the update operation is $\tilde{O}(k)$. The sketches in the buckets except the bucket numbered $\phi(l)$ are left unchanged.

The main observation of the paper is that the hash partitioning of the sketch Y into $\{Y_b\}_{b\in[B]}$ reduces the variance of Y significantly, while maintaining that $\mathbf{E}[Y] = F_k$. This is stated in the lemma below.

Lemma 3. *Let* $B \leq 2n^{1-\frac{1}{k}}$. *Then,* $\mathbf{Var}[Y] = O(F_k^2 n^{k-2}/B^{k-1})$.

A hash table organization of the sketches is normally used to reduce the time complexity of processing each stream update [6, 8]. However, for $k > 2$, the hash table organization of the sketches has the additional effect of reducing the variance.

Finally, we keep s_1 independent copies $Y[0], \ldots, Y[s_1 - 1]$ of the variable Y. The average of these variables is denoted by $\bar{Y}$; thus $\mathbf{Var}[\bar{Y}] = (1/s_1)\mathbf{Var}[Y]$. The result of the paper is summarized below, which states that $\bar{Y}$ estimates F_k to within an accuracy factor of $(1 \pm \epsilon)$ with constant probability greater than $1/2$ (at least $2/3$).

Theorem 1. *Let* $n^{1-\frac{1}{k-1}} \leq B \leq 2 \cdot n^{1-\frac{1}{k-1}}$ *and* $s_1 = 6 \cdot 2^k \cdot k^{3k}/\epsilon^2$. *Then,* $\mathbf{Pr}\left\{|\bar{Y} - F_k| > \epsilon F_k\right\} \leq 1/3$.

The space usage of the algorithm is therefore $\tilde{O}(B \cdot s_1) = O(n^{1-\frac{1}{k-1}})$ bits, since a logarithmic overhead is required to store each sketch Z_b. To boost the confidence of the answer to at least $1 - 2^{-\Omega(s_2)}$, a standard technique of returning the median value among s_2 such average estimates can be used, as shown in [1, 2].

The algorithm assumes that the number of buckets in the hash table is B, where, $n^{1-\frac{1}{k-1}} \leq B \leq 2 \cdot n^{1-\frac{1}{k-1}}$. Since, in general, the number of distinct items in the stream is not known in advance, one possible method that can be used is as follows. First estimate n to within a factor of $(1 \pm \frac{1}{8})$ using an algorithm for estimating F_0, such as [10, 1, 2, 4]. This can be done with high probability, in space $O(\log N)$. Keep $2 \log N + 4$ group of (independent) hash tables, such that the ith group uses $B_i = \lceil 2^{i/2} \rceil$ buckets. Each group of the hash tables uses the data structure described earlier. At the time of inference, first n is estimated as $\hat{n}$, and, then, we choose a hash table group indexed by i such that $i = 2 \cdot \lceil (1 - \frac{1}{k-1}) \log(8 \cdot \hat{n}/7) \rceil$. This ensures that the hash table size B_i satisfies $n^{1-\frac{1}{k-1}} \leq B_i \leq 2 \cdot n^{1-\frac{1}{k-1}}$, with high probability. Since, the number of hash table groups is $2 \cdot \log N$, this construction adds an overhead in terms of both space complexity and update time complexity by a factor of $2 \cdot \log N$. In the remainder of the paper, we assume that n is known exactly, with the understanding that this assumption can be alleviated as described.

3 Analysis

The jth frequency moment of the set of elements that map to bucket b under the hash function ϕ, is a random variable denoted by $F_{j,b}$. Thus, $F_{j,b} = \sum_l f_l^j y_{l,b}$. Further, since every element in the stream hashes to exactly one bucket, $\sum_b F_{j,b} = F_j$. We define $h_{l,b}$, for $l \in \{0, 1, \ldots, N-1\}$ and $b \in [B]$ to be $h_{l,b} = f_l \cdot y_{l,b}$. Thus, $F_{j,b} = \sum_l h_{l,b}^j$, for $j \geq 1$.

Notation: Marginal Expectations. The random variables, $Y, \{Y_b\}_{b \in B}$ are functions of two families of random variables, namely, $\mathbf{x} = \{x_l\}_{l \in \{0,1,\dots,N-1\}}$, used to generate the random roots of unity, and $\mathbf{y} = \{y_{l,b}\}, l \in \{0, 1, \dots, N-1\}$ and $b \in [B]$, used to map elements to buckets in the hash table. Our independence assumptions imply that these two families are mutually independent (i.e., their seeds use independent random bits), that is, $\mathbf{Pr}\{\mathbf{x} = \mathbf{u}$ and $\mathbf{y} = \mathbf{v}\} = \mathbf{Pr}\{\mathbf{x} = \mathbf{u}\} \cdot \mathbf{Pr}\{\mathbf{y} = \mathbf{v}\}$ Let $W = W(\mathbf{x}, \mathbf{y})$ be a random variable that is a function of the random variables in $\mathbf{x}$ and $\mathbf{y}$. For a fixed random choice of $\mathbf{y} = \mathbf{y}_0$, $\mathbf{E}_\mathbf{x}[W]$ denotes the marginal expectation of W as a function of y. That is, $\mathbf{E}_\mathbf{x}[W] = \sum_\mathbf{u} W(\mathbf{u}, \mathbf{y}_0)\mathbf{Pr}\{\mathbf{x} = \mathbf{u}\}$. It follows that $\mathbf{E}[W] = \mathbf{E}_\mathbf{y}[\mathbf{E}_\mathbf{x}[W]]$.

Overview of the Analysis. The main steps in the proof of Theorem 1 are as follows. In Section 3.1, we show that $\mathbf{E}_\mathbf{x}[Y] = F_k$. In Section 3.2, we show that $\mathbf{E}[\mathrm{Re}\, Z^k] \leq k^{2k} F_2^k$. In Section 3.3, using the above result, we show that $\mathbf{E}_\mathbf{x}[Y^2] \leq k^{2k} \sum_b F_{2,b}^k$. Section 3.4 shows that $\mathbf{E}_\mathbf{y}[F_{2,b}^k] \leq (2/B + 2^k \cdot n^{k-2}/B^k) F_k^2$ and also concludes the proof of Theorem 1. Finally, we conclude in Section 4.

Notation: Multinomial Expansion. Let X be defined as $X = \sum_{l \in \{0,1,\dots,N-1\}} a_l$, where, $a_l \geq 0$, for $l \in \{0, 1, \dots, N-1\}$. Then, X^k can be written as

$$X^k = \sum_{s=1}^{k} \sum_{e_1 + \dots e_s = k, e_1 > 0, \dots, e_s > 0} \binom{k}{e_1 e_2 \cdots e_s} \sum_{l_1 < l_2 < \dots < l_s} a_{l_1}^{e_1} a_{l_2}^{e_j} \cdots a_{l_s}^{e_s}$$

where, s is the number of distinct terms in the product and e_i is the exponent of the ith product term. The indices l_i are therefore necessarily distinct, $l_i \in \{0, 1, \dots, N-1\}, i = 1, 2, \dots, s$. For easy reference, the above equation is written and used in the following form.

$$X^k = \sum_{s,e:Q(e,s)} C(\mathbf{e}) \sum_{1:R(e,1,s)} \left(\prod_{j=1}^{s} a_{l_j}^{e_j}\right) . \tag{2}$$

where, $Q(\mathbf{e}, s) \equiv 1 \leq s \leq k$ and $\mathbf{e} = (e_1, e_2, \dots, e_s)$ is s-dimensional and $\sum_{j=1}^{s} e_j = k$; $R(\mathbf{e}, 1, s) \equiv \mathbf{1} = (l_1, l_2, \dots, l_s)$ is s-dimensional and $0 \leq l_1 < l_2 < \dots < l_s \leq N-1$; and the multinomial coefficient $C(\mathbf{e}) = \binom{k}{e_1,\dots,e_s}$. In this notation, the following inequality holds .

$$\sum_{1:R(e,1,s)} \prod_{j=1}^{s} a_{l_j}^{e_j} \leq \prod_{j=1}^{s} (\sum_l a_l^{e_j}) . \tag{3}$$

By setting $n = k$, and $a_1 = a_2 = \dots = a_k = 1$, we obtain,

$$k^k = \sum_{e,s} C(\mathbf{e}) \binom{k}{s} > \sum_{e,s} C(\mathbf{e}).$$

By squaring the above equation on both sides, we obtain that $k^{2k} = (\sum_{\mathbf{e},s} C(\mathbf{e})\binom{k}{s})^2 > \sum_{\mathbf{e},s} C^2(\mathbf{e})$. We therefore have the following inequalities.

$$\sum_{e,s} C(\mathbf{e}) < k^k, \quad \sum_{e,s} C^2(\mathbf{e}) < k^{2k} . \tag{4}$$

3.1 Expectation

In this section, we show that $\mathbf{E}[\mathrm{Re}\,Z^k] = F_k$, thereby proving Lemma 1, and that $\mathbf{E}_\mathbf{x}[Y] = F_k$.

Proof (of Lemma 1). Since the family of variables x_l's is k-wise independent, therefore

$$\mathbf{E}\Big[\prod_{j=1}^{s} x_{l_j}^{e_j}\Big] = \prod_{j=1}^{s} \mathbf{E}[x_{l_j}^{e_j}] \ .$$

Applying equation (2) to $Z^k = (\sum_l f_l x_l)^k$ and using linearity of expectation and k-wise independence property of x_l's, we obtain

$$\mathbf{E}[Z^k] = \sum_{s,e:Q(e,s)} C(e) \sum_{l:R(e,l,s)} \Big(\prod_{j=1}^{s} f_{l_j}^{e_j}\Big)\Big(\prod_{j=1}^{s} \mathbf{E}[x_{l_j}^{e_j}]\Big) \ .$$

Using equation (1), we note that the term $\big(\prod_{j=1}^{s} \mathbf{E}[x_{l_j}^{e_j}]\big) = 0$, if $s > 1$, since in this case, $e_j < k$, for each $j = 1, \ldots, s$. Thus, the above summation reduces to

$$\mathbf{E}[Z^k] = \sum_l f_l^k = F_k \ .$$

Since F_k is real, $\mathbf{E}[\mathrm{Re}\,Z^k]$ is also F_k, proving Lemma 1. $\qquad\square$

Lemma 4. *Suppose that the family of random variables $\{x_l\}$ is k-wise independent. Then,* $\mathbf{E}_\mathbf{x}[Y_b] = F_{k,b}$ *and* $\mathbf{E}_\mathbf{x}[Y] = \mathbf{E}[Y] = F_k$.

Proof. We first show that $\mathbf{E}_\mathbf{x}[Y_b] = F_{k,b}$. $\mathbf{E}_\mathbf{x}[Z_b^k] = \mathbf{E}_\mathbf{x}[(\sum_l f_l y_{l,b} x_l)^k] = \mathbf{E}_\mathbf{x}[(\sum_l h_{l,b} x_l)^k]$, by letting $h_{l,b} = f_l \cdot y_{l,b}$. By an argument analogous to the proof of Lemma 1, we obtain $\mathbf{E}_\mathbf{x}[(\sum_l h_{l,b} x_l)^k] = \sum_l h_{l,b}^k = \sum_l f_l^k y_{l,b}^k = \sum_l f_{l,k}^k y_{l,b} = F_{k,b}$, (since $y_{l,b}$'s are binary variables). Since $F_{k,b}$ is always real, $\mathbf{E}_\mathbf{x}[Y_b] = \mathbf{E}_\mathbf{x}[\mathrm{Re}\,Z_b^k] = F_{k,b}$. Finally, $\mathbf{E}_\mathbf{x}[Y] = \mathbf{E}_\mathbf{x}[\sum_b Y_b] = \sum_b \mathbf{E}_\mathbf{x}[Y_b] = \sum_b F_{k,b} = F_k$, since each element is hashed to exactly one bucket. Further, $\mathbf{E}[Y] = \mathbf{E}_\mathbf{y}[\mathbf{E}_\mathbf{x}[Y]] = \mathbf{E}_\mathbf{y}[F_k] = F_k$. $\qquad\square$

3.2 Variance of Re Z^k

In this section, we estimate the variance of $\mathrm{Re}\,Z^k$ and derive some simple corollaries.

Lemma 5. *Let* $W = \mathrm{Re}\,(\sum_l a_l x_l)^k$. *Then,* $\mathbf{Var}[W] \leq k^{2k}(\sum_l a_l^2)^k$.

Proof. Let $X = (\sum_l a_l x_l)^k$. Then, $\mathbf{Var}[W] = \mathbf{E}[W^2] - (\mathbf{E}[W])^2 \leq \mathbf{E}[X\bar{X}] - (\mathbf{E}[W])^2$. Using equation (2), for $X, \bar{X}$, we obtain the following.

$$X = \sum_{s,e:Q(e,s)} C(e) \sum_{l:R(e,l,s)} \Big(\prod_{j=1}^{s} a_{l_j}^{e_j}\Big) \cdot \Big(\prod_{j=1}^{s} x_{l_j}^{e_j}\Big)$$

$$\bar{X} = \sum_{t,g:Q(g,t)} C(g) \sum_{l:R(g,m,t)} \Big(\prod_{j'=1}^{t} a_{m_{j'}}^{g_{j'}}\Big) \cdot \Big(\prod_{j'=1}^{t} \bar{x}_{m_{j'}}^{g_{j'}}\Big)$$

Multiplying the above two equations, we obtain

$$X \cdot \bar{X} = \sum_{s,e:Q(e,s)} \sum_{t,g:Q(g,t)} C(e) \cdot C(g) \sum_{l:R(e,l,s)} \sum_{l:R(g,m,t)}$$

$$(\prod_{j=1}^{s} a_{l_j}^{e_j}) \cdot (\prod_{j'=1}^{t} a_{m_{j'}}^{g_{j'}}) \cdot (\prod_{j=1}^{s} x_{l_j}^{e_j}) \cdot (\prod_{j'=1}^{t} \bar{x}_{m_{j'}}^{g_{j'}}).$$

The general form of the product of random variables that arises in the multinomial expansion of $X\bar{X}$ is $(\prod_{j=1}^{s} x_{l_j}^{e_j})(\prod_{j'=1}^{t} \bar{x}_{m_{j'}}^{g_{j'}})$. Since the random variables x_l's are $2k$-wise independent, using equation (1), it follows that,

$$\mathbf{E}\left[\prod_{j=1}^{s} x_{l_j}^{e_j} \prod_{j'=1}^{t} \bar{x}_{m_{j'}}^{g_j}\right] = \begin{cases} 1 & \text{if } s = t = 1, e_1 = g_1 = k \\ 1 & \text{if } s = t, t > 1, \mathbf{e} = \mathbf{g} \text{ and } \mathbf{l} = \mathbf{m}, \\ 0 & \text{otherwise.} \end{cases}$$

This directly yields the following.

$$\mathbf{E}[X\bar{X}] = \sum_{e,s:Q(e,s)} C^2(e) \sum_{l:R(e,l,s)} \prod_{j=1}^{s} a_{l_j}^{2e_j}$$

$$\leq \sum_{e,s} C^2(e) \prod_{j=1}^{s} (\sum_l a_l^{2e_j}), \quad \text{by equation (3)}$$

$$\leq \sum_{e,s} C^2(e) \prod_{j=1}^{s} (\sum_l a_l^2)^{e_j}, \quad \text{since } \sum_l a_l^{2e_j} \leq (\sum_l a_l^2)^{e_j}$$

$$= (\sum_l a_l^2)^k (\sum_{e,s} C^2(e)), \quad \text{since } \sum_{j=1}^{s} e_j = k$$

$$\leq (\sum_l a_l^2)^k \cdot k^{2k}, \quad \text{by equation (4).} \quad \Box$$

By letting $a_l = f_l, l \in \{0, 1, 2 \ldots, N-1\}$, Lemma 5 yields

$$\mathbf{Var}[\text{Re } Z^k] = \mathbf{Var}[\text{Re } (\sum_l f_l x_l)^k] \leq k^{2k} F_2^k, \tag{5}$$

which is the statement of Lemma 2. By letting $a_l = h_{l,b} = f_l \cdot y_{l,b}$, where, b is a fixed bucket index, and $l \in \{0, 1, 2 \ldots, N-1\}$, yields the following equation.

$$\mathbf{E}_{\mathbf{x}}[Y_b^2] \leq k^{2k} F_{2,b}^k, \quad \text{for } b \in [B]. \tag{6}$$

3.3 Var$[Y]$: Vanishing of Cross-Bucket Terms

We now consider the problem of obtaining an upper bound on $\mathbf{Var}[Y]$. Note that $\mathbf{Var}[Y] = \mathbf{E}_{\mathbf{y}}[\mathbf{E}_{\mathbf{x}}[Y^2]] - (\mathbf{E}_{\mathbf{y}}[\mathbf{E}_{\mathbf{x}}[Y]])^2$. ¿From Lemma 4, $\mathbf{E}_{\mathbf{y}}[\mathbf{E}_{\mathbf{x}}[Y]] = F_k$. Thus,

$$\mathbf{Var}[Y] = \mathbf{E}_{\mathbf{y}}[\mathbf{E}_{\mathbf{x}}[Y^2]] - F_k^2. \tag{7}$$

Lemma 6. $\mathrm{Var}\big[Y\big] \leq k^{2k} \sum_b \mathbf{E_y}\big[F_{2,b}^k\big]$, *assuming independence assumption I.*

Proof. $\mathbf{E_x}\big[Y^2\big] = \mathbf{E_x}\big[(\sum_b Y_b)^2\big] = \mathbf{E_x}\big[\sum_b Y_b^2 + \sum_{a\neq b} Y_a Y_b\big] = \sum_b \mathbf{E_x}\big[Y_b^2\big] + \sum_{a\neq b} \mathbf{E_x}\big[Y_a Y_b\big]$.

We now consider $\mathbf{E_x}\big[Y_a Y_b\big]$, for $a \neq b$. Recall that $Y_a = \mathrm{Re}\ Z_a^k$ (and analogously, Y_b is defined). For any two complex numbers z, w, $(\mathrm{Re}\ z)(\mathrm{Re}\ w) = (1/2)\mathrm{Re}\ (z(w + \bar{w}))$. Thus, $Y_a Y_b = (\mathrm{Re}\ Z_a^k)(\mathrm{Re}\ Z_b^k) = (1/2)\mathrm{Re}\ (Z_a^k Z_b^k + Z_a^k \overline{Z_b^k})$.

Let us first consider $\mathbf{E_x}\big[Z_a^k Z_b^k\big]$. The general term involving product of random variables is $(\prod_{j=1}^{s} f_{l_j}^{e_j}) \cdot (\prod_{j'=1}^{t} f_{m_j}^{g_{j'}}) \cdot (\prod_{j=1}^{s} y_{l_j,a} \cdot x_{l_j}^{e_j}) \cdot (\prod_{j'=1}^{t} y_{m_{j'},b} \cdot x_{m_{j'}}^{g_{j'}})$. Consider the last two product terms in the above expression, that is, $(\prod_{j=1}^{s} y_{l_j,a} \cdot x_{l_j}^{e_j}) \cdot (\prod_{j'=1}^{t} y_{m_{j'},b} \cdot x_{m_{j'}}^{g_{j'}})$. For any $1 \leq j \leq s$ and $1 \leq j' \leq t$, it is not possible that $l_j = m_{j'}$, that is, the same element whose index is given by $l_j = m_{j'}$ cannot simultaneously hash to two distinct buckets, a and b (recall that $a \neq b$). By $2k$-wise independence, we therefore obtain that the only way the above product term can be non zero (i.e., 1) on expectation, is that $s = t = 1$ and therefore, $e_1 = k$ and $g_1 = k$. Thus, we have $\mathbf{E}\big[Z_a^k Z_b^k\big] = \sum_{l,m} h_{l,a}^k h_{m,b}^k = F_{k,a} F_{k,b}$.

Using the same observation, it can be argued that $\mathbf{E_x}\big[Z_a^k \overline{Z_b^k}\big] = F_{k,a} F_{k,b}$. It follows that $\mathbf{E_x}\big[(1/2)(Z_a^k Z_b^k + Z_a^k \overline{Z_b^k}))\big] = F_{k,a} F_{k,b}$, which is a real number. Therefore $\mathbf{E_x}\big[\mathrm{Re}\ (1/2)(Z_a^k Z_b^k + Z_a^k \overline{Z_b^k})\big] = F_{k,a} F_{k,b} = \mathbf{E_x}\big[Y_a Y_b\big]$.

By equation (7), $\mathrm{Var}\big[Y\big] = \mathbf{E_y}\big[\mathbf{E_x}\big[Y^2\big]\big] - F_k^2$. Further, from Lemma 4, $F_k = \mathbf{E_y}\big[\sum_b F_{k,b}\big]$. We therefore have,

$$\mathrm{Var}\big[Y\big] = \mathbf{E_y}\Big[\mathbf{E_x}\big[Y^2\big] - \Big(\sum_b F_{k,b}\Big)^2\Big]$$

$$= \mathbf{E_y}\Big[\sum_b \mathbf{E_x}\big[Y_b^2\big] + \sum_{a\neq b} \mathbf{E_x}\big[Y_a Y_b\big] - \Big(\sum_b F_{k,b}\Big)^2\Big]$$

$$= \mathbf{E_y}\Big[\sum_b \mathbf{E_x}\big[Y_b^2\big] + \sum_{a\neq b} F_{k,a} F_{k,b} - \Big(\sum_b F_{k,b}\Big)^2\Big], \quad \text{by above argument}$$

$$= \mathbf{E_y}\Big[\sum_b \mathbf{E_x}\big[Y_b^2\big] - \sum_b F_{k,b}^2\Big] \leq \mathbf{E_y}\Big[\sum_b \mathbf{E_x}\big[Y_b^2\big]\Big]$$

$$\leq \mathbf{E_y}\Big[\sum_b k^{2k} F_{2,b}^k\Big], \quad \text{by equation (6)} \qquad \square$$

3.4   Calculation of $\mathbf{E}\big[F_{2,b}^k\big]$

Given a t-dimensional vector $\mathbf{e} = (e_1, \dots, e_t)$ such that $e_i > 0$, for $1 \leq i \leq t$ and $\sum_{j=1}^{t} e_j = k$, we define the function $\psi(\mathbf{e})$ as follows. Without loss of generality, let the indices e_j be arranged in non-decreasing order. Let $r = r(\mathbf{e})$ denote the largest index such that $e_r < k/2$. Then, we define the function $\phi(e)$ as follows.

$$\psi(\mathbf{e}) = n^{\sum_{j=1}^{r}(1-2e_j/k)}/B^t$$

The motivation of this definition stems from its use in the following lemma.

Lemma 7. *Suppose* $\sum_{j=1}^{t} e_j = k$ *and* $e_j > 0$, *for* $j = 1, \ldots, t$. *Then,* $\prod_{j=1}^{t} F_{2e_j} \le \psi(\mathbf{e}) \cdot F_k^2 \cdot B^t$.

Proof. From [1,2], $F_j \le n^{1-j/k} F_k^{j/k}$, if $j < k$ and $F_j \le F_k^{j/k}$, if $j > k$. Thus,

$$\prod_{j=1}^{t} F_{2e_j} = \left(\prod_{j=1}^{r} F_{2e_j} \right) \left(\prod_{j=r+1}^{t} F_{2e_j} \right) = \left(\prod_{j=1}^{r} n^{1-2e_j/k} F_k^{2e_j/k} \right) \left(\prod_{j=r+1}^{t} F_k^{2e_j/k} \right)$$

$$= n^{\sum_{j=1}^{r}(1-2e_j/k)} F_k^{\sum_{j=1}^{t} 2e_j/k} = \psi(\mathbf{e}) \cdot B^t \cdot F_k^2, \quad \text{since} \sum_{j} e_j = k. \square$$

The function ψ satisfies the following property that we use later.

Lemma 8. *If* $B < 2 \cdot n^{1-\frac{1}{k}}$, *then,* $\psi(\mathbf{e}) \le \max \left(2/B, 2^k \cdot n^{k-2}/B^k \right)$.

Proof. Let $\mathbf{e}$ be a t-dimensional vector. If $t = 1$, $\psi(\mathbf{e}) = 1/B$. If $t = r$, then $\psi(\mathbf{e}) = n^{t-2}/B^t \le 2^k \cdot n^{k-2}/B^k$. If $t \ge r + 2$, then $\psi(\mathbf{e}) = (2^t/B^t) \cdot n^{t-((t-r)+\sum 2e_j/k)} < 2^t \cdot n^{t-2}/B^t \le 2^k \cdot n^{k-2}/B^k$. Finally, let $t = r + 1$. Then,

$$\psi(\mathbf{e}) = 2^t \cdot n^{t-1-2\sum e_j/k}/B^t \le 2^t \cdot n^{t-1-2(t-1)/k}/B^t,$$

since $\sum_{j=1}^{r} e_j \ge r = t - 1$. Thus,

$$\psi(\mathbf{e}) \le \psi(\mathbf{e})(2 \cdot n^{1-\frac{1}{k}}/B)^{k-t} \le (2^t \cdot n^{t-1-2(t-1)/k}/B^t)(2 \cdot n^{1-\frac{1}{k}}/B)^{k-t} =$$
$$2^k \cdot n^{k-2-(t-2)/k}/B^k \le 2^k \cdot n^{k-2}/B^k .$$

where, the first inequality follows from the assumption that $B < n^{1-\frac{1}{k}}$ and the second inequality follows because $t \ge 2$. $\square$

Lemma 9. *Let* $B < 2 \cdot n^{1-\frac{1}{k}}$. *Then,* $\mathbf{E}\left[F_{2,b}^k \right] < k^k F_k^2 (2/B + 2^k \cdot n^{k-2}/B^k)$.

Proof. For a fixed b, the variables $y_{l,b}$ are k-wise independent. $F_{2,b}$ is a linear function of $y_{l,b}$. Thus, $F_{2,b}^k$ is a symmetric multinomial of degree k, as follows.

$$F_{2,b}^k = \left(\sum_{l} f_l^2 y_{l,b} \right)^k = \sum_{s,e} C(\mathbf{e}) \sum_{l_1 < l_2 < \cdots l_s} f_{l_1}^{2e_1} \cdots f_{l_s}^{2e_s} y_{l_1,b} \cdot y_{l_2,b} \cdot y_{l_s,b} .$$

Taking expectations, and using k-wise independence of the $y_{l,b}$'s, we have,

$$\mathbf{E}\big[F_{2,b}^k\big] = \sum_{s,\mathbf{e}} C(\mathbf{e}) \sum_{l_1 < l_2 < \cdots < l_s} f_{l_1}^{2e_1} \cdots f_{l_j}^{2e_j} \mathbf{E}\big[y_{l_1,b} \cdot y_{l_2,b} \cdot y_{l_s,b}\big]$$

$$= \sum_{s,\mathbf{e}} C(\mathbf{e}) \sum_{l_1 < l_2 < \cdots < l_s} f_{l_1}^{2e_1} \cdots f_{l_j}^{2e_j} \mathbf{E}\big[y_{l_1,b}\big] \cdot \mathbf{E}\big[y_{l_2,b}\big] \cdots \mathbf{E}\big[y_{l_s,b}\big]$$

$$= \sum_{s,\mathbf{e}} C(\mathbf{e}) \sum_{l_1 < l_2 < \cdots < l_s} f_{l_1}^{2e_1} \cdots f_{l_j}^{2e_j} \frac{1}{B^s}, \quad \text{since, } \mathbf{E}\big[y_{l_j,b}\big] = \frac{1}{B}$$

$$\leq \sum_{s,\mathbf{e}} C(\mathbf{e}) \cdot (1/B^s) \cdot \prod_{j=1}^{s} F_{2e_j} \leq \sum_{s,\mathbf{e}} C(\mathbf{e}) \cdot \psi(\mathbf{e}) \cdot F_k^2, \quad \text{by Lemma 7}$$

$$\leq \sum_{s,\mathbf{e}} C(\mathbf{e}) \cdot F_k^2 \cdot (2/B + 2^k \cdot n^{k-2}/B^k), \quad \text{by Lemma 8}$$

$$\leq k^k \cdot F_k^2 \cdot (2/B + 2^k \cdot n^{k-2}/B^k), \quad \text{since, } \sum_{s,\mathbf{e}} C(\mathbf{e}) < k^k \qquad \Box$$

Combining the result of Lemma 6 with Lemma 9, we obtain the following bound on $\mathbf{Var}[Y]$.

$$\mathbf{Var}[Y] \leq k^{3k} \cdot F_k^2 \cdot (2 + 2^k \cdot n^{k-2}/B^{k-1}) \tag{8}$$

Recall that $\bar{Y}$ is the average of s_1 independent estimators, each calculating Y. The main theorem of the paper now follows simply.

Proof (of Theorem 1). By Chebychev's inequality, $\mathbf{Pr}\big\{|\bar{Y} - F_k| > \epsilon F_k\big\} < \mathbf{Var}[\bar{Y}]/(\epsilon^2 F_k^2)$. Substituting Equation (8), we have $\mathbf{Var}[\bar{Y}]/(\epsilon^2 \cdot F_k^2) \leq 1/3$. $\qquad \Box$

4 Conclusions

The paper presents a method for estimating the kth frequency moment, for $k > 2$, of data streams with general update operations. The algorithm has space complexity $\tilde{O}(n^{1-\frac{1}{k-1}}))$ and is based on constructing random linear combinations using randomly chosen k^{th} roots of unity. A gap remains between the lower bound for this problem, namely, $O(n^{1-2/k})$, for $k > 2$, as proved in [3,5] and the complexity of a known algorithm for this problem.

References

1. Noga Alon, Yossi Matias, and Mario Szegedy. "The Space Complexity of Approximating the Frequency Moments". In *Proceedings of the 28th Annual ACM Symposium on the Theory of Computing STOC, 1996*, pages 20–29, Philadelphia, Pennsylvania, May 1996.
2. Noga Alon, Yossi Matias, and Mario Szegedy. "The space complexity of approximating frequency moments". *Journal of Computer Systems and Sciences*, 58(1):137–147, 1998.
3. Ziv Bar-Yossef, T.S. Jayram, Ravi Kumar, and D. Sivakumar. "An information statistics approach to data stream and communication complexity". In *Proceedings of the 34th ACM Symposium on Theory of Computing (STOC), 2002*, pages 209–218, Princeton, NJ, 2002.

4. Ziv Bar-Yossef, T.S. Jayram, Ravi Kumar, D. Sivakumar, and Luca Trevisan. "Counting distinct elements in a data stream". In *Proceedings of the 6th International Workshop on Randomization and Approximation Techniques in Computer Science, RANDOM 2002*, Cambridge, MA, 2002.
5. Amit Chakrabarti, Subhash Khot, and Xiaodong Sun. "Near-Optimal Lower Bounds on the Multi-Party Communication Complexity of Set Disjointness". In *Proceedings of the 18th Annual IEEE Conference on Computational Complexity, CCC 2003*, Aarhus, Denmark, 2003.
6. Moses Charikar, Kevin Chen, and Martin Farach-Colton. "Finding frequent items in data streams". In *Proceedings of the 29th International Colloquium on Automata Languages and Programming*, 2002.
7. Don Coppersmith and Ravi Kumar. "An improved data stream algorithm for estimating frequency moments". In *Proceedings of the Fifteenth ACM SIAM Symposium on Discrete Algorithms*, New Orleans, LA, 2004.
8. G. Cormode and S. Muthukrishnan. "What's Hot and What's Not: Tracking Most Frequent Items Dynamically". In *Proceedings of the Twentysecond ACM SIGACT-SIGMOD-SIGART Symposium on Principles of Database Systems*, San Diego, California, May 2003.
9. Joan Feigenbaum, Sampath Kannan, Martin Strauss, and Mahesh Viswanathan. "An Approximate L^1-Difference Algorithm for Massive Data Streams". In *Proceedings of the 40th Annual IEEE Symposium on Foundations of Computer Science*, New York, NY, October 1999.
10. Philippe Flajolet and G.N. Martin. "Probabilistic Counting Algorithms for Database Applications". *Journal of Computer Systems and Sciences*, 31(2):182–209, 1985.
11. Sumit Ganguly. "A bifocal technique for estimating frequency moments over data streams". *Manuscript*, April 2004.
12. Sumit Ganguly, Minos Garofalakis, and Rajeev Rastogi. "Processing Set Expressions over Continuous Update Streams". In *Proceedings of the 2003 ACM SIGMOD International Conference on Management of Data*, San Diego, CA, 2003.
13. Piotr Indyk. "Stable Distributions, Pseudo Random Generators, Embeddings and Data Stream Computation". In *Proceedings of the 41st Annual IEEE Symposium on Foundations of Computer Science*, pages 189–197, Redondo Beach, CA, November 2000.
14. M. Saks and X. Sun. "Space lower bounds for distance approximation in the data stream model". In *Proceedings of the 34th ACM Symposium on Theory of Computing (STOC), 2002*, 2002.

Fooling Parity Tests with Parity Gates

Dan Gutfreund[1,*] and Emanuele Viola[2,**]

[1] School of Computer Science and Engineering
The Hebrew University of Jerusalem, Israel, 91904
danig@cs.huji.ac.il
[2] Division of Engineering and Applied Sciences
Harvard University, Cambridge, MA 02138
viola@eecs.harvard.edu

Abstract. We study the complexity of computing k-wise independent and ϵ-biased generators $G : \{0,1\}^n \to \{0,1\}^m$. Specifically, we refer to the complexity of computing G *explicitly*, i.e. given $x \in \{0,1\}^n$ and $i \in \{0,1\}^{\log m}$ computing the i-th output bit of $G(x)$. [MNT90] show that constant depth circuits of size poly(n) cannot explicitly compute k-wise independent and ϵ-biased generators with seed length $n \leq 2^{\log^{o(1)} m}$.
In this work we show that DLOGTIME-uniform constant depth circuits of size poly(n) *with parity gates* can explicitly compute k-wise independent and ϵ-biased generators with seed length n roughly $\log m \lll 2^{\log^{o(1)} m}$. In some cases the seed length of our generators is optimal up to constant factors, and in general up to polynomial factors. To obtain our results, we show a new construction of combinatorial designs, and we also show how to compute, in DLOGTIME-uniform AC_0, random walks of length $\log^c n$ over certain expander graphs of size 2^n.

1   Introduction

The notion of *pseudorandom generators* (PRGs) is central to the fields of Computational Complexity and Cryptography. Informally, a PRG is an efficient deterministic procedure that maps a short *seed* to a long output, such that certain tests are *fooled* by the PRG, i.e. they cannot distinguish between the output of the PRG (over a random input) and the uniform distribution. More formally, a PRG $G : \{0,1\}^n \to \{0,1\}^m$ fools a test $M : \{0,1\}^m \to \{0,1\}$ if

$$\left| \Pr_{y \in \{0,1\}^m}[M(y) = 1] - \Pr_{x \in \{0,1\}^n}[M(G(x)) = 1] \right| \leq \epsilon.$$

One popular class of tests $M : \{0,1\}^m \to \{0,1\}$ are *parity* ones (sometimes called linear tests). These are the tests that, given $z \in \{0,1\}^m$, only XOR fixed subsets of the bits of z. A generator is called ϵ-*biased* if it fools such tests.

---

* Research supported in part by the Leibniz Center, the Israel Foundation of Science, a US-Israel Binational research grant, and an EU Information Technologies grant (IST-FP5).
** Research supported by NSF grant CCR-0133096 and US-Israel BSF grant 2002246.

K. Jansen et al. (Eds.): APPROX and RANDOM 2004, LNCS 3122, pp. 381–392, 2004.
© Springer-Verlag Berlin Heidelberg 2004

Another popular type of tests are those whose value, given $z \in \{0,1\}^m$, depends only on k fixed bits of z, but on those bits is arbitrary. A generator is called (ϵ, k)-*wise independent* if it fools such tests. An important special case of (ϵ, k)-wise independent generators are those where $\epsilon = 0$, i.e. every k fixed output bits of the generator are uniform over $\{0,1\}^k$. Such generators are called k-wise independent. ϵ-biased and k-wise independent generators have found several applications in Complexity Theory and Cryptography. For a discussion of these generators we refer the reader to the excellent book by Goldreich [Gol99].

In this paper we study the following general question: *what are the minimal computational resources needed to compute k-wise and ϵ-biased generators?* Throughout this work, when we refer to the complexity of a generator G, we actually refer to the complexity of computing the i-th bit of $G(x)$ given a seed $x \in \{0,1\}^n$ and an index $i \in \{0,1\}^{\log m}$. This is a more refined notion of complexity, which is especially adequate for generators having logarithmic seed length (i.e. $n \approx \log m$), as is the case with k-wise independent and ϵ-biased generators (see below). Generators for which we can efficiently compute the i-th output bit of $G(x)$, given $x \in \{0,1\}^n$ and $i \in \{0,1\}^{\log m}$, are called *explicit*. Explicitness plays a crucial role in many applications where only some portion of $G(x)$ is needed at each time, and the application runs in time polynomial in the *seed length* of the generator (rather than in its output length). Such applications range from constructing hash functions to Probabilistically Checkable Proofs.

k-wise independent and ϵ-biased generators, $G : \{0,1\}^n \to \{0,1\}^m$, explicitly computable in time $\text{poly}(n)$ are known with seed length logarithmic in the output length, i.e. $n = O(\log m)$ [CG89,ABI86,NN90,AGHP92] (we ignore here other parameters, we will be more accurate later). On the other hand, Mansour, Nisan and Tiwari [MNT90] showed[1] that any k-wise independent or ϵ-biased generator $G : \{0,1\}^n \to \{0,1\}^m$ explicitly computable by constant depth circuits (AC_0) of size $\text{poly}(n)$ must have seed length n at least $2^{\log^{\Omega(1)} m} \gg \log m$.

In this paper we ask the following technical question: Can constant depth circuits *with parity gates* ($AC_0[\oplus]$) of size $\text{poly}(n)$ explicitly compute k-wise independent and ϵ-biased generators, $G : \{0,1\}^n \to \{0,1\}^m$, with seed length $n \approx \log m$? (Where parity is the function $\oplus(x) := \sum_i x_i \mod 2$).

We give an affirmative answer by showing several constructions of such generators explicitly computed by $AC_0[\oplus]$ circuits of size $\text{poly}(n)$ (our results are discussed below). All our circuits are DLOGTIME-uniform. Informally this means that each gate (resp. edge) in the circuit can be specified in time linear *in the name* of the gate (resp. edge). DLOGTIME-uniformity is the strongest notion of uniformity found generally applicable, and gives nice characterizations of AC_0, $AC_0[\oplus]$ [BIS90]. *In this paper whenever we say that a circuit class is uniform we always mean that it is DLOGTIME-uniform.* Note that uniform $AC_0[\oplus]$ is strictly contained in $L :=$ logarithmic space. Therefore all our generators are in particular explicitly computable in space $O(\log n)$. For background on circuit complexity we refer the reader to the excellent book by Vollmer [Vol99].

---

[1] In [MNT90] this negative result is stated for hash functions only, but their techniques apply to k-wise independent and ϵ-biased generators as well.

Table 1. Our main results.

Generators : $\{0,1\}^n \rightarrow \{0,1\}^m$ explicitly computable by DLOGTIME-uniform $AC_0[\oplus]$ circuits of size poly(n)			
Seed length	Type	Limitations	Reference
$n = O(k^3 \log m)$	k-wise	$k = O(1)$	Theorem 2 (1)
$n = O(k^2 \log(m) \log(k \cdot \log m))$	k-wise	-	Theorem 2 (2)
$n = O(\log m + \log 1/\epsilon)$	ϵ-biased	$\epsilon = \Omega(1/2^{\log^c \log(m)})$	Theorem 4 (1)
$n = O(\log m \log 1/\epsilon)$	ϵ-biased	-	Theorem 4 (2)
$n = O(k + \log \log m + \log(1/\epsilon))$	(ϵ, k)-wise	$k = O(\log^c \log(m))$, $\epsilon = \Omega(1/2^{\log^c \log(m)})$	Omitted due to space restrictions

1.1   Our Results

Our main results are summarized in Table 1. We now discuss some of the generators in Table 1.

k-Wise Independent Generators in $AC_0[\oplus]$: k-wise independent generators explicitly computable in time poly(n) are known with seed length $n = O(k \log m)$ [CG89,ABI86] which is optimal up to constant factors (a lower bound of $(k \log m)/2$ was proven in [CGH+85]).

To understand the difficulty of implementing these generators in $AC_0[\oplus]$ let us discuss a construction from [CG89,ABI86]. For simplicity of exposition, assume that $m = 2^h$ for some integer h. Let $GF(2^h)$ be the field of size 2^h, then for every k, the generator $G : \{0,1\}^n \rightarrow \{0,1\}^m$ defined as

$$G(a_0, a_1, \ldots, a_{k-1})_i := \sum_{j<k} a_j i^j, \text{ where } a_0, a_1, \ldots, a_{k-1}, i \in GF(2^h), \quad (1)$$

is a k-wise independent generator[2] with seed length $n = kh = k \log m$.

For concreteness, in the following discussion let us fix k at most poly(h), so that poly(n) = poly(h). To compute the generator in Equation 1 we must first find a representation of $GF(2^h)$ (e.g., an irreducible polynomial of degree h over $GF(2)$). We do not know how to find a representation of $GF(2^h)$ in *uniform* $AC_0[\oplus]$ circuits of size poly(h) for *every* given h. And even if one has such a representation we do not know *in general* how to compute field operations in uniform $AC_0[\oplus]$. On the other hand, it is known that for every h of the form $2 \cdot 3^l$ (for some l) the polynomial $x^h + x^{h/2} + 1$ is irreducible over GF(2) (see e.g. [vL99], Theorem 1.1.28). Furthermore, it can be shown that using this *specific* representation of $GF(2^h)$ (for $h = 2 \cdot 3^l$) it is possible to compute field operations over $GF(2^h)$ in uniform $AC_0[\oplus]$ circuits of size poly(h). (Eric Allender (personal communication, June 2004) pointed out field multiplication to us. More involved techniques seem to give field exponentiation with a slightly growing

---

[2] This generator outputs non-Boolean random variables. To get Boolean random variables we can, say, take the least significant bit of $G(x)_i$.

exponent of at most poly(h)). However, these results are bound to a specific field representation, and thus are somewhat unsatisfactory.

In this work we exhibit alternative constructions of k-wise independent generators computable in uniform $AC_0[\oplus]$.

Techniques: Our k-wise independent generators are based on *combinatorial designs*. A combinatorial design is a collection of m subsets $S_1, \ldots, S_m$ of $\{1 \ldots n\}$ with small pairwise intersections. We obtain the i-th output bit of the generator by taking the parity of the bits of the seed indexed by S_i. Roughly speaking, the small intersection size of the sets will guarantee the k-wise independence of the variables. For the case $k = O(1)$, we give a new construction of combinatorial designs with $n = O(\log m)$, which is computable in $AC_0[\oplus]$ (Lemma 3). Following an idea from [HR03] (credited to S. Vadhan), we obtain this construction by concatenating Reed-Solomon codes with a design construction from [Vio04].

To compute our design constructions (specifically the Reed-Solomon code) we work over finite fields as well (as does the classic k-wise independent generator in Equation 1). The difference is that in our constructions we need finite fields of size exponentially smaller, i.e. poly(n) (as opposed to 2^n) . Representations of such fields can be found by brute force (even in uniform AC_0), and results by Agrawal et. al. [AAI+01] show how to perform field operations over fields of size poly(n) by uniform AC_0 circuits of size poly(n).

The question rises whether there is a construction of k-wise independent generators that does not use finite fields of growing size. We argue (see the remark at the end of Section 3) that such a construction is possible using our approach based on combinatorial designs, however we only get a generator explicitly computable in P-uniform $AC_0[\oplus]$ (rather than DLOGTIME-uniform).

ϵ-*Biased Generators in* $AC_0[\oplus]$: Similar to k-wise independent generators, ϵ-biased generators explicitly computable in time poly(n) are known with seed length $n = O(\log m + \log 1/\epsilon)$ [NN90,AGHP92] which is optimal up to constant factors [AGHP92].

Alon et. al. [AGHP92] give three simple constructions, but none of them seems to be implementable in uniform $AC_0[\oplus]$: they either seem to be inherently sequential, or they need large primes, or they need exponentiation over finite fields of size 2^n with exponent of n bits, which does not seem to be computable in uniform $AC_0[\oplus]$ circuits of size poly(n).

Prior to their work, Naor and Naor [NN90] gave another construction. We show that their construction (as long as $\epsilon \geq 1/2^{\log^c(\log m)}$ for some fixed constant c) can be implemented in uniform $AC_0[\oplus]$ (for $\epsilon \leq 1/2^{\log^c(\log m)}$ we obtain seed length $O((\log m)(\log 1/\epsilon))$ with a slight modification of the construction in [NN90]).

Techniques: The construction of [NN90] goes through a few stages. In particular it uses as a component a 7-wise independent generator. By the discussion above, we have such a generator in uniform $AC_0[\oplus]$. [NN90] also needs to compute random walks on an expander graph. We show that, for every fixed c, random walks of length $\log^c n$ on the Margulis expander [Mar73] of size 2^n, can be computed by uniform AC_0 circuits of size poly(n) (where the depth of the circuit depends on c). We also argue that for this specific expander, our result is tight. Bar-Yossef, Goldreich and Wigderson [ZBYW99] show how to compute walks on the same expander but in an *online* model of computation. Their result is incomparable to ours. Ajtai [Ajt93] shows how to compute in uniform AC_0 of size $\text{poly}(n)$ walks of length $\log n$ on expander graphs of size exponentially smaller, i.e. n.

Almost k-Wise Independent Generators in $AC_0[\oplus]$: Using the approach of [NN90] that combines k-wise independent and ϵ-biased generators to get almost k-wise independent generators, we can use our constructions to obtain an (ϵ, k)-wise independent generator $G : \{0,1\}^n \rightarrow \{0,1\}^m$ explicitly computable by uniform $AC_0[\oplus]$ circuits of size $\text{poly}(n, \log m)$. Its seed length is $O(k + \log \log m + \log(1/\epsilon))$ as long as $k = O(\log^c \log m)$ and $\epsilon = \Omega(1/2^{\log^c \log m})$ (for any fixed constant $c \geq 1$). In this range of parameters, this seed length matches that of the best known constructions [NN90,AGHP92] up to constant factors. Due to space limitations this construction is omitted.

Organization: In Section 2 we discuss some preliminaries. In Section 3 we show the connection between combinatorial designs and k-wise independent generators, and we exhibit our construction of combinatorial designs and our constructions of k-wise independent generators in uniform $AC_0[\oplus]$. In Section 4 we give our ϵ-biased generators in uniform $AC_0[\oplus]$. This includes our results about computing walks on expander graphs.

2 Preliminaries

We now define k-wise independent and ϵ-biased generators. Denote the set $\{1, \ldots, m\}$ by $[m]$. For $I \subseteq [m]$ and $G(x) \in \{0,1\}^m$ we denote by $G(x)|_I \in \{0,1\}^{|I|}$ the projection of $G(x)$ on the bits specified by I. Recall that $\oplus$ is the *parity* function, i.e. $\bigoplus_{i \in I}(x) := \sum_{i \in I} x_i \bmod 2$.

Definition 1. *Let $G : \{0,1\}^n \rightarrow \{0,1\}^m$ be a generator.*

- *G is (ϵ, k)-wise independent if for every $M : \{0,1\}^k \rightarrow \{0,1\}$ and $I \subseteq [m]$ such that $|I| \leq k$: $\left| \Pr_{y \in \{0,1\}^k}[M(y) = 1] - \Pr_{x \in \{0,1\}^n}[M(G(x)|_I) = 1] \right| \leq \epsilon$.*

- *G is k-wise independent if it is $(0, k)$-wise independent.*

- *G is ϵ-biased if for every $I \subseteq [m]$: $\left| \Pr_{x \in \{0,1\}^n}[\bigoplus_{i \in I} G(x)_i = 0] - \frac{1}{2} \right| \leq \epsilon$.*

We now define the circuit classes of interest in this paper. AC_0 is the class of constant depth circuits with $\neg, \vee$ and $\wedge$ gates, where $\vee$ and $\wedge$ have unbounded fan-in. $AC_0[\oplus]$ is the class of constant depth circuits with $\neg, \vee, \wedge$ and $\oplus$ gates, where $\vee, \wedge$ and $\oplus$ have unbounded fan-in. A family of circuits $\{C_n\}$ is DLOGTIME-uniform, if there is a Turing machine M running in linear time that decides *the direct connection language of $\{C_n\}$*, which is the language of tuples (t, a, b, n), such that b and a are names of gates in C_n, b is a child of a, and a is of type t [BIS90,Vol99]. *In this paper uniform always means DLOGTIME-uniform.*

Definition 2. *A generator* $G : \{0,1\}^n \to \{0,1\}^m$ *is explicitly computable by uniform* $AC_0[\oplus]$ *(resp.* AC_0*) circuits of size* g*, if there is a uniform* $AC_0[\oplus]$ *(resp.* AC_0*) circuit* C *of size* g *such that* $C(x,i) = G(x)_i$ *for all* $r \in \{0,1\}^n$ *and* $i \in \{0,1\}^{\log m}$*, where* $G(r)_i$ *is the* i*-th output bit of* G*.*

In [MNT90] they essentially prove the following negative result on the ability of AC_0 circuits to explicitly compute (ϵ, k)-wise independent and ϵ-biased generators.

Theorem 1 ([MNT90]). *Fix any constant* $\epsilon < 1/2$*. Let* $G : \{0,1\}^n \to \{0,1\}^m$ *be a generator either* $(\epsilon, 2)$*-wise independent or* ϵ*-biased. Let* C *be a circuit of size* g *and depth* d *such that* $C(r,i) = G(r)_i$ *for every* $r \in \{0,1\}^n, i \in \{0,1\}^{\log m}$*. Then* $\log^{d-1} g \geq \Omega(\log m)$*.*

In some of our constructions we make use of the following result from [AAI+01] about field operations in uniform $AC_0[\oplus]$. We denote by $GF(2^t)$ the field with 2^t elements, and we identify these elements with bit strings of length t.

Lemma 1 ([AAI+01], Theorem 3.2 and proof of Theorem 1.1). *Let* $t = O(\log n)$*. There is a uniform* $AC_0[\oplus]$ *circuit of size* $\mathrm{poly}(n)$ *such that given a polynomial* $p(x) := \sum_{i=0}^{k} a_i x^i$ *of degree* $k = \mathrm{poly}(n)$ *over* $GF(2^t)$*, and* $b \in GF(2^t)$*, computes* $p(b)$*.*

3 k-Wise Independent Generators from Designs

In this section we present a general approach that gives k-wise independent generators from *combinatorial designs*. First we define combinatorial designs. Then we show how to get k-wise independent generators from combinatorial designs. We then turn to the problem of efficiently constructing designs with good parameters.

Definition 3. *[NW94] A* (l, d)*-design of size* m *over a universe of size* s *is a family* $S = (S_1, \ldots, S_m)$ *of subsets of* $\{1, \ldots, s\}$ *that satisfies: (1) for every* i*,* $|S_i| = l$*, and (2) for every* $i \neq j$*,* $|S_i \cap S_j| \leq d$*.*

We now show how to get k-wise independent generators from combinatorial designs.

Lemma 2. *Let* $S = (S_1, \ldots, S_m)$ *be a* $((k + 1)d, d)$*-design of size* m *over a universe of size* n*. Define the generator* $G_S : \{0,1\}^n \to \{0,1\}^m$ *as:* $G_S(r)_i := \bigoplus_{j \in S_i} r_j$*. Then* G_S *is a* k*-wise independent generator.*

Proof. Fix k output bits $i_1, \ldots, i_k$. We show that $G_S(x)_{i_k}$ is uniform over $\{0,1\}$ and independent from $G_S(x)_{i_1}, \ldots, G_S(x)_{i_{k-1}}$. By definition, $G_S(x)_{i_k}$ is the parity of the bits in x indexed by S_{i_k}. Since S is a $((k + 1)d, d)$-design there is $e \in S_{i_k}$ such that $e \notin \bigcup_{0<j<k} S_{i_j}$. Thus the value of $G_S(x)_{i_k}$ is independent from $G_S(x)_{i_1}, \ldots, G_S(x)_{i_{k-1}}$, because its parity includes a bit that is independent from the bits in the parities of $G_S(x)_{i_1}, \ldots, G_S(x)_{i_{k-1}}$. $\qquad\square$

3.1   Combinatorial Designs Computable in Uniform $AC_0[\oplus]$

We now describe a new construction of combinatorial designs of size m with universe size $O(\log m)$. In [HR03] (Section 5.3) it was suggested (following a suggestion of Salil Vadhan) to combine error-correcting codes with combinatorial designs to achieve new designs with good parameters. While their construction does not achieve the parameters that we need, we can use their idea to obtain designs with the desired parameters combining Reed-Solomon codes with a design construction from [Vio04].

Lemma 3. *For every constant $c > 1$ and large enough m there is a family S of $(c^2 \log m, 2c \log m)$-designs of size m over a universe of size $50 \cdot c^3 \log m$. Moreover, there is a uniform $AC_0[\oplus]$ circuit of size $\mathrm{poly}(\log m)$, such that given $i \in \{0,1\}^{\log m}$ computes the characteristic vector of S_i.*

Lemma 3 uses as a component a family of designs with exponentially smaller parameters computable in AC_0 [Vio04].

Lemma 4 ([Vio04]). *For every constant $c > 1$, and for every large enough n there is a family S of $(c \log n, \log n)$-designs of size n over a universe of size $50 \cdot c^2 \log n$. Moreover, there is a uniform AC_0 circuit of size $\mathrm{poly}(n)$, such that given $i \in \{0,1\}^{\log n}$ computes the characteristic vector of S_i.*

Proof (of Lemma 3). First we describe the construction, then we show it is a design with the claimed parameters and then we study its complexity.

Construction: The idea is 'combining' a Reed-Solomon code with the designs given by Lemma 4. Let n be such that $n \log n = c \log m$. Fix a field $\mathbf{F}$ of size n (without loss of generality we assume there is such a field). Let $h := n/c$. Let z be a bit string $z = a_0 \ldots a_h$ where each a_i has $\log(n)$ bits. Define the polynomial p_z over $\mathbf{F}$ as $p_z := \sum_{i=0}^{h} a_i x^i$. Let $b_1, \ldots, b_n$ be an enumeration of all elements of $\mathbf{F}$.

Now consider a family $(D_1, \ldots, D_n)$ of $(c \log n, \log n)$ designs of size n over a universe of size $O(c^2 \log n)$ as guaranteed by Lemma 4.

Then $S = (S_1, \ldots, S_{2^{h \log n}})$ is defined as follows: The characteristic vector of S_z is $D_{p_z(b_1)} \cdots D_{p_z(b_n)}$. Namely it is a concatenation of n characteristic vectors of sets from D.

Analysis: S has $2^{h \log n} = m$ sets, each of size $cn \log n = c^2 \log m$, and the universe size is $O(c^2 n \log n) = O(c^3 \log m)$. We now bound the intersection size. Consider $z \neq z'$. Since p_z and $p_{z'}$ are polynomials of degree at most h, there are at most h distinct $b \in \mathbf{F}$ such that $p_z(b) = p_{z'}(b)$. Whenever $p_z(b) \neq p_{z'}(b)$, we have that $D_{p_z(b)}$ and $D_{p_{z'}(b)}$ are distinct sets in the design D, and thus their intersection is at most $\log n$. Therefore

$$|S_z \cap S_{z'}| \leq hc \log n + (n - h) \log n \leq 2n \log n = 2c \log m$$

Complexity: By Lemma 1, computing $p_z(b)$ can be done by uniform $AC_0[\oplus]$ circuits of size $\mathrm{poly}(n) = \mathrm{poly}(\log m)$, and D can be computed in uniform AC_0 circuits of size $\mathrm{poly}(n) = \mathrm{poly}(\log m)$ by Lemma 4. $\square$

The construction above gives k-wise independent generators with $k = O(1)$ (see Theorem 2). For $k = \omega(1)$ we use as a component a celebrated design construction by Nisan and Wigderson [NW94] which we now state without proof. The complexity of computing this construction follows from Lemma 1.

Lemma 5 ([NW94]). *For every integers ℓ, m such that $\log m \leq \ell \leq m$, there is a $(\ell, \log m)$-design of size m over a universe of size $O(\ell^2)$. Moreover, there is a uniform $AC_0[\oplus]$ circuit of size $\text{poly}(\ell)$, such that given $i \in \{0,1\}^{\log m}$ computes the characteristic vector of S_i.*

We now state our k-wise independent generators.

Theorem 2. *For every large enough m, there is a k-wise independent generator $G : \{0,1\}^n \to \{0,1\}^m$, that is explicitly computable by uniform $AC_0[\oplus]$ circuits of size $\text{poly}(n)$, where,*

1. *$k \geq 2$ is a fixed constant, and $n = O(k^3 \log m)$. Or,*
2. *k is any function (of m), and $n = O(k^2 \log(m) \log(k \cdot \log m))$.*

Proof. For Item (1) we plug the design construction from Lemma 4 (with $c := 2(k+1)$) into Lemma 2. Item (2) is obtained as follows. Consider a $((k+1) \log m, \log m)$-design S of size m over a universe of size $O(k^2 \log^2 m)$ as guaranteed by Lemma 5. By Lemma 2, G_S is a k-wise independent generator, however the seed length of G_S is $n = O(k^2 \log^2 m)$. To reduce the seed length, suppose we have a $(k(k+1) \log m)$-wise independent generator $G' : \{0,1\}^{n'} \to \{0,1\}^n$. We claim that $G \circ G' : \{0,1\}^{n'} \to \{0,1\}^m$ is still a k-wise independent generator. This is because every k output bits of G depend on at most $k(k+1) \log m$ output bits of $G'(x)$. Since G' is $(k(k+1) \log m)$-wise independent, these bits will by uniformly and independently distributed. Now, using for G' the generator of [CG89,ABI86] (Section 1.1, Equation 1) we have $n' = O(k \cdot (k+1) \log(m) \cdot (\log k + \log(\log m))) = O(k^2 \log(m) \log(k \cdot \log m))$. Finally, by Lemma 1, G' is computable by uniform $AC_0[\oplus]$ circuits of size $\text{poly}(n')$. $\qquad\square$

A "Combinatorial" Construction. We note that our approach allows for a construction of k-wise independent generators (for $k = O(1)$) that does not use finite fields of growing size. This is obtained combining, as in Lemma 3, the designs from Lemma 4 with an error correcting code. For the latter, we take an expander code [SS96] based on the Margulis expander discussed in Section 4.1. The resulting generators match the parameters of Theorem 2 (1), and they are computable in P-uniform $AC_0[\oplus]$ circuits of size $\text{poly}(n)$.

4 ϵ-Bias in $AC_0[\oplus]$

In this section we describe our constructions of ϵ-biased generators in uniform $AC_0[\oplus]$. We obtain our constructions by exhibiting uniform $AC_0[\oplus]$ implementations of the ϵ-biased generator due to Naor and Naor [NN90]. Due to space restrictions we do not describe the ϵ-biased generator in [NN90] here. We only point out that it is built combining 7-wise independent generators, 2-wise independent generators and *random walks on expander graphs*. Using our k-wise independent generators in uniform $AC_0[\oplus]$ (Theorem 2), all that is left to do is computing walks on expander graphs in uniform $AC_0[\oplus]$. We now formally state what random walks we need to get ϵ-biased generators.

A family of graphs $\{G_N\}_N$ is a family of d-regular expander graphs if there is a constant $\lambda < d$ such that for every N the graph G_N has N nodes, is d-regular, and the absolute value of the second eigenvalue of its adjacency matrix is at most λ. To get an ϵ-biased generator we need random walks of length $l = O(\log 1/\epsilon)$ on expander graphs of size $2^{O(n)}$. We show in the next section that for every fixed c we can compute walks of length $\log^c n$ on certain expander graphs of size 2^n in uniform AC_0 circuits of size poly(n).

Theorem 3. *There is a family $\{G_N\}_N$ of 8-regular expander graphs such that for every c there is a uniform AC_0 circuit of size poly(n) that, given $v \in G_{2^n}$ and a path w of length $\log^c n$, computes the node $v' \in G_{2^n}$ reached starting from v and walking according to w.*

Using the expander walks in Theorem 3 we get, for $\epsilon \geq 1/2^{\log^c(\log m)}$, an ϵ-biased generator with seed length optimal up to constant factors [AGHP92]. For smaller ϵ, we replace the random walk on the expander with random (independent) nodes in the graph, and obtain a generator with larger seed.

Theorem 4. *For every large enough m, there is an ϵ-biased generator G_ϵ : $\{0,1\}^n \to \{0,1\}^m$, explicitly computable by uniform $AC_0[\oplus]$ circuits of size poly(n), where*

1. $n = O(\log m + \log(1/\epsilon))$ *and* $\epsilon = 1/2^{\log^c(\log m)}$, *for any constant $c > 0$. Or,*
2. $n = O((\log m)\log(1/\epsilon))$ *and* $\epsilon = \epsilon(m)$ *is arbitrary.*

4.1 Expander Walks in AC_0

In this section we prove Theorem 3, i.e. we show that there is an expander graph of size $N = 2^n$ where random walks of length $l = O(\log^c n)$ can be computed by uniform AC_0 circuits of size poly(n) (for every fixed constant c; the depth of the circuit depends on c). We use an expander construction due to Margulis [Mar73] (the needed expansion property was proved later in [GG81,JM87]), which we now recall.

Let $m := \sqrt{N}$ (we assume without loss of generality that m is a power of 2). The vertex set of G_N is $Z_m \times Z_m$, where Z_m is the ring of the integers modulo m. Each vertex v is a pair $v = (x, y)$ where $x, y \in Z_m$. For matrices T_1, T_2 and vectors b_1, b_2 defined below, each vertex $v \in G_N$ is connected to $T_1 v, T_1 v + b_1, T_2 v, T_2 v + b_2$ and the four inverses of these operations.

Theorem 5. *[Mar73,GG81,JM87] The family $\{G_N\}_N$ with $T_1 := \begin{pmatrix} 1 & 1 \\ 0 & 1 \end{pmatrix}$, $T_2 := \begin{pmatrix} 1 & 0 \\ 1 & 1 \end{pmatrix}$, $b_1 := \begin{pmatrix} 1 \\ 0 \end{pmatrix}$ and $b_2 := \begin{pmatrix} 0 \\ 1 \end{pmatrix}$ is a family of 8-regular expander graphs (the absolute value of the second eigenvalue of the adjacency matrix is $5\sqrt{2} < 8$).*

We now prove Theorem 3, with the family $\{G_N\}_N$ of expander graphs from Theorem 5. That is, we show that walks of length $\log^c n$ on G_{2^n} can be computed in uniform AC_0 circuits of size poly(n). We note that this result is tight, i.e. there is no $AC_0[\oplus]$ circuit (uniform or not) of size poly(n) that computes random walks of length $\log^{\omega(1)} n$ on G_{2^n}. This is because computing $\sum_i x_i$ given $x \in \{0,1\}^l$ (which cannot be done in $AC_0[\oplus]$ for $l = \log^{\omega(1)} n$) can be AC_0 reduced to the problem of computing random walks of length $O(l)$ over G_{2^n}. (Proof sketch: Given x, replace '1' in x with a step along the edge associated with T_1, replace '0' in x with a self-loop. Call x' the string thus obtained. Now start at vertex $(0,1)$, and compute a walk according to x'. It is easy to see that the first coordinate of the ending node is $\sum_i x_i$). However this negative result relies on the particular expander graph and on its representation. We do not know if $AC_0[\oplus]$ (or AC_0) circuits of size poly(n) can compute random walks of length $\log^{\omega(1)} n$ on *some* expander on $\omega(n)$ vertices.

We now turn to the proof of Theorem 3. By definition of G_N, there are 8 matrices $\tilde{T}_1, \ldots, \tilde{T}_8$ with constant size entries and 8 vectors $\tilde{b}_1, \ldots, \tilde{b}_8$ with constant size entries such that the set of neighbors of a vertex v are $\{\tilde{T}_i v + \tilde{b}_i : i \leq 8\}$. Thus computing the random walk translates to computing

$$v' = A_l(\ldots(A_3(A_2(A_1 v + a_1) + a_2) + a_3)\ldots) + a_l$$

where for every i, $A_i \in \{\tilde{T}_i : i \leq 8\}$ and $a_i \in \{\tilde{b}_i : i \leq 8\}$. We write this as

$$v' = Av + A' \text{ where } A := A_l \ldots A_2 A_1 \text{ and } A' := A_l \ldots A_2 a_1 + A_l \ldots A_3 a_2 + \ldots + a_l.$$

So we are left with the following tasks: computing the matrixes A, A' and then computing $Av + A'$. We now show how to solve these problems in uniform AC_0. First note that since the matrixes A_i's have constant size entries, the matrixes A and A' have entries of size at most $O(l) = \log^c n$ bits for some c.

To compute $Av + A'$ given A, A' and v we use the facts that (1) sum of two n-bit integers is in AC_0 and (2) multiplication of a n-bit integer by a $\log^c n$-bit integer is in AC_0. For (1) see e.g. [Vol99], Theorem 1.20, for (2) see e.g. [Vol99], Theorem 1.21 (this latter theorem shows multiplication of a n-bit integer by a $\log n$-bit integer. The same techniques give (2)). These circuits can be easily shown to be uniform.

We now show how to compute A, the same techniques give A'.

Lemma 6. *For every fixed constant c there is a uniform AC_0 circuit of size* poly(n) *that, given $l = \log^c n$, 2×2 matrixes $A_1, \ldots, A_l$ with constant size entries, computes $A = \prod_{i \leq l} A_i$.*

Proof. Instead of proving the lemma directly, it is convenient to show that the product of n (as opposed to l) given matrices $A_1, \ldots, A_n$ with constant size entries can be computed in space $O(\log n)$, and then appeal to the following lemma (that can be obtained combining results in [Nep70] and in [BIS90], details omitted).

Lemma 7 ([Nep70] + [BIS90]). *Let L be a language computable in logarithmic space. Then for every constant $c \geq 1$ there is a uniform AC_0 circuit of size $\text{poly}(n)$ that, given x of size $\log^c n$, correctly decides whether x is in L.*

Suppose we are given n matrices $A_1, \ldots, A_n$ with constant size entries. The idea is to first compute $\prod_{i \leq n} A_i$ in Chinese Remainder Representation (CRR) and then convert the CRR to binary (using a result by Chiu, Davida and Litow [CDL01]). More formally, note that every entry of $\prod_{i \leq k} A_i$, for every $k \leq n$, will be at most d^n, for some constant d. By the Chinese Remainder Theorem each number $x \leq d^n$ is uniquely determined by its residues modulo $\text{poly}(n)$ primes, each of length $O(\log n)$. The Prime Number Theorem guarantees that there will be more than enough primes of that length. We refer to such a representation of a number x (i.e. as a list of residues modulo primes) as the CRR of x. Note that to find the primes for the CRR we search among the integers of size $O(\log n)$. This clearly can be done in logarithmic space.

We compute the product of the n matrices modulo one prime at a time, reusing the same space for different primes. To compute the product modulo one prime first note that transforming a matrix in CRR is easy because the matrixes have constant size entries. Each matrix multiplication is a simple sum (because the matrices have constant size entries) and therefore can be computed in logarithmic space. The machine operates in space $O(\log n)$ because it only needs to store a constant number of residues modulo a prime of length $O(\log n)$.

All that is left to do is to convert the product matrix from CRR to binary, and this can be done in space $O(\log n)$ by a result in [CDL01]. □

Acknowledgments

We thank the following people for helpful discussions and valuable suggestions: Eric Allender, Michael Ben-Or, Yonatan Bilu, Oded Goldreich, Alex Healy, Troy Lee, Eyal Rozenman, Ronen Shaltiel, Salil Vadhan, and Avi Wigderson.

References

[AAI+01]   M. Agrawal, E. Allender, R. Impagliazzo, T. Pitassi, and S. Rudich. Reducing the complexity of reductions. *Comput. Complexity*, 10(2):117–138, 2001.

[ABI86]   N. Alon, L. Babai, and A. Itai. A fast and simple randomized algorithm for the maximal independent set problem. *J. of algorithms*, 7:567–583, 1986.

[AGHP92]   N. Alon, O. Goldreich, J. Håstad, and R. Peralta. Simple constructions of almost k-wise independent random variables. *Random Structures Algorithms*, 3(3):289–304, 1992.

[Ajt93]   M. Ajtai. Approximate counting with uniform constant-depth circuits. In *Advances in computational complexity theory (New Brunswick, NJ, 1990)*, pages 1–20. Amer. Math. Soc., Providence, RI, 1993.

[BIS90]   D.A.M. Barrington, N. Immerman, and H. Straubing. On uniformity within NC^1. *J. Comput. System Sci.*, 41(3):274–306, 1990.

[CDL01]   A. Chiu, G. Davida, and B. Litow. Division in logspace-uniform nc^1. *RAIRO Theoretical Informatics and Applications*, 35:259–276, 2001.

[CG89]   B. Chor and O. Goldreich. On the power of two-point based sampling. *Journal of Complexity*, 5(1):96–106, March 1989.

[CGH+85]   B. Chor, O. Goldreich, J. Hastad, J. Friedman, S. Rudich, and R. Smolensky. The bit extraction problem and t-resilient functions. In *26th Annual Symposium on Foundations of Computer Science*, pages 396–407, Portland, Oregon, 21–23 October 1985. IEEE.

[GG81]   O. Gabber and Z. Galil. Explicit constructions of linear size superconcentrators. *JCSS*, 22:407–420, 1981.

[Gol99]   O. Goldreich. *Modern cryptography, probabilistic proofs and pseudorandomness*, volume 17 of *Algorithms and Combinatorics*. Springer-Verlag, Berlin, 1999.

[HR03]   T. Hartman and R. Raz. On the distribution of the number of roots of polynomials and explicit weak designs. *Random Structures & Algorithms*, 23(3):235–263, 2003.

[JM87]   S. Jimbo and A. Maruoka. Expanders obtained from affine transformations. *Combinatorica*, 7(4):343–355, 1987.

[Mar73]   G. A. Margulis. Explicit construction of concentrator. *Problems Inform. Transmission*, 9:325–332, 1973.

[MNT90]   Yishay Mansour, Noam Nisan, and Prasoon Tiwari. The computational complexity of universal hashing. In *Proceedings of the 22nd Annual ACM Symposium on Theory of Computing (May 14–16 1990: Baltimore, MD, USA)*, pages 235–243, New York, NY 10036, USA, 1990. ACM Press.

[Nep70]   V.A. Nepomnjaščiĭ. Rudimentary predicates and turing calculations. *Soviet Mathematics-Doklady*, 11(6):1462–1465, 1970.

[NN90]   J. Naor and M. Naor. Small-bias probability spaces: efficient constructions and applications. In *Proceedings of the Twenty-Second Annual ACM Symposium on the Theory of Computing*, pages 213–223, 1990.

[NW94]   N. Nisan and A. Wigderson. Hardness vs randomness. *Journal of Computer and System Sciences*, 49(2):149–167, October 1994.

[SS96]   M. Sipser and D.A. Spielman. Expander codes. *IEEE Trans. Inform. Theory*, 42:1710–1722, 1996.

[Vio04]   E. Viola. The complexity of constructing pseudorandom generators from hard functions. Technical Report TR04-020, Electronic Colloquium on Computational Complexity, 2004. http://www.eccc.uni-trier.de/eccc. To appear in Journal of Computational Complexity.

[vL99]   J. H. van Lint. *Introduction to coding theory*, volume 86 of *Graduate Texts in Mathematics*. Springer-Verlag, Berlin, third edition, 1999.

[Vol99]   H. Vollmer. *Introduction to circuit complexity*. Springer-Verlag, Berlin, 1999.

[ZBYW99]   O. Goldreich Z. Bar-Yossef and A. Wigderson. Deterministic amplification of space bounded probabilistic algorithms. In *Proceedings of the 14nd Annual IEEE Conference on Computational Complexity*, pages 188–198, 1999.

Distribution-Free Connectivity Testing

Shirley Halevy[1] and Eyal Kushilevitz[2]

[1] Department of Computer Science, Technion, Haifa 3200, Israel
`shirleyh@cs.technion.ac.il`
[2] Department of Computer Science, Technion, Haifa 3200, Israel
`eyalk@cs.technion.ac.il`

Abstract. We consider distribution-free property-testing of graph connectivity. In this setting of property testing, the distance between functions is measured with respect to a *fixed but unknown* distribution D on the domain, and the testing algorithms have an oracle access to random sampling from the domain according to this distribution D. This notion of distribution-free testing was previously defined, and testers were shown for very few properties. However, no distribution-free property testing algorithm was known for any graph property.

We present the first distribution-free testing algorithms for one of the central properties in this area - graph connectivity (specifically, the problem is mainly interesting in the case of sparse graphs). We introduce three testing models for sparse graphs: (1) a model for bounded-degree graphs, (2) a model for graphs with a bound on the total number of edges (both models were already considered in the context of uniform distribution testing), and (3) a model which is a combination of the two previous testing models; i.e., bounded-degree graphs with a bound on the total number of edges. We prove that connectivity can be tested in each of these testing models, in a distribution-free manner, using a number of queries independent of the size of the graph. This is done by providing a new analysis to previously known connectivity testers (from "standard", uniform distribution property-testing) and by introducing some new testers.

1 Introduction

The classical notion of *decision problems* requires an algorithm to distinguish objects having some property $\mathcal{P}$ from those objects which do not have the property. *Property testing* is a relaxation of decision problems, where algorithms are only required to distinguish objects having the property $\mathcal{P}$ from those which are at least "ϵ-far" from every such object. The main goal of property testing is to avoid "reading" the whole object (which requires complexity at least linear in the size of its representation); i.e., to make the decision by reading a small (possibly, selected at random) fraction of the input (e.g., a fraction of size polynomial in $1/\epsilon$ and poly-logarithmic in the size of the representation) and still having a good (say, at least 2/3) probability of success.

The notion of property testing was introduced by Rubinfeld and Sudan [19] and since then attracted a considerable amount of attention. Property testing

K. Jansen et al. (Eds.): APPROX and RANDOM 2004, LNCS 3122, pp. 393–404, 2004.
© Springer-Verlag Berlin Heidelberg 2004

algorithms (or *testers*) were introduced for problems in graph theory (e.g. [1, 9, 11, 17]), monotonicity testing (e.g. [5, 8, 14]) and other properties (e.g. [2, 4, 16]; the reader is referred to surveys by Ron [18], Goldreich [7], and Fischer [6] for a presentation of some of this work, including some connections between property testing and other areas). In these papers, the distance between an object and the property is measured by the fraction of the domain that has to be modified in order for the object to obtain the property. For example, when dealing with graphs, this distance is measured by the number of edges that should be added or removed so that the given graph posses the property.

However, it is natural to consider situations where not all edges are equivalent. That is, removal or addition of some edges may be more expensive than others, depending on various factors of the problem. We are interested in taking these differences between the edges into consideration when measuring the distance between a given graph and the property. In other words, we wish to put different weights (probabilities) on different elements of the domain. This notion of testing, termed *distribution-free testing*, has been previously introduced by [9] and some *negative* results for distribution-free testing of certain graph properties were proved (see below); In [13] the authors presented the first (non-trivial) distribution-free testers for two properties: monotonicity and low-degree polynomials. This raises the question whether distribution-free testers can be found for problems in other areas, and specifically graph properties, where the concept of non-uniform weights for the points of the domain seems natural.

In this paper, we study distribution-free testing of graph properties, and more specifically the connectivity property. In this case, we wish to distinguish, with high probability, connected graphs from those that are far from any such graph. The distance between two graphs is measured with respect to some fixed but unknown distribution D. Testing of graph properties in general, and connectivity in particular, are among the most central problems studied in the field of property testing (with respect to the uniform distribution). Below we survey some of the relevant previous work.

Graph Representations and Previous Results: In the study of graph properties (e.g. [1, 3, 9–12, 15, 17]), most previous work distinguish between two basic models - one is more suitable for dense graphs while the other is more suitable for sparse graphs. In the first model, a graph is represented by its adjacency matrix or by a function $f : V \times V \rightarrow \{0, 1\}$, specifying for every $u, v \in V$ whether there is an edge between u and v. The distance between two graphs in this model, is measured by the fraction of entries in the matrix that should be altered in order to transform one graph to the other. This model was first considered (in the context of property-testing) by Goldreich, Goldwasser and Ron in [9], who proved that a variety of graph-partition problems can be tested with query complexity $poly(\frac{1}{\epsilon})$ (which is independent of the size of the graph). Testing in this graph model was further studied, for example, in [1], where a characterization is given of the first order properties that are testable using a number of queries independent of n.

Though representation as an adjacency matrix seems very natural, in the context of testing it is only relevant when dealing with dense graphs; that is, graphs whose number of edges is $\Omega(n^2)$. This is because every two sparse graphs (that contain $o(n^2)$ edges) are close in the adjacency matrix representation. Hence, another model is needed to deal with sparse graphs. Property-testing for sparse graphs was first considered by Goldreich and Ron [11], who defined the following representation for *bounded-degree* graphs. Let $G = (V, E)$ be an undirected graph with bounded-degree d (i.e., every vertex in V has at most d neighbors) and assume that, for every vertex $v \in V$, there is an ordering among the neighbors of v. The graph G is viewed as a function $f : V \times \{1, \ldots d\} \to V \cup \{\perp\}$ where $f(v, i)$ is set to be u if the ith neighbor of v is u, and is set to be $\perp$ if v has less than i neighbors. Notice that, since the ordering of the neighbors of each vertex is arbitrary, a graph G may have more than one function representing it. In this representation, the distance between two graphs is measured by the fraction of f values that have to be changed in order to transform one function into the other. [11] shows, for example, that in this graph representation, connectivity, k-edge connectivity, k-vertex connectivity and cycle freeness are testable using a constant (i.e., a function of ϵ^{-1}, but independent of n and d) number of queries, and that planarity is testable using $O(d^4 \cdot \epsilon^{-1})$ queries. Properties of graphs represented as incidence lists of length d have also been studied in [3] and [10].

A different representation was presented by Parnas and Ron [17] for testing properties of all sparse graphs, rather than only the bounded-degree ones. In this case, the graphs are represented by incidence lists of varying length. For each vertex v the tester can query both the degree of v and the $i'th$ neighbor of v. The distance between two graphs is then measured by the fraction of edge modifications necessary to obtain the property, defined with respect to an upper bound m on the number of edges in the graph. They present a testing algorithm for graph diameters, in this representation. In addition, they explained how many of the results proved in [11] for the bounded-degree case, can be transformed to this representation. Among these properties are the testing algorithms for k-connectivity. This graph representation was further studied in [15], where the testing model is adjusted to deal with dense graphs as well, by allowing the tester to query whether there exists an edge in the graph between two given vertices. The authors study testing of graph bipartiteness in this model.

Connectivity Testing: Connectivity is a central property in sparse graphs; as mentioned above, it has been previously studied in the context of testing with respect to the uniform distribution. However, although extending the testing models for graph problems to the distribution-free setting seems natural, the problem of connectivity testing in the distribution-free setting has not been dealt with before. Moreover, as mentioned, the only known results for distribution-free testing of graph properties in general are impossibility results for graphs represented by adjacency matrix [9]. They proved that it is impossible to test a variety of partition problems (for which they showed testers with respect to the uniform distribution) in a distribution-free manner. The generalization of the adjacency matrix model to a distribution-free one, is straightforward; on the other hand, every graph is close to be connected in this representation with respect to any distribution[1], hence testing connectivity is trivial. However, unlike the dense case, when dealing with representations for sparse graphs, their generalization to the distribution-free case is not so straightforward, and the testing is not obvious.

Our Contributions. We consider three representations of sparse graphs in the distribution-free setting. These representations are generalizations of the ones presented previously for sparse graphs in the context of uniform distribution testing. We show that in each of these representations, connectivity can be tested in a distribution-free manner, using a number of queries independent of the size of the graph.

- The first representation is for bounded-degree graphs [11]. We show that it is possible to test graph connectivity, in this representation, in a distribution-free manner, using a number of queries which is polynomial in $\frac{1}{\epsilon}$. The test is similar to the one presented in [11] for the uniform setting, however the analysis required for the distribution-free case is different (Section 3).
- The second representation is for graphs with a bound m on the total number of edges [17]. We show that in this representation, it is possible to test connectivity, in a distribution-free manner, whenever $m \geq n(1+c)$ for some constant c using a poly$(\frac{1}{\epsilon})$ number of queries (Subsection 4.1) .
- We then combine the two previous testing models, and deal with bounded-degree graphs with at most m edges. This model was not studied before even in the uniform setting. Hence, we first present a uniform distribution tester for this model for every m; then, using the distribution-free tester for the previous model, we prove that there exists a distribution-free tester for graph connectivity in this model whenever $m \geq n(1+c)$, for some constant c (Subsection 4.2).

2   Definitions

In this section, we define the notions of being ϵ-far from a property $\mathcal{P}$ with respect to a given distribution D, and of distribution-free testing. Denote the range of functions in question by $\mathcal{A}$, and denote by $[n]$ the set $\{1, \ldots, n\}$.

Definition 1. *Let D be a distribution over a set $\mathcal{X}$. The D-distance between functions $f, g : \mathcal{X} \to \mathcal{A}$ is $dist_D(f, g) \stackrel{def}{=} \mathrm{Pr}_{x \sim D}\{f(x) \neq g(x)\}$.*
The D-distance of a function f from a property $\mathcal{P}$ (i.e., the class of functions satisfying the property $\mathcal{P}$) is $dist_D(f, \mathcal{P}) \stackrel{def}{=} \min_{g \in \mathcal{P}} dist_D(f, g)$. We say that f is (ϵ, D)-far from a property $\mathcal{P}$ if $dist_D(f, \mathcal{P}) \geq \epsilon$.

When the distribution in question is the uniform distribution over $\mathcal{X}$, we either use U instead of D or (if clear from the context) we omit any reference to the distribution.

---

[1] Under any distribution measure on $V \times V$, the total probability of the path with minimal probability is at most $\frac{1}{n}$.

Next, we define the notion of distribution-free tester for a property $\mathcal{P}$. Note that the type of queries possible for the tester is determined by the model (as is the case in the literature on testing with respect to the uniform distribution).

Definition 2. *A distribution-free tester for a property $\mathcal{P}$ is a probabilistic oracle machine M, which is given a distance parameter $\epsilon > 0$, and an oracle access to an arbitrary function $f : \mathcal{X} \to \mathcal{A}$ and to sampling of a fixed but unknown distribution D over $\mathcal{X}$, and satisfies the following two conditions:*
1. *If f satisfies $\mathcal{P}$, then $\Pr\{M^{f,D} = Accept\} = 1$.*
2. *If f is (ϵ, D)-far from $\mathcal{P}$, then $\Pr\{M^{f,D} = Accept\} \leq \frac{1}{3}$.*

Note that a more general definition of testers that allows two-sided errors is not needed here; all our testers, like many previously-known testers, have one-sided error and always accept any function that satisfies the property $\mathcal{P}$ in question.

The definition of a uniform distribution tester for a property $\mathcal{P}$ can be derived from the above definition by omitting the sampling oracle (since the tester can sample in the uniform distribution by itself) and by measuring the distance with respect to the uniform distribution.

Finally, notice that since the distribution D in question is arbitrary, it is possible that there are two distinct functions f and g such that $dist_D(f, g) = 0$. Moreover, it is possible that $f \notin \mathcal{P}$ and $g \in \mathcal{P}$. Since the notion of testing is meant to be a relaxation of the notion of decision problems, it is required that the algorithm accepts functions that satisfy $\mathcal{P}$, but may reject functions that have distance 0 from $\mathcal{P}$ and do not satisfy $\mathcal{P}$. In addition, note that the algorithm is allowed to query the value of the input function also in points with probability 0 (which is also the case with membership queries in learning theory)[2]. This definition of distribution-free testing was introduced in [9, Definition 2.1].

3   Connectivity Testing of Bounded-Degree Graphs

The representation for bounded-degree graphs which is the subject of this section is a generalization of [11] to deal with arbitrary probability distributions. First, we generalize the notion of a function $f_G : V \times [d] \to V \cup \{\bot\}$ that represents a degree d (undirected) graph G. Since the d outgoing edges of a vertex v may have different probabilities, it is essential to allow $f_G(v, i)$ to equal $\bot$ even if $f_G(v, i+1) \neq \bot$ (i.e., we cannot assume that a node of out-degree $d' \leq d$ uses the first d' entries in its incidence list).

Definition 3. *A function $f_G : V \times [d] \to V \cup \{\bot\}$ represents a graph $G = (V, E)$, if the following holds:*
(a) for every edge $(u, v) \in E$ there exist unique i_u and i_v such that $f_G(u, i_u) = v$ and $f_G(v, i_v) = u$, and
(b) for every vertex $v \in V$ and $i \in [d]$, if there exists no neighbor u of v such that $f_G(v, i) = u$, then $f_G(v, i) = \bot$. In this case, we say that the pair (v, i) is free in f_G.

---

[2] It is not known whether membership queries are essential in general for testing even in the uniform case (see [18]); such a result is known only for specific problems such as monotonicity testing (see [5]).

Denote by $\mathcal{P}^d$ the class of functions f that represent connected graphs with bounded-degree d. Given a probability measure $D : V \times [d] \to [0,1]$ and a function f_G that represents a graph G, we are interested in the D-distance of f_G from $\mathcal{P}^d$. Hence, we examine possible ways to transform f_G into a function $f_{G'}$ that represents a connected graph G' with degree d (i.e., $f_{G'} \in \mathcal{P}^d$). To do so, we have to connect all the connected components of G; that is, if the connected components of G are $C_1, \ldots, C_k$, then we wish to add an edge between C_i and C_{i+1} while, at the same time, preventing the graph's degree from exceeding d, and keeping the connectivity of each of the connected components. In addition, we want the total probability of the modified (i.e., added and removed) edges to remain small. Therefore, we look, in any connected component C_i of G, either for an edge that is unnecessary to the connectivity, or for two distinct free pairs (v, i_v) and (u, i_u). In other words, we find a list of possible connecting points between the connected components of G. We define more accurately the notion of such a list. For this purpose, we need the following definition.

Definition 4. *Let $G = (V, E)$ be a graph, let $e = (u, v) \in E$ be an edge, and let C be a connected component of G containing e. We say that e is* redundant, *if removing e from E does not affect the connectivity of C.*

The following definition formally states the idea of a list of possible connections between the connected components of a graph. The $m'th$ edge in this list, (u_m, v_m), will connect the components C_m and C_{m+1}.

Definition 5. *Let f_G be a function that represents a graph G with bounded-degree d, and let $C_1, \ldots, C_k$ be the connected components of G. We say that a list of k quad-tuples $L = ((u_1, i_1), (v_1, j_1)), \ldots, ((u_k, i_k), (v_k, j_k))$ connects f_G if the following holds, for every $m \in [k]$:*
1. $u_m \in C_m$ and $v_m \in C_{m+1}$ (in case $m = k$, then $m + 1$ refers to 1).
2. $(v_m, j_m) \neq (u_{m+1}, i_{m+1})$.
3. one of the following holds: (a) (u_{m+1}, i_{m+1}) and (v_m, j_m) are free in f_G; or (b) (u_{m+1}, v_m) is redundant in C_{m+1}, and $f_G(u_{m+1}, i_{m+1}) = v_m$, $f_G(v_m, j_m) = u_{m+1}$ [3].

Indeed, given a list L that connects f_G, we can construct from f_G a function $f_{G'}$, that represents a connected degree d graph G', by setting $f_{G'}(u_m, i_m) = v_m$ and $f_{G'}(v_m, j_m) = u_m$ (notice that the graph G' is obtained from G by removing at most one redundant edge from each connected component, thereby not damaging the component's connectivity).

Let f_G be a function that represents a graph G. For every edge $(u, v) \in E$, define the D-probability of the edge (u, v) to be the total probability of its endpoints with respect to D (it can be seen as the cost of a change in the edge (u, v), since a change in the edge (u, v) requires a change in the incidence lists of both u and v); equivalently, let i_u and i_v be such that $f_G(u, i_u) =$

---

[3] The above definition actually describes a cycle connecting all the connected components of the graph. It is possible to define a path from C_1 to C_k; however, in such a case, not all the quad-tuples are symmetric, causing the definition to be slightly more complicated.

v and $f_G(v, i_v) = u$, then the D-probability of the edge (u,v) is $D(u,v) \overset{\text{def}}{=} D(u, i_u) + D(v, i_v)$. In addition, define the D-probability of a vertex $v \in V$ to be $D(v) \overset{\text{def}}{=} \Sigma_{i \in [d]} D(v, i)$ and the D-probability of a list L to be $D(L) \overset{\text{def}}{=} \Sigma_m (D(u_m, i_m) + D(v_m, j_m))$.

Observation 1: If a list L connects f_G, then the D-distance of f_G from $\mathcal{P}^d$ is at most $D(L)$.

We present an $O(\epsilon^{-2})$ distribution-free tester for connectivity of bounded-degree graphs that, given access to random sampling of $V \times [d]$ according to the distribution D and to membership queries of a function f_G, distinguishes, with probability at least $\frac{2}{3}$, between the case that f_G represents a connected graph with bounded-degree d (i.e., $f_G \in \mathcal{P}^d$), and the case that f_G is (ϵ, D)-far from $\mathcal{P}^d$. For lack of space, we deal here only with the case where d is at least 3. An algorithm for the special case $d = 2$ can be constructed using similar arguments and will appear in the full version of the paper.

The tester is similar to the one presented in [11] for the uniform distribution, in the sense that it also looks for small connected components in G. However, while the original analysis is based on the fact that the number of small connected components in a graph G' which is far from being connected, is big, this claim no longer holds when dealing with arbitrary distributions[4]. Hence, a whole new analysis is required for the distribution-free case. A natural generalization of the tester for the uniform case may seem to be seeking for connected components where the total probability of their vertices is small (note that, there is no correlation between the number of vertices in a connected component and their total probability). However, there are some drawbacks to this approach. First of all, in the distribution-free setting we have no knowledge of the actual probability D of the sampled points, and are only allowed to sample the domain according to D; hence, we are only able to estimate their probability. In addition, the size of such components may be very large, therefore finding out whether two vertices lie in the same component may not be possible in time independent of the size of the graph. Thus, a different generalization is required.

Algorithm - connectivity(ϵ, d)

Repeat $\frac{4}{\epsilon}$ times:

- Choose, using the sampling oracle, $(v, i) \sim D$.
- Perform BFS starting from v until $\frac{96}{\epsilon d}$ vertices have been reached, or no new vertex can be reached. If the search was ended since no new vertex can be reached, return **FAIL**[5].

return **PASS**

---

[4] Consider for example a graph G that consists of two connected components, one of size $|V| - 1$ and the other containing a single vertex v_0, and the distribution D is set to be $D(v_0, i) = \frac{1}{d}$ for every $1 \le i \le d$. In this case, G is $(\frac{1}{d}, D)$-far from being connected, while it contains only one small connected component

[5] Assume $|V| \ge \frac{96}{\epsilon d}$. If $|V| < \frac{96}{\epsilon d}$, then it is possible to *decide* whether the graph is connected in query complexity $O(\epsilon^{-1})$.

To prove the correctness of this algorithm, we use the following notation. Let f_G be a function that represents a graph G, and let C be a connected component of G; denote by n_C the number of vertices in C and define $w_C = \sum_{v \in C} D(v)$ (i.e., the total probability of all the vertices in C).

The next lemma states that if f_G is (ϵ, D)-far from $\mathcal{P}^d$, then the total weight of the small connected components in the graph G is at least $\frac{\epsilon}{2}$. Denote by $S_D(f_G)$ the total probability of connected components in G whose size is smaller than $\frac{96}{\epsilon d}$; i.e., $S_D(f_G) \stackrel{\text{def}}{=} \sum_{C : n_C < \frac{96}{\epsilon d}} w_C$.

Lemma 1. *Let f_G be a function (ϵ, D)-far from $\mathcal{P}^d$. Then, $S_D(f_G) \geq \frac{\epsilon}{2}$.*

To prove the lemma, we show that if $S_D(f_G) < \frac{\epsilon}{2}$, then it is possible to construct a list L that connects f_G, such that $D(L) < \epsilon$. Then, by Observation 1, we deduce that $dist_D(f_G, \mathcal{P}^d) < \epsilon$. The full proof is omitted and will appear in the full version of the paper.

We can now prove the correctness of the algorithm *connectivity*.

Theorem 1. *Algorithm connectivity(ϵ, d) is a distribution-free tester for connectivity of bounded-degree graphs with degree $d \geq 3$; its query complexity is $O(\epsilon^{-2})$.*

Proof. By the definition of a function representing a graph, in every stage of the BFS the algorithm looks for the next neighbor of the current vertex, which causes the BFS step to cost $O(d)$ queries (in the worst case). Since there are $O(\frac{\epsilon^{-2}}{d})$ BFS steps, the query complexity is as required. By the definition of the algorithm, if $f_G \in \mathcal{P}^d$, then it is accepted by the algorithm with probability 1. Let f_G be (ϵ, D)-far from $\mathcal{P}^d$. By Lemma 1, $S_D(f_G) \geq \frac{\epsilon}{2}$. Hence, the probability of the algorithm to randomly pick a pair (v, i) such that $v \in C$ and C is a connected component of size less than $\frac{96}{\epsilon d}$, is at least $\frac{\epsilon}{2}$. Therefore, the probability that the algorithm fails to find a small component is at most $(1 - \frac{\epsilon}{2})^{\frac{4}{\epsilon}} \leq \frac{1}{e}^2 \leq \frac{1}{3}$.   □

4   Alternative Testing Models for Sparse Graphs

In this section, we consider additional testing models for sparse graphs, as discussed in the Introduction. For each of these testing models, we define the class of functions that represent graphs, the queries that the tester is allowed to ask, and prove the existence of a distribution-free tester in that model. Section 4.1 deals with graphs with a bounded number of edges. Then, Section 4.2 deals with bounded-degree graphs with a bound also on the total number of edges.

4.1   Graphs with a Bound on the Number of Edges

The testing model described in this section is a generalization of the model presented in [17] for testing with respect to the uniform distribution: There is no upper bound on the degree, only a bound m on the total number of edges in the graph; as in the uniform model, the tester can query the degree of a given vertex v or the $i'th$ neighbor of v. As explained in [17], connectivity can be tested with respect to the uniform distribution.

To generalize this model to the distribution-free setting, we view a graph $G = (V, E)$ as a function $f : [m] \rightarrow \binom{V}{2} \bigcup \{\perp\}$, indicating for each $i \in [m]$ the $i'th$ entry in the edge list if such an edge exists, or $\perp$ if that entry is empty. As before, a graph G can have more than one representation, due to different orderings of the edge list. Denote by $\mathcal{P}_m$ the class of functions that represent connected graphs with at most m edges.

Note that, the differences between the weights of the entries in the list are more significant when removing edges (in order to make room for new edges) and less relevant when talking about edges additions (where one can choose the cheapest entry).

The next lemma, shows that if $m \geq (1 + c)n$ for some constant c, then it is possible to use any tester known in this model for the uniform distribution, to construct a distribution-free tester.

Lemma 2. *Let $m \geq (1 + c)n$, for some constant c. Then, there exists a distribution-free tester for $\mathcal{P}_m$ with query complexity of $O(\epsilon^{-3})$.*

To prove the lemma, we show that if f_G is ϵ-far from $\mathcal{P}_m$ with respect to any distribution D, then it is also $g(\epsilon)$-far from $\mathcal{P}_m$ with respect to the uniform distribution, when $g(\epsilon)$ is a linear function in ϵ. (proof omitted.)

4.2 Graphs with a Bound on the Number of Edges and on the Degree

Next, we consider a combination of the two previous testing models. One may think of graphs where there are at most m edges and the degree of each vertex is at most d. This model can be seen as an intermediate model between the above two models, since it contains all bounded-degree graphs and is contained in the set of all graphs with a bound on the total number of edges. As in the previous model, graphs are represented by functions $f : [m] \rightarrow \binom{V}{2} \bigcup \{\perp\}$, and the distribution is defined over the list of edges, i.e. the set $[m]$. However, in this case we are only interested in functions that represent degree-d graphs with at most m edges. Denote the class of these functions by $\mathcal{P}_m^d$. As in the previous model, the tester is able to query the $i'th$ neighbor of a given vertex v.

Though we believe that this is a very natural testing model for sparse graphs, it was not considered before even in the uniform setting. We prove existence of both uniform and distribution-free testers for $\mathcal{P}_m^d$, using similar approaches. When dealing with the uniform distribution, the distance between two functions is measured as in [17]; that is, by the fraction of edge modifications necessary to transform the graph into a connected one, measured with respect to m.

Lemma 3. *There exists a uniform tester for $\mathcal{P}_m^d$ with query complexity $poly(\frac{1}{\epsilon})$.*

To prove the above lemma, we show that if a function f_G is ϵ-close to $\mathcal{P}_m$, then it is 3ϵ-close to $\mathcal{P}_m^d$ (note that both testing models use the same graph representation). It follows, that if f_G is ϵ-far from $\mathcal{P}_m^d$, it is $\frac{\epsilon}{3}$-far from $\mathcal{P}_m$. Hence, the existence of a uniform tester for $\mathcal{P}_m^d$ follows from the existence of such a tester for $\mathcal{P}_m$ [17]. The full details are omitted for lack of space.

We now turn to the distribution-free setting. Assume that $d \geq 3$; the proof for $d = 2$ is similar.

Theorem 2. *Let $m \geq (1+c)n$, for some constant c. There exists a distribution-free tester for $\mathcal{P}_m^d$ with query complexity $poly(\frac{1}{\epsilon}, d)$.*

To construct the distribution-free tester, we want to use the tester presented for bounded-degree graphs in Section 3. Hence, we first need a way to transform graphs in our representation to the bounded-degree representation. We describe how, given a function $f_G : [m] \rightarrow \binom{V}{2} \bigcup \{\bot\}$ that represents a graph G with degree at most d, we transform it to a function $T(f_G) : V \times [d] \rightarrow V \bigcup \{\bot\}$ that represents the same graph G. The transformation is as follows: for every $v \in V$, assume an ordering on the edges outgoing from v (i.e., the same ordering according to which the tester's queries are answered). For every $i \in [d]$, define the value of $T(f_G)(v, i) = u$ if u is the i'th neighbor of v and $\bot$ if v has less than i neighbors. Note that necessarily if $T(f)(v, i) = \bot$, then $T(f)(v, i+1) = \bot$.

Note that, given access to queries for the function f_G, one can simulate queries for the function $T(f_G)$ (this is due to the choice of the same ordering on the outgoing edges of a vertex v).

However, in order to use the distribution-free tester for bounded-degree graphs, we need to also find a way to translate distribution measures from our model to the bounded-degree model.

Given the functions f_G and $T(f_G)$, we describe a transformation from the distribution D defined over $[m]$ to a distribution $T(D, f_G)$ defined over $V \times [d]$. Define W to be the total probability of all non-empty entries in $[m]$ under the distribution D. That is, $W = \sum_{i:f(i)\neq\bot} D(i)$. The distribution $T(D, f_G)$ is now defined as follows. For every $v \in V$ and $i \in [d]$:

• If $T(f_G)(v, i) = u$ and (v, u) is the j'th edge in the list, define $T(D, f_G)(v, i) = D(j)/2W$. If $W = 0$, then $T(D, f_G)$ is defined to be the uniform distribution over all non-empty entries in $[m]$.

• Otherwise, define $T(D, f_G)(v, i) = 0$.

Note that, sampling of $V \times [d]$ with respect to the distribution $T(D, f_G)$ using sampling of $[m]$ according to the distribution D is not possible. This is due to the possibility to sample empty entries in $[m]$, that cannot be translated to a pair (v, i) in $V \times [d]$.

Sampling of $T(D, f_G)$: Set $c = 8$.

• Sample, at most $\frac{8c}{\epsilon} \log \frac{1}{\epsilon}$ times, the distribution D until the sampled value i is such that $f(i) \neq \bot$.

• If all $\frac{8c}{\epsilon} \log \frac{1}{\epsilon}$ samples were of empty entries, return FAIL.

• Otherwise, let i be the sampled non-empty entry, and assume $f(i) = (v, u)$. Find the location of u in v's neighbors list, denote it by i_v. Similarly, find v's place in u's neighbors list, denoted by i_u. Then, with probability $\frac{1}{2}$ return (v, i_v), otherwise return (u, i_u).

Each sampling of $T(D, f_G)$, using the above sampling procedure, requires $poly(\frac{1}{\epsilon}, d)$ queries. In addition, if the sampling procedure indeed returns a value (v, i), then $(v, i) \sim T(D, f_G)$. However, there is also the possibility that the sampling procedure returns FAIL, which does not happen in direct sampling of $T(D, f_G)$. Let A be a connectivity distribution-free testing algorithm in the bounded-degree model. Denote by A' the algorithm obtained from A by adding the following rule: if any sampling query of A, during its run on a function f, was answered FAIL, then A accepts f. We will show that we can still use the tester after this modification for our construction.

Assume for now, that we actually have an ability to sample $T(D, f_G)$. One may wonder whether the distribution-free tester for $\mathcal{P}^d$, that was presented in Section 3, can be used as a tester for $\mathcal{P}^d_m$ as well. It is easy to see, however, that in some cases the function $T(f_G)$ is very close to $\mathcal{P}^d$ with respect to $T(D, f_G)$, while f_G is far from $\mathcal{P}^d_m$ with respect to D [6]. Therefore, a different test is required. The distribution-free tester for $\mathcal{P}^d_m$ is as follows:

Algorithm Connectivity$_m(\epsilon, d)$:
Repeat twice:
 Run connectivity'(ϵ, d) (obtained from the algorithm presented in
  Section 3) on $T(f_G)$ using the sampling procedure for $T(D, f_G)$ with
  distance $\frac{\epsilon}{2}$.
 Run the distribution-free tester for $\mathcal{P}_m$ on f_G and D with distance $\frac{\epsilon}{2}$.
 return PASS

Theorem 2 now follows immediately from the correctness of the above tester. This, in turn, is based on Lemma 4 and Lemma 5 below, and the existence of the distribution-free testers for $\mathcal{P}_m$ (Section 4.1). As in Section 3, we denote by $S_D(f)$ the total weight of the small connected components in G. That is, the total weight of connected components of size less than $\frac{96}{\epsilon d}$. Lemma 4 below shows that if the total weight of small connected components in G, measured with respect to D, is large then, with high probability (over the possible runs of the tester), the function f_G fails to pass the first stage of the above tester. Lemma 5 then shows that if the total weight of small connected components in G is small, and the function is far from $\mathcal{P}^d_m$, then with high probability, over the possible runs of the tester, it fails to pass the second stage of the above tester.

Lemma 4. _Let $f_G : [m] \to \binom{V}{2} \bigcup \{\perp\}$ be a function and D a distribution over $[m]$. If $S_D(f_G) > \frac{\epsilon}{4}$, then the probability that f_G passes the first stage of the tester is at most $\frac{2}{9}$._

The proof of this Lemma uses the following observation, which is based on the analysis of the tester for $\mathcal{P}^d$ (Section 3).

Observation 2: Let f_G be a function in the bounded-degree model. If $S_D(f_G) > \frac{\epsilon}{2}$, then the probability that f passes the distribution-free test presented in Section 3, with respect to a distribution D over $V \times [d]$ with distance parameter ϵ, is at most $\frac{1}{3}$.

---

[6] Let $m = nd/4$, let D be the uniform distribution over $[m]$, and consider the following graph G on n nodes. G consists of $n/2$ isolated vertices and a connected component of size $n/2$ that contains $nd/4$ edges. It is easy to see that $dist_D(f_G, \mathcal{P}^d_m) \geq 1/d$, while $dist_{T(D,f_G)}(T(f_G), \mathcal{P}^d) = 4/nd$.

Lemma 5. *Let* $f : [m] \rightarrow \binom{V}{2} \bigcup \{\perp\}$ *be a function and* D *a distribution over* $[m]$. *If* f *is* $\frac{\epsilon}{2}$-*close to* $\mathcal{P}_m$ *with respect to* D, *and* $S_D(f_G) \leq \frac{\epsilon}{4}$, *then* f *is* ϵ-*close to* $\mathcal{P}_m^d$.

References

1. N. Alon, E. Fischer, M. Krivelevich, and M. szegedy, *Efficient testing of large graphs. FOCS 1999*, pp. 656–666.
2. M. Blum, M. Luby, and R. Rubinfeld, *Self testing/correcting with applications to numerical problems, Journal of Computer and System Sceince* 47:549–595, 1993.
3. A. Bogdanov, K. Obata, and L. Trevisan, *A lower bound for testing 3-colorability in bounded-degree graphs. FOCS 2002*, pp. 93-102.
4. A. Czumaj and C. Sohler, *Testing hypergraph coloring, ICALP 2001*, 493–505.
5. E. Ergün, S. Kannan, R. Kumar, R. Rubinfeld, and M. Viswanathan, *Spot-checkers, Journal of Computing and System Science*, 60:717–751, 2000 (a preliminary version appeared in STOC 1998).
6. E. Fischer, *The art of uninformed decisions: A primer to property testing, The Computational Complexity Column of The bulletin of the European Association for Theoretical Computer Science*, 75:97–126, 2001.
7. O. Goldreich, *Combinatorial property testing – a survey, In: Randomized Methods in Algorithms Design* (P. Pardalos, S. Rajasekaran and J. Rolim eds.), AMS-DIMACS pages 45–61, 1998.
8. O. Goldreich, S. Goldwasser, E. Lehman, D. Ron, and A. Samorodnitsky, *Testing Monotonicity, Combinatorica*, 20(3):301–337, 2000 (a preliminary version appeared in FOCS 1998).
9. O. Goldreich, S. Goldwasser, and D. Ron, *Property testing and its connection to learning and approximation, Journal of the ACM*, 45(4):653–750, 1998 (a preliminary version appeared in FOCS 1996).
10. O. Goldreich and D. Ron, *On testing expansion in bounded-degree graphs. In Electronic Colloquium on Computational Complexity* 7(20), 2000.
11. O. Goldreich and D. Ron, *Property testing in bounded-degree graphs. STOC 1997*, pp. 406–415.
12. O. Goldreich and L. Trevisan, *Three theorems regarding testing graph properties. FOCS 2001*, pp. 302–317.
13. S. Halevy and E. Kushilevitz, *Distribution-free property testing. RANDOM-APPROX 2003*, pp. 341–353.
14. S. Halevy and E. Kushilevitz, *Testing monotonicity over graph products. To appear in ICALP 2004*.
15. T. Kaufman, M. Krivelevich, and D. Ron, *Tight bounds for testing bipartiteness in general graphs. RANDOM-APPROX 2003*, pp. 341–353.
16. Y. Kohayakawa, B. Nagle, and V. Rodl, *Efficient testing of hypergraphs. ICALP 2002*, pp. 1017–1028.
17. M. Parnas, and D. Ron, *Testing the diameter of graphs, RANDOM-APPROX 1999*, pp.85–96.
18. D. Ron, *Property testing (a tutorial), In: Handbook of Randomized Computing* (S.Rajasekaran, P. M. Pardalos, J. H. Reif and J. D. P. Rolin eds), Kluwer Press (2001).
19. R. Rubinfeld and M. Sudan, *Robust characterization of polynomials with applications to program testing, SIAM Journal of Computing*, 25(2):252–271, 1996. (first appeared as a technical report, Cornell University, 1993).

Testing the Independence Number
of Hypergraphs

Michael Langberg[*]

Department of Computer Science, California Institute of Technology
Pasadena, California 91125
mikel@cs.caltech.edu

Abstract. A k-uniform hypergraph G of size n is said to be ε-far from having an independent set of size ρn if one must remove at least εn^k edges of G in order for the remaining hypergraph to have an independent set of size ρn. In this work, we present a natural *property testing* algorithm that distinguishes between hypergraphs which have an independent set of size $\geq \rho n$ and hypergraphs which are ε-far from having an independent set of size ρn. Our algorithm is natural in the sense that we sample $\simeq c(k)\frac{\rho^{2k}}{\varepsilon^3}$ random vertices of G, and according to the independence number of the hypergraph induced by this sample, we distinguish between the two cases above. Here $c(k)$ depends on k alone (*e.g.* the sample size is independent of n). To the best of our knowledge, property testing of the independence number of hypergraphs has not been addressed in the past.

1 Introduction

A k-uniform hypergraph is a hypergraph $G = (V, E)$ in which each (hyper) edge is of size exactly k. An independent set I in G, is a subset of vertices that do not include any edges (*i.e.* there does not exist an edge $\{v_1, \ldots, v_k\} \in E$ for which $v_i \in I$ for all $i \in \{1, \ldots, k\}$). The size of the maximum independent set in G is denoted by $\alpha(G)$, and referred to as the independence number of G. Consider a k-uniform hypergraph G of (vertex) size n, with a maximum independent set I of size $\alpha(G)$. Let H be a random subgraph of G of size s (*i.e.* H is the subgraph induced by a random subset of vertices of G of size s). In this work we study the independence number of H.

For example, if it is the case that $\alpha(G) \geq \rho n$ for some parameter ρ, it is not hard to verify that with constant probability $\alpha(H)$ will be at least ρs. This follows from the fact that the expected size of $H \cap I$ is at least ρs. However, if $\alpha(G) < \rho n$, and this is our only assumption on G, we cannot hope for a positive bound on the probability that $\alpha(H) < \rho s$ (unless s is equal to n).

Thus, in the latter case, we strengthen our assumption on G to hypergraphs G which not only satisfy $\alpha(G) < \rho n$ but are also *far* from having an independent

---

[*] Part of this work was done while visiting the School of Computer Science, Tel Aviv University, Tel Aviv 69978, Israel.

K. Jansen et al. (Eds.): APPROX and RANDOM 2004, LNCS 3122, pp. 405–416, 2004.

set of size ρn (we defer defining the exact notion of "far" until later in this discussion). That is, given a hypergraph G which is *far* from having an independent set of size ρn, we study the minimal value of s for which (with high probability) a random subgraph of size s does not have an independent set of size ρs. This question (and many other closely related ones) have been studied in (2-uniform hyper) graphs in [GGR98] under the title of *property testing*.

Property Testing: Let $\mathcal{C}$ be a class of objects, and $\mathcal{P}$ a property of objects from $\mathcal{C}$. Property testing addresses the problem of distinguishing between elements $c \in \mathcal{C}$ which have the property $\mathcal{P}$ and elements that are *far* from having the property $\mathcal{P}$. The aim is to construct efficient (randomized) distinguishing algorithms that sample the given element c in relatively few places. The notion of property testing was first presented by Rubinfeld and Sudan in [RS96] where the testing of algebraic properties of functions was addressed. Goldreich, Goldwasser, and Ron [GGR98] later initiated the study of combinatorial objects in the context of property testing. In their work they studied (2-uniform) graphs and considered several fundamental combinatorial graph properties related to the independence number, chromatic number, size of maximum cut, and size of the maximum bisection of these graphs. Since, many papers have addressed the notion of property testing, both in the context of functions and in the combinatorial setting (*e.g.* see surveys [Gol98,Ron01,Fis01]).

Property testing of hypergraphs has also been studied in the past. Czumaj and Sohler [CS01] initiated this line of study when analyzing the property of being ℓ colorable. Colorability, and other properties of k-uniform hypergraphs (that can be phrased as a Max-k-CNF formula) were also studied in [AdlVKK02,AS03]. In this work, we consider testing the independence number of hypergraphs. To the best of our knowledge, this property has not been addressed in the past (in the context of hypergraphs).

Testing the Independence Number: Goldreich, Goldwasser, and Ron [GGR98] study property testing of the independence number of (2-uniform) graphs. In [GGR98] a graph G of size n is said to be ε-far from having an independent set of size ρn if any set of size ρn in G has at least εn^2 induced edges. It was shown in [GGR98] that if G is ε-far from having an independent set of size ρn then with high probability a random subgraph of size $s = \frac{c \log{(1/\varepsilon)}\rho}{\varepsilon^4}$, for a sufficiently large constant c, does not have an independent set of size ρs. The sample size s was later improved in [FLS02] to $c\frac{\rho^4}{\varepsilon^3} \log{\left(\frac{\rho}{\varepsilon}\right)}$ (again c is a sufficiently large constant). It is not hard to verify that this implies a (two-sided error) property testing algorithm for the independence number of G. Namely, given a graph G, one may sample a random subgraph H of G of size s, and exhaustively compute $\alpha(H)$. On one hand, if G happens to have an independent set of size ρn, then with some constant probability p_1 the independence number of H will be at least ρs (recall that the expected value of $\alpha(H)$ is at least ρs). On the other hand, if G is ε-far from having an independent set of size ρn then, as mentioned above, with at most some small probability p_2 (which is set to be smaller than p_1) it is the case that $\alpha(H) \geq \rho s$.

Our Results: In this work, we address the study of hypergraphs in the above context. Namely, given a k-uniform hypergraph $G = (V, E)$ of size n, we show that if one must remove at least εn^k edges of G in order for the remaining hypergraph to have an independent set of size ρn (*i.e.* G is ε-far from having an independent set of size ρn) then a random subgraph H of G of size $s \simeq c(k) \frac{\varrho^{2k}}{\varepsilon^3}$ satisfies $\alpha(H) < \rho s$ with high probability. Here $c(k)$ depends on k alone, which implies that the sample size s is independent of n. We also show a lower bound for the size of s of value $c'(k) \frac{\varrho^{2k-1}}{\varepsilon^2}$ (where $c'(k)$ is yet an additional function which depends on k alone).

Definition 1 *Let A be a subset of V. Define $E(A)$ to be the set of edges in the hypergraph induced by A.*

Definition 2 *Let $\rho < 1$. A k-uniform hypergraph $G = (V, E)$ is said to be $\langle \rho, \varepsilon \rangle$-connected iff every subset A of V of size ρn satisfies $|E(A)| \geq \varepsilon n^k$ (i.e. the number of edges in the subgraph induced by A is greater than εn^k).*

Theorem 3. *Let G be a k-uniform hypergraph. Let H be a random sample of G of size $s \geq c 2^k k! \frac{\varrho^{2k}}{\varepsilon^3} \log \left(\frac{\varrho}{\varepsilon} \right)$ for a large constant c.*

1. *If G has an independent set of size ρn, then with probability $\geq 1/4$ the subgraph H will have an independent set of size ρs.*
2. *If G is $\langle \rho, \varepsilon \rangle$-connected then with probability $\leq 1/20$ the subgraph H will have an independent set of size ρs.*

A few remarks are in place. First of all, in the above theorem we consider only hypergraphs which are k-uniform. Theorem 3 can be extended to hypergraphs in which any edge is of size at most k. This follows from the fact that any sufficiently large hypergraph which is $\langle \rho, \varepsilon \rangle$-connected with edges of size k and smaller contains a subgraph (on the same vertex set) which is k-uniform and $\langle \rho, \varepsilon/2 \rangle$-connected. Secondly, throughout this work we analyze the properties of random subsets H which are assumed to be *small*. Namely, we assume that the value of s (as stated in Theorem 3) and the parameters ρ, ε and n satisfy (a) $s < c\sqrt{n}$ and (b) $s < c\rho n$ for a sufficiently small constant c. These assumptions typically hold in the context of property testing. Finally, the constants $1/4$, and $1/20$ presented above are not tight (we have not made an attempt to find the tightest possible constants).

Proof Techniques: Given a k-uniform hypergraph G with vertex size n which is $\langle \rho, \varepsilon \rangle$-connected, the bulk of our work addresses the study of the minimal value of s for which a random subgraph H of G satisfies $\alpha(H) < \rho s$ with high probability. This is done by analyzing the probability that a random subset H of G of size s satisfies $\alpha(H) < \rho s$ (as a function of s, ρ and ε). Our proof structure is strongly based on that appearing in (the journal version) of [FLS02], and proceeds as follows. Given a sample size s, we start by bounding the probability that a random subset R of G of size $\ell \geq \rho s$ is an independent set. As any subset R of H is random in G, one may bound the probability that $\alpha(H) \geq \rho s$ using the standard union bound on all subsets R of H of size $\geq \rho s$. We analyze this naive strategy and show that it only suffices to bound the probability of a slightly stronger condition than the condition $\alpha(H) \geq \rho s$. Namely, using this scheme, we bound the probability for which $\alpha(H) \geq 3\rho s$ (instead of exactly ρs). The naive strategy is then enhanced in order to bound the probability that $\alpha(H) > \delta \rho s$ as a function of δ for any $\delta > 1$. This now suffices to bound the probability that $\alpha(H) \geq \rho s$ (using a few additional ideas).

Applying the proof technique of [FLS02] to the case of k-uniform hypergraphs involves several difficulties. Roughly speaking, these are overcome by considering for each vertex v in the given graph G, a weighted set system of all subsets of V which *share* an edge with v. This set system plays the role of the standard "set of neighbors" considered when studying (two-uniform) graphs. Specifically, for all vertices v we consider the set system $\mathcal{A}_v$ consisting of sets $\alpha = \{u_1, \ldots, u_i\}$ for which there exists an edge e in G which includes both the vertex v and the set α. The weight applied to each subset is fixed to be a function of the subset size. Ideas along this line have been used in the past in the study of hypergraphs (*e.g.* [AS03]).

Organization: The remainder of the paper is organized as follows. In Section 2 we present the proof of Theorem 3. In Section 3 we present our lower bounds on the sample size s. As mentioned above, our proof uses many of the arguments presented in [FLS02]. During our presentation, we repeat quite a few of these arguments, in aim to make the paper comprehensible. Due to space limitations, some of our results appear without detailed proof. Most of the details can be found in an extended version of this paper [Lan03].

2   Proof of Theorem 3

In this section, we prove Theorem 3. The proof of Theorem 3 (1) is straightforward and (due to space limitations) is omitted. The proof of Theorem 3 (2) follows.

2.1   The Naive Scheme

Let $G = (V, E)$ be a $\langle \rho, \varepsilon \rangle$-connected k-uniform hypergraph. In this section we study the probability that a random subset R of V of size ℓ is an independent set. We then use this result to bound the probability that a random subset H of G of size s has a large independent set.

We would like to bound (from above) the probability that R induces an independent set. Let $\{r_1, \ldots, r_\ell\}$ be the vertices of R. Consider choosing the vertices of R one by one, such that at each step the random subset chosen so far is $R_i = \{r_1, \ldots, r_i\}$ and the vertex r_{i+1} is chosen from $V \setminus R_i$. Assume that at some stage R_i is an independent set. We would like to show (with high probability) that after adding the remaining vertices $\{r_{i+1}, \ldots, r_\ell\}$ to R_i, the final set R will not be an independent set.

The vertices in $V \setminus R_i$ that *cannot* be added to R_i are exactly the vertices v that share an edge with some $k - 1$ vertices in R_i (namely, these vertices will result in a set R_{i+1} that is no longer independent). Let $N(R_i) = N_i$ be the set of such vertices in $V \setminus R_i$, and let $I(R_i) = I_i$ be $V \setminus N_i$. Consider the next random vertex $r_{i+1} \in R$. If r_{i+1} is chosen from N_i then it cannot be added to R_i, and we view this round as a success regarding the set R_i. Otherwise, r_{i+1} happens to be in I_i and can be added to R_i. But if the addition of r_{i+1} to R_i happens to add many vertices to N_i, we also view this round as a successful round regarding R_i.

Motivated by the discussion above, we continue with the following definitions. For each subset R_i we define the following weighted set systems of subsets of V. Let $RES(R_i, 1)$ (for *restrict*) be the set of singletons $\{v\}$ that share an edge with vertices in R_i. Namely, $\{v\} \in RES(R_i, 1)$ iff there exists vertices $\{w_1, \ldots, w_{k-1}\}$ in R_i such that there is an edge $\{v, w_1, \ldots, w_{k-1}\}$ in E. $RES(R_i, 1)$ is exactly the set N_i defined above. Define the weight of each element in $RES(R_i, 1)$ as n^{k-2}. Let $RES(R_i, 2)$ be the set of *pairs* of vertices v_1, v_2 which (together) share an edge with vertices in R_i. Namely, $\{v_1, v_2\} \in RES(R_i, 2)$ iff there exists vertices $\{w_1, \ldots, w_{k-2}\}$ in R_i such that there is an edge $\{v_1, v_2, w_1, \ldots, w_{k-2}\}$ in E. Define the weight of each element in $RES(R_i, 2)$ as n^{k-3}. Similarly for each $j \in \{1, \ldots, k - 1\}$ let $RES(R_i, j)$ be the set of subsets $\{v_1, \ldots, v_j\}$ of V of size j that share an edge with vertices in R_i. Define the weight of each element in $RES(R_i, j)$ as n^{k-j-1}. Finally let $RES_i = RES(R_i)$ be the union of the sets $RES(R_i, j)$ where $j \in \{1, \ldots, k - 1\}$. Let $\|RES(R_i, j)\|$ ($\|RES_i\|$) denote the weight of elements in $RES(R_i, j)$ (RES_i). Notice that $\|RES(R_i, j)\| \leq \binom{n}{j} n^{k-j-1} \leq n^{k-1}$ and that $\|RES_i\| \leq \sum_{j=1}^{k-1} \binom{n}{j} n^{k-j-1} \leq k n^{k-1}$.

Definition 4 *Let the normalized degree w.r.t. R_i of a vertex $v \in V$ be the amount on which v restricts upon R_i:*

$$d_v(R_i) = \|RES(R_i \cup \{v\})\| - \|RES(R_i)\| = \|RES(R_i \cup \{v\}) \setminus RES(R_i)\|.$$

In the above notice that $RES(R_i) \subseteq RES(R_i \cup \{v\})$. We call a vertex v in V heavy with respect to R_i (or R_i-heavy for short) if $d_v(R_i) \geq \frac{1}{2^k (k-3)!} \frac{\varepsilon}{\rho} n^{k-1}$.

Each subset R_i of V now defines the following partition (LI_i, HI_i, N_i) of V and the set RES_i. Let RES_i and N_i be as defined as above. Let $I_i = V \setminus N_i$. I_i is now partitioned into two parts: vertices in I_i with *low* normalized degree (w.r.t. R_i), denoted as the set LI_i, and vertices with *high* normalized degree, denoted as HI_i. Namely LI_i is defined to be the ρn vertices of I_i with minimal normalized degree and HI_i is defined to be the remaining vertices of I_i. Ties are broken arbitrarily or in *favor* of vertices in R_i (namely, vertices in R_i are placed in LI_i before other vertices of identical degree). If it is the case that $|I_i| \leq \rho n$ then LI_i is defined to be I_i, and HI_i is defined to be empty.

We define the partition corresponding to $R_0 = \phi$ as (LI_0, HI_0, N_0), where LI_0 are the ρn vertices of G of minimal normalized degree, HI_0 are the remaining vertices of G, and N_0 is empty. RES_0 is also defined to be empty.

Notice, using this notation, that the subset R_i is an independent set iff $R_i \subseteq I_i$. Moreover, in this case $R_i \subseteq LI_i$ (all vertices of R_i have normalized degree 0).

Furthermore, each vertex r_i in an independent set $R = R_\ell = \{r_1, \ldots, r_\ell\}$ satisfies $r_i \in I_{i-1}$.

We are now ready to bound the probability that a random subset $R = \{r_1, \ldots, r_\ell\}$ of G is independent. Let $R_i = \{r_1, \ldots, r_i\}$, and let (LI_i, HI_i, N_i) be the corresponding partition of V defined by R_i. Consider the case in which R is an independent set. As mentioned above, this happens iff for every i the vertex r_i is chosen to be in $I_{i-1} = LI_{i-1} \cup HI_{i-1}$. We would like to show that this happens with small probability (if ℓ is large enough).

Consider the set RES_i as we proceed in the choice of vertices in R. Initially, the subset RES_0 is of weight 0, and it gets larger and larger as we proceed in the choice of vertices in R. Each vertex in $r_i \in HI_{i-1}$ increases the weight of RES_{i-1} substantially, while each vertex in LI_{i-1} may only slightly change the weight of RES_{i-1}. In the following, we show that there cannot be many vertices $r_i \in R$ that happen to fall into HI_{i-1}. We thus turn to consider vertices r_i that fall in LI_{i-1} (there are almost ℓ such vertices). The size of LI_i is bounded by ρn. Hence, the probability that $r_i \in LI_i$ is bounded by ρ (by our definitions $R_{i-1} \subseteq LI_{i-1}$ and the vertex r_i is random in $V \setminus R_{i-1}$). This implies that the probability that R is an independent set is roughly bounded by ρ^ℓ. Details follow.

Lemma 1. *Let G be a $\langle \rho, \varepsilon \rangle$-connected hypergraph. Let $R = \{r_1, \ldots, r_\ell\}$ be a set in G. The number of vertices r_i which satisfy $r_i \in HI_{i-1}$ is bounded by $t = \frac{2^k k! \rho}{\varepsilon}$.*

Proof. We start with the following claim which is the main technical contribution of this work.

Claim 5 *Let R_i be as defined above, and let (LI_i, HI_i, N_i) and RES_i be its corresponding partition and set system. Let $I_i = LI_i \cup HI_i$. If G is $\langle \rho, \varepsilon \rangle$-connected then, every vertex in HI_i is R_i-heavy.*

Proof. Assume that $LI_i = \rho n$ (otherwise HI_i is empty and the claim holds). By our assumptions on G we have that LI_i induces at least εn^k edges. Let $E(LI_i)$ be the set of these edges, and let m be the size of $E(LI_i)$. To simplify our notation, let $LI = LI_i$, $HI = HI_i$, $N = N_i$ and $R = R_i$. Now that the index i is free, we use it in the following new context. For $i \in \{1, \ldots, k-1\}$ consider the following sequence of weighted i-uniform hypergraphs $H_i = (LI, E_i)$ which all have vertex set LI. The edge set E_i is defined to be all subsets of size i of edges in $E(LI)$ (here and throughout our work we consider a (hyper) edge of size k as a subset of vertices of size k). For example, each edge in $E(LI)$ induces $\binom{k}{2}$ edges in H_2, $\binom{k}{3}$ edges in H_3, and so on. The weight of an edge e_i in H_i is equal to the number of edges e in $E(LI)$ which satisfy $e_i \subseteq e$.

Recall the sets $RES(R, j)$ and $RES(R)$. Our goal is to prove the existence of a vertex $v \in LI$ which is R-heavy (this will imply our assertion). Namely a vertex v which when added to R will significantly increase the weight of $RES(R)$. Formally we are looking for a vertex $v \in LI$ which satisfies $\|RES(R \cup \{v\})\| - \|RES(R)\| \geq \frac{1}{2^k (k-3)!} \frac{\varepsilon}{\rho} n^{k-1}$. Notice that the weight of a subset of vertices with respect to $RES(R)$ may (and usually does) differ from its weight with respect to H_i. To avoid confusion we denote the weight of a subset e with respect to

$RES(R)$ as $w_R(e)$. Recall that every (partial) edge e of size i in $RES(R)$ has weight $w_R(e) = n^{k-i-1}$. The weight of an edge e in H_i will be marked as $w_{H_i}(e)$.

For each i we now consider the weight (with respect to H_i) of the edges in H_i which are in $RES(R)$. Denote this weight as res_i. Notice that $res_1 = 0$ (by our construction $LI \cap N = \phi$, which implies that singletons $\{v\} \subseteq LI$ are not in $RES(R)$). We consider the following cases.

Case 1: We start by assuming that $res_{k-1} \leq (k-1)m$. As the total weight of edges in H_{k-1} is km, we conclude that the weight of edges in H_{k-1} that are not in $RES(R)$ is at least m. Let v be a vertex in LI. We say that an edge e in H_{k-1} is a v-edge if there exists an edge in $E(LI)$ consisting of the union of e and $\{v\}$. Notice that an edge e in H_{k-1} of weight $w_{H_{k-1}}(e)$ is actually a v edge for $w_{H_{k-1}}(e)$ distinct vertices v. As each edge in H_{k-1} is a v-edge for some vertex v in LI, we have the existence of a vertex v in LI with at least $\frac{m}{\rho n}$ (distinct) v-edges that are not in $RES(R)$. The weight of each such v-edge in $RES(R)$ is 1. Furthermore, each such v-edge appears in $RES(R \cup \{v\})$ implying that

$$\|RES(R \cup \{v\})\| - \|RES(R)\| \geq \frac{m}{\rho n} \geq \frac{\varepsilon}{\rho} n^{k-1}$$

which in turn implies the v is R-heavy.

Cases 2 to $k-1$: Starting at $i = k-1$ and iteratively continuing until $i = 2$ consider the following cases. From the previous step, we may assume that $res_i > \frac{2m}{2^{k-i}} \frac{(k-1)}{(k-i)!}$. We now also assume that $res_{i-1} \leq \frac{m}{2^{k-i}} \frac{(k-1)}{(k-i+1)!}$. Consider an edge $e = (v_1, \ldots, v_i)$ in H_i of weight $w_{H_i}(e)$ which contributes to res_i. Each such edge induces i edges in H_{i-1}: $\{e'_1, \ldots, e'_i\}$ (each edge obtained by removing one vertex from e). We are interested in bounding (by below) the weight of edges e as above with corresponding edges e'_j which are not in $RES(R)$ (here $j \in \{1, \ldots, i\}$).

By our construction, the weight of any edge e' in H_{i-1} equals the number of edges in G that include e'. It is not hard to verify that this equals $\frac{1}{k-i+1}$ times the weight of edges in H_i which include e'. We conclude that the weight of edges e in H_i which are in $RES(R)$ with some corresponding edge e' in H_{i-1} which is also in $RES(R)$ is bounded by $res_{i-1}(k-i+1)$. This leaves us with

$$res_i - (k-i+1)res_{i-1} > \frac{m}{2^{k-i}} \frac{(k-1)}{(k-i)!}$$

edges e which contribute to res_i for which e'_j for $j \in \{1, \ldots, i\}$ are not in $RES(R)$.

We now conclude the existence of a vertex $v \in LI$ which is adjacent to (*i.e.* is included in) at least the weight of $\frac{im}{2^{k-i}\rho n} \frac{(k-1)}{(k-i)!}$ edges in H_i which appear in $RES(R)$ such that their corresponding edges in H_{i-1} are not in $RES(R)$. The weight of each edge in H_i is bounded by $\binom{\rho n}{k-i} \leq (\rho n)^{k-i}$. Thus there are at least $\frac{im}{2^{k-i}(\rho n)^{k-i+1}} \frac{(k-1)}{(k-i)!}$ distinct edges in H_i adjacent to v with corresponding edges in H_{i-1} that are not in $RES(R)$.

We will now show that v is R-heavy. Consider one of the distinct edges $e = (v, v_1, \ldots, v_{i-1})$ as discussed above. By our assumption the set $\{v_1, \ldots, v_{i-1}\}$ is not in $RES(R)$. The edge e appears in $RES(R)$ implying that there exists vertices $\{w_1, \ldots, w_{k-i}\}$ (in R) such that $(v, v_1, \ldots, v_{i-1}, w_1, \ldots, w_{k-i})$ is an edge in the original hypergraph G. This in turn implies that $\{v_1, \ldots, v_{i-1}\}$ will be included in $RES(R \cup \{v\})$. As each set of size $i - 1$ in $RES(R)$ has weight $w_R = n^{k-i}$, and there are at least $\frac{im}{2^{k-i}(\rho n)^{k-i+1}} \frac{(k-1)}{(k-i)!}$ distinct edges of interest adjacent to v, we conclude that

$$\|RES(R \cup \{v\})\| - \|RES(R)\| \geq \frac{1}{2^{k-i}} \frac{(k-1)}{(k-i)!} \frac{\varepsilon}{\rho} n^{k-1}.$$

Now to prove our lemma, consider the subsets $R_i = \{r_1, \ldots, r_i\}$ and their corresponding partitions (LI_i, HI_i, N_i). Let $I_i = LI_i \cup HI_i$. We would like to bound the number of vertices r_i that are in HI_{i-1}. Consider a vertex r_i in HI_i. By Claim 5, its normalized degree w.r.t. R_i is at least $\frac{1}{2^k(k-3)!} \frac{\varepsilon}{\rho} n^{k-1}$. $RES(\phi)$ is initially empty, and for any i, $RES(R_i)$ is of weight at most $\sum_{i=2}^{k-1} \binom{n}{i} n^{k-i-1} \leq kn^{k-1}$. Each vertex $r_i \in HI_{i-1}$ increases $\|RES(R_{i-1})\|$ by at least $\frac{1}{2^k(k-3)!} \frac{\varepsilon}{\rho} n^{k-1}$. We conclude that there are at most $\frac{2^k k! \rho}{\varepsilon}$ vertices r_i in R which are in HI_{i-1}.

Let δ be a large constant. "Plugging" Lemma 1 into the proof appearing in (the journal version of) [FLS02], we are now able to bound the probability that a random subset H of G of size s has an independent set of size $> \delta \rho s$. This is done in two steps, first we bound the probability that a random subset R of G is independent. Then, using the union bound, we obtain the result mentioned for H. In the upcoming Section 2.2 we present the refined proof technique of [FLS02] (which *gets rid* of the parameter δ) applied to the case of hypergraphs. Namely, we will bound the probability that a random subset H of G of size s has an independent set of size $\geq \rho s$.

Lemma 2 ([FLS02]). *Let G be a $\langle \rho, \varepsilon \rangle$-connected hypergraph. Let t be as in Lemma 1. Let $\ell \geq 2t$. The probability that ℓ random vertices of G induce an independent set is at most*

$$\rho^\ell \left(\frac{e\ell}{t\rho} \right)^t$$

Corollary 6 ([FLS02]) *Let G be a $\langle \rho, \varepsilon \rangle$-connected hypergraph. Let t be as in Lemma 1. Let H be a random sample of G of size s. Let $\delta > e$, and let c be a sufficiently large constant. If $s \geq ct\frac{\log(1/\rho)}{\rho}$ then the probability that $\alpha(H) > \delta \rho s$ is at most $\left(\frac{e}{\delta} \right)^{\Omega(\delta \rho s)}$.*

2.2 An Enhanced Analysis

Recall the proof technique from Section 2.1. We started by analyzing the probability that a subset R of H of size ℓ is an independent set. Afterwards we bounded the probability that $\alpha(H) > \delta\rho n$ (using the standard union bound on all subsets R of H of size greater than $\ell = \delta\rho n$). In this section we enhance the first part of this scheme by analyzing the probability that a subset R of H of size ℓ is a *maximum* independent set in H (rather than just an independent set of H). Then, as before, using the standard union bound on all large subsets R of H, we bound the probability that $\alpha(H) \geq \rho s$. We show that taking the maximality property of R into account will suffice to prove Theorem 3 (2).

Let $H = \{h_1, \ldots, h_s\}$ be s random vertices in G. We would like to analyze the probability that a given subset R of H of size ℓ is a *maximum* independent set. Recall (Section 2.1), that the probability that R is an independent set is bounded by approximately ρ^ℓ. An independent set R is a maximum independent set in H only if adding any other vertex in H to R will yield a set which is no longer independent. Let $R = R_\ell$ be an independent set, and let $(LI_\ell, HI_\ell, N_\ell)$ be the partition (as defined in Section 2.1) corresponding to R. Consider an additional random vertex h from H. The probability that $R \cup h$ is no longer an independent set is approximately $|N_\ell|/n$ (here we assume that $|R|$ is small compared to n). The probability that for every $h \in H \setminus R$ the subset $R \cup \{h\}$ is no longer independent is thus $\simeq (|N_\ell|/n)^{s-\ell}$. Hence, the probability that a given subset R of H of size ℓ is a *maximum* independent set is bounded by approximately $\rho^\ell (|N_\ell|/n)^{s-\ell}$. This value is substantially smaller than ρ^ℓ iff $|N_\ell|$ is substantially smaller than n. We conclude that it is in our favor to somehow ensure that $|N_\ell|$ is not too large. We do this in an artificial manner.

Let $R = \{r_1, \ldots, r_\ell\}$ be an independent set, let $R_i = \{r_1, \ldots, r_i\}$, and let (LI_i, HI_i, N_i) be the partition (as defined in Section 2.1) corresponding to R_i. Let RES_i be the set system corresponding to R_i. Roughly speaking, in Section 2.1, every time a vertex r_i was chosen, the set system RES_i and N_i was updated. If r_i was chosen in HI_{i-1}, then RES_{i-1} increased substantially and N_{i-1} potentially also grew substantially, and if r_i was chosen in LI_{i-1}, both RES_{i-1} and N_{i-1} only slightly changed. We would like to change the definition of the partition (LI_i, HI_i, N_i) and of RES_i corresponding to R_i as to ensure that N_i is always substantially smaller than n. This cannot be done unless we relax the definition of N_i. Recall that N_i was defined (in Section 2.1) to be the set of vertices v which share an edge with vertices in R_i. Specifically, N_i is the set of vertices v for which there exists vertices $\{w_1, \ldots, w_{k-1}\}$ in R_i such that there is an edge $\{v, w_1, \ldots, w_{k-1}\}$ in E. In this section N_i will only include a subset of these vertices (a subset which is substantially smaller than n). Namely, in our new definitions RES_{i-1} and N_{i-1} will be changed only if r_i was chosen in HI_{i-1}. In the case in which $r_i \in LI_{i-1} \cup N_{i-1}$, we do not change N_{i-1} at all. As we will see, such a definition will imply that $|N_i| \leq (1 - \rho)s$, which will now suffice for our proof.

A New Partition and Set System: Let $H = \{h_1, \ldots, h_s\}$ be a subset of V. Let $R_i = \{r_1, \ldots, r_i\}$ be a subset of H of size i. In the previous section, the subset R_i defined a partition (LI_i, HI_i, N_i) of V and a set system RES_i. In this section, for each i we will define a subset $\hat{R}_i$ of R_i, and a new partition and set system. The new partition and set system corresponding to R_i will be defined similarly to those defined in the previous section with the exception that $\hat{R}_i$ will play the role that R_i played previously. The set $\hat{R}_i$, the new partition (LI_i, HI_i, N_i), and the set system RES_i are defined as follows (as before, let $I_i = LI_i \cup HI_i$).

1. Initially $R_0 = \hat{R}_0 = \phi$, LI_0 is the ρn vertices in V of minimal normalized degree w.r.t. $\hat{R}_i$, $HI_0 = V \setminus LI_0$, and $N_0 = \phi$. In the above, ties are broken by an assumed ordering on the vertices in V. $RES_0 = RES(\hat{R}_0)$ is defined to be empty.

2. Let $\hat{R}_i$, (LI_i, HI_i, N_i) and RES_i be the sets corresponding to R_i, let r_{i+1} be a new random vertex. Let $R_{i+1} = R_i \cup \{r_{i+1}\}$, we now define the sets $\hat{R}_{i+1}$, $(LI_{i+1}, HI_{i+1}, N_{i+1})$ and RES_{i+1}. Let $N(r_{i+1})$ be the set of vertices which share an edge with vertices in $\hat{R}_i \cup r_{i+1}$. We consider the following cases: (a) If $r_{i+1} \in LI_i$ or $r_{i+1} \in N_i$ then the sets corresponding to R_{i+1} will be exactly those corresponding to R_i. Namely, $\hat{R}_{i+1} = \hat{R}_i$, $LI_{i+1} = LI_i$, $HI_{i+1} = HI_i$, $N_{i+1} = N_i$, and RES_{i+1} will be defined as RES_i. (b) If $r_{i+1} \in HI_i$ then we consider two sub-cases: (b1) If $|N_i \cup N(r_{i+1})| \le (1-\rho)n$, then $\hat{R}_{i+1} = \hat{R}_i \cup \{r_{i+1}\}$, $RES_{i+1} = RES(\hat{R}_{i+1})$, and LI_{i+1}, HI_{i+1}, N_{i+1}, are defined as in Section 2.1. Namely, $N_{i+1} = N_i \cup N(r_{i+1})$. I_{i+1} is defined to be $V \setminus N_{i+1}$. LI_{i+1} is defined to be the ρn vertices of I_{i+1} with minimal normalized degree w.r.t. $\hat{R}_{i+1}$. Finally, HI_{i+1} is defined to be the remaining vertices of I_{i+1}. Ties are broken by the assumed ordering on V. (b2) If $|N_i \cup N(r_{i+1})| > (1 - \rho)n$, then let $\hat{N}(r_{i+1})$ be the first (according to the assumed ordering on V) $(1 - \rho)n - |N_i|$ vertices in $N(r_{i+1})$ and set $N_{i+1} = N_i \cup \hat{N}(r_{i+1})$. Furthermore, set LI_{i+1} to be the remaining ρn vertices of G, and HI_i to be empty. Notice that in this case $|N_{i+1}|$ is of size exactly $(1 - \rho)n$. Finally, let $\hat{R}_{i+1} = \hat{R}_i$ and $RES_{i+1} = RES_i$.

A few remarks are in place. First of all it is not hard to verify that the definition above implies $|N_i| \le (1-\rho)n$ for all $i \in \{1, \ldots, \ell\}$. Secondly, due to the iterative definition of our new partition, the sets corresponding to the subsets R_i depend strongly on the specific ordering of the vertices in R_i. Namely, in contrast to the partitions (set systems) used in Section 2.1, a single subset R with two different orderings may yield two different partitions (set systems). For this reason, in the remainder of this section, we will assume that the vertices of H are chosen one by one. This will imply an ordering on H and on any subset R of H. The partitions we will study will correspond to these orderings only.

Finally, in Section 2.1, an (ordered) subset $R = \{r_1, \ldots, r_\ell\}$ was independent iff $\forall i \ r_i \in I_{i-1}$ (according to the definition of I_{i-1} appearing in Section 2.1). In this section, if R is independent then it still holds that $\forall i \ r_i \in I_{i-1}$. However, it may be the case that $\forall i \ r_i \in I_{i-1}$ but R is not an independent set. In the remainder of this section, we call ordered subsets R for which $\forall i \ r_i \in I_{i-1}$ *free* sets. We analyze the probability that a random ordered subset H of V of size s does not have any free sets R of size larger then ρs. This implies, that H does not include any independent sets of size ρs.

We now turn to prove Theorem 3 (2). Roughly speaking, we start by analyzing the probability that a random subset R is a free set. We then analyze the probability that a given subset R in H is a maximum free set. Finally, we use the union bound on all subsets R of H of size $\geq \rho s$ to obtain our results.

We start by stating the following lemma which is analogous to Lemma 1 from Section 2.1. The main difference between the lemma below (and its proof), and that of the previous section is in the definition of the partition (LI_i, HI_i, N_i), and the set system RES_i. Proof of the lemma is omitted.

Lemma 3. *Let G be a $\langle \rho, \varepsilon \rangle$-connected hypergraph. Let $R = \{r_1, \ldots, r_\ell\}$ be an ordered set in G of size ℓ. The number of vertices r_i which satisfy $r_i \in HI_{i-1}$ is bounded by $t = \frac{2^k k! \rho}{\varepsilon}$.*

We now address the probability that a random subset R of H is a maximum free set. We then use the union bound on all subsets R of H of size $\geq \rho s$ (and a few additional ideas) to obtain our main result. Due to space limitations, the proof of the following Lemma and Theorem is omitted. The interested reader can find their proofs in [Lan03] (or proofs of similar nature in [FLS02]).

Lemma 4. *Let G be a $\langle \rho, \varepsilon \rangle$-connected hypergraph. Let t be as in Lemma 3. Let $\ell \geq 2t$. Let H be an ordered random sample of G of size $s \geq \ell$. The probability that a given subset R of H of size ℓ is a maximum free set is at most*

$$\rho^\ell \left(\frac{e\ell}{t\rho} \right)^t (1 - \rho)^{s-\ell}$$

Theorem 3 (2). *Let G be a $\langle \rho, \varepsilon \rangle$-connected hypergraph. Let t be as in Lemma 3. Let H be a random sample of G of size s. Let c be a sufficiently large constant. If $s \geq ct \frac{\rho^{2k-1}}{\varepsilon^2} \log \left(\frac{\rho}{\varepsilon} \right)$ then the probability that H has an independent set of size $\geq \rho s$ is at most $e^{-\Omega(t)}$.*

3   Lower Bounds on the Sample Size s

Roughly speaking, Theorem 3 (2) states that given a $\langle \rho, \varepsilon \rangle$-connected hypergraph G, a random sample H of G of size s proportional to $2^k k! \frac{\rho^{2k}}{\varepsilon^3}$ (or larger) will not have an independent set of size ρs (with high probability). We now continue to study the minimal value of s for which $\alpha(H) < \rho s$ with high probability, and present a lower bound of $\frac{\rho^{2k-1}}{4(k!)^2 \varepsilon^2}$ on the size of s.

Lemma 5. *Let n be a sufficiently large constant. Let $\rho > 0$ and $\varepsilon > 0$ satisfy (a) $\varepsilon << \frac{\rho^k}{2k!}$, (b) $\rho^{2k-1}/\varepsilon^2 << n$ and (c) $k^2 << \rho n$ (here $a << b$ if for a sufficiently small constant $\delta > 0$ it holds that $a < \delta b$). There exists a graph G on n vertices for which G is $\langle \rho, \varepsilon \rangle$-connected, and with probability $\geq 1/20$ a random subgraph H of size $s = \frac{\rho^{2k-1}}{4(k!)^2 \varepsilon^2}$ will have an independent set of size ρs.*

Proof. Consider the k-uniform hypergraph $G = (V, E)$ in which (a) $|V| = n$, (b) V consists of two disjoint sets A and $V \setminus A$, where A is of size $(1 - \frac{2k! \varepsilon}{\rho^k})\rho n$, and (c) the edge set E of G consists of all subsets of V of size k except those included in A (namely, A is an independent set). On one hand, every subset of size ρn in G induces a subgraph with at least εn^k edges (implying that G is $\langle \rho, \varepsilon \rangle$-connected). On the other, let H be a random subset of V obtained by picking each vertex independently with probability $\frac{1}{n} \frac{\rho^{2k-1}}{4(k!)^2 \varepsilon^2}$. The expected size of H is $s = \frac{\rho^{2k-1}}{4(k!)^2 \varepsilon^2}$. In the following, we assume H is exactly of size s, minor modifications in the proof are needed if this assumption is not made. The set $H \cap A$ is an independent set in the subgraph induced by H. The expected size of $H \cap A$ is $(1 - \frac{2k! \varepsilon}{\rho^k}) \rho s$. Let $N(0, 1)$ denote a standard normal variable. It can be seen using the central limit theorem (for example [Fel66]) that for our choice of parameters, the probability that $|H \cap A|$ is greater than $\left(1 - \frac{2k! \varepsilon}{\rho^k}\right) \rho s + \sqrt{\rho s}$ is at least $\Pr\left[N(0, 1) > 3/2\right] \geq \frac{1}{20}$. In such a case the size of $H \cap A$ will be greater than $(1 - \frac{2k! \varepsilon}{\rho^k}) \rho s + \sqrt{\rho s} = \rho s$ for our value of s. Hence implying the lemma.

References

[AdlVKK02] N. Alon, W. Fernandez de la Vega, R. Kannan, and M. Karpinski. Random sampling and approximation of Max-CSP problems. *In Proceddings of STOC*, pages 232–239, 2002.

[AS03] N. Alon and A. Shapira. Testing satisfyability. *Journal of Algorithms*, 47:87–103, 2003.

[CS01] A. Czumaj and C. Sohler. Testing hypergraph coloring. *In Proceddings of ICALP*, pages 493–505, 2001.

[Fel66] W. Feller. *An introduction to probability theory and its applications*, volume 2. John Wiley & Sons, 1966.

[Fis01] E. Fisher. The art of uninformed decisions: A primer to property testing. *The Computational Complexity Column of The Bulletin of the European Association for Theoretical Computer Science*, 75:97–126, 2001.

[FLS02] U. Feige, M. Langberg, and G. Schechtman. Graphs with tiny vector chromatic numbers and huge chromatic numbers. *In proceedings of 43rd annual Symposium on Foundations of Computer Science*, pages 283–292, 2002. *To appear in SIAM Journal on Computing, manuscript availiable at http://www.cs.caltech.edu/~mikel.*

[GGR98] O. Goldreich, S. Goldwasser, and D. Ron. Property testing and its connection to learning and approximation. *Journal of ACM*, 45(4):653–750, 1998.

[Gol98] O. Goldreich. Combinatorial property testing - a survey. *Randomization Methods in Algorithm Design (P. Pardalos, S. Rajasekaran and J. Rolim eds.), AMS-DIMACS*, pages 45–60, 1998.

[Lan03] M. Langberg. Testing the independence number of hypergraphs. *Electronic Colloquium on Computational Complexity (ECCC))*, TR03-076, 2003.

[Ron01] D. Ron. Property testing (a tutorial). *In Handbook of Randomized Computing (S. Rajasekaran, P. M. Pardalos, J. H. Reif and J. D. P. Rolim eds), Kluwer Press*, 2001.

[RS96] R. Rubinfeld and M. Sudan. Robust characterization of polynomials with applications to program testing. *SIAM Journal of Computing*, 25:252–271, 1996.

A Note on Approximate Counting for k-DNF

Luca Trevisan[*]

Computer Science Division, U.C. Berkeley
luca@cs.berkeley.edu

Abstract. We describe a deterministic algorithm that, for constant k, given a k-DNF or k-CNF formula φ and a parameter ε, runs in time linear in the size of φ and polynomial in $1/\varepsilon$ (but doubly exponential in k) and returns an estimate of the fraction of satisfying assignments for φ up to an additive error ε. This improves over previous polynomial (but super-linear) time algorithms. The algorithm uses a simple recursive procedure and it is not based on derandomization techniques. It is similar to an algorithm by Hirsch for the related problem of solving k-SAT under the promise that an ε-fraction of the assignments are satisfying. Our analysis is different from (and somewhat simpler than) Hirsch's.

We also note that the argument that we use in the analysis of the algorithm gives a proof of a result of Luby and Velickovic that every k-CNF is "fooled" by every δ-biased distribution, with $\delta = 1/2^{O(k2^k)}$.

1 Introduction

We consider the following problem: given a k-CNF formula φ and a parameter ε, approximate within an *additive* error ε the fraction of satisfying assignments for φ [1].

The problem is easy to solve using randomization: just generate $O(1/\varepsilon^2)$ assignments at random and then output the fraction of assignments in the sample that satisfies φ, and the question is whether efficient *deterministic* algorithms exist.

We also consider the related problem of finding a satisfying assignment for φ under the promise that an ε fraction of assignments are satisfying. Again, we are interested in deterministic algorithms, and the problem is easy to solve probabilistically, since after picking $O(1/\varepsilon)$ assignments at random it is likely that one of them satisfies the formula.

One can consider the approximate counting problem as the problem of *derandomizing two-sided* error algorithms implemented by *depth-two* circuits. The

---

[*] Supported by NSF grant CCR-9984703, a Sloan Research Fellowship and an Okawa Foundation Grant.

[1] Note that an algorithm achieving additive approximation ε for k-CNF immediately implies an algorithm achieving the same additive approximation for k-DNF. Also, achieving multiplicative approximation $(1 + \varepsilon)$ for k-DNF reduces to achieving additive approximation $\varepsilon 2^{-k}$, since a satisfiable k-DNF is satisfied by at least a $1/2^k$ fraction of assignments.

K. Jansen et al. (Eds.): APPROX and RANDOM 2004, LNCS 3122, pp. 417–425, 2004.

problem of finding a satisfying assignment for φ under the promise that that there is a large number of such assignments can be seen as the problem of de-randomizing *one-sided* error algorithms implemented by depth-two circuits.

These problems were first studied by Ajtai and Wigderson [AW89]. Using derandomization techniques (specifically, t-wise independence) they give an algorithm for the counting problem running in time $O(n^{k^2} + 2^{(\log(1/\varepsilon))^{2^k}})$ and an algorithm for the satisfiability problem running in time $O(n^{k2^k} \log(1/\varepsilon))$. They also give sub-exponential time algorithm for the counting problem for functions computed by AC^0 circuits[2].

The algorithm of Ajtai and Wigderson for k-CNF could be improved by using *almost* t-wise independent distributions, for example the small bias distributions of [NN93], instead of distributions that are perfectly t-wise independent. For constant ε, this would improve the running time to roughly $n \cdot (\log n)^{O(1)} \cdot 2^{k^{O(k^2)}}$ for both the approximate counting problem and the satisfiability problem. Almost t-wise independent distributions were introduced after the publication of [AW89].

Nisan [Nis91] and Nisan and Wigderson [NW94] construct a pseudorandom generator that fools constant-depth circuits and that has poly-logarithmic seed length. As a consequence, they achieve $n^{(\log n)^{O(1)}}$ time algorithms for the counting and satisfiability problems for AC^0 circuits.

Luby, Velickovic and Wigderson [LVW93] optimize the constructions of Nisan and Wigderson [Nis91,NW94] to the case of depth-2 circuits, thus solving the counting and satisfiability problem in time $n^{O((\log n)^3)}$ for general CNF and DNF. Luby and Velickovic [LV96] show how to reduce an arbitrary CNF and DNF to a formula in a simplified format, and show that the counting and satisfiability problems can be solved in polynomial time for k-CNF even if $k = O((\log n)^{1/8})$ is more than a constant. Luby and Velickovic [LV96] also also present an improved derandomization of general CNF and DNF that runs in slightly super-polynomial time $n^{O(2^{\sqrt{\log \log n}})}$. Finally, Luby and Velickovic also show that every k-CNF is "fooled" by every distribution that is close to being $t(k)$-wise independent, where $t(k) = 2^{O(k2^k)}$. (A distribution X "fools" a function $f()$ if the probability that $f(X) = 1$ is approximately the same as the probability that $f()$ equals 1 under the uniform distribution.)

Hirsch [Hir98] shows how to solve the satisfiability problem for k-CNF in time $O(Lk(2/\varepsilon)^{B(k)})$, where $L \leq nk$ is the size of the formula and $B(k)$ is a function for which a closed formula is not given, but that seems to grow exponentially in k. Hirsch's algorithm does not use derandomization techniques.

In this paper, we show how to solve the approximate counting problem and the satisfiability problem in time $O(L(1/\varepsilon)^{(\ln 4)k2^k})$.

Our algorithm is based on the following simple observation: given a k-CNF φ, then for every fixed c, either we can efficiently find a set of $\leq kc$ variables

---

[2] An AC^0 circuit is a circuit of constant depth and polynomial size with unbounded fan-in AND and OR gates. A CNF formula is a depth-two AC^0 circuit, and so is a DNF formula.

that hits all the clauses, or we can efficiently find $> c$ clauses over disjoint sets of variables. In the former case, we can try all assignments to those variables, and recurse on each assignment, thus reducing our problem to 2^{kc} problems on $(k-1)$-CNF instances; in the latter case, less than a $(1 - 1/2^k)^c$ fraction of assignments can satisfy φ, and thus 0 is an approximation within an additive error $(1 - 1/2^k)^c$ of the fraction of satisfying assignments for φ. Fixing c to be $2^k \ln 1/\varepsilon$ gives us the main result.

Using the same recursive approach adopted in our algorithm, we give a slightly different proof of the result of Luby and Velickovic about k-CNF being fooled by almost t-wise independen distributions. We show that every k-CNF is well approximated, in a certain technical sense, by a decision tree of depth $t = O(k2^k)$, and it is well known that functions that are well approximated (in the above technical sense) by a decision tree of depth t cannot distinguish the uniform distribution from a distribution that is approximately t-wise independent.

2   The Recursive Algorithm

We describe the algorithm only for the case of k-CNF. As discussed in the introduction, an algorithm for k-DNF is an immediate corollary of the algorithm for k-CNF.

We use the following simple fact.

Lemma 1. *There is an algorithm that, on input a k-CNF formula φ and a parameter t, runs in time linear in the size of φ and then it returns either a set of t clauses over disjoint sets of variables, or a set S of at most $k(t-1)$ variables such that every clause in the formula contains at least one variable from S.*

Proof. (Sketch) Consider the k-uniform hypergraph H that has a vertex for every variable and an hyperedge for every clause. It is easy to find a maximal matching in H in linear time, that is, a set S of clauses over disjoint sets of variables and such that every other clause in φ shares some variables with some clause in S. If $|S| \geq t$, then we return t of the clauses in S. Otherwise, we return the set of $\leq k \cdot |S| \leq k(t-1)$ variables that occur in the clauses of S. Such a set of variables clearly "hits" all the clauses of φ. $\square$

The algorithm works as follows: given φ and ε,

- If φ is a 1-CNF, that is, it is just an AND of literals, then we output 0 if there are two inconsistent literals and 2^{-c} where c is the number of distinct literals, otherwise. This procedure is exact and can be implemented in linear time.
- Otherwise, we let t be the smallest integer such that $(1 - 1/2^k)^{t+1} < \varepsilon$, so that $t \leq 2^k(\ln 1/\varepsilon)$, and we run the algorithm of Lemma 1 on φ with parameter $t+1$.

- If the algorithm of Lemma 1 finds $t+1$ clauses $C_1, \ldots, C_{t+1}$ over disjoint sets of variables, then it is clear that the probability that φ is satisfied by random assignment is at most ε, and we return the value 0 as our approximation.
- If the algorithm of Lemma 1 finds a set V of at most $tk \leq k2^k \ln(1/\varepsilon)$ variables that hit all the clauses, then, for every assignment a to the variables V, define φ_a to be the formula obtained from φ by substituting the assigment into the variables. Note that φ_a is a $(k-1)$-CNF formula. We recurse on each of the φ_a with parameter ε, and take the average of the results. Assuming that each recursive call returns an ε additive approximation, the algorithm returns an ε additive approximation.

If we denote by $T(L, k)$ the running time of the algorithm for a k-CNF instance of size L, then we have

$$T(L, 1) = O(L)$$

and

$$T(L, k) \leq O(L) + 2^{k(\ln(1/\varepsilon))2^k} T(L, k-1)$$

which solves to $T(L, k) = O(L \cdot 2^{2k(\ln 1/\varepsilon)2^k}) = O(L(1/\varepsilon)^{(\ln 4)k2^k})$.

For the promise problem of finding a satisfying assignment under the promise that an ε fraction of assignments are satisfiable, we essentially use the same recursive algorithm. When we are down to 1-CNF, we find a satisfying assignment or fail if the instance is unsatisfiable. (Indeed, we can stop at 2-CNF.) In the recursive step, we fail if t is such that $(1 - 1/2^k)^t < \varepsilon$. The analysis of the running time is the same, and it is clear that at least one of the recursive branches produces a satisfying assignment.

Hirsch's algorithm is similar to the above sketch of the algorithm for the satisfiability promise problem, except that a different greedy strategy is used to pick the variables in V. The analysis is slightly different and somewhat more difficult.

3   Pseudorandomness Against k-CNF Formulae

3.1   Some Technical Preliminaries

We begin this section with a few technical definitions.

We denote by U_n the uniform distribution over $\{0, 1\}^n$. If $f : \{0, 1\}^n \to \{0, 1\}$ is a function and X is a distribution over $\{0, 1\}^n$, then we say that X ε-*fools* f if

$$|\mathbf{Pr}[f(U_n) = 1] - \mathbf{Pr}[f(X) = 1]| \leq \varepsilon$$

If $\mathcal{F}$ is a collection of functions, then we say that a distribution X ε-fools $\mathcal{F}$ if X ε-fools every function $f \in \mathcal{F}$.

Our goal will be to find a distribution X that ε-fools the class of k-CNF formulae over n variables, and that is uniform over an efficiently constructable support of polynomial size. Then, given a k-CNF formula f, we can approximate $\mathbf{Pr}[f(U_n) = 1]$ by computing the close value $\mathbf{Pr}[f(X) = 1]$, and we compute the latter by enumerating all the polynomially many elements in the support of X, and applying $f()$ to each of them. We will show that ε-biased distributions, defined below, can be used towards such goal.

We say that a distribution X over $\{0,1\}^n$ is ε-biased [NN93] if for every subset $S \subseteq \{1, \ldots, n\}$ we have

$$\frac{1}{2} - \varepsilon \leq \mathbf{Pr}\left[\bigoplus_{i \in S} x_i = 1\right] \leq \frac{1}{2} + \varepsilon$$

Equivalently, we can say that a distribution is ε-biased if it ε-fools every linear function. (Where, of course, we mean *linear* over the field $GF(2)$.)

Theorem 1 ([NN93,AGHP92]). *For every ε, and n, there is an ε-biased distribution over $\{0,1\}^n$ that is uniform over a support of size polynomial in n and $1/\varepsilon$. Furthermore, the support can be constructed in time polynomial in n and $1/\varepsilon$.*

We say that a distribution $X = (X_1 \cdots X_n)$ over $\{0,1\}^n$, where each X_i is unbiased, is *k-wise independent* if every k of the random variables $X_1, \ldots, X_n$ are mutually independent. Equivalently, a distribution is k-wise independent if it 0-fools the class of functions that depend only on k or fewer input variables.

We say that X is *ε-close to k-wise independent* if for every function $g : \{0,1\}^n \to \{0,1\}$ that depends on k or fewer inputs we have

$$|\mathbf{Pr}[g(U_n) = 1] - \mathbf{Pr}[g(X) = 1]| \leq \varepsilon$$

that is, if X ε-fools the class of functions that depend on at most k inputs.

We say that X is *ε-close to k-wise independent in ℓ_∞ norm* if for every $t \leq k$, for every t indices $i_1, \ldots, i_t$ in $\{1, \ldots, n\}$ and for every t values $a_1, \ldots, a_t \in \{0,1\}$ we have

$$\frac{1}{2^t} - \varepsilon \leq \mathbf{Pr}[X_{i_1} = a_1 \wedge \cdots \wedge X_{i_t} = a_t] \leq \frac{1}{2^t} + \varepsilon$$

that is, if X ε-fools the class of functions that can be expressed as checking that a subset of at most k bits of the input equals a particular sequence of values.

The following connection between the notions that we have described is well known.

Lemma 2. *Let X be an ε-biased distribution over $\{0,1\}^n$. Then, for every k, X is $\varepsilon \cdot 2^{k/2}$-close to k-wise independent, and also 2ε-close to k-wise independent in ℓ_∞ norm.*

3.2 ε-Biased Distribution and k-CNF Fomulae

In this section iwe give a proof of the following theorem.

Theorem 2 ([LV96]). *There are functions* $t(k,\varepsilon) = O(k \cdot 2^k \cdot \log(1/\varepsilon))$ *and* $\delta(k,\varepsilon) = 1/2^{O(k \cdot 2^k \cdot \log(1/\varepsilon))}$ *such that the following happens.*

Let $f : \{0,1\}^n \to \{0,1\}$ *be a function defined by a k-CNF formula and let X be a distribution over $\{0,1\}^n$ that is $\delta(k,\varepsilon)$-close to $t(k,\varepsilon)$-wise independent in ℓ_∞ norm. Then X ε-fools f.*

The application of Theorem 2 to ε-biased distributions is immediate.

Corollary 1. *There is a function* $\delta'(k,\varepsilon) = 1/2^{O(k \cdot 2^k \cdot \log(1/\varepsilon))}$ *such that if X is a $\delta'(k,\varepsilon)$-biased distributions, then X ε-fools every function computed by a k-CNF formula.*

The rest of this section is devoted to the proof of Theorem 2.

Let $f : \{0,1\}^n \to \{0,1\}$ be a function defined by a k-CNF formula φ over variables $x_1, \ldots, x_n$, and consider a *decision tree* over the variables $x_1, \ldots, x_n$. Every leaf of the decision tree (indeed, every node of the decision tree) defines a *restriction*, that is, an assignment to a subset of the variables $x_1, \ldots, x_n$. If a leaf v is at distance t from the root, then it defines an assignment to t variables; if we pick a random assignment and then apply the decision tree to it, there is a probability $1/2^t$ that we reach the leaf v. In general, for a vertex v at distance t from the root we define the *probability* of v to be $1/2^t$, and for a set of vertices such that none of them is an ancestor of any other we define the probability of the set as the sum of the probabilities of the individual vertices.

Lemma 3. *Let* $f : \{0,1\}^n \to \{0,1\}$ *be the function defined by a k-CNF formula φ and $\varepsilon > 0$. Let t be an integer such that $(1 - 1/2^k)^t \le \varepsilon$. Then there is a decision tree of depth at most tk such that: either (i) all the leaves define restrictions relative to which f is a constant, or (ii) all the leaves, except possibly a set of probability at most ε, define restrictions relative to which φ becomes a $(k-1)$-CNF.*

Proof. We apply Lemma 1 to φ with parameter t. Then we either find t clauses over disjoint variables or $k(t-1)$ variables that hit all the clauses.

In the former case, consider the decision tree that reads all the $\le kt$ variables that occur in the t clauses. All but an ε fraction of the leafs of the decision tree correspond to restrictions relative to which φ is zero, and, in particular, is constant.

In the latter case, consider the decision tree that reads all the $\le kt$ variables returned by the algorithm. All the leafs of the decision tree correspond to restrictions relative to which φ is a $(k-1)$-CNF. □

We now compose the construction.

Lemma 4. *There is a function* $t(k,\varepsilon) = O(k2^k \ln(1/\varepsilon))$ *such that for every $f : \{0,1\}^n \to \{0,1\}$ defined by a k-CNF formula and for every $\varepsilon > 0$ there is a decision tree of depth at most $t(k,\varepsilon)$ such that all the leaves, except possibly a subset of probability ε, correspond to a restriction relative to which f is a constant.*

Proof. We prove the theorem by induction. Lemma 3 proves the theorem for $k = 1$ and $t(1, \varepsilon) = \log_2 1/\varepsilon$. For general k, start by constructing the decision tree for f as in Lemma 3. If all but an ε fraction of the leaves of the tree make f become a constant, then we are done. Otherwise, every leaf of the tree defines a restriction relative to which f is a $(k-1)$-CNF, and we can apply the induction hypothesis to get decision trees for each of these $(k-1)$-CNF.

This argument proves the theorem for every function $t()$ that satisfies $t(1, \varepsilon) = \log_2 1/\varepsilon$ and $t(k, \varepsilon) \geq k2^k \ln(1/\varepsilon) + t(k-1, \varepsilon)$. In particular, the theorem is true for $t(k, \varepsilon) = 2k2^k \ln(1/\varepsilon)$. $\qquad\square$

To prove Theorem 2 we now only need the following simple last step, which is well known.

Lemma 5. *Let $f : \{0,1\}^n \rightarrow \{0,1\}$ be a function and T be a decision tree of depth t such that all but an ε fraction of the leaves of T define a restriction relative to which f is constant. Let X be a distribution that is δ-close to t-wise independent in ℓ_∞ norm. Then*

$$|\mathbf{Pr}[f(U_n) = 1] - \mathbf{Pr}[f(X) = 1]| \leq \varepsilon + \delta \cdot 2^t$$

Proof. We may assume withouth loss of generality that T has 2^t leaves, all at distance t from the root. (Otherwise, from leaves that are closer to the root, we read additional variables until we reach distance t. This does not change the properties of T assumed in the Lemma.) Let S be the set of leaves of T that define a restriction relative to which f is the constant 1. Then we have

$$\mathbf{Pr}[f(U_n) = 1] \leq \frac{|S|}{2^t} + \varepsilon$$

If we sample an assignment according to X, we see that for each leaf of T there is a probability at least $1/2^t - \delta$ that the assignment is consistent with the leaf. In each of these event, f evaluates to one and, moreover, all these events are disjoint. We deduce

$$\mathbf{Pr}[f(X) = 1] \geq \frac{|Z|}{2^t} - \delta \cdot |Z|$$

and

$$\mathbf{Pr}[f(U_n) = 1] - \mathbf{Pr}[f(X) = 1] \leq \varepsilon + \delta \cdot 2^t$$

Similarly, we can prove

$$\mathbf{Pr}[f(X) = 1] - \mathbf{Pr}[f(U_n) = 1] \leq \varepsilon + \delta \cdot 2^t$$

$\qquad\square$

4   Perspective

The current body of work on derandomization (see [Kab02] for a survey) strongly suggests that every problem (including search problems and promise problems) that is solvable probabilistically in polynomial time can also be solved deterministically in polynomial time. It is then a natural research program to look for deterministic polynomial time algorithms for all the interesting problems for which only probabilistic polynomial time algorithms are known.

After the discovery of a deterministic polynomial time algorithm for testing primality [AKS02], the most interesting algorithms to derandomize are now the identity test for low-degree polynomials and the approximate counting algorithms based on the Markov-Chain Monte-Carlo approach. Kabanets and Impagliazzo [KI03] show that derandomizing the polynomial identity test algorithm for general arithmetic circuits implies the proof of circuit lower bounds that may be beyond our current proof techniques. It is not clear whether there are similar inherent difficulties in derandomizing approximate counting algorithms such as, say, the Permanent approximation algorithm of [JSV01].

The problem of approximately counting the number of satisfying assignments for a given circuit up to a small *additive* error is clearly precisely the same problem as derandomizing every promise-BPP problem. (In particular, such an approximate counting algorithm would imply that $NEXP \not\subseteq P/poly$.) It seems possible, however, to derandomize at least bounded-depth circuits, and, at the very least, depth-two circuits in polynomial time using current techniques. In this paper we note that a special case of this problem can be solved by a simple divide-and-conquer algorithm, without using derandomization techniques.

References

[AGHP92] N. Alon, O. Goldreich, J. Håstad, and R. Peralta. Simple constructions of almost k-wise independent random variables. *Random Structures and Algorithms*, 3(3):289–304, 1992.

[AKS02]   Manindra Agrawal, Neeraj Kayal, and Nitin Saxena. PRIMES is in P. Manuscript, 2002.

[AW89]   Miklos Ajtai and Avi Wigderson. Deterministic simulation of probabilistic constand-depth circuits. *Advances in Computing Research - Randomness and Computation*, 5:199–223, 1989. Preliminary version in *Proc. of FOCS'85*.

[Hir98]   Edward A. Hirsch. A fast deterministic algorithm for formulas that have many satisfying assignments. *Journal of the IGPL*, 6(1):59–71, 1998.

[JSV01]   M. Jerrum, A. Sinclair, and E. Vigoda. A polynomial time approximation algorithm for the permanent of a matrix with non-negative entries. In *Proceedings of the 33rd ACM Symposium on Theory of Computing*, pages 712–721, 2001.

[Kab02]   Valentine Kabanets. Derandomization: A brief overview. *Bulletin of the European Association for Theoretical Computer Science*, 76:88–103, 2002.

[KI03]   Valentine Kabanets and Russell Impagliazzo. Derandomizing polynomial identity tests means proving circuit lower bounds. In *Proceedings of the 35th ACM Symposium on Theory of Computing*, pages 355–364, 2003.

[LV96]   Michael Luby and Boban Velickovic. On deterministic approximation of DNF. *Algorithmica*, 16(4/5):415– 433, 1996.

[LVW93]   Michael Luby, Boban Velickovic, and Avi Wigderson. Deterministic approx-
          imate counting of depth-2 circuits. In *Proceedings of the 2nd ISTCS*, pages
          18–24, 1993.

[Nis91]   N. Nisan. Pseudorandom bits for constant depth circuits. *Combinatorica*,
          12(4):63–70, 1991.

[NN93]   J. Naor and M. Naor. Small-bias probability spaces: efficient constructions
          and applications. *SIAM Journal on Computing*, 22(4):838–856, 1993.

[NW94]   N. Nisan and A. Wigderson. Hardness vs randomness. *Journal of Computer
          and System Sciences*, 49:149–167, 1994. Preliminary version in *Proc. of
          FOCS'88*.

Author Index

Lecture Notes in Computer Science

For information about Vols. 1–3062

please contact your bookseller or Springer

Vol. 3114: R. Alur, D.A. Peled (Eds.), Computer Aided Verification. XII, 536 pages. 2004.

Vol. 3113: J. Karhumäki, H. Maurer, G. Paun, G. Rozenberg (Eds.), Theory Is Forever. X, 283 pages. 2004.

Vol. 3112: H. Williams, L. MacKinnon (Eds.), Key Technologies for Data Management. XII, 265 pages. 2004.

Vol. 3111: T. Hagerup, J. Katajainen (Eds.), Algorithm Theory - SWAT 2004. XI, 506 pages. 2004.

Vol. 3110: A. Juels (Ed.), Financial Cryptography. XI, 281 pages. 2004.

Vol. 3109: S.C. Sahinalp, S. Muthukrishnan, U. Dogrusoz (Eds.), Combinatorial Pattern Matching. XII, 486 pages. 2004.

Vol. 3108: H. Wang, J. Pieprzyk, V. Varadharajan (Eds.), Information Security and Privacy. XII, 494 pages. 2004.

Vol. 3107: J. Bosch, C. Krueger (Eds.), Software Reuse: Methods, Techniques and Tools. XI, 339 pages. 2004.

Vol. 3106: K.-Y. Chwa, J.I. Munro (Eds.), Computing and Combinatorics. XIII, 474 pages. 2004.

Vol. 3105: S. Göbel, U. Spierling, A. Hoffmann, I. Iurgel, O. Schneider, J. Dechau, A. Feix (Eds.), Technologies for Interactive Digital Storytelling and Entertainment. XVI, 304 pages. 2004.

Vol. 3104: R. Kralovic, O. Sykora (Eds.), Structural Information and Communication Complexity. X, 303 pages. 2004.

Vol. 3103: K. Deb, e. al. (Eds.), Genetic and Evolutionary Computation – GECCO 2004. XLIX, 1439 pages. 2004.

Vol. 3102: K. Deb, e. al. (Eds.), Genetic and Evolutionary Computation – GECCO 2004. L, 1445 pages. 2004.

Vol. 3101: M. Masoodian, S. Jones, B. Rogers (Eds.), Computer Human Interaction. XIV, 694 pages. 2004.

Vol. 3100: J.F. Peters, A. Skowron, J.W. Grzymała-Busse, B. Kostek, R.W. Świniarski, M.S. Szczuka (Eds.), Transactions on Rough Sets I. X, 405 pages. 2004.

Vol. 3099: J. Cortadella, W. Reisig (Eds.), Applications and Theory of Petri Nets 2004. XI, 505 pages. 2004.

Vol. 3098: J. Desel, W. Reisig, G. Rozenberg (Eds.), Lectures on Concurrency and Petri Nets. VIII, 849 pages. 2004.

Vol. 3097: D. Basin, M. Rusinowitch (Eds.), Automated Reasoning. XII, 493 pages. 2004. (Subseries LNAI).

Vol. 3096: G. Melnik, H. Holz (Eds.), Advances in Learning Software Organizations. X, 173 pages. 2004.

Vol. 3095: C. Bussler, D. Fensel, M.E. Orlowska, J. Yang (Eds.), Web Services, E-Business, and the Semantic Web. X, 147 pages. 2004.

Vol. 3094: A. Nürnberger, M. Detyniecki (Eds.), Adaptive Multimedia Retrieval. VIII, 229 pages. 2004.

Vol. 3093: S.K. Katsikas, S. Gritzalis, J. Lopez (Eds.), Public Key Infrastructure. XIII, 380 pages. 2004.

Vol. 3092: J. Eckstein, H. Baumeister (Eds.), Extreme Programming and Agile Processes in Software Engineering. XVI, 358 pages. 2004.

Vol. 3091: V. van Oostrom (Ed.), Rewriting Techniques and Applications. X, 313 pages. 2004.

Vol. 3089: M. Jakobsson, M. Yung, J. Zhou (Eds.), Applied Cryptography and Network Security. XIV, 510 pages. 2004.

Vol. 3087: D. Maltoni, A.K. Jain (Eds.), Biometric Authentication. XIII, 343 pages. 2004.

Vol. 3086: M. Odersky (Ed.), ECOOP 2004 – Object-Oriented Programming. XIII, 611 pages. 2004.

Vol. 3085: S. Berardi, M. Coppo, F. Damiani (Eds.), Types for Proofs and Programs. X, 409 pages. 2004.

Vol. 3084: A. Persson, J. Stirna (Eds.), Advanced Information Systems Engineering. XIV, 596 pages. 2004.

Vol. 3083: W. Emmerich, A.L. Wolf (Eds.), Component Deployment. X, 249 pages. 2004.

Vol. 3080: J. Desel, B. Pernici, M. Weske (Eds.), Business Process Management. X, 307 pages. 2004.

Vol. 3079: Z. Mammeri, P. Lorenz (Eds.), High Speed Networks and Multimedia Communications. XVIII, 1103 pages. 2004.

Vol. 3078: S. Cotin, D.N. Metaxas (Eds.), Medical Simulation. XVI, 296 pages. 2004.

Vol. 3077: F. Roli, J. Kittler, T. Windeatt (Eds.), Multiple Classifier Systems. XII, 386 pages. 2004.

Vol. 3076: D. Buell (Ed.), Algorithmic Number Theory. XI, 451 pages. 2004.

Vol. 3075: W. Lenski (Ed.), Logic versus Approximation. IX, 205 pages. 2004.

Vol. 3074: B. Kuijpers, P. Revesz (Eds.), Constraint Databases and Applications. XII, 181 pages. 2004.

Vol. 3073: H. Chen, R. Moore, D.D. Zeng, J. Leavitt (Eds.), Intelligence and Security Informatics. XV, 536 pages. 2004.

Vol. 3072: D. Zhang, A.K. Jain (Eds.), Biometric Authentication. XVII, 800 pages. 2004.

Vol. 3071: A. Omicini, P. Petta, J. Pitt (Eds.), Engineering Societies in the Agents World. XIII, 409 pages. 2004. (Subseries LNAI).

Vol. 3070: L. Rutkowski, J. Siekmann, R. Tadeusiewicz, L.A. Zadeh (Eds.), Artificial Intelligence and Soft Computing - ICAISC 2004. XXV, 1208 pages. 2004. (Subseries LNAI).

Vol. 3068: E. André, L. Dybkjær, W. Minker, P. Heisterkamp (Eds.), Affective Dialogue Systems. XII, 324 pages. 2004. (Subseries LNAI).

Vol. 3067: M. Dastani, J. Dix, A. El Fallah-Seghrouchni (Eds.), Programming Multi-Agent Systems. X, 221 pages. 2004. (Subseries LNAI).

Vol. 3066: S. Tsumoto, R. Słowiński, J. Komorowski, J.W. Grzymała-Busse (Eds.), Rough Sets and Current Trends in Computing. XX, 853 pages. 2004. (Subseries LNAI).

Vol. 3065: A. Lomuscio, D. Nute (Eds.), Deontic Logic in Computer Science. X, 275 pages. 2004. (Subseries LNAI).

Vol. 3064: D. Bienstock, G. Nemhauser (Eds.), Integer Programming and Combinatorial Optimization. XI, 445 pages. 2004.

Vol. 3063: A. Llamosí, A. Strohmeier (Eds.), Reliable Software Technologies - Ada-Europe 2004. XIII, 333 pages. 2004.